LAW AND NEUROSCIENCE
CURRENT LEGAL ISSUES 2010

VOLUME 13

CURRENT LEGAL PUBLICATIONS

Editor-in-Chief
MICHAEL D.A. FREEMAN

Editorial Board

SIR JOHN BAKER	N. LACEY
E. BARENDT	A. LE SUEUR
I.H. DENNIS	A.D.E. LEWIS
D. GALLIGAN	E. MCKENDRICK
DAME HAZEL GENN	R. MOKAL
SIR BOB HEPPLE	R.W. RIDEOUT
J. HODDER	P. SANDS
J. JOWELL	LORD STEYN
LORD WOOLF OF BARNES	

Law and Neuroscience

Current Legal Issues 2010

VOLUME 13

Edited by
MICHAEL FREEMAN F.B.A.
Professor of English Law
University College London

This book has been printed digitally and produced in a standard specification in order to ensure its continuing availability

OXFORD
UNIVERSITY PRESS

Great Clarendon Street, Oxford OX2 6DP
United Kingdom

Oxford University Press is a department of the University of Oxford.
It furthers the University's objective of excellence in research, scholarship,
and education by publishing worldwide. Oxford is a registered trade mark of
Oxford University Press in the UK and in certain other countries

© Oxford University Press 2011

The moral rights of the authors have been asserted
Database right Oxford University Press (maker)

Reprinted 2012

All rights reserved. No part of this publication may be reproduced,
stored in a retrieval system, or transmitted, in any form or by any means,
without the prior permission in writing of Oxford University Press,
or as expressly permitted by law, or under terms agreed with the appropriate
reprographics rights organization. Enquiries concerning reproduction
outside the scope of the above should be sent to the Rights Department,
Oxford University Press, at the address above

You must not circulate this book in any other form
and you must impose this same condition on any acquirer

British Library Cataloguing in Publication Data
Data available

Library of Congress Cataloging in Publication Data
Data available

ISBN 978-0-19-959984-4

Printed and bound by CPI Group (UK) Ltd, Croydon, CR0 4YY

Preface

The essays in this volume are the product of UCL's 13th international interdisciplinary colloquium held in July 2009. We had previously held a conference on this subject in February 2008, which has been published as *Law, Mind and Brain*, edited by myself and Oliver Goodenough (Farnham, Ashgate, 2009).

Neuroscience offers many challenges to the lawyer, and these are taken up in this volume. Both the value and limits of neuroscience to the discipline and practice of law are explored, and there is, appropriately, a note of healthy scepticism. The book brings together many of the leading thinkers on the interdiscipline and thus offers a source for jurists, neuroscientists, practising lawyers (particularly those in the criminal law), and policy-makers. The range of subjects covered is wide. There are several papers on neuroimaging, in particular in relation to criminal responsibility and in relation to evidence. There are papers on juvenile justice, on tort, in particular on emotional harm. There are papers also on end-of-life decisions, for example on the PVS condition. And decisions on the beginning of life are also considered. There is discussion of religion, of the right to silence, and how jurors process information. Papers consider such questions as empathy, and on conflicts between our moral intuitions and legal doctrine. Papers also offer historical insights.

This colloquium and volume could not have been put together without the assistance of Professor Semir Zeki, Professor of Neuroesthetics at UCL. He also gave a public lecture at the colloquium, which unfortunately we are not able to include in this volume. As ever Lisa Penfold provided amazing support, as did Jacqui Bennett and Deborah Burns. I am grateful to all of these people.

It is with some irony that I record that as the volume was going to press I was diagnosed with a neurological disease—Parkinson's. This will not stop me producing further colloquia and volumes in the series. The 2010 one is on 'Law and Childhood Studies' (5 and 6 July 2010), and the next projected one is 'Law and Language' in July 2011. Further information on 'Law and Language' can be obtained from me (michael.freeman@ucl.ac.uk) or Dr Fiona Smith (fiona.m.smith@ucl.ac.uk).

Michael Freeman

April 2010

Contents

List of Contributors	xi
1. Introduction: Law and the Brain *Michael Freeman*	1
2. What Neuroscience Can (and Cannot) Tell Us about Criminal Responsibility *Walter Glannon*	13
3. *Mens Rea*, Logic, and the Brain *Gert-Jan Lokhorst*	29
4. Indeterminism and Control: An Approach to the Problem of Luck *John Martin Fischer*	41
5. Neuroscience and Criminal Responsibility: Proving 'Can't Help Himself' as a Narrow Bar to Criminal Liability *Henry T. Greely*	61
6. Madness, Badness, and Neuroimaging-Based Responsibility Assessments *Nicole A. Vincent*	79
7. Brain Images as Evidence in the Criminal Law *Adina L. Roskies and Walter Sinnott-Armstrong*	97
8. The Neural Correlates of Third-Party Punishment *Joshua W. Buckholtz, Christopher L. Asplund, Paul E. Dux, David H. Zald, John C. Gore, Owen D. Jones, and René Marois*	115
9. Law, Neuroscience, and Criminal Culpability *Lisa Claydon*	141
10. How (Some) Criminals Are Made *Theodore Y. Blumoff*	171
11. Neuroscience and Penal Law: Ineffectiveness of the Penal Systems and Flawed Perception of the Under-Evaluation of Behaviour Constituting Crime. The Particular Case of Crimes Regarding Intangible Goods *David Terracina*	193
12. Neuroscience and Emotional Harm in Tort Law: Rethinking the American Approach to Free-Standing Emotional Distress Claims *Betsy J. Grey*	203

13. Neuroscience and Ideology: Why Science Can Never Supply a Complete Answer for Adolescent Immaturity — 231
June Carbone

14. Adolescent Brain Science and Juvenile Justice — 255
Terry A. Maroney

15. The Neuroscience of Cruelty as Brain Damage: Legal Framings of Capacity and Ethical Issues in the Neurorehabilitation of Motor Neurone Disease and Behavioural Variant Frontotemporal Dementia — 283
Robin Mackenzie and Mohamed Sakel

16. The Carmentis Machine: Legal and Ethical Issues in the Use of Neuroimaging to Guide Treatment Withdrawal in Newborn Infants — 309
Dominic Wilkinson and Charles Foster

17. The Right to Silence Protects Mental Control — 335
Dov Fox

18. Minds Apart: Severe Brain Injury, Citizenship, and Civil Rights — 367
Joseph J. Fins

19. Reciprocity and Neuroscience in Public Health Law — 385
A. M. Viens

20. Pathways to Persuasion: How Neuroscience Can Inform the Study and Practice of Law — 395
Cheryl Boudreau, Seana Coulson, and Mathew D. McCubbins

21. The Juridical Role of Emotions in the Decisional Process of Popular Juries — 407
Laura Capraro

22. Possible Legal Implications of Neural Mechanisms Underlying Ethical Behaviour — 419
Donald Pfaff and Sandra Sherman

23. What Hobbes Left Out: The Neuroscience of Compassion and its Implications for a New Common-wealth — 433
James D. Duffy

24. Neuroscience and the Free Exercise of Religion — 449
Steven Goldberg

25. Steps toward a Constructivist and Coherentist Theory of Judicial Reasoning in Civil Law Tradition — 459
Enrique Cáceres

26. Evolutionary Jurisprudence: The End of the Naturalistic Fallacy and the Beginning of Natural Reform? — 483
Morris B. Hoffman

27. The History of Scientific and Clinical Images in Mid-to-Late Nineteenth-Century American Legal Culture: Implications for Contemporary Law and Neuroscience 505
    Daniel S. Goldberg

28. Lost in Translation? An Essay on Law and Neuroscience 529
    Stephen J. Morse

Index 563

List of Contributors

Christopher L. Asplund, Department of Psychology, Neuroscience Graduate Program of Vanderbilt University.

Theodore Y. Blumoff is Professor of Law, Mercer University, Macon, GA.

Cheryl Boudreau is Assistant Professor in the Department of Political Science at the University of California, Davis.

Joshua W. Buckholtz is PhD candidate in Neuroscience at the Vanderbilt Brain Institute and Department of Psychology, Vanderbilt University.

Enrique Cáceres, Institute for Legal Research, National Autonomous University of Mexico.

Laura Capraro is Assistant Professor in Criminal Procedure at the University of Rome 'Tor Vergata'.

June Carbone holds the Edward A. Smith/Missouri Chair of Law, the Constitution and Society, at the University of Missouri, Kansas City.

Lisa Claydon is Principal Lecturer in Law, Department of Law, Bristol Law School, University of the West of England.

Seana Coulson is Associate Professor in the Department of Cognitive Science at the University of California, San Diego.

James D. Duffy is Professor of Psychiatry, University of Texas, M. D. Anderson Cancer Center, Fellow, McGovern Center for Health, Healing, and the Human Spirit.

Paul E. Dux, Department of Psychology of Vanderbilt University.

Joseph J. Fins, MD, FACP is Chief of the Division of Medical Ethics at Weill Cornell Medical College where he serves as Professor of Medicine, Professor of Public Health and Professor of Medicine in Psychiatry; Attending Physician and Director of Medical Ethics at New York-Presbyterian Weill Cornell Medical Center and member of the Adjunct Faculty of Rockefeller University and Senior Attending Physician at The Rockefeller University Hospital.

John Martin Fischer is Distinguished Professor of Philosophy, University of California, Riverside, and University of California President's Chair (2001–2010).

Charles Foster, The Ethox Centre, Department of Public Health and Primary Health Care, University of Oxford; Barrister, Outer Temple Chambers, London WC2.

Dov Fox is Law Clerk to Judge Stephen Reinhardt of the United States Court of Appeals for the Ninth Circuit.

Michael Freeman is Professor of English Law, University College London.

Walter Glannon is Associate Professor of Philosophy at the University of Calgary.

List of Contributors

Daniel. S. Goldberg, JD, PhD, is Assistant Professor in the Department of Bioethics & Interdisciplinary Studies at the Brody School of Medicine, East Carolina University.

The late **Steven Goldberg**. Until his death in 2010, Steven Goldberg was James and Catherine Denny Professor of Law, Georgetown University Law Center, Washington, DC.

John C. Gore, Institute of Imaging Science, Center for Integrative and Cognitive Neurosciences, Departments of Radiology and Radiological Sciences and Biomedical Engineering of Vanderbilt University.

Henry T. Greely is Deane F. and Kate Edelman Johnson Professor of Law; Professor, by courtesy, of Genetics, Stanford University.

Betsy J. Grey is Professor of Law and Faculty Fellow, Center for Law, Science & Innovation at the Sandra Day O'Connor College of Law at Arizona State University.

Morris B. Hoffman is a District Judge, State of Colorado, Second Judicial District (Denver); Adjunct Professor of Law, University of Colorado; Member, MacArthur Foundation Law and Neuroscience Project; Research Fellow, Gruter Institute for Law and Behavioral Research.

Owen D. Jones is New York Alumni Chancellor's Chair in Law and Professor of Biology at Vanderbilt University; Director of the MacArthur Foundation Law and Neuroscience Project.

Gert-Jan Lokhorst is senior researcher in the Section of Philosophy of the Faculty of Technology, Policy and Management at the Delft University of Technology in the Netherlands.

Mathew D. McCubbins is Provost Professor of Business, Law and Political Economy in the Marshall School of Business, Gould School of Law, and Department of Political Science at the University of Southern California.

Robin Mackenzie is Director, Medical Law & Ethics, Kent Law School, University of Kent.

René Marois is Associate Professor of Psychology at Vanderbilt University.

Terry Maroney is an Associate Professor of Law at the Vanderbilt University Law School.

Stephen J. Morse is Ferdinand Wakeman Hubbell Professor of Law & Professor of Psychology and Law in Psychiatry, University of Pennsylvania Law School and School of Medicine.

Donald Pfaff is Professor and Head of the Laboratory of Neurobiology and Behavior at The Rockefeller University, a Fellow of the New York Academy of Sciences, a member of the Advisory Board of the National Academy of Sciences, a Fellow of the American Academy of Arts and Sciences, and serves on the editorial boards of numerous scholarly journals.

Adina L. Roskies is Assistant Professor, Department of Philosophy, Dartmouth College.

Mohamed Sakel is Director, Neurorehabilitation Services and Director of Research and Development, East Kent University Hospital Trust.

Sandra Sherman is Assistant Director of the Fordham Intellectual Property Law Institute, and formerly Professor at the University of Arkansas and Georgia State University; Fellow of the National Endowment for the Humanities and the Andrew Mellon Foundation; Visiting

Fellow of Lucy Cavendish College, Cambridge University; and the Institute for Research in the Humanities at the University of Wisconsin-Madison; Visiting Scholar at Harvard.

Walter Sinnott-Armstrong is Chauncey Stillman Professor in Practical Ethics in the Department of Philosophy and the Kenan Institute for Ethics at Duke University.

David Terracina is Assistant Professor in Criminal Law at the University of Rome Tor Vergata.

A. M. Viens, School of Law, Queen Mary, University of London and Joint Centre for Bioethics, University of Toronto.

Nicole A. Vincent is a postdoctoral researcher in the Philosophy Department at Delft University of Technology in the Netherlands.

Dominic Wilkinson, The Ethox Centre, Department of Public Health and Primary Health Care, the University of Oxford; Oxford Uehiro Centre for Practical Ethics, the University of Oxford.

David H. Zald, Department of Psychology, Center for Integrative and Cognitive Neurosciences of Vanderbilt University.

1

Introduction: Law and the Brain

*Michael Freeman**

The human brain is the most complicated device in the universe.[1] It weighs just 1.2 kg and contains one hundred billion nerve cells.

Its importance has been recognized since early civilizations.[2] Alcmaeon of Croton,[3] a follower of Pythagoras, is believed to have been one of the first to have realized that the brain is the likely centre of the intellect.

The human brain has been a focus of medico-legal debates since the late 1960s. It was then that efforts were made to formulate an alternative definition of death, one centred on brain function rather than heart-and-lung function.[4] It was the development of mechanical ventilation and the emergence of heart transplantation that necessitated this new definition of death. The focus is now on the brain in the living human, and it is this on which this volume concentrates.

We are seeing the emergence of 'neurolaw'. In a sense this is only what we have wanted to be able to do since ancient times. As long ago as Hippocrates[5] scientists were searching for the source of our behaviour. He developed a humoral theory, explaining that there were four determinants of temperament. Of course, the Greeks believed that there were four elements (earth, air, water, and fire), and this may have been at the root of his rather implausible theory.[6] Hippocrates' influence can be observed four centuries later in the theory of Galen.[7] His knowledge of the human body, in part acquired when he was physician at a school for gladiators, enabled him to build on earlier ideas. He linked the presence in the brain of three fluid-filled cavities (or ventricles) with the division of mental faculties into imagination, reason, and memory. In Galen's view, the primary function of the brain is to distribute vital fluid from the ventricles through the nerves to the muscles and

---

* Professor of English Law, University College London.
[1] A good introduction is Michael O'Shea, *The Brain* (Oxford, Oxford University Press, 2005).
[2] Though not by the Egyptians: ibid. 12.
[3] Born in 535 BCE.
[4] See J. L. Bernat, 'The Biophilosophical Basis of Whole-Brain Death' (2004) 19(2) *Social Philosophy and Policy* 324–42.
[5] See L. G. Panourias et al., 'Hippocrates: A Pioneer in the Treatment of Head Injuries' (2005) 57 *Neurosurgery* 181.
[6] See R. Mazzolini, 'Schemes and Models of the Thinking Machine' in P. Corsi (ed.), *The Enchanted Loom: Chapters in the History of Neuroscience* (New York, Oxford University Press, 1991).
[7] Claudius Galenus of Pergamum (131–201 CE).

organs, in this way controlling bodily activity. Galen was influential for centuries. He clearly left his imprint on Leonardo da Vinci, some of whose early drawings picture the brain crudely as having three cavities. But subsequently Leonardo was to make significant observations on the brain and its ventricles. He determined the shape and extent of the brain's cavities. Even so, he failed to throw off Galen's influence and failed to appreciate the importance of the solid tissue of the brain. With the Renaissance there emerged a new understanding of the brain. In the seventeenth century, the French philosopher René Descartes wrote of the brain as a machine: he understood it as based on the principles of hydraulics. He accepted that even with a complete understanding of the brain we would still not completely understand behaviour. Descartes did not locate cognitive processes in the brain's fluid-filled ventricles, but he made no attempt to assign functions to specific brain structures, with the exception of the pineal gland. Where Descartes was a pioneer was in comparing the workings of the brain with complex hydraulic machines. This is somewhat reminiscent of today's discussions of the way computers acquire, process, and store information.

The term 'neurology' was coined by Descartes's near contemporary, Thomas Willis. He was the first to argue that solid cerebral tissue has important functions. But he did not break with the past entirely: he considered that fluid-flow was the key to understanding brain function. Nevertheless, he stressed the importance of solid cerebral tissues and showed that nervous function depends on the flow of blood to them.

Our understanding of the brain increased in the eighteenth century with the discovery of electricity. Luigi Galvani discovered the importance of electricity to the operation of the nervous system. But it was not until the middle of the nineteenth century that the ability of nerves and muscles to generate rapidly propagating electrical impulses was confirmed by a German physiologist, Emil Du Bois-Reymond.

Our understanding of the brain could have taken a blind alley.

In the eighteenth and nineteenth centuries, the pseudo-sciences of physiognomy and phrenology developed. The former tends to be associated today with Cesare Lombroso, an Italian proto-criminologist who tried to explain criminal behaviour in biological terms.[8] Phrenology began with the studies of Gall and Spurzheim: their work related brain functions (and character traits) to protuberances on the skull. It provoked considerable interest at the time.[9] Like Lombroso, Gall studied criminals.[10] Cranioscopy was to become standard within criminology. Carl Wernicke, the nineteenth-century German neurologist, was the first to advance the theory that there are 'centres' in the brain. Though not

---

[8] See M. Gibson, *Born to Crime: Cesare Lombroso and the Origins of Biological Criminology* (Westport, CT, Praeger, 2002).
[9] It is completely discredited by the findings of neuropsychology.
[10] See C. Pogliano, 'Between Form and Function: A New Science of Man' in op. cit., note 6, 144–203.

without his critics, Wernicke's concept of localization is at the root of brain imaging techniques such as fMRI still today.[11]

Phrenology was very popular, and did not die out until early in the twentieth century. In fact the British Phrenological Society only came to an end in 1967. But we now know that many discrete mental functions are localized to particular parts of the brain. There is some irony that though phrenology's claim to read the mind from bumps on the head has been soundly refuted, its premise has been vindicated.

The true foundations of neuroscience are found in the work of Santiago Ramon y Cajal.[12] He was the first to recognize the cellular nature of the brain and of its mental functions. Most neuroanatomists, well into the twentieth century, believed that the brain, unlike other organs, was not composed of cells at all. Of course, brain cells are not like the cells in other parts of the body. Neurons are diverse in morphology. So, when viewed through a microscope the brain appears to consist of a tangled morass (a reticulum) without the distinct cell-defining boundaries so evident in other tissues.

To understand the brain, science had to identify the functional components of its microscopic structure. The Italian anatomist, Camillo Golgi,[13] was able to highlight the morphology of very few neurons in any particular region of the brain. It was a staining method, and it was particularly useful because it enabled individual neurons to be viewed without being blocked by the mass of neighbouring cells. It revealed individual neurons as dark silver-impregnated silhouettes. Golgi offered the key to a new set of scientifically testable ideas about how the brain works. Ramon y Cajal, with the help of Golgi's staining method, connected two important propositions: that the neuron is a cell, and neurons are structurally polarized with respect to function. For the first time the workings of the brain were explicitly associated with the functions of physical structures at a microscopic level. As Michael O' Shea explains:

> Cajal concluded that a neuron's function must be concerned with the movement and processing of information in the brain. He could only guess about the form in which information might be encoded... In a stroke of genius, however, he postulated that it would be sensible for the components of function to impose directionality on information flow (or streaming as he called it). So he proposed that information flows in one direction, from an input region to an output region. The neuron's cell body and its shorter processes, known as dendrites, perform input functions. Information then travels along the longest extension from the cell body, known as the axon, to the output region—the terminals of the axon and its branches that contact the input dendrites and cell body of another neuron.[14]

By the middle of the twentieth century neuroscience had become the fastest growing discipline in the history of scientific endeavour. By the end of that century

---

[11] See G. Fernandez et al., 'Intrasubject Reproducibility of Resurgical Language Lateralization and Mapping Using fMRI' (2003) 60 *Neurology* 969.
[12] 1852–1934. See, further, S. Finger, *Origins of Neuroscience: A History of Explorations into Brain Function* (Oxford, Oxford University Press, 2001).
[13] 1843–1926.
[14] Op. cit., note 1, 21.

we had a more or less complete understanding, in molecular detail, of how neurons generate electrical and chemical signals.

The discovery of the X-ray in 1895 (by Roentgen) is a watershed event. For the first time it was possible to see into the body without undertaking surgery. The first arteriogram/angiogram in the brain was performed in 1927; the first recording of electrical brain activity (electroencephalography, or EEG) was made in Germany just after the end of the First World War.[15] By the 1970s computers enabled us to make topographic maps of brain electrical activity. Quantitative EEG (QEEG) became possible and we had brain maps comparing a patient with groups of patients.

Neuroimaging began with the discovery of computerized axial tomography by Hounsfield and Cormack in 1972 (they were awarded the Nobel Prize for this discovery in 1979).[16] As Hounsfield explained in his Nobel lecture:

Computer tomography... measures the attenuation of x-ray beams passing through sections of the body from hundreds of different angles, and then, from the evidence of these measurements, a computer is able to reconstruct pictures of the body's interior. Pictures are based on the separate examination of a series of contiguous cross sections, as though we looked at the body separated into a series of thin 'slices'. By doing so, we virtually obtained total three-dimensional information about the body.[17]

Hounsfield's initial focus was the detection of small tumours in the body, but he found he was able to perform a brain scan on a patient who had a frontal lobe brain tumour. Swiftly, technology enabled the development of CT scanning machines. Though initially criticized—there was concern about their cost—they have come to revolutionize neuroscience and, of course, medicine. It has also had an impact on law, and this will increase. Thirteen years ago Jennifer Kulynych noted:[18]

It is now common for a psychiatrist to refer to the physiological state of an individual's brain when evaluating a mental disorder. Moreover, such evaluation increasingly includes a reference to neuroimages. In a legal proceeding, the visual impact of such neuroimages is hard to overstate.

How much more so now! And this is not to underestimate its dangers: that it will, for example, license abuse of detainees in the war against terrorism.[19]

In 1848, John Harlow's observations[20] of Phineas Gage,[21] a railroad worker whose prefrontal cortex was extensively damaged in an industrial accident, provided the first evidence that prefrontal cortex damage could be linked to behavioural

---

[15] By Hans Berger, a psychiatrist in Jena.
[16] See R. L. Eisenberg, *Radiology: An Illustrated History* (St Louis, Mosby, 1992) 467.
[17] Ibid.
[18] 'Psychiatric Neuroimaging Evidence: A High-Tech Crystal Ball?' (1997) 29 *Stanford Law Review* 1249, 1251.
[19] See J. H. Marks, 'Interrogational Neuroimaging in Counterterrorism: A "No-Brainer" or a Human Rights Hazard?' (2007) 33 *American Journal of Law and Medicine* 483–500.
[20] Reported in (1848) 13 *Boston Medical & Surgical Journal* 390.
[21] See M. MacMillan, *An Odd Kind of Fame: Stories of Phineas Gage* (Cambridge, MA, MIT Press, 2000).

dysfunction. Gage had been reliable and courteous, but was now a coarse, unstable, and antisocial individual. He was never able to work again (though physically fit to do so). Harlow's observations are supported by modern neuroimaging studies of patients with frontal lobe injuries.[22] Research on violent offenders has shown hypersensitivity in the amygdala and diminished activity in the prefrontal cortex as compared with control subjects.[23] Brain-imaging showed that such individuals had significantly smaller (11 per cent) prefrontal cortexes than normal individuals.[24] The smaller the volume of the prefrontal cortex, the greater was the tendency towards aggressive and antisocial behaviour.[25]

It is hardly surprising that the implications of this have been grasped by lawyers (and philosophers).[26] This has led to a healthy debate, which can be sampled in the papers in this volume. It has also led defence lawyers, in particular in the US in death penalty cases, to advance such evidence in mitigation. An example is the trial (and appeal) of Eddie Lee Sexton in Florida.[27] Evidence was adduced that brain scans suggested that Sexton, a convicted murderer, had a diminished level of self-control due to dysfunction in his prefrontal cortex.[28]

It is not surprising that the study of law has begun to draw insights from this gushing fountain of new knowledge. It was less than a decade ago that the combination of neuroscience with law first began to be mooted in academic circles at meetings[29] and in writings.[30] The Neuroethics Society was founded as recently as 2006. And now the field is bursting open. Some have rightly raised cautionary warnings that we should not get too carried away and lose our intellectual footing in the flood of discovery.[31] But a flood it is becoming. UCL's Law and Neuroscience

---

[22] For example, J. Grafman et al., 'Frontal Lobe Injuries, Violence and Aggression: A Report of the Vietnam Head Injury Study' (1996) 46 (5) *Neurology* 1231–8.

[23] See A. Raine et al., 'Reduced Prefrontal Gray Matter Volume and Reduced Autonomic Activity in Anti-Social Personality Disorder' (2000) 57 *Archives of General Psychiatry* 121.

[24] Ibid, 121.

[25] M. Brower and B. Price, 'Neuropsychiatry of Frontal Lobe Dysfunction in Violent and Criminal Behaviour: A Critical Review' (2001) 71 *Neurosurgery and Psychiatry* 725.

[26] H.-L. Kroeber, 'The Historical Debate on Brain and Legal Responsibility—Revisited' (2007) 25 *Behavioral Science and the Law* 252; L. S. Khoshbin and S. Khoshbin, 'Imaging the Mind, Minding the Image' (2007) 33 (2–3) *American Journal of Law and Medicine* 171–92.

[27] *Sexton v State* 775 So 2d 923 (Fla 2000), and 16–17, below.

[28] See also *Roper v Simmons* 543 US 551 (2005) and 241, 255, 540, below.

[29] See L. A. Frolik, Report: Seventh Annual Teaching Seminar: 'Neurobiology, Human Behavior and the Law', Squaw Valley, CA, 17–21 June 1998; K. Wermke and O. Goodenough, Report: Workshop: 'Law and Neuroscience' Humboldt University, Berlin, 19 February 2001, available at <http://www.gruter.org/index.php?option=com_content&task=view&id=145&Itemid=88888947>; S. Taha and K. Nagel, Report: Workshop: 'Neurological Basis for Justice' University of California at San Francisco, CA, 16 November 2001, available at <http://www.gruter.org/index.php?option=com_content&task=view& id=154& Itemid=88888947>; S. E. Hyman, 'Neuroethics: At age 5, field continues to evolve', available at <http://www.dana.org/news/publications/details.aspx?id=5850>.

[30] See O. R. Goodenough, 'Mapping Cortical Areas Associated with Legal Reasoning and with Moral Intuition' 41 *Jurimetrics* 429–42; B. Garland (ed.), *Neuroscience and the Law, Brain, Mind and the Scales of Justice* (Washington DC, Dana Press, 2004); S. Zeki and O. R. Goodenough (eds.), Theme Issue 'Law and the Brain' (2004) 359 *Phil. Trans. R. Soc. Lond.* 1661–890.

[31] See S. J. Morse, 'New Neuroscience, Old Problems: Legal Implications of Brain Science' (2004) 6 *Cerebrum* 81–90. And see Morse in this volume at 529.

colloquium, and the papers in this volume, are examples of this growing ferment and important contributions to it.

A good study to set the scene is Libet et al.'s now classic, published in *Brain* in 1983.[32] This showed that the conscious intention to act upon a decision lagged 300–500 milliseconds behind the unconscious brain activity that lead to intentions. Does this mean that our brain knows our decisions before we do, or at least before we become conscious of them?[33] This would appear to problematize free will. But, as Stephen Morse points out,[34] these findings simply demonstrate that unconscious brain events precede conscious experience. This is precisely what one would expect of the normal functioning of the brain. It does not mean that intentionality does not play a causal role in our actions, since there is more to decision-making and actions. Libet concedes that, despite the unconscious brain activity, people can 'veto' the act choosing not to carry it out—this is another form of mental act that is key, a causal role in what actions are ultimately taken.[35] Furthermore, the assertion that all our actions arise from unconscious brain events cannot accommodate the 'phenomenon of making up or changing one's mind',[36] especially when one is faced with two possible courses of action at the same time.

Whether what scientists know about our brains, coupled with brain imaging technology, is sufficient to threaten free will as we know it remains contentious. For example, an fMRI scan showing an abnormally small amygdala cannot tell us whether a violent offender lacked the capacity for impulse control, or whether he possessed but failed to exercise it. Glannon points out that it is unclear how images showing differences in amygaloid volume could conclusively explain why one person could, and another not, act rationally.[37] This is because brain imaging provides only 'one window of many into the multiple influences on behaviour'.[38] Behavioural decisions are influenced not just by the brain, but by other variables as well, social, cultural, genetic, and endocrinological.[39] As Morse puts it,[40] while the same neurological processes may 'produce' the same delusional beliefs in all people with the processes, the 'delusional contact' and resultant behaviour of delusional, thirteenth-century subcontinental Indians will surely differ from that of delusional late twentieth-century Americans. As Gazzaniga observes, 'no pixel in a brain will ever be able to show culpability or non-culpability'.[41]

---

[32] B. Libet et al., 'Time of Conscious Intention to Act in Relation To Onset of Cerebral Activity (Readiness Potential); The Unconscious Initiaton of a Freely Voluntary Act' (1983) 106 *Brain* 623.
[33] And see M. Gazzaniga, 'Facts, Fictions, and the Future of Neuroethics' in Judy Illes (ed.), *Neuroethics* (New York, Oxford University Press, 2006) 145.
[34] 'New Neuroscience, Old Problems', op. cit., note 31.
[35] B. Libet, 'Do We Have Free Will?' (1999) 6 *Journal of Consciousness Studies* 50.
[36] Per W. Glannon, *Bioethics and the Brain* (Oxford, Oxford University Press, 2007) 55.
[37] Ibid. 60.
[38] Per D. Mobbs et al., 'Law, Responsibility and the Brain' in M. Freeman and O. Goodenough (eds.), *Law, Mind and Brain* (Farnham, Ashgate, 2009) 20.
[39] Ibid.
[40] 'Brain and Blame' (1996) 84 *Georgetown Law Journal* 534.
[41] *The Ethical Brain* (New York, Dana Press, 2005) 100.

There are many with prefrontal cortex damage and with amygdaloid abnormalities who do not commit crimes. And brain defects are not observed in all violent criminals. There are even studies which show that PFC damage can decrease antisocial behaviour,[42] and others which show that damage to the amygdala can result in both increased[43] and decreased[44] aggression. Just as it would be wrong to expect full localization of criminality genetically, so it would be misleading to assume there was neurological localization of it.[45] Our brain structures are not immutable; they are susceptible to change and are adaptable.[46] Some brain systems can compensate for others that have been damaged and can effectively perform the same tasks.[47] So, a lone brain scan showing a neurological aberration is not irrefutable proof that a person cannot control his behaviour.

Much of the focus of this volume is on these issues, but there is much else besides. The volume begins with Walter Glannon, whose book *Bioethics and the Brain* has already been cited.[48] Glannon presents us with eight cases to frame and discuss the question of how neuroscience, in the form of neuroimaging, can inform evaluations of people's actions in the criminal law. The discussion supports the view that neuroscience can inform but not determine judgments of criminal responsibility. The cases he presents suggest that brain imaging may be more useful in assessing judgments of criminal negligence, less useful in cases of impulsive behaviour and psychopathy, and least useful in judgments of criminal intent. We should not fall prey to what Morse calls 'brain overclaim syndrome'.[49]

Gert-Jan Lokhorst discusses *mens rea*[50] and modern neuroscience with the help of conceptual work in artificial intelligence and the law. He discusses the concept of *mens rea* using the tools of modern formal logic, and then discusses the concept in relation to neuroscience. He reveals in this way something about the scope and limits of neuroscience in law.

In the third essay in this volume John Martin Fischer addresses determinism and the problem of luck. His closely reasoned analysis repays careful study: it cannot adequately be summarized in the space available.

Greely focuses on the claim that the criminal defendant 'can't help himself', asking specifically 'how could such a claim be proven?'. He points out that persons with Tourette Syndrome were once burnt at the stake. Greely concludes that for a defendant to mount the defence that he 'can't help himself', there must be specific

---

[42] J. M. Ellenbogen et al., 'Ventromedial Frontal Lobe Trauma' (2005) 64 *Neurology* 759.
[43] J. E. Le Doux, *The Emotional Brain: The Mysterious Underpinnings of Emotional Life* (New York, Simon & Schuster, 1996).
[44] R. J. Blair, 'The Roles of Orbital Frontal Cortex in the Modulation of Anti-Social Behavior' (2004) 55 *Brain and Cognition* 121.
[45] Op. cit., note 38, 21.
[46] See L. R. Tancredi, *Hardwired Behavior: What Neuroscience Reveals about Morality* (New York, Cambridge University Press, 2005) 43.
[47] There is 'brain plasticity'. See ibid. 42–5.
[48] Op. cit. note 36.
[49] 'Brain Overclaim Syndrome and Criminal Responsibility: A Diagnostic Note' (2006) 3 *Ohio State Journal of Criminal Law* 397.
[50] On which see *R v G* [2003] UKHL 50 at [32] per Lord Bingham.

proof that ties some characteristics of that defendant (a condition, whether genetic, brain-based or behavioural) that correlates extremely strongly with the criminal behaviour in question. He cannot see it being successfully argued very often. He cites Branner Syndrome, and coprolalia (as part of Tourette Syndrome) as examples where proof may exonerate. He refers to Nicole Vincent's comment on his paper that the number of such cases may expand with developments in neuroscience and human genetics. We must wait and see.

Nicole Vincent in her paper poses the interesting question whether lacking the mental capacity for moral agency excuses or condemns further. Heidi Maibom, in a recent article,[51] has argued for the latter, so that such evidence would enhance the prosecution's case. Marga Reimer,[52] also in a recent article, claims that such evidence both increases and condemns. Vincent's argument is that once we distinguish condemnation of people 'for who they are' from 'what they do', and realize that each of these two types of condemnation plays a role at a *different* stage in a criminal trial, we will see that at the guilt determination stage such evidence clearly favours the defence. She rejects the claims of Maibom and Reimer.

Adina Roskies and Walter Sinnott-Armstrong examine the value of brain images as evidence in the criminal law, specifically of the US. Do they pass muster under the Federal Rules of Evidence? They conclude that brain images are as confusing and misleading in trials as in reported experiments, that their 'moderate dangers' outweigh their minimal probative value. Thus, they fail the balancing test in FRE 403 and should not be admitted into trials.

The chapter that follows, by Buckholtz et al., is rather different from those thus far. It examines the decision-maker rather than the defendant. Legal decision-making in criminal contexts include the essential functions performed by impartial 'third parties', assessing responsibility and determining an appropriate punishment. In order to explore the neural underpinnings of these processes, subjects were scanned with fMRI while they determined the appropriate punishment for crimes that varied in perpetrator responsibility and crime severity. Activity within regions linked to effective processing (amygdala, medial prefrontal, and posterior cingulate cortex) predicted punishment magnitude for a range of criminal scenarios. By contrast, activity in right dorsolateral prefrontal cortex distinguished between scenarios on the basis of criminal responsibility, suggesting that it plays a key role in third-party punishment. The same prefrontal region has previously been shown to be involved in punishing unfair economic behaviour in two-party interactions, raising the possibility that the cognitive processes supporting third-party decision-making and second-party economic norm enforcement may be supported by a common neural mechanism in human prefrontal cortex.

Lisa Claydon's paper also addresses the relationship between explanations of human behaviour drawn from neuroscientific imaging and the approach adopted by the criminal justice system to the establishment of guilt. Unlike the bulk of the other papers in this collection, the focus is on England and Wales, where hitherto

---

[51] H. L. Maibom, 'The Mad, the Bad and the Psychopath' (2008) 1 *Neuroethics* 167.
[52] M. Reimer, 'Psychopathy without (the language of) Disorder' (2008) 1 *Neuroethics* 185.

less use has been made of these developments. Claydon discusses some well-known cases, Brian Thomas (who killed his wife in his sleep) and Sally Clark, who was the victim of a serious statistical miscalculation. She points to new legislative developments that impose more mechanistic definitions of accountability—in particular to a new definition of diminished responsibility.[53]

Blumoff asks how it is that a minor deficit in a brain enzyme can significantly affect predisposition to violent action. He discusses moral luck, and he focuses on abused children. The MAOA (monoamine oxidase type A), a brain enzyme responsible for inhibiting neurotransmitters associated with aggression and the facts of constitutive and circumstantial moral luck make a strong case, he argues, for the impact of genetic/environment interactions (G&E) on violent behaviour. He suggests strategies to tackle this: minimize early childhood abuse—he does not say, but perhaps it is too obvious, make it unlawful to hit children as 29 countries have now done; and use the tools of rehabilitation to mitigate the effects of neurotransmitter deficits as part of the treatment of those who commit crime. Neither of these strategies should be controversial.

For David Terracina, cognitive neuroscience may help us—in his area in particular the Italian legal system—to understand why the penal system is ineffective.

With Betsy Grey's paper we turn to the law of tort, and in particular to emotional harm. It is her view that advances in neuroimaging challenge the traditional US doctrine. In particular, these advances may result in a 'more probative avenue to verify claims for injury related to mental distress'. Advances are also thought by her to improve proof on the subjective element of the tort— whether the claimant did actually suffer harm. The paper discusses in some detail also the different approaches of the English legal system and that in US states on this question.

The next two papers are about juvenile justice.[54] The relevance of neuroscience to this became apparent when the constitutionality of the death penalty for juveniles in the US was challenged in *Roper v Simmons*.[55] This Supreme Court case is the backdrop for both papers. June Carbone offers us a detailed analysis of the Supreme Court judgments. She shows that the existence of neuroscience had the flimsiest of impacts on the judgments. And, as Terry Maroney points out, the court remarked that 'any parent knows' that teenagers are immature. Maroney's own paper examines the value of neuroscience beyond the *Simmons* decision. She asks whether it is necessary to draw on neuroscience to reconfigure juvenile justice. Adolescent brain science, she concludes, does not provide an independent basis to recommit to traditional juvenile justice values. Rather it reinforces the wisdom of so doing.

---

[53] Coroners and Justice Act 2009, s. 53.
[54] See also K. H. Federle and P. Skendelas, 'Thinking Like A Child: Legal Implications of Recent Developments in Brain Research for Juvenile Offenders' and C. Ross, 'A Stable Paradigm: Revisiting Capacity, Vulnerability and the Rights Claims of Adolescents' in M. Freeman and O. Goodenough (eds.), *Law, Mind and Brain* (Farnham, Ashgate, 2009) 199 and 183, respectively.
[55] See note 28.

Mackenzie and Sakel explore ethical and legal issues arising from the impact of neuroscience of decision-making and emotion on end-of-life decisions of patients with certain neurodegenerative disorders, such as motor neurone disease (known in the US as amyotrophic lateral sclerosis).

End-of-life decisions are also the focus of Joseph Fins's paper, which discusses the controversies surrounding the diagnosis of the PVS condition. This raises profound ethical and legal issues. The *Bland* case[56] in England continues to provoke criticism: the turning off of AHN as an omission, not an act, to avoid its classification as murder; the removal of food and water as the removal of treatment, not basic care; the discussion of best interests (what interests does a PVS patient have?) etc. It is known that some patients wake up: is this because of misdiagnosis, or is it recovery? And is PVS the thin end of the wedge? What are the dangers of its extension to analogous conditions, and not-so analogous ones, 'locked-in syndrome', for example? Fins is equally critical of the leading US case of *Quinlan*.[57] He points to numerous flaws in the court's reasoning. He points out that many patients diagnosed as being vegetative may be in a minimally conscious state, a condition in which there is definite, albeit intermittent and episodic, evidence of awareness of self, others or the environment. Such patients may say words or phrases and gesture or show evidence of memory, attention, and intention. He cites a recent study of 41 patients labelled as vegetative which find that 41 per cent were in fact in MCS. And he observes that 'sequestering' these patients in the chronic care sector is concerning as we learn more about mechanisms of brain recovery and potential therapeutic interventions.

The following paper (by Wilkinson and Foster) examines issues at the beginning of life. They acknowledge that neuroscience has not been thought to raise particularly novel problems for newborn infants, and ask whether, as neuroimaging techniques improve, legal judgments about the permissibility of withholding and withdrawing life support from severely damaged neonates will take account of its insights. They conduct a thought experiment—they call it the 'Carmentis Machine'. This will be able to predict accurately future impairments in newborn infants. They consider what implications the development of this machine will have for the law and for practice.

Dov Fox's paper argues that the right to silence (the US Fifth Amendment) protects a right of mental control. Brain imaging brings the moral and legal significance of mental control into sharp relief. Fox argues that the state is prohibited from extracting a suspect's thoughts without her meaningful consent or from making use of a suspect's compelled recall or recognition to lay criminal blame upon her.

There are many policy questions raised by Fins. Viens shows in the next paper that neuroscience has implications too for public health questions. He shows how neuroscience can (and will) be used within the formulation and enforcement of

---

[56] [1993] AC 789.
[57] *In The Matter of Karen Quinlan* 355 A 2d 647 (1976).

public health law. He focuses on public health law as it pertains to pandemic prevention and response.[58]

For Boudreau and her colleagues also neuroscience can inform the study and practice of law. Of their many findings, the one with the most implications for the practice of law relates to how jurors process information.

Capraro's paper too focuses on the jury, in particular on the contribution that neuroscience can make to our understanding of the relationship between emotions and decision-making.

Pfaff and Sherman argue that humans are 'wired for reciprocity'. Recent work in neuroscience points to a physical/hormonal basis for moral reciprocity. They ask: 'Is there room for empathy in a courtroom dealing with issues that are usually left to actuaries?' Should courts take account of it? And what of the person not 'wired' like others who does harm? Are the findings of neuroscience equally relevant to the criminal law as to the civil law of torts?

James Duffy's paper also addresses empathy. He looks to a future drawn not by the legacy of Thomas Hobbes, but rather one influenced by the Dalai Lama. He emphasizes the role of contemplation in creating a new commonwealth built on compassion, not fear.

The next paper, by Steven Goldberg, focuses on religion. Would respect for religious belief erode if religion came to be seen as merely a byproduct of biology and, he adds, not a very attractive byproduct? But, for two reasons neuroscience is not likely to undermine the social and legal status of religion. First, because of the philosophical hurdles that need to be surmounted—upon which there is hardly a beginning. And, recently, because the challenges are not new—they can be traced back 100 years or more—and have had little or no impact. But this is only 'half the story'. He sees neuroscience playing a role in law and religion disputes in courts, for example where parents object to medicine for their children on religious grounds. I must say that I, for one, remain to be convinced that this will happen.

The paper of the Mexican scholar, Enrique Cáceres, presents a theoretical model of judicial reasoning that attempts to integrate partially offered explanations by three different paradigms: ones drawn from legal philosophy, legal epistemology, and artificial intelligence and law. He claims an epistemic evaluation is more effective if it considers the pragmatic use of the system of investigation, and does not rely solely on a set of rules out of touch with the practices of its users.

The next paper is by the American judge, Morris Hoffman. It asks what we should do when our moral institutions conflict with existing legal doctrine, a dilemma that on any test judges must themselves face from time to time. But his particular concern is the dissonance between the standard murder rule and our intuitions of moral blameworthiness. He offers some principles to assist us in resolving this dilemma.

The penultimate paper is an excursion into legal, social, and medical history. In it Daniel Goldberg argues that an understanding of this has important implications

---

[58] See, further, the two volumes by M.Freeman (ed.), *The Ethics of Public Health* (Farnham, Ashgate, 2010).

for evaluating the scope and significance of neuroimaging evidence today. He notes that discussions of neuroscientific evidence and the law have tended to be historical. We can learn from the nineteenth-century experience with photographs and the early twentieth-century use of the X-ray.

The volume concludes with Stephen Morse cautioning against putting too much faith in the contribution that neuroscience can make to legal doctrine, practice, and theory. He draws entirely on criminal law, but believes what he says is generalizable to other legal contexts. He sees neuroscience as 'simply the most recent mechanistic causal science that appears deterministically to explain behaviour'. The volume thus ends, as it began, on a sceptical note. Whether this is an appropriate judgement to make, or to make at this juncture, must be assessed by the many studies that now exist, some of which can be found in this collection of papers. Of one thing we can be sure: the debates about the relationship between law and neuroscience are here for a long time yet.

2

What Neuroscience Can (and Cannot) Tell Us about Criminal Responsibility

*Walter Glannon**

2.1 Introduction

Criminal responsibility presupposes the capacity to respond to reasons against actions that make one liable to punishment.[1] This capacity enables one to restrain one's impulses and form and execute intentions in actions. It is by virtue of our rational capacity that we can control our behaviour. Until recently, whether or to what extent one had this capacity was inferred entirely from one's behaviour. The neurobiological basis of our actions remained elusive, making it difficult to draw valid inferences from brain to mind to action. The increasing use of structural and especially functional neuroimaging in the criminal law has revealed an association between some types of criminal behaviour and dysfunction in brain regions mediating practical and moral reasoning and decision-making. By showing correlations between brain dysfunction and impaired cognitive, conative, and affective processing, neuroimaging might support claims of mitigation or excuse from criminal responsibility for certain actions.

There are, however, limitations in what neuroimaging can tell us about brain function and its relation to motivational states and behaviour. Because of these limitations, the interpretation of brain images and their legal significance are fraught with uncertainty. What further contributes to this uncertainty is that empirical data from brain scans are assessed against a background of normative considerations regarding how persons are expected to control their behaviour. There is disagreement about the extent to which we have and exercise this control.

I present six cases to frame and discuss the general question of how neuroscience, in the form of neuroimaging, can inform evaluations of people's actions in the criminal

---

* Associate Professor of Philosophy, University of Calgary. The author is grateful to the other participants in the Law and Neuroscience Colloquium at University College London, 6–7 July 2009, for helpful comments on an earlier version of this paper.

[1] H. L. A. Hart defines this capacity as 'the ability to understand what conduct legal rules or morality require, to deliberate and reach decisions concerning those requirements, and to conform to decisions when made': *Punishment and Responsibility* (Oxford, Clarendon Press, 1968) 227.

law. The first two cases involve the issue of whether measuring function in the prefrontal cortex can determine that an individual had or lacked the requisite intention to be held responsible for sexual assault and first degree murder. The third and fourth cases involves the issue of whether an underactive prefrontal cortex and overactive amygdala make one incapable of controlling violent impulses. If they impair or undermine this control, then these brain features could constitute a mitigating or excusing condition for the actions that result from them. The fifth case raises the question of whether dysfunction in the brain regions mediating the capacity for empathy and emotional responses to other people demonstrates that psychopaths are not responsible for their behaviour. I also consider whether an abnormal cortical-limbic pathway in the psychopath's brain makes him unable to control his aggressive impulses. The sixth case considers whether appeals to neuroimaging can elucidate the role of memory in behaviour and resolve disputes about criminal negligence. Discussion of these actual and hypothetical cases will support the view that neuroscience can inform but not determine judgments of criminal responsibility. Even when scans show significant dysfunction in regions of the brain associated with reasoning and decision-making, they at most will supplement and not supplant behavioural evidence in legal judgments of responsibility, mitigation, and excuse.

2.2 Criminal Intent

For one to be criminally liable for sexual assault as well as first degree murder, one must have formed and executed an intention to perform the criminal act. The *actus reus* ('bad act') must be related in the relevant causal respects to the *mens rea* ('guilty mind'[2]). The intention to perform a harmful or wrongful action issues in that action. In a 2007 case in the US, defendant Peter Braunstein was charged with kidnapping, sexual abuse, burglary, robbery, and arson for a series of actions committed on Halloween in 2005. Positron emission tomography (PET) scans of the defendant's brain were introduced by the defence in an attempt to show that he was unable to plan the sexual assault of a former co-worker.[3] The defendant's not guilty plea was not based on the insanity defence. He did not claim that he was suffering from a defect of reason making him unable to appreciate the wrongfulness of his conduct. Instead, the not guilty plea was based on the claim that a brain abnormality made him incapable of forming an intention or plan to commit the crime.[4] He lacked the necessary *mens rea* to be criminally responsible and liable for

---

[2] See M. S. Moore, *Causation and Responsibility* (New York, Oxford University Press, 2009).
[3] A. Hartocollis, 'Attack not Disputed at Trial, just Intent of the Attacker' (2007) *New York Times*, 23 May.
[4] P. S. Appelbaum, 'Through a Glass Darkly: Functional Neuroimaging Evidence Enters the Courtroom' (2009) 60 *Psychiatric Services* 21–3; P. S. Appelbaum, 'Insanity, Guilty Minds, and Psychiatric Testimony' (2006) 57 *Psychiatric Services* 1370–2. See also M. Bratman, *Intention, Plans, and Practical Reason* (Cambridge, MA, Harvard University Press, 1987); A. Mele, *Springs of Action: Understanding Intentional Behavior* (New York, Oxford University Press, 1992); and R. A. Duff, *Intention, Agency and Criminal Liability: Philosophy of Action and the Criminal Law* (Oxford, Blackwell, 1990).

it. The PET scans were introduced as evidence that the defendant had dysfunctional frontal lobes, the executive part of the brain regulating personality, planning, decision-making, and moral judgement. A psychiatrist testifying for the defence stated: 'If I saw this scan without knowing anything else... I would say this person has changes in personality, will have difficulty planning, making executive judgments, and thinking ahead.' Under cross-examination, the psychiatrist responded affirmatively to the question: 'Was the defendant completely unable to plan?'[5]

Presumably, the PET scans were behind the change in the psychiatrist's assessment from the weaker claim that the defendant had 'difficulty planning' to the stronger claim that he was 'completely unable to plan'. But this distinction points to one of the principal limitations of functional brain imaging. Except perhaps in cases of severe brain dysfunction, PET or fMRI scans of the prefrontal cortex alone cannot determine whether an individual lacked the capacity to plan a crime, whether he had less than full capacity to plan and thus had difficulty in forming and executing his intention, or whether he had full capacity and executed his intention straightaway. This threefold distinction is more pertinent to the issue of impulse control, and I will return to it in the next section. In the present case, neither the stronger claim that the defendant lacked the capacity to plan the assault nor the weaker claim that he had difficulty planning it withstand scrutiny. The elaborate scheme he devised to assault the victim, which consisted in not one but a series of actions, and the duration of the assault, clearly show that he had full capacity to plan and execute his criminal behaviour. The defendant dressed as a firefighter and set off a smoke bomb in the lobby of his victim's apartment building. To gain access to her apartment, Braunstein knocked on her door and claimed that he had come to inspect the smoke damage. After she let him into her apartment, he used chloroform to render her unconscious and then committed the sexual assault over a period of 13 hours. His behaviour demonstrated that any brain abnormality displayed on the PET scans did not interfere with his cognitive capacity to plan and act. It played no role in his behaviour and thus provided no grounds for reversing the charge and conviction that his criminal course of action was intentional.

In most cases, the claim that one was unable to plan a criminal act one performed is not defensible. It would have to be confirmed by the individual's pattern of behaviour over time. The idea that one was unable to plan a criminal act would have to be consistent with a general pattern of behaviour symptomatic of an inability to plan or difficulty in planning. One would have difficulty performing many cognitive tasks in everyday life. Holding a job and managing one's finances would be a challenge. The defence claimed that Braunstein was a paranoid schizophrenic and had recently walked away from a successful career in journalism. Abandoning his journalistic career may have been an irrational decision. But it is not necessarily a symptom of paranoid schizophrenia or a general inability to plan.

---

[5] A. Hartocollis, 'In Support of Sex Attacker's Insanity Plea, a Look at his Brain' (2007) *New York Times*, 11 May. Contrary to what the reporter claims in the title of this article, neither the defendant nor his defence attorney invoked a plea of insanity. Appelbaum (2009), note 4, above, points this out in citing the two comments by the testifying psychiatrist.

Just because a decision is irrational does not mean that it was coerced or compelled by a dysfunctional brain and that the individual making the decision had no control of the motivational states that led to it. It is possible that dysfunction in the frontal-parietal network could cause an acute episode of psychosis, in which case one could not know what one was doing and in this regard could not control one's actions. This might constitute an excusing condition from responsibility for sexual assault. But this type of dysfunction and behaviour is not consistent with the behaviour of the defendant in committing the assault. Given the elaborate nature of the interconnected set of his actions, any claim that his behaviour was caused by acute psychosis would be implausible. Braunstein's behaviour undermined the defence attorney's claim that his actions were performed in a 'vague, improvisational haze', and that he did not know 'what he was going to do from one moment to the next'.[6]

It is surprising that Braunstein's attorney did not argue explicitly for the insanity defence, since the characterization of his action conceivably could meet the cognitive criterion of the defence according to the M'Naghten Rules and the Model Penal Code.[7] 'Vague, improvisational haze' suggests that Braunstein had a defect of reason preventing him from knowing what he was doing or of appreciating the wrongfulness of his actions. This could rule out criminal intent. Nevertheless, the defendant's behaviour indicated that he had no difficulty planning the crime. The interconnected set of actions leading up to and including the assault were evidence of the fact that considerable planning went into it. The PET scans introduced to support the defence's claim that Braunstein was unable to form an intention to commit the crime were refuted by his behaviour, and the assault charge was sustained. The testifying psychiatrist's initial weaker claim that the PET scan indicated difficulty in planning was not sustainable in court either. Braunstein was sentenced to eighteen years to life for his crime. Even if the scans displayed frontal lobe dysfunction correlating with cognitive impairment involving difficulty in planning, this by itself would not be evidence that Braunstein lacked criminal intent. One may have difficulty planning and executing a criminal act and still perform it. What matters for criminal responsibility is that one executed the relevant intention, regardless of whether one did it with ease or difficulty.

The ruling in another case reinforces the idea that behavioural evidence alone may be sufficient to establish the necessary aggravating factors and uphold a conviction of first degree murder. In a 2000 Florida case, at a sentencing hearing, a psychologist presented brain scans suggesting that the defendant, Eddie Lee Sexton, had a diminished level of self-control due to dysfunction in his prefrontal cortex.[8] This evidence was introduced to support the defendant's claim that he was suffering from extreme mental disturbance when he committed the crime and that

---

[6] Hartocollis (2007), notes 3 and 5, above.

[7] *M'Naghten Case*, 8 Eng Rep 718, 722 (1843), cited in the *Report of the Committee on Mentally Abnormal Offenders* (London, Her Majesty's Stationery Office, 1975) 217. *Model Penal Code*: Official Draft and Commentaries (Philadelphia, American Law Institute, 1985) 4.01.

[8] *Sexton v State* 775 So 2d 923 (Fla 2000).

this should be a mitigating factor against the death sentence. Yet the trial court imposed the death penalty on the grounds that aggravating factors outweighed any evidence of diminished self-control. The defendant's behaviour displayed a carefully planned course of action in which he manipulated his mentally disabled son to kill his son-in-law, who knew of Sexton's involvement in another homicide. The behavioural evidence that the crime was 'cold, calculating, and premeditated' not only outweighed but discredited the neuroscientific evidence suggesting diminished responsibility. The Florida Supreme Court upheld the death sentence, noting that 'nothing in the expert testimony presented [including the brain scans] suggests that Sexton was incapable of planning the murder and manipulating his children to assist with the murder'.[9] Even in cases where imaging displaying brain dysfunction might be used to support the claim that an individual lacked the capacity to intend to kill and thereby change a conviction from first degree murder to second degree murder or manslaughter, behavioural evidence would be needed to confirm the neuroscientific findings.

2.3 Violent Impulses

Suppose that a couple is having an argument. One partner criticizes and blames the other for everything that has gone wrong in their disintegrating relationship. The one who is blamed becomes violently angry and strangles and kills the other. Charged with second degree murder, the defence for the accused claims that his action was an instance of a general inability to control violent impulses. He undergoes PET and fMRI scans showing abnormally high activity in limbic areas such as the amygdala, which regulates emotions such as fear and anger. The imaging also shows abnormally low activity in the prefrontal cortex, which regulates the inhibition of impulses. These particular findings are consistent with general findings from studies conducted by a number of neuroscientists. They have shown that many impulsive murderers have reduced activity in the prefrontal cortex and increased activity in the amygdala.[10] Since normal activity (measured in terms of blood flow and glucose metabolism) in these brain regions is necessary for one to control one's impulses, the images in this hypothetical case suggest that the accused had significantly impaired impulse control. This ought to be a mitigating, if not an

---

[9] Ibid. 934.
[10] R. J. Davidson, K. M. Putnam, and C. L. Larson, 'Dysfunction in the Neural Circuitry of Emotion Regulation—A Possible Prelude to Violence' (2000) 289 *Science* 591–4; A. Raine, T. Lencz, S. Bihrle, L. LaCasse, and P. Colletti, 'Reduced Prefrontal Grey Matter Volume and Reduced Autonomic Activity in Antisocial Personality Disorder' (2000) 58 *Archives of General Psychiatry* 119–27; A. S. New et al., 'Blunted Prefrontal Cortical 18 Fluorodeoxyglucose Positron Emission Tomography Response to Meta-Chlorophenylpiperazine in Impulsive Aggression' (2002) 59 *Archives of General Psychiatry* 621–9; S. W. Anderson, A. Bechara, H. Damasio, D. Tranel, and A. Damasio, 'Impairment of Social and Moral Behavior Related to Early Damage in Human Prefrontal Cortex' (1999) 2 *Nature Neuroscience* 1032–7; and A. Damasio, 'Neuroscience and Ethics: Intersections' (2007) 7(1) *The American Journal of Bioethics: AJOB Neuroscience* 3–7.

excusing, condition that could absolve the accused from responsibility for killing his partner.

It would appear that impulse control is one area where neuroimaging would be particularly helpful in evaluating criminal behaviour. Yet it is in this area where the limitations of neuroimaging are most obvious. I will cite six such limitations, which is by no means an exhaustive list.[11]

First, brain scans may establish correlations between neurobiological abnormalities and criminal behaviour. But correlation is not causation. In some cases, the correlation between a brain abnormality and criminal behaviour may be strong enough to establish a causal connection and excuse a person from responsibility for that behaviour. For example, if loss of the capacity for impulse control and decision-making correlated with the presence of a tumour in the prefrontal region of the brain, and this capacity was restored upon removal of the tumour, then this would suggest a causal connection between the tumour and the criminal behaviour.[12] In most cases, there are only correlations between brain abnormalities and criminal behaviour. So, one could not claim that a particular brain abnormality caused a particular action.

Second, those who appeal to neuroimaging to claim that brain dysfunction can excuse one from responsibility for criminal actions tend to focus on a specific region of the brain that mediates cognitive functions associated with reasoning and decision-making. Yet these functions are not usually located in a single region of the brain but are distributed across complex interacting neural networks. The relevant brain–mind connections are not one–one but many–many.

Third, given the artificial setting of a PET or fMRI scanner, brain regions activated when one is asked to perform a cognitive task might not be activated in the same way as they were when one performed a criminal act. One may be able to rehearse one's responses in the affectively 'cool' scanning scenario, which would be very different from the emotionally heightened state that issued in the earlier impulsive act. Attempts to elicit an emotional reaction in the scanner through photos or verbal cues would not replicate the emotional reaction that led to the crime because it could not replicate the same environmental, neurobiological, and psychological triggers of the act. Because the individual could know ahead of time which tasks he or she would be asked to perform and the type of photos shown to elicit an emotional response, the amygdala and other regions of the brain mediating his emotions would likely not be as active as they were during the commission of the crime.

---

[11] D. Mobbs, H. C. Lau, O. D. Jones, and C. Frith, 'Law, Responsibility, and the Brain' (2007) 5(4) *PLoS Biology* 693–700. See also Appelbaum (2007), note 4, above.
[12] This occurred in the case of an American teacher in Virginia whose paedophilia was traced to a meningioma pressing on his orbitofrontal cortex. Removal of the tumour resolved the paedophilia. This behaviour returned with a new meningioma in the same brain region and resolved again when the second tumour was removed. See J. M. Burns and R. H. Swerdlaw, 'Right Orbitofrontal Tumor with Pedophilia Symptom and Constructional Apraxia' (2003) 62 *Archives of Neurology* 437–40. See also S. Batts, 'Brain Lesions and their Implications in Criminal Responsibility' (2009) 27 *Behavioral Sciences & the Law* 261–72; and J. J. Knabb, R. K. Welsh, J. G. Ziebell, and K. S. Reimer, 'Neuroscience, Moral Reasoning, and the Law' (2009) 27 *Behavioral Sciences & the Law* 219–36.

Fourth, a brain scan would not be able to determine whether the individual lacked the capacity to inhibit the impulse to strangle and kill his partner, whether he had some capacity to inhibit it but had difficulty doing so, or whether he had the capacity to inhibit the impulse but simply failed to at the time of his action.[13] The degree to which one has or lacks the capacity to control one's behaviour can only be assessed by observing that behaviour over time. Brain images may be necessary to confirm this assessment. But it cannot be done on the basis of brain imaging alone.

Fifth, dysfunction in the prefrontal cortex would not rule out the possibility of other regions in the frontal lobes taking over some of its inhibitory mechanisms and enabling one to have some control of one's impulses. Neuroplasticity, the ability of nerve cells to modify their activity in response to change, might enable other regions of the brain to take over the tasks associated with regions that have become dysfunctional. Given that the brain becomes less plastic as we age, however, rewiring or bypassing dysfunctional circuits in the prefrontal area may be more likely in a younger person and less likely in an older one. This might suggest that the first individual would have a greater degree of control of his impulses than the second and might be held to a higher standard of responsibility. Admittedly, this is speculative. It also involves making questionable inferences from groups to individuals, which neglects the fact that no two people's brains respond in the same way to injury. Accordingly, this should remain an open question. In any case, a determination of greater or lesser control on the basis of age would have to be confirmed by observing the individual's behaviour.

Sixth, brain dysfunction and mental impairment come in degrees. There is no empirical measure that could establish a threshold at or over which one is responsible and under which one is not responsible. As an imperfect empirical measure of brain function and its relation to behaviour, neuroimaging cannot tell us what degree of dysfunction in different regions of the brain can excuse one from criminal responsibility for an action.

There will be disagreement within the legal community about what degree of brain dysfunction is enough to impair responsibility-relevant mental functions and support mitigation or excuse. Empirical considerations regarding information about the brain cannot be isolated from normative considerations regarding the behavioural and legal significance of that information. What degree of control an individual has of his mental states and behaviour, and how much control is necessary for responsibility, will be influenced by social expectations about how individuals should act in conforming their conduct to social rules and the requirements of the law.

---

[13] R. Sapolsky, 'The Frontal Cortex and the Criminal Justice System' (2004) 359 *Philosophical Transactions of the Royal Society of London* 1787–809; B. McSherry, 'Criminal Responsibility, Fleeting States of Mental Impairment, and the Power of Self-Control' (2004) 27 *International Journal of Law and Psychiatry* 224–57; N. Vincent, 'Neuroimaging and Responsibility Assessments' (2009) 2(2) *Neuroethics* 39–54; and N. Vincent, 'Responsibility, Dysfunction, and Capacity' (2008) 1(3) *Neuroethics* 199–204.

A significant degree of prefrontal damage might support the claim that one could not control one's impulses. Even here, though, a general pattern of impulsive behaviour would be needed to support this claim. The execution of the violent impulse that resulted in the death of the partner would be one manifestation of this behaviour. If the individual's inability to control his impulse on this occasion was caused by a dysfunctional prefrontal cortex, then presumably this dysfunction would also manifest in other incidents of violent impulsive action. Other than perhaps an acute psychotic episode, brain dysfunction disrupting cognitive and affective processing would likely not manifest in mental impairment and irrational or unlawful behaviour in only one incident. In addition, there would have to be consensus within the medical and legal community in interpreting imaging data and evaluating the significance of the correlation of the brain damage to the behaviour in order to argue that the individual was not responsible. Nevertheless, a certain degree of dysfunction in cortical-limbic pathways could make it difficult for an individual to fully engage his cognitive capacities and control his behaviour. Neuroimaging displaying this dysfunction might support a claim of impairment in restraining impulses and thus might be a mitigating factor in judgments of criminal responsibility for harmful impulsive acts. But imaging alone would not be able to establish this. Behavioural evidence would be needed as well.

The fact that neuroimaging data may be open to different interpretations presents a number of challenges for the use of imaging in criminal cases. If brain scans cannot objectively determine that an individual has or lacks the requisite degree of control of his or her behaviour, then scans have minimal probative value for criminal responsibility. Any probative value they might have is outweighed by the potential for misinterpretation and prejudice in determining the facts of the case.[14] Prosecution and defence could present ambiguous imaging data to insufficiently cautious judges and juries in biased ways. Moreover, in countering a claim from the defence that brain dysfunction impaired the cognitive processing of the accused, prosecution could appeal to neuroplasticity to argue that dysfunction in a particular region of the brain associated with cognitive processes would not necessarily impair or undermine this processing. Appeals to age could complicate matters further. Prosecution could argue that a forty-year-old with some damage to frontal regions may have recovered some inhibitory cognitive functions thanks to neuroplasticity. At the same time, defence could exploit the fact that the frontal lobes are not fully developed until early adulthood in arguing that a seventeen-year-old could not be fully responsible for a criminal act because he lacked the inhibitory mechanisms necessary for responsibility.

This was one of the arguments presented in the US case of *Roper v Simmons*.[15] At issue was whether Christopher Simmons, who twelve years earlier had murdered a woman when he was seventeen, should receive the death penalty for his crime. The American Medical Association filed an *amicus curiae* brief based on imaging studies

---

[14] W. Sinnott-Armstrong, A. Roskies, T. Brown, and E. Murphy, 'Brain Images as Legal Evidence' (2009) 5(3) *Episteme* 359–73.

[15] *Roper v Simmons* 543 US 551 (2005).

of adolescent brains. Drawing from these studies, the Association stated: 'To a degree never before understood, scans can now determine that adolescents are immature not only to the observer's naked eye, but in the very fibers of their brains.'[16] The US Supreme Court ruled 5-to-4 that the death penalty was unconstitutional as punishment for murder by a juvenile and upheld the sentence of life imprisonment. The imaging studies did not influence so much as confirm the Supreme Court majority opinion by Justice Anthony Kennedy, which was based largely on accepted views of developmental psychology. The ruling in this case underscores the following point. Just because brain scans confirm that adolescents are comparatively less mature in their reasoning and decision-making than adults does not imply that they cannot be responsible to any degree for their actions.

Unlike cases involving the question of criminal intent, brain imaging data may play a greater role in cases where responsibility hinges on the question of impulse control. This is because the correlations between images of brain dysfunction and behaviour are stronger in the latter than they are in the former. Yet behavioural evidence, normative evaluations of behaviour, and normative evaluations of imaging data are more significant than imaging alone. A functional brain scan showing an underactive prefrontal cortex or overactive amygdala by itself will not be diagnostic of a loss of impulse control or cognitive control of one's behaviour.

2.4 Psychopathy

A man has been bullying a younger co-worker with verbal taunts for several weeks. He lures his colleague to his apartment, claiming that he is having a party with friends. As soon as the younger person arrives, the bully kills him by stabbing him with a knife and then disposes of his body. After he is charged with first degree murder, a psychologist interviewing the defendant notices that he has a number of psychopathic traits. The defendant undergoes PET and fMRI scans, which show that he has abnormal activity in the pathway between the prefrontal cortex and the amygdala. During the trial, his defence attorney presents data from neuroimaging studies showing that this brain abnormality is common among psychopaths. He then presents images from the brain of the defendant to argue that his psychopathy constitutes an excusing condition and that he could not be criminally responsible for the murder.

Psychopathy can be described as a mental disorder involving impaired capacity for empathy and remorse, as well as impaired responsiveness to fear-inducing stimuli and poor behaviour control. Psychopaths are impaired in their capacity for both practical and moral reasoning. These characteristics explain the difficulty psychopaths have in considering the interests, needs, and rights of others, to deliberate, plan, and choose rationally, and conform to social norms. There have been a number of imaging studies revealing the neurobiological underpinning of psychopathy. James Blair and colleagues note that 'there are strong reasons to

[16] Ibid., No. 03–633.

believe that individuals with psychopathy present with orbital/ventrolateral frontal cortex dysfunction as well as amygdala dysfunction'.[17] These brain abnormalities desensitize the psychopath to normal emotional responses. Further, Blair et al. state that 'these impairments interfere with socialization, such that the individual does not learn to avoid actions that cause harm to other individuals'.[18]

Another region of the brain that has been implicated in psychopathy is the anterior cingulate cortex, which mediates the capacity for empathy. A normal empathic response to other persons and a normal fear response to threatening situations are necessary to inhibit the performance of harmful and wrongful actions. These responses inhibit these actions in part by enabling a person to feel regret or remorse for them, which become anticipatory once one forms a memory of these feelings. The prefrontal cortex, amygdala, and anterior cingulate constitute a cortical-limbic network that mediates the capacity for practical and moral reasoning. This capacity, which depends on the capacity for empathic and fearful responses, is necessary for one to internalize moral reasons for or against certain actions. It enables one to recognize and respond appropriately to these reasons and to consider how one's actions affect others.[19] A more recent imaging study conducted by Andrea Glenn, Adrian Raine, and Robert Schug has shown that psychopaths have reduced activity in all of the regions comprising the moral neural circuit. This includes the amygdala, medial prefrontal cortex, anterior and posterior cingulate, and angular gyrus.[20] Such reduced activity interferes with the capacity for the interpersonal interactions central to moral behaviour. In so far as these capacities and responses are necessary for one to be criminally responsible for one's actions, and the psychopath lacks these capacities, it seems to follow that he is not responsible for any criminal acts he commits. Or, if he does not lack but is impaired in these capacities, then the impairment constitutes a mitigating factor and thus he at most can be only partly responsible.

Not everyone agrees with this argument. Psychologist Robert Hare, who has studied psychopaths for years and is the author of the well-known *Hare Psychopathy Checklist*, asserts that psychopathy is an aggravating rather than a mitigating factor in determining criminal responsibility.[21] Although Hare's assertion is based mainly

---

[17] R. J. R. Blair, D. Mitchell, and K. Blair, *The Psychopath: Emotion and the Brain* (Malden, Blackwell, 2005); R. J. R. Blair, 'Neurobiological Basis of Psychopathy' (2003) 182 *British Journal of Psychiatry* 5–7; R. J. R. Blair, 'The Amygdala and Ventromedial Prefrontal Cortex in Morality and Psychopathy' (2007) 11 *Trends in Cognitive Sciences* 387–92; R. J. R. Blair, 'The Cognitive Neuroscience of Psychopathy and Implications for Judgments of Responsibility' (2008) 1(3) *Neuroethics* 149–57; and K. A. Kiel, A. M. Smith, R. D. Hare et al., 'Limbic Abnormalities in Affective Processing by Criminal Psychopaths as Revealed by Functional Magnetic Resonance Imaging' (2001) 50 *Biological Psychiatry* 677–84.

[18] Blair et al., *The Psychopath* 13.

[19] John Martin Fischer and Mark Ravizza develop a theory of moderate reasons-responsiveness that consists of strong receptivity and weak reactivity to reasons for or against actions in *Responsibility and Control: A Theory of Moral Responsibility* (New York, Cambridge University Press, 1998).

[20] A. L. Glenn, A. Raine, and R. A. Schug, 'The Neural Correlates of Moral Decision-Making in Psychopathy' (2009) 14 *Molecular Psychiatry* 5–6.

[21] R. Hare, 'Psychopaths and their Nature' in T. Millon et al. (eds.), *Psychopathy: Antisocial, Criminal, and Violent Behavior* (New York, Guilford Press, 1998) 188–212; and R. Hare, *The Hare Psychopathy Checklist*, second edition (Toronto, Multi-Health Systems, 2003). See also note 17, above.

on behavioural traits, he has been involved in imaging studies showing limbic abnormalities in psychopaths. As noted, studies by Blair and others show that psychopaths have some brain dysfunction. Yet, as in cases involving questions about impulse control, brain dysfunction comes in degrees. It is not clear what degree of dysfunction is enough for the psychopath to be impaired in or lack the cognitive and affective capacities necessary to respond to reasons against harmful and wrongful actions and to be criminally responsible for them. Here too normative considerations involving evaluation of the psychopath's behaviour and how scans of his brain are interpreted will influence judgments about responsibility. Imaging data alone will not determine that the psychopath lacks the capacity to respond to moral and legal reasons against certain actions. Being impaired in a cognitive capacity is not equivalent to lacking that capacity altogether.

Severe dysfunction in the form of a highly underactive or inactive amygdala or anterior cingulate cortex might impair the psychopath's cognitive and affective processing to such a degree that he would not be able to internalize and respond to reasons against performing criminal acts. Damage to the anterior cingulate could make him lose his capacity for empathy and result in desensitized responses to other persons. The brain dysfunction would deprive the psychopath of his rational and moral capacity and would be grounds for excuse from responsibility.[22] Curiously, like psychopaths, individuals with autism spectrum disorders are impaired in their capacity for empathy. One hypothesis for this impairment is that individuals with autism have dysfunctional mirror neuron systems in cortical and limbic regions. Unlike psychopaths, those with autism do not deliberately harm others. Perhaps differences in the structure and function of the anterior cingulate between these two groups might go some way toward explaining why one group deliberately commits harmful acts while the other does not. I cannot adequately address this issue here. But the fact that psychopaths and persons with autism have impaired capacity for empathy suggests that this type of impairment alone is not enough to excuse psychopaths from responsibility for their actions. There would have to be agreement within the legal community that severe dysfunction in the anterior cingulate or other brain regions clearly deprive the psychopath of an empathic capacity. In these cases, and even more so in cases of moderate brain dysfunction, imaging alone would not be a sufficient basis on which to conclude that psychopaths could not be responsible for their behaviour.

Let us now consider the relation between a dysfunctional amygdala and the psychopath's impaired response to fear-inducing stimuli. Because the amygdala mediates primitive emotions such as fear, it may not play as great a role as the anterior cingulate in regulating cognitive-affective states and responsiveness to moral reasons. But its regulation of fearful responses probably plays at least as significant a role as the anterior cingulate in the inhibitory mechanisms that enable us to refrain from performing harmful actions. Appeals to neuroplasticity might not be helpful in arguing for the responsibility of the psychopath despite a dysfunctional amygdala.

---

[22] S. Morse, 'Psychopathy and Criminal Responsibility' (2008) 1(3) *Neuroethics* 205–12. W. Glannon, 'Moral Responsibility and the Psychopath' (2008) 1(3) *Neuroethics* 158–66.

Through its regulation of fear and its role in the fight-or-flight response to external threats, this brain region is critical to survival. This function might not be taken over so readily by other regions. More primitive hardwired functions in the brain and the unconscious mental states they generate and sustain might not be so amenable to transfer from one region to another. Yet most imaging studies of psychopaths have shown not severe but moderate dysfunction in cortical-limbic pathways. It is also important to point out that one study twelve years ago by Adrian Raine and colleagues using PET scans of the brains of impulsive murderers and predatory psychopaths yielded different results from those in Blair's studies. Compared to normal controls, the impulsive murderers showed reduced activation in the bilateral prefrontal cortex and enhanced activation in limbic areas such as the amygdala. The predatory psychopaths had relatively normal prefrontal function and increased activity in the amygdala and hippocampus.[23] As Dean Mobbs et al. put it: 'These results suggest that predatory psychopaths are able to regulate their impulses, in contrast to impulsive murderers, who lack the prefrontal "inhibitory" machinery that stops them from committing violent transgressions.'[24] While I have emphasized the limitations of imaging to support claims of mitigation or excuse, if anything this study supports the claim that psychopaths have enough control of their behaviour to be responsible for it. It is also noteworthy that Glenn et al. did not conclude from their imaging study that dysfunction in brain regions mediating moral decision-making decisively shows that psychopaths have no control of their behaviour. Instead, they concluded that this dysfunction 'may *partly* explain a complex social problem—the psychopath'.[25]

The characterization of psychopaths as 'predatory' is significant because it suggests that they are not as deficient in instrumental and moral reasoning as some might think. Blair et al. point out that psychopaths present with highly elevated levels of both instrumental and reactive aggression.[26] The first type of aggression is purposeful and goal-directed. In so far as it is instrumental, the aggression is planned and aimed at a specific goal. Bullying is an example of instrumental aggression. When he engages in instrumental aggression such as bullying, the psychopath identifies the person to whom the aggression is directed as a potential victim, as a vulnerable individual whom he desires to manipulate for his selfish ends. Given that his instrumental aggression is deliberate and he knows that a particular individual will be victimized and harmed by his actions, he is capable of knowing to some degree that this is a moral and criminally liable transgression. He is capable of knowing that it is more than just a conventional transgression, where there is no identifiable victim who might be harmed by the action. If the psychopath is capable of knowing that his instrumental aggression is a moral and legal transgression, then it seems that he would also be capable of

---

[23] A. Raine, J. R. Meloy, S. Bihrle et al., 'Reduced Prefrontal and Increased Subcortical Brain Functioning Assessed Using Positron Emission Tomography in Predatory and Affective Murderers' (1998) 16 *Behavioral Sciences & the Law* 319–32.
[24] Note 11, above.
[25] Note 20, above, 6, emphasis added.
[26] *The Psychopath*, note 17, above, 12–13.

responding to moral and legal reasons against the transgression. Imaging showing dysfunction in prefrontal and limbic brain regions of psychopaths may support the claim that they are impaired in their capacity for moral reasoning, which may make them less than fully responsible for their behaviour. But impaired capacity is not equivalent to a lack of capacity. In spite of any brain abnormalities, the instrumental aggression of psychopaths suggests that they have enough cognitive capacity to respond to moral and legal reasons against their harmful actions. Thus they can be at least partly criminally responsible for them. They may not be fully responsible; but they should not be excused either.

2.5 Criminal Negligence

I have argued that neuroimaging displaying brain abnormalities by itself cannot determine that an individual was so mentally impaired that he or she could not be criminally responsible for a particular action. Yet imaging may have more weight in judgments about responsibility when there are strong correlations between moderately severe to severe brain dysfunction and mental impairment. Earlier, I noted that in some cases there may be a causal connection between a brain abnormality and criminal behaviour. Such a connection could excuse a person from responsibility for that behaviour. Another type of case where imaging may support an excuse is when it shows a brain abnormality correlating with an unintentional omission of one person resulting in the death of another. Damage to brain regions mediating the formation of episodic memories of recent events may significantly impair one's capacity for short-term working memory and one's capacity for reasoning and decision-making. It may result in negligence with criminal implications.

Suppose that a father leaves his young son unattended at the edge of an outdoor swimming pool when he goes inside his house to answer a phone call. The conversation lasts about an hour, and during this time he is unaware of where he left his son. As soon as he hangs up the phone, he goes outside and sees him floating in the pool. He tries to resuscitate him to no avail because he is already dead. The father is charged with criminal negligence causing death for the consequence of his child drowning. He claims that he could not recall leaving his child by the pool and that this is one instance of a recent pattern of forgetful behaviour. After explaining that he had a herpes encephalitis infection several months ago, his defence attorney suspects that his memory impairment may not be just psychogenic but have an organic cause beyond his cognitive control. Structural CT and MRI scans of his brain show significant bilateral damage to his hippocampi and adjacent structures in the medial temporal lobes. These structures are necessary for declarative memory, including episodic memory of events one has experienced. Damage to these brain structures can interfere with the process of consolidation and result in anterograde amnesia, the inability to form and retain new memories. It is unlikely that subcortical systems mediating non-declarative procedural memory, such as the striatum and cerebellum, would be able to take over or compensate for the functions lost by the hippocampus. Episodic and procedural memory systems serve distinct purposes. Episodic memory involves consciously knowing *when* an event occurred. Procedural memory involves unconsciously knowing *how* to do something, such as ride a bicycle or drive an automobile. It seems probable that the earlier viral infection of the accused resulted in significant damage to his hippocampi and severely impaired his capacity to form new episodic memories.[27]

In the hypothetical case at hand, if the hippocampal damage was extensive, then this could support the claim that the father was not responsible for his son's drowning. His amnesia made him unable to form and retain a memory of leaving his son unattended by the pool. This in turn undermined his capacity to reason that leaving a young child unattended by a swimming pool entailed a high probability of the child drowning. Ordinarily, he would have been able to draw this inference and not leave his child out of his sight for so long. This would have prevented the tragic consequence. Yet he was unable to do this because he had no memory of where he left his child. Given his damaged and dysfunctional hippocampus and anterograde amnesia, he was not responsible for the consequence of the drowning because he was not responsible for failing to recall where he left his child. The first state of affairs is causally connected to the second. In fact, it is inaccurate to describe this case as one involving an omission, since the father could not have failed to recall something of which he had no memory.

This argument is more persuasive than the argument that he was suffering from information overload at the time, with too many cognitive tasks to plan and execute.[28] This could have led to a temporary retrieval block of his episodic memory of leaving his son unattended. But unless the memory block was strongly correlated with a detectable brain dysfunction, it would not have much force as an excusing condition. He may have had difficulty with his working memory while executing a number of cognitive tasks within a compressed interval of time. But this would do little to support the claim that he was unable to remember leaving his son beside the pool. Although one may have difficulty recalling an event, one may be able to recall it with some effort. In that case, it could still be within one's cognitive control. Unlike non-declarative memory, declarative memory is not involuntary and thus is to a certain degree within our capacity for conscious recall.[29] Yet anterograde amnesia that severely impaired his episodic memory could have put the critical event outside of the father's cognitive control of the sequence of events resulting in the drowning. Episodic memory and its retrieval in prefrontal-mediated working memory are necessary for the practical and moral reasoning and decision-making

---

[27] This is precisely what befell British musicologist Clive Wearing, as discussed by O. Sacks, 'The Abyss: Music and Amnesia' (2007) *The New Yorker*, 24 September, 100–11.

[28] T. Klingberg, *The Overflowing Brain: Information Overload and the Limits of the Working Memory* (trans. N. Betteridge) (Oxford, Oxford University Press, 2008).

[29] D. Schacter, *Searching for Memory: The Brain, the Mind, and the Past* (New York, Basic Books, 1996); D. Schacter and E. Scarry (eds.), *Memory, Brain, and Belief* (Cambridge, MA, Harvard University Press, 2000); D. Schacter and E. Tulving (eds.), *Memory Systems* (Cambridge, MA, MIT Press, 1994); and R. F. Thompson and S. A. Madigan, *Memory: The Key to Consciousness* (Princeton, Princeton University Press, 2005) ch. 5, 'Amnesia'.

that provide an agent with this control. Individuals with anterograde amnesia are able to retain any newly formed episodic memories for only a very brief period of time. This clearly would not be enough for the father to have the requisite cognitive control to prevent the harmful consequence from occurring. Still, the evidence of hippocampal damage from the brain scans as an explanation for his behaviour would be only one part of the argument for excuse and by itself would not be sufficient to absolve the father of responsibility. His inability to recall leaving his child beside the pool would have to be one instance in a consistent pattern of forgetful behaviour over a period of time. It is only within this pattern that the information about his brain from CT and MRI scans would be legally significant. As in the other cases discussed earlier, imaging showing structural and functional brain abnormalities would have to be evaluated together with behavioural evidence to suitably inform legal judgments of responsibility for actions, omissions, and their consequences.

2.6 Conclusion

What can neuroscience, in the form of neuroimaging, tell us about criminal responsibility? Structural and functional imaging can reveal abnormalities in regions of the brain mediating cognitive, conative, and affective processing. These abnormalities can confirm a neurobiological basis of mental impairment resulting in criminal behaviour. In this respect, neuroimaging can inform judgments of criminal responsibility. Yet this technology alone cannot determine whether one had enough mental capacity to respond to reasons against harmful and wrongful actions, to restrain one's impulses, and be aware of the circumstances of one's actions. Even in cases where there is a causal connection between a brain abnormality and criminal behaviour, whether one had the requisite capacity will be based primarily on behavioural evidence and only secondarily on imaging data. Neuroimaging should not supplant but supplement behaviour as the basis of these judgments. In most cases, brain scans do not establish causal connections between brain states and behaviour but only correlations between them. Moreover, brain scans cannot determine that a person lacks the capacity to control his or her behaviour, has limited capacity to do this, or has full capacity but fails to exercise it on one or more occasions. Also, there will be differences in the interpretation of the psychological and legal significance of brain imaging data between and among different individuals and groups. This will not be decided by appeal just to empirical considerations but also to normative considerations regarding how individuals can be expected to respond to reasons in conforming to social rules and the requirements of the law. The cases I have presented suggest that brain imaging may be more useful in assessing judgments of criminal negligence, less useful in judgments of psychopathy and impulsive behaviour, and least useful in judgments of criminal intent.

To claim that neuroimaging alone can tell us that a person lacked the necessary cognitive control to be criminally responsible for an action and its consequences is to fall prey to what legal scholar Stephen Morse calls 'brain overclaim syndrome'.[30] It displays an oversimplified view of the relation between the brain and the mind, and between the brain and behaviour. Further refinement of neuroimaging may result in it having a greater supplementary role in elucidating the complexity of these relations and the processes through which a person's mental states issue in actions. But it will not obviate the need for behavioural evidence in shaping judgments of criminal responsibility. As Morse puts it:

> In principle, no amount of increased causal understanding of behavior, from any form of science, threatens the law's notion of responsibility unless it shows definitively that we humans (or some subset of us) are not intentional, minimally rational, actors. And no information about biological or social causes can show this directly. It would have to be demonstrated behaviorally.[31]

Another legal scholar, Rebecca Dresser, expresses this same sentiment in a way that captures the gist of my discussion: 'Proof that a biological (or environmental) condition influenced someone's behavior does not necessarily defeat the legal presumption that most people are sufficiently rational to be held accountable for their actions.'[32] If neuroscientific techniques reveal brain abnormalities that might impair one's capacity to respond to reasons, then there will be questions about whether or to what extent one retains this capacity and thus the capacity to be criminally responsible. But given the limitations of brain imaging, neuroscience alone will not resolve these questions.

---

[30] S. Morse, 'Brain Overclaim Syndrome and Criminal Responsibility: A Diagnostic Note' (2006) 3 *Ohio State Journal of Criminal Law* 397–412.

[31] S. Morse, 'New Neuroscience, Old Problems' in B. Garland (ed.), *Neuroscience and the Law: Brain, Mind, and the Scales of Justice* (New York, Dana Press, 2004), 157–98.

[32] R. Dresser, 'Neuroscience's Uncertain Threat to Criminal Law' (2008) 38(6) *Hastings Center Report* 9–10.

3

Mens Rea, Logic, and the Brain

*Gert-Jan Lokhorst**

In this paper, we discuss some recent work in the artificial intelligence and law community on the logic of *mens rea* ascriptions. We extend this work to give an account of quantified mens rea ascriptions in the sense in which they have been discussed in the legal literature. We study the consequences of this work for neuroscience. We point out that it is logically impossible to derive *de re* ascriptions of *mens rea*—as opposed to *de dicto* ascriptions—from neuroscientific evidence. In other words: the distinction between wide and narrow scopes in propositional attitude ascriptions, of which logicians have been aware for a long time, has dramatic consequences for the scope and limits of neuroscience in areas that are of interest to the law.

3.1 Introduction

The term *mens rea* (literally 'guilty mind') comes from the English legal precept *Actus non facit reum nisi mens rea sit* ('an act does not make someone guilty unless his mind is also guilty'). This phrase sounds clearer than it is. This is especially evident to those who have not been trained in the Anglo-American tradition. In the Germanic legal family, for example, nothing corresponds to the term *mens rea* and German legal scholars have gone to extraordinary lengths to explain this notion in terms that they can understand.[1] Conversely, the German legal system has concepts such as *dolus eventualis* that are entirely foreign to legal scholars versed in the common law.[2] Given these difficulties among humans, it is hardly surprising that researchers in the field of artificial intelligence and law face even greater difficulties when they set out to explicate concepts like *mens rea* in languages that can be

---

* Senior reseacher in the Section of Philosophy of the Faculty of Technology, Policy and Management at the Delft University of Technology in the Netherlands. The author wishes to thank Bjørn Jespersen, Roberto Ciuni, and Rosja Mastop for their useful comments.

[1] Hermann Mannheim, '*Mens Rea* in German and English Criminal Law' part 1 (1935) 17 *Journal of Comparative Legislation and International Law 3rd Series* 82; part 2 (1935) 17 ibid. 236; part 3 (1936) 18 ibid. 78.

[2] Greg Taylor, 'Concepts of Intention in German Criminal Law' (2004) 24 *Oxford Journal of Legal Studies* 99. Greg Taylor, 'The Intention Debate in German Criminal Law' (2004) 17 *Ratio Juris* 346.

processed by robotic judges that have not been raised in any legal tradition whatsoever. The first steps towards conceptual clarification in this area have been set, however, and we shall briefly sketch them below.

We are not so much interested in the concept of *mens rea* in itself as in its possible relations to modern neuroscience. Several papers have already been published about this.[3] The present contribution is quite different from these papers because we shall discuss the topic of *mens rea* and modern neuroscience with the help of the conceptual work in artificial intelligence and law that we have just mentioned. In other words, we shall analyse the concept of *mens rea* using the tools of modern formal logic and then discuss this concept in relation to modern neuroscience. Our investigations will reveal that the logical analysis of *mens rea* has significant implications for the scope and limits of neuroscience in the area of the law.

3.2 Logic

In the beginning of the twentieth century, logic was mainly concerned with negation (not), disjunction (or), conjunction (and) and quantification (the quantifiers 'for all' and 'for some'). In the second half of the twentieth century, logic expanded and began to address modality (necessity and possibility) and similar notions as well. The latter are usually taken to include temporal notions, epistemic and doxastic notions (knowledge and belief), deontic notions (obligation, permission and prohibition), and operators having to do with causality, ability, and agency, such as 'the agent sees to it that A'. Much of this work is now being applied in computer science and artificial intelligence research.

The field to which we refer is vast and rapidly expanding, and we cannot hope to summarize it; just to give an indication, the first edition of the *Handbook of Philosophical Logic* (1983–9) consisted of four volumes of about 800 pages each, while the second edition (2001–) is intended to comprise some eighteen volumes.

From the point of view of the law and disciplines such as artificial intelligence and the law the most exciting thing of these developments is that larger and larger fragments of language become amenable to rigorous formal analysis, with the prospect of possible mechanization in the near future. In other words, Leibniz's dream of resolving differences of opinion by means of calculation seems to come closer and closer to realization. As he wrote,

If we had it [a *characteristica universalis*], we should be able to reason in metaphysics and morals in much the same way as in geometry and analysis. If controversies were to arise, there would be no more need of disputation between two philosophers than between two

---

[3] See, for example, Stephen J. Morse, 'Inevitable *Mens Rea*' (2003) 27 *Harvard Journal of Law and Public Policy* 51; Stephen J. Morse, 'The Non-Problem of Free Will in Forensic Psychiatry and Psychology' (2007) 25 *Behavioral Sciences and the Law* 203; Stephen J. Morse and Morris B. Hoffman, 'The Uneasy Entente Between Legal Insanity and *Mens Rea*: Beyond Clark v. Arizona' (2007) 97 *Journal of Criminal Law and Criminology* 1071.

accountants (*Computistas*). For it would suffice to take their pencils in their hands, to sit down to their slates (*abacos*), and to say to each other: Let us calculate (*Calculemus*).[4]

3.3 Seeing To It That

To give a flavour of some of the literature in the area of logic, we quote some impressive results achieved by Belnap and Perloff.[5] They discuss the following conjecture:

If an agent is morally responsible for doing something, then the agent could have done otherwise.

Because this is a complex claim, they split it up into two parts:

1) If an agent is morally responsible for doing something, then the agent did it.
2) If an agent did something, then the agent could have done otherwise.

Are these claims right? Belnap and Perloff answer as follows:

Our view is that the armoury of ordinary language is inadequate to the task of deciding such questions. We need the weaponry provided by something like *stit* theory, a theory that is careful in placing agents and their doings in relation to the causal structure of our world.

By *stit* theory, Belnap and Perloff mean a theory about the logic of 'seeing to it that', its syntax, semantics, axiomatization, and so forth. The 500-page book by Belnap, Perloff, and Xu, *Facing the Future*, presents such a theory in all its details.[6] The theory is completely rigorous and the book is full of complicated mathematics, but the ideas on which the theory is based are simple and intuitive, so that the proofs can often be given by referring to schematic diagrams. The authors assume that agents make choices against a background of branching time; in other words, time has a tree-like structure and the future is open.

Illustrating the formal developments with numerous attractive pictures, Belnap and Perloff are able to prove the following results:

1) 9B1. Claim: What an agent might have done is different from what he or she could have done. Evaluation: True.
2) 9B2. Claim: What an agent could have done is different from what he or she might have done. Evaluation: False.

---

[4] Gottfried Wilhelm Leibniz, as quoted by Edward N. Zalta, 'Computational Metaphysics', at <http://mally.stanford.edu/cm/>. Leibniz's own words were as follows: '*Quo facto quando orientur controversiae, non magis disputatione opus erit inter duos philosophos, quam inter duos Computistas. Sufficiet enim calamos in manus sumere sedereque ad abacos, et sibi mutuo (accito si placet amico) dicere: calculemus*'. Gottfried Wilhelm Leibniz, *Sämtliche Schriften und Briefe*, series 6, vol. 4 (Berlin, Akademie Verlag, 1999) 913.
[5] Nuel Belnap and Michael Perloff, 'The Way of the Agent' (1992) 51 *Studia Logica* 463, reworked into ch. 9 of Nuel Belnap, Michael Perloff, and Ming Xu, *Facing the Future: Agents and Choices in Our Indeterminist World* (New York, Oxford University Press, 2001). We quote from the latter.
[6] Belnap, Perloff, and Xu (2001), op. cit. A similar book is John F. Horty, *Agency and Deontic Logic* (New York, Oxford University Press, 2001).

3) 9C. Claim: If yon fellow sees to some state of affairs, then it might have been that the state of affairs not obtain—at that very instant. Evaluation: True.

4) 9D. Claim: If an agent does something, then it might have been otherwise; that is, he or she might not have done it. Evaluation: True.

5) 9E. Claim: 'The fact that a person could not have avoided doing something is a sufficient condition of his having done it.'[7] Evaluation: 'In spite of the plausibility derived from thick and interesting stories, attention to austere form suggests that there is no reading of Frankfurt's conjecture on which it is both interesting and true.'[8]

6) 9F. Claim: That we are responsible for some state of affairs implies that it must have been possible for us to have been responsible for its absence. Evaluation: False.

7) 9G. Claim: If an agent saw to something, he or she could have refrained from seeing to it. Evaluation: False.

8) 9J. Claim: If an agent is morally responsible for doing something, then the agent could have done otherwise. Evaluation: True.

In other words, Harry Frankfurt's famous paper in which he attempted to refute 9J was wrong.[9]

3.4 A Logical Analysis of *Mens Rea*

Treading in the footsteps of Belnap and his collaborators, the Dutch computer scientist Jan Broersen has recently sketched how the logical analysis of *stit* can be used to shed light on the concepts of 'knowingly doing' and *mens rea* and in this way be applied in the area of the law.[10]

In his formal language, Broersen uses the following basic symbols:

- O, read as: 'it is obligatory that',
- [a xstit]A, read as: 'the agent sees to it that A is the case in the next state',
- L, read as: 'it is necessary that',
- ~, read as: 'it is not the case that',
- K_a, read as: 'the agent knows that',
- A→B, read as: 'if A then B', and
- V, read as: 'there is a violation'.

---

[7] Harry Frankfurt, 'Alternate Possibilities and Moral Responsibility' (1969) 66 *Journal of Philosophy* 828.
[8] Belnap, Perloff, and Xu (2001), op. cit. 262.
[9] Harry Frankfurt (1969), op. cit.
[10] Jan Broersen, 'Deontic Epistemic *stit* Logic Distinguishing Modes of *Mens Rea*' (2010) *Journal of Applied Logic*, to appear. An earlier paper covering some of the same terrain is Jan Broersen, 'First Steps in the *stit*-Logic Analysis of Intentional Action' (2009) in *Proceedings of the ESSLLI 2009 Workshop on Logical Methods for Social Concepts (LMSC'09)*.

Broersen reads $K_a[a\ xstit]A$ as 'the agent knowingly does A'. He proposes the following disambiguations of the concept of *mens rea* in terms of his formal symbols:

- $O_1 K_a[a\ xstit]A = L(\sim K_a[a\ xstit]A \rightarrow [a\ xstit]V)$
- $O_2 K_a[a\ xstit]A = L(K_a \sim [a\ xstit]A \rightarrow [a\ xstit]V)$
- $O_3 K_a[a\ xstit]A = L(K_a[a\ xstit] \sim A \rightarrow [a\ xstit]V)$

Broersen gives the following explanation of these notions; he uses the relative pronoun 'it' rather than 'he' or 'she' because he is interested in artificial rather than natural agents.[11]

The first operator, that is $O_1 K_a[a\ xstit]A$, captures the mens rea mode of 'recklessly'. Here the agent has to knowingly see to it that A obtains, since otherwise there will be a violation. The third operator, that is $O_3 K_a[a\ xstit]A$, captures the mens rea mode of 'knowingly'. Here there is only a violation if the agent knowingly sees to it that the opposite of the lawful outcome A obtains. Finally, the second operator, that is $O_2 K_a[a\ xstit]A$ defines a mode of mens rea in between 'recklessly' and 'knowingly'. It says that the agent is liable if it knowingly refrains from doing. So, on the one hand, there is an aspect of recklessness: if the agent knowingly omits to do something, a violation occurs, because omitting may have an undesirable consequence. On the other hand, if omitting is seen as a form of doing, we can also say that this expresses that there is a violation if the agent knowingly 'does' the omission which is inexcusable for this level of mens rea.

Broersen defines a semantical interpretation for these notions in terms of branching time models in which the future is open and agents make choices. Broersen's semantics are somewhat different from those of Belnap and his collaborators, with the result that it is easy to find an axiomatization that is sound and complete with respect to the semantics.[12] It is also easy to prove that the logic is *decidable* (validity and invalidity can be determined by calculation), so that whenever there is a dispute about the validity of claims made in this rudimentary language, the opposite parties 'can sit down and calculate without further ado to see who is right', in the manner described by Leibniz.

We might mention a few weak points of Broersen's work,[13] but we prefer to skip over these here as he did and continue roughly along the same lines.

---

[11] Broersen (unpublished manuscript 2009), op. cit. We have corrected a few typos, made some stylistic changes and omitted a sentence.

[12] For this, one may use the SQEMA algorithm, which has been implemented in a computer program with an online interface at <http://www.fmi.uni-sofia.bg/fmi/logic/sqema/> and <http://dimiter.slavi.biz/sqema/>.

[13] For example, Broersen uses a crude concept of knowledge, which applies to 'logical saints' rather than real humans; the concept of knowledge used in legal definitions of *mens rea* is similar to the notion of conviction studied in epistemic logic, rather than that of knowledge; more attention should be paid to the notion of concurrence because this is usually regarded as a *conditio sine qua non* of *mens rea*; furthermore, concepts such as likelihood, causality, intentionality, volition, and expectation play important roles in the legal terminology but are missing from Broersen's account. The traditional requirement of concurrence is more problematical than one might think because mental states and events cannot be timed accurately (Daniel C. Dennett, *Freedom Evolves* (New York, Viking Press, 2003) ch. 8) and are neither necessary nor sufficient for voluntary action (M. R. Bennett and P. M. S. Hacker, *Philosophical Foundations of Neuroscience* (Oxford, Blackwell, 2003) ch. 8).

3.5 Quantifiers and Scope Distinctions

All the work we have mentioned up to this point is in the area of propositional logic and does not address the quantifiers (the universal quantifier 'for all' and the existential quantifier 'for some'). We should like to remove this limitation and shall therefore extend the work we have described with quantification.

We shall mainly discuss the issue of *scope distinctions*, which is important in the context of the distinction between *de re* and *de dicto* propositional attitude ascriptions. The importance of this subject in the context of the law was recognized a long time ago.[14] We shall see that it is also important when one wants to assess the scope and limits (in the sense of relevance and applicability) of neuroscience.

The literature about scope distinctions is huge in itself.[15] The highlights are as follows. The whole topic started with Russell's theory of definite descriptions,[16] which has been described as 'one of the Archimedean points in the philosophy of language during the past 100 years'.[17] Russell discussed sentences such as 'the present king of France is bald'. This sentence *seems* to refer to a non-existent entity, the present king of France; but, if so, how should we assess its truth-value—and what are non-existent entities anyway? Russell suggested that a sentence such as 'the present king of France is bald' is not about a non-existent entity and can be analyzed as:

For some x: (i) x is a present king of France, (ii) for all y: if y is a present king of France, then y=x, and (iii) x is bald.

Matters become interesting when we consider the sentence 'the present king of France is not bald'. This sentence can be analyzed in two ways:

1) For some x: x is the present king of France and it is not the case that x is bald.
2) It is not the case that for some x: x is the present king of France and x is bald.

The difference between these analyses is as follows:

- In (1), 'not' has a narrow scope; this is a *de re* ascription.
- In (2), 'not' has a wide scope; this is a *de dicto* ascription.

It is obvious that (1) and (2) have different meanings. (1) is false because there is no present king of France and (2) is true because there is no present king of France. Similarly, 'the present king of France does not exist' is false when taken *de re* and true when taken *de dicto*.

---

[14] Deborah M. Weiss, 'Scope, Mistake, and Impossibility: The Philosophy of Language and Problems of *Mens Rea*' (1983) 83 *Columbia Law Review* 1029; Burkhard Schäfer, 'Leśniewski-Quantifiers and Modal Arguments in Legal Discourse' (1998) 6 *Logic and Logical Philosophy* 133.

[15] For some references, see Peter Ludlow 'Descriptions' in Edward N. Zalta (ed.), *Stanford Encyclopedia of Philosophy* (<http://plato.stanford.edu/>, 2007); and Thomas McKay and Michael Nelson (2005) 'Propositional Attitude Reports', ibid.

[16] Bertrand Russell, 'On Denoting' (1905) 14 *Mind* 479.

[17] Peter Ludlow (2007), op. cit.

Fifty years after Russell, Smullyan made a similar observation about scope differences in modal contexts.[18] For example, 'the number of planets is necessarily eight' can be analyzed in two ways, where x ranges over natural numbers and N is the number of planets:

1) For some x: x = N and it is necessary that x = 8.
2) It is necessary that for some x: x = N and x = 8.

The difference between these analyses is as follows:

- In (1), the necessity operator has a narrow scope; this is a *de re* ascription.
- In (2), the necessity operator has a wide scope; this is a *de dicto* ascription.

It is obvious that (1) and (2) have different meanings. (1) is true because there happen to be eight planets,[19] and (2) is false because it is not necessary that there are eight planets.[20]

A couple of years after Smullyan, Quine noted that propositional attitude ascriptions are, in this respect, similar to modal statements.[21] He pointed out that 'John believes that Ortcutt is a spy' can be analyzed as follows:

1) For some x: x = Ortcutt and John believes that x is a spy.
2) John believes that for some x: x = Ortcutt and x is a spy.

The difference between these analyses is as follows:

- In (1), belief has a narrow scope; this is a *de re* ascription.
- In (2), belief has a wide scope; this is a *de dicto* ascription.

As Quine pointed out, (1) and (2) are not equivalent. One may believe that Ortcutt (whoever he is) is a spy without believing of the actual Ortcutt that he is a spy because one is mistaken about the identity of Ortcutt. In other words, belief *de re* presupposes knowledge of the identity of Ortcutt, while belief *de dicto* does not; instead, it only presupposes knowledge of a condition for being Ortcutt.[22]

It is useful to introduce some notation to represent scope differences in a perspicuous way. We define '[a]Fa' as 'for some x: x=a and Fx'. The *de re* sentences mentioned above are of the form [a]XFa, where X is the operator of negation,

---

[18] A. Smullyan, 'Modality and Description' (1948) 13 *Journal of Symbolic Logic* 483.
[19] Recall that Pluto was reclassified as a dwarf planet (hence a non-planet) in 2006.
[20] These sentences about the planets illustrate an important difference between *de re* and *de dicto* sentences: *de re* sentences are 'extensional' and 'referentially transparent' wheras *de dicto* sentences are 'intensional' and 'referentially opaque'. By 'extensional' and 'referentially transparent' we mean that only the denotations (references) of terms count. Substituting terms with the same reference for one another does not affect the truth-value. (1) is extensional: if M=N, then (for some x: x=N & L(x=8)) if and only if (for some x: x=M & L(x=8)). On the other hand, (2) is not: N=8, but L(N=8) is false and L(8=8) is true.
[21] W. V. Quine, 'Quantifiers and Propositional Attitudes' (1956) 53 *The Journal of Philosophy* 177.
[22] This is a condition that is relative to the agent. It does not depend on how those who ascribe beliefs to agents individuate things in the world, but on how the agent individuates things. We shall come back to this below.

necessity, or belief. The *de dicto* sentences are of the form X[a]Fa, where X indicates negation, necessity, or belief.

3.6 *Mens Rea* and Scope Distinctions

If we enrich Broersen's language of 'knowingly see to it that' with quantifiers, we can make scope distinctions similar to those we have just discussed. For example, 'the agent knowingly kills a policeman' can be analyzed in two ways:

1) For some x: x is a policeman and the agent knowingly sees to it that x is not alive. Formally: for some x: Px and K_a[a xstit]Fx.
2) The agent knowingly sees to it that for some x: x is a policeman and x is not alive. Formally: K_a[a xstit] for some x: Px and Fx.

The difference between these analyses is as follows:

- In 1), 'knowingly sees to it that' has a narrow scope; this is a *de re* ascription.
- In 2), 'knowingly sees to it that' has a wide scope; this is a *de dicto* ascription.

In general, we can and must distinguish between:

1) [b]K_a[a xstit]Fb,
2) K_a[b][a xstit]Fb, and
3) K_a[a xstit][b]Fb,

where F is a predicate that applies to a term such as b. These three formulas are formally similar to Broersen's formulas:

1) ~K_a[a xstit]A,
2) K_a~[a xstit]A, and
3) K_a[a xstit]~A,

respectively, which we have encountered above. Broersen defined three 'modes of *mens rea*' in terms of the latter three formulas; we can similarly define a number of modes of *mens rea* in terms of the former three formulas. Since there are four positions in each of the three formulas at which can insert the negation symbol (for example, at each of the four points marked with a '#' in '#[b]#K_a#[a xstit]#Fb', in the case of first formula), there are twelve corresponding modes of *mens rea*. If we insert yet another single occurrence of a one-place operator (such as necessity or belief), then there are sixty possibilities, and so on: there is therefore an infinite number of modes of *mens rea*. We could list them (alphabetically, for example), just as Broersen did, but there seems little point in doing so.

The distinctions that we have made are important. For example, we find the following informal distinctions in the Model Penal Code's definition of *mens rea*:[23]

---

[23] M. D. Dubber, *Criminal Law: Model Penal Code* (New York, Foundation Press, 2002).

1) Purposely: express purpose to commit a specific crime against *a particular person*.
2) Knowingly: knowledge that one's actions would certainly result in a crime against *someone*, but did not specifically intend to commit that crime against *the particular victim* which one is accused of injuring.

The first notion is *de re* (the agent wants to kill b, whatever properties b may have), the second one is *de dicto* (the agent wants to kill whoever is an F, whatever individual—if any—may be an F).

In a recent case in the Netherlands, a man knowingly shot a policewoman. He received a reduced sentence because he did not know of her that she was a policewoman. (Or at least the court could not prove that he knew this.) He knowingly shot the policewoman in the *de re* sense, but not in the *de dicto* sense. Many discussions of similar cases—for example, a Gettier-type[24] case in which a man was accused of receiving stolen goods and another case in which a man unknowingly married his mother—are to be found in the logico-legal literature.[25]

Burkhard Schäfer has claimed that the German and Anglo-American legal systems behave differently vis-à-vis *de dicto* and *de re* because the former is inclined towards *de dicto* interpretations whereas the latter is more favourably disposed towards the *de re* stance.[26] He has also claimed that this difference is gradually diminishing because the Anglo-American system is slowly moving towards the German system (not the other way round).[27] This may be true, but we cannot go into this here. This is a subject for comparative criminal law.

3.7 Implications for Neuroscience

Propositional attitude ascriptions have the following important property: *de re* ascriptions are *relational*, *de dicto* ascriptions are not. A *de re* ascription is made true by the fact that two individuals are related to each other in a certain way, whereas a *de dicto* ascription is made true by the fact that one individual (the agent to whom the attitude is ascribed) has a certain property. This is evident from the logical form of *de re* and *de dicto* ascriptions. Let us examine the Ortcutt sentences again:

1) For some x: x = Ortcutt and the agent believes that x is a spy.
2) The agent believes that for some x: x = Ortcutt and x is a spy.

Sentence (1) refers to two individuals, Ortcutt and the agent to whom the belief is ascribed, but sentence (2) only refers to a property of the agent (namely that he is in the state of having some particular belief). If Ortcutt did not exist, (1) would be false, but (2) might still be true. Children have all kinds of beliefs about Santa

---

[24] Edmund L. Gettier, 'Is Justified True Belief Knowledge?' (1963) 23 *Analysis* 121.
[25] Weiss (1983) and Schäfer (1998), op. cit.
[26] Schäfer (1998), op. cit.
[27] Schäfer (1998), op. cit.

Claus, even though Santa Claus does not exist. The sentence 'John believes that the present king of France is bald' appears to refer to the present king of France but does not really do so because there is no present king of France. It is made true or false by John's properties, not those of the present king of France.[28]

Similar remarks apply to the 'knowingly killing' scenario. An agent may fail to kill the king of France knowingly not because he failed to kill the person whom he took to be the king of France, but because the king of France was just dethroned and has not yet been succeeded. All one's *de re* beliefs about the king of France cease to be true as soon as there is no longer a king of France, but all one's *de dicto* beliefs about the king of France may stay the same regardless of whether there is a king or not.

This is interesting from the point of view of neuroscience because it indicates that *de re* attitudes can never be read off from brain scans, although *de dicto* attitudes might. A *de dicto* propositional attitude ascription ascribes a property to an agent that depends only on that agent. For example, the sentence 'the agent believes that for some x: x = Ortcutt and x is a spy' ascribes the property *believing that Ortcutt is a spy* to the agent. It could in principle be ascertained on the basis of neuroscientific evidence whether the agent has that property or not. To put it crudely: we only have to know whether the 'Ortcutt area' and 'spy area' are jointly active or not. The real Ortcutt (if there is any) has nothing to do with the truth-value of this sentence.

But the following *de re* sentence is not only about the agent, but also about Ortcutt; it says that Ortcutt and the agent are related in a certain way:

For some x: x = Ortcutt and the agent believes that x is a spy.

It is not sufficient to study the agent's brain to determine whether Ortcutt and the agent are related in the relevant way. The relationship can only hold if Ortcutt exists; but it cannot be inferred from brain scans whether Ortcutt—or anything else, for that matter—exists or not. Brain scans will never reveal whether Santa Claus exists 'out there', in the real world. Similarly, it is not sufficient to study Ortcutt by himself to determine whether anybody suspects him of being a spy.

To summarize: *de dicto* beliefs could possibly depend on, or be systematically correlated with, brain activity, but *de re* beliefs cannot. The latter can vary widely with external circumstances which are not internal to the brain. Like fatherhood, having a certain *de re* belief is an *intrinsically relational* property. A person can become a father without being aware of it or undergoing any internal change. We may in the same way lose all *de re* knowledge of any given thing without having any

---

[28] Recall what we have said above about the sentence 'the present king of France does not exist'. It seems to refer to the present king of France but does not really do so because there is no present king of France. If it is true, it is true in virtue of the logical properties of negation, in the sense that the negation of a false claim—in this case the claim that the world contains exactly one individual that is presently the king of France and that whoever is presently the king of France exists—is true. In the same way, 'the present king of France' is believed to exist' is, if true, true by virtue of what people believe; it is true by virtue of *their* properties, not those of the present king of France. Only if there were a present king of France could people *de re* believe of him that he exists—or deny this (falsely).

special experience or undergoing any internal change. Such a 'purely *de re* amnesia' would not be detectable by a brain scanner. Even total *de re* amnesia could go unnoticed. *De re* attitudes cannot possibly be discovered on the basis of brain scans alone.

If we are interested in the *mens rea in cerebro reo*, it is important to keep in mind that only the *de dicto* type of *mens rea* can be detected on the basis of studies of the *cerebrum*. By the same token, it will never be justified to ascribe (or withhold) *de re* culpability on the evidence of nothing but brain scans.[29]

There are alternative accounts of quantification in contexts such as those we have discussed, but these do not seem to affect our conclusions. We give only two examples. First, some have argued that *de dicto* ascriptions are relational. In sentences like 'John believes that the present king of France is bald', 'the present king of France' is said to refer to an 'intensional object' which does not exist but 'subsists'. If they are right, then our remark about the limitations of neuroscience applies to those ascriptions too: when assessing the truth of a relational statement, evidently both *relata* have to be taken into account. Second, Hintikka has argued that epistemic logic needs two kinds of quantifiers: one type to refer to physical objects, and another one to refer to objects as they are individuated by the knower.[30] If this is correct—which has been doubted[31]—we need to make all the points we have made twice: once for each type of quantifiers.

There is more to be said about logic and neuroscience,[32] but nothing seems as relevant to the concerns of the law as the quantificational and scope issues that we have discussed and that have already been discussed for different reasons by the logico-legal community.[33]

3.8 Conclusion

We have tried to identify an area at the intersection of logic, neuroscience, and the law, and we have not only found one, as we have demonstrated above, but even made a discovery in that area: namely that the logical distinction between scope differences in propositional attitude ascriptions has significant consequences for the scope and limits (in the sense of relevance and applicability) of neuroscience in areas that are of interest to the law.

---

[29] This may have greater consequences for the Anglo-American legal system than for the German system, if the former is indeed more oriented towards *de re*, as was suggested above, following Schäfer (1998), op. cit.
[30] Jaakko Hintikka, 'The Cartesian *Cogito*, Epistemic Logic and Neuroscience: Some Surprising Interrelations' (1990) 83 *Synthese* 133. Jaakko Hintikka and John Symons, 'Systems of Visual Identification in Neuroscience: Lessons from Epistemic Logic' (2003) 70 *Philosophy of Science* 89.
[31] Romane Clark, 'Old Foundations for a Logic of Perception' (1976) 33 *Synthese* 75. Gregory Mellema, 'Multiple Quantifiers and Hintikka's Logic of Perception' (1982) 11 *Philosophia* 95.
[32] Ahti-Veikko Pietarinen, 'What do Epistemic Logic and Cognitive Science Have to do with Each Other?' (2003) 4 *Cognitive Systems Research* 169.
[33] Weiss (1983) and Schäfer (1998), op. cit.

4

Indeterminism and Control: An Approach to the Problem of Luck

*John Martin Fischer**

4.1 The Dilemma of Determinism

William James' famous 'Dilemma of Determinism' can be formulated as follows:

1) Either causal determinism is true, or it is false.

2) If causal determinism is true, then I *have* to act as I do, and thus I am not morally responsible for my actions.

3) If causal determinism is false (in a relevant way), then how I act is a matter of luck, and thus I am not morally responsible for my actions.

Therefore:

4) I am not morally responsible for my actions.[1]

Alternatively, the dilemma could be put as follows:

1) Either causal determinism is true, or it is false.

2*) If causal determinism is true, then I cannot do otherwise, and thus I am not morally responsible for my actions.

3*) If causal determinism is false (in a relevant way, i.e. in the sequences leading to my behaviour), then my actions are not appropriately connected to my

---

* Distinguished Professor of Philosophy, University of California, Riverside, and University of California President's Chair (2001–10). The author has benefited from giving versions of this paper at the Law and Neuroscience Conference, Faculty of Laws, University College, London; University of California, San Diego; and as the Bo Clark Lecture, Department of Philosophy, Indiana University. He is extremely grateful for very helpful comments on a previous versions of this paper by Neal Tognazzini, Christopher Franklin, Ben Mitchell-Yellin, Justin Coates, Garrett Pendergraft, Patrick Todd, Paul Hoffman, Randloph Clarke, Anthony Brueckner, Michael McKenna, Dana Nelkin, Michael Hardimon, David Brink, and Tim O'Connor.

[1] William James, 'The Dilemma of Determinism,' originally an address to Harvard Divinity Students (in Lowell Lecture Hall, Harvard University), published in *The Unitarian Review* (September 1884); reprinted in William James, *The Will to Believe and Other Essays in Popular Philosophy* (New York, Longmans Green and Co., 1907) 145–83.

prior states (that is 'my actions' are not in a genuine sense *my* actions), and thus I am not morally responsible for my actions.

Therefore:

4) I am not morally responsible for my actions.

The picture behind the Dilemma of Determinism might be thought to be roughly this. To be morally responsible, I must be in control of my behaviour. But there are at least two components of the relevant kind of control (or perhaps at least two requirements for such control). First, in order to be in control, it must not be the case that I *have* to do what I actually do; that is, I must have freedom to choose and to do otherwise. Second, the action must be an 'outflowing' of *me*—it must be genuinely *my* action in some important sense.[2]

When I am in control of my behaviour, in the way required for moral responsibility, I select from among various genuinely open paths into the future through my free choice. Again, there are at least two crucial ideas here: I must *select* from among *various open alternatives*, and *I* must select. As the two formulations of the Dilemma of Determinism indicate, causal determinism would seem to challenge (at least) the notion that I select from among various genuinely open paths into the future, and indeterminism appears to challenge the idea that it is genuinely *I* who selects (or, perhaps, makes a difference).

In this paper I shall focus primarily on the 'indeterministic horn', that is, premises 3 and 3*. But I shall begin by considering the 'deterministic horn', (premises 2 and 2*). The proper analysis of the deterministic horn will be illuminating with respect to the indeterministic horn. I shall argue that neither the second premises of the parallel arguments nor the third premises are true, and thus the argument is unsound for two separate reasons. I shall further argue that similar considerations help to establish the failure of *both* the deterministic and indeterministic horns of the dilemma. Not only are the worries similar at a deep level, but the appropriate replies are also based on similar insights.

4.2 The Deterministic Horn

4.2.1 The argument for the second premise

Why exactly does it follow from causal determinism that an agent does not have freedom to do otherwise? And why would it follow from an agent's lacking freedom to do otherwise that he is not morally responsible for his actions?

There are various ways to seek to establish that causal determinism implies that we never have the freedom to do other than we actually do. Perhaps the most salient

---

[2] Timothy O'Connor calls the relevant notion 'agent control'. As he puts it, agent control is 'the manner in which a particular piece of behaviour is connected to, controlled by, or an "utflowing of" the agent': Timothy O'Connor, *Persons and Causes: The Metaphysics of Free Will* (New York, Oxford University Press, 2000) 23.

such argument has been dubbed the 'Consequence Argument' by Peter Van Inwagen.[3] The argument gets its name from the fact that, under causal determinism, every choice and action is the consequence of the past together with the laws of nature. More specifically, if causal determinism is true, every choice and action is the result of a causally deterministic causal sequence. The argument has it that if this is so, we don't ever have a choice about what we do; that is, if causal determinism is true, we never have it in our power to do otherwise than we actually do. As Van Inwagen puts it:

> If determinism is true, then our acts are the consequences of the laws of nature and events in the remote past. But it is not up to us what went on before we were born, and neither is it up to us what the laws of nature are. Therefore, the consequence of these things (including our present acts) are not up to us.[4]

Van Inwagen (and others) have given various more formal versions of this intuitive argument.[5] Here is a still-quite-informal presentation of the bare bones of the argument. Suppose that causal determinism is true. Given a standard definition of causal determinism, it follows that my current choice to raise my coffee cup (caffeinated, fortunately) is entailed by true propositions about the past and laws of nature. Thus, if I were free (just prior to my actual choice to pick up my coffee cup) to choose (say) to listen to my wife's admonition to drink less coffee and to refrain from picking up the cup, then I must have been free so to act that the past would have been different or the natural laws would have been different. But intuitively the past is 'fixed' and out of my control and so are the natural laws. Thus, I cannot now do anything that is such that, if I were to do it, the past would have been different (say, the Spanish Armada never would have been defeated or Napoleon would have been victorious at Waterloo). Similarly, I cannot ever so act that the natural laws would be different (say, some things would travel faster than the speed of light). It appears to follow that, despite the natural impression I have that I am sometimes genuinely free to choose and do otherwise, it turns out that I cannot choose to refrain from picking up the coffee cup (although I will not try that line on my wife) and indeed I am never free to choose and do otherwise, on the assumption of causal determinism.

The Consequence Argument is a 'sceptical argument' that challenges our common-sense view that we are often free to do otherwise. As with other sceptical arguments, it gets its bite from intuitively plausible ingredients, such as the idea that the past and natural laws are fixed and out of our control. Of course, one needs to be considerably more careful in developing a definition of causal determinism and in specifying the relevant senses in which the past and natural laws are indeed

---

[3] Peter Van Inwagen, 'The Incompatibility of Free Will and Determinism' (1975) 27 *Philosophical Studies* 185–99; and *An Essay on Free Will* (Oxford, Clarendon Press, 1983).
[4] Van Inwagen, *An Essay on Free Will* 56.
[5] For just a small selection, see: ibid 55–105; Carl Ginet, *On Action* (Cambridge, Cambridge University Press, 1990); John Martin Fischer, *The Metaphysics of Free Will: An Essay on Control* (Oxford, Blackwell Publishers, 1994); and J. Howard Sobel, *Puzzles for the Will* (Toronto, University of Toronto Press, 1998).

fixed; but I am confident that the intuitive ideas can be crystallized in such a way as to yield a powerful and highly plausible argument for the incompatibility of causal determinism and freedom to do otherwise. If the Consequence Argument is indeed sound, and if moral responsibility requires freedom to do otherwise, then it would follow that the premises that present the Deterministic Horn of the Dilemma of Determinisms 2 and 2* are true.

4.2.2 A reply to the argument for the second premise

As developed above, there are two steps in the argument for premises 2 and 2*. First, the Consequence Argument seeks to establish that if causal determinism is true, then I cannot choose and do otherwise. Second, the 'Principle of Alternative Possibilities' (PAP) is invoked, according to which moral responsibility requires precisely the freedom to choose and do otherwise.[6] Each step of the argument can be called into question.

Some philosophers have rejected the Consequence Argument. Philosophers such as Keith Lehrer, Terence Horgan, and John Perry have denied that the past is fixed in any sense strong enough to yield the incompatibilistic result.[7] They are willing to say that the past is fixed in *one* sense; that is, no one can initiate a causal chain extending into—and altering—the past. But they are quick to point out that this point, in itself, does not yield a result strong enough to get incompatibilism via the Consequence Argument. What would be strong enough is the contention that no one can ever so act that the past would have been different from what it actually was; and this, they say, is simply not true. One might think it true by not distinguishing it from the uncontroversial sense in which the past is fixed.

Other philosophers, such as David Lewis, have rejected the fixity of the natural laws.[8] (Actually, David Lewis would reject both the fixity of the past and the natural laws.) Lewis is willing to say that the natural laws are fixed in *one* sense; as he puts it, we are not free to *violate* a natural law. But it does not follow that we are never free so to act that some proposition expressing a natural law *would not* have expressed a natural law. Lewis argues that the uncontroversial truth is not enough to support the Consequence Argument, and the contention that would be enough is false.

These debates are highly contentious, and I have argued that they end up in what I have called 'Dialectical Stalemates'.[9] I prefer to address the second step of the argument for premises 2 and 2*. That is, although it is useful to note that the

---

[6] The principle is so-called by Harry Frankfurt in 'Alternate Possibilities and Moral Responsibility' (1969) 66 *Journal of Philosophy* 829–39. Actually, Frankfurt dubs the principle, 'The Principle of Alternate Possibilities'.

[7] See, for example, Keith Lehrer, ' "Can" in Theory and Practice: A possible Worlds Analysis' in Myles Brand and Douglas Walton (eds), *Action Theory: Proceedings of the Winnipeg Conference on Human Action* (Dordrecht, D. Reidel, 1976) 241–70; Terence Horgan, ' "Could", Possible Worlds, and Moral Responsibility' (1979) 17 *Southern Journal of Philosophy* 345–58; and John Perry, 'Compatibilist Options' in Joseph Keim Campbell, Michael O'Rourke, and David Shier (eds), *Freedom and Determinism* (Cambridge, MA, Bradford/MIT Press, 2004) 231–54.

[8] David Lewis, 'Are We Free to Break the Laws?' (1981) 47 *Theoria* 113–21.

[9] Fischer, *The Metaphysics of Free Will: An Essay on Control* 83–5.

Consequence Argument is not entirely persuasive, I find it more appealing to attack the contention that moral responsibility requires freedom to choose and do otherwise (PAP).

Again, there are various routes to a rejection of PAP. I have focused primarily on the so-called 'Frankfurt-style' counter-examples to PAP—examples that were originally introduced in contemporary philosophy by Harry Frankfurt and that purport to impugn PAP.[10] Here is an updated version of a Frankfurt example:

Because Black dares to hope that the Democrats finally have a good chance of winning the White House, the benevolent but elderly neurosurgeon, Black, has come out of retirement to participate in yet another philosophical example.[11] (After all, what would these thought-experiments be without the venerable *eminence grise*—or should it be *noire*?) He has secretly inserted a chip in Jones's brain that enables Black to monitor and control Jones's activities. Black can exercise this control through a sophisticated computer which he has programmed so that, among other things, it monitors Jones's voting behavior. If Jones were to show any inclination to vote for McCain (or, let us say, anyone other than Obama), then the computer, through the chip in Jones's brain, would intervene to assure that he actually decides to vote for Obama and does so vote. But if Jones decides on his own to vote for Obama (as Black, the old progressive, would prefer), the computer does nothing but continue to monitor—without affecting—the goings-on in Jones's head.

Now suppose that Jones decides to vote for Obama on his own, just as he would have if Black had not inserted the chip in his head. It seems, upon first thinking about this case, that Jones can be held morally responsible for this choice and act of voting for Obama, although he could not have chosen otherwise and he could not have done otherwise.[12]

Notoriously (in some quarters), there has been much discussion of the Frankfurt-style examples and their implications in the years subsequent to the publication of Frankfurt's (1969) paper, 'Alternate Possibilities and Moral Responsibility'. I certainly will not go into the details here. For my purposes here it is enough to distill a few (admittedly controversial) lessons. First, it seems to me that Black's presence (as described in the example), perhaps together with other features, makes it the case that Jones cannot choose or do other than he actually does. Further, it seems to me that Black's presence (in the context of those other features) is *irrelevant* to Jones's moral responsibility. Both of these contentions would need to be defended, and, perhaps, qualified.[13] But for the purposes of this paper, I will take them as working hypotheses.

It might be helpful to have before us Frankfurt's statements on behalf of the contention that Black's presence is irrelevant to Jones's moral responsibility:

---

[10] Frankfurt, 'Alternate Possibilities and Moral Responsibility'. For a selection of papers on Frankfurt-style examples, see David Widerker and Michael McKenna (eds.), *Moral Responsibility and Alternative Possibilities: Essays on the Importance of Alternative Possibilities* (Aldershot, Ashgate, 2003).
[11] Whew! Black was right; obviously, the specifics of the case are now out of date, but what matters is the structure of the example.
[12] I originally presented such an example in John Martin Fischer, 'Responsibility and Control' (1982) 79 *Journal of Philosophy* 24–40.
[13] For a recent effort along these lines, see John Martin Fischer, 'The Frankfurt Cases: The Moral of the Stories' (2010) 119 *Philosophical Review* 315–36.

The fact that a person could not have avoided doing something is a sufficient condition of his having done it. But, as some of my examples show, this fact may play no role whatever in the explanation of why he did it. It may not figure at all among the circumstances that actually brought it about that he did what he did, so that his action is to be accounted for on another basis entirely... Now if someone had no alternative to performing a certain action but did not perform it because he was unable to do otherwise, then he would have performed exactly the same action even if he *could* have done otherwise. The circumstances that made it impossible for him to do otherwise could have been subtracted from the situation without affecting what happened or why it happened in any way. Whatever it was that actually led the person to do what he did, or that made him do it, would have led him to do it or made him do it even if it had been possible for him to do something else instead... When a fact is in this way irrelevant to the problem of accounting for a person's action it seems quite gratuitous to assign it any weight in the assessment of his moral responsibility.[14]

Although I recognize that further argumentation would be required to convince sceptics, I agree with Frankfurt's intuition that it would be 'quite gratuitous' to assign any weight to Black's presence in assessing Jones's moral responsibility. After all, Black did not play any role in the 'actual sequence'—the actual causal pathway to Jones's choice and action; Black's device, although present, is *untriggered*. I think that the Frankfurt-style examples help to provide motivation for an 'actual-sequence' approach to moral responsibility, according to which moral responsibility attributions depend on (possibly dispositional or modal) features of the actual sequence, rather than on the availability of genuinely open alternative possibilities. The mere presence of certain sorts of *untriggered ensurers* (such as Black's device) rules out alternative possibilities without in any way affecting the actual sequence that issues in the relevant behaviour. In my view, this is the fundamental reason to reject the deterministic horn of the dilemma, as expressed by premises 2 and 2*. Of course, there might well be *other* reasons (not captured in these premises) to suppose that causal determinism rules out moral responsibility; here I am focusing primarily upon the specific reasons featured in the premises of the Dilemma of Determinism as presented above.[15]

4.3 The Indeterministic Horn

4.3.1 The argument for the third premise

Above I suggested that the intuitive idea of control that underwrites moral responsibility requires that an agent selects from among genuinely open alternatives and thereby makes a difference through his free choice (and action). The Consequence Argument calls into question whether an agent can select from among open alternatives, if causal determinism is true. But if causal determinism is false (in the relevant way), an interestingly similar argument—The Rollback Argument—calls

---

[14] Frankfurt, 'Alternate Possibilities and Moral Responsibility' 836–7.
[15] For a discussion of the 'sourcehood' worry, see John Martin Fischer, 'Compatibilism' in John Martin Fischer et al., *Four Views on Free Will* (Oxford, Blackwell Publishers, 2007) 44–84, esp. 61–71.

Indeterminism and Control: An Approach to the Problem of Luck

into question whether it is *the agent* who makes the difference. In order to have the requisite control, the choices and actions must flow from the *agent* in the right way—it must be genuinely *the agent's* action. And, as suggested above, in order for the choice and action to be genuinely the agent's, they must be suitably related to the agent's prior mental states. If it is purely a matter of chance or luck that an individual chooses as he does, then the choice in question would not appear to be related to the agent's prior mental states—including his desires, beliefs, intentions, and acceptances of values—in the right way. And, according to the Rollback Argument, if causal indeterminism is true, then an agent's choices are—in a sense to be specified—purely a matter of luck. Thus, the Consequence Argument and the Rollback Argument play parallel roles in purporting to support the second and third premises.

More specifically, the Rollback Argument seeks to explain precisely *why* it would follow from causal indeterminism (in the relevant places) that the choices and actions would not be an outflowing of the agent in the required sense; that is, it seeks to pinpoint the reason why, if indeterminism were to obtain, the required *relationship* between the agent's prior mental states and his choice (and action) would not be present. For ease of discussion, let us call this relationship, the 'responsibility-grounding relationship'.

Peter Van Inwagen gives a particularly clear presentation of the Rollback Argument, and thus I ask the reader's patience with the following lengthy quotation:

> Let us suppose undetermined free acts occur. Suppose, for example, that in some difficult situation Alice was faced with a choice between lying and telling the truth and that she freely chose to tell the truth—or, what is the same thing, she seriously considered telling the truth, seriously considered lying, told the truth, and was able to tell the lie she had been contemplating. And let us suppose that...Alice's telling the truth, being a free act, was therefore undetermined. Now suppose that immediately after Alice told the truth, God caused the universe to revert to precisely its state one minute before Alice told the truth (let us call the first moment the universe was in this state 't1' and the second moment the universe was in this state 't2' and then let things go forward again. What would have happened the second time? What would have happened after t2? Would she have lied or would she have told the truth? Since Alice's 'original' decision, her decision to tell the truth, was undetermined—since it was undetermined whether she would lie or tell the truth—her 'second' decision would also be undetermined, and this question can therefore have no answer; or it can have no answer but, 'Well, although she would either have told the truth or lied, it's not the case that she would have told the truth and it's not the case that she would have lied; lying is not what she would have done and telling the truth is not what she would have done. One can say only that she *might have lied* and she *might have told the truth.*'
>
> Now let us suppose that God *a thousand times* caused the universe to revert to exactly the state it was in at *t1* (and let us suppose that we are somehow suitably placed, metaphysically speaking, to observe the whole sequence of 'replays'). What would have happened? What should we expect to observe? Well, again, we can't say what would have happened, but we can say what would *probably* have happened: sometimes Alice would have lied and sometimes she would have told the truth. As the number of 'replays' increases, we observers shall—almost certainly—observe the ratio of the outcome 'truth' to the outcome 'lie' settling down to, converging on, some value...[L]et us imagine the simplest case: we

observe that Alice tells the truth in about half the replays and lies in about half the replays. If, after 100 replays, Alice has told the truth 53 times and has lied 48 times, we'd begin strongly to suspect that the figures after a 1,000 replays would look something like this: Alice has told the truth 493 times and has lied 508 times. Let us suppose that these are indeed the figures after 1,000 replays. Is it not true that as we watch the number of replays increase, we shall become convinced that what will happen on the *next* reply is a matter of chance?[16]

Van Inwagen goes on to state that our inclination to say that what will happen on the 'next replay' is a matter of chance will not be different, if we change the probabilities that Alice will tell the truth (as long as this objective probability is under 100 per cent). Van Inwagen adds, 'Nothing we could possibly learn, nothing God knows, it would seem, should lead us to distrust our initial inclination to say that the outcome of the next replay will be a matter of chance.'[17] Van Inwagen goes on to conclude from the fact that the action is a matter of chance that Alice lacked a kind of dual ability: the ability to tell the truth and the ability to lie. It might be said that Alice lacked the ability to guarantee (in advance) which action would occur, where this ability included the power to guarantee either that she would lie or that she would not lie.

It should be noted that Van Inwagen himself does not accept that the Rollback Argument decisively shows that indeterminism implies that the responsibility-grounding relationship fails to hold (and thus that premises 3 and 3* are true). Rather, he thinks that the Rollback Argument presents a significant challenge to the coherence of libertarianism—a challenge that he thinks can be met. Van Inwagen does not see *how* the challenge can be met, so he must accept what he calls the 'puzzling' view that indeterminism does not rule out the responsibility-grounding relationship, rather than the 'inconceivable' view that causal determinism is compatible with freedom to do otherwise. Van Inwagen famously says, 'I must choose between the puzzling and the inconceivable. I choose the puzzling.'[18] In the discussion in the previous section of this paper, I sought to call into question the contention that the compatibility of causal determinism and certain notions of freedom and moral responsibility is inconceivable. In the following I shall seek to diminish at least some of Van Inwagen's puzzlement.

It is interesting to note that, not only do the Consequence Argument and the Rollback Argument play similar roles in putatively supporting the relevant premises in the Dilemma of Determinism, but they can be seen to share an abstract form. The problem raised by the Consequence Argument is that if one were to roll back the universe to the relevant prior instant a thousand (or a million or . . .) times, the agent would *always* make the same choice. This calls into question the notion that the agent has the power to *select* from among *more than one* path that are *genuinely*

---

[16] Peter Van Inwagen, 'Free Will Remains a Mystery', reprinted in John Martin Fischer (ed.), *Critical Concepts of in Philosophy: Free Will Volume IV* (London, Routledge, 2005) 173–92; the quotation is from 187–8, and citations of page numbers below are from the Fischer volume.
[17] Van Inwagn, 'Free Will Remains a Mystery' 189.
[18] Van Inwagen, *An Essay on Free Will* 150.

available to him. And, as we have seen above, the problem raised by the Rollback Argument is that if one were to roll back the universe to the relevant prior instant a thousand (or a million or...) times, the agent would *not* always make the same choice. This calls into question the notion that the *agent* has it in his power to make the relevant selection; that is it calls into question whether the responsibility-grounding relationship obtains. The Rollback and Consequence Arguments can thus be seen to share a common structure; but they are, as it were, 'inverses' or mirror images of each other. And, of course, they target different aspects of the basic idea that when I am morally responsible, it is *I* who *select* from among various open options.[19]

4.3.2 Critique of the argument for the third premise

Recall the (reformulated) third premise of the Dilemma of Determinism:

> 3*. If causal determinism is false (in a relevant way, i.e. in the sequences leading to my behavior), then my actions are not appropriately connected to my prior states (that is 'my actions' are not in a genuine sense *my* actions), and thus I am not morally responsible for my actions.

Note that the Rollback Argument is supposed to establish that (or give a reason for) the contention that if causal determinism is false, then my actions are not appropriately connected to my prior states; that is, the Rollback Argument is supposed to show exactly why the responsibility-grounding relationship would not obtain, if causal determinism were false.

I now wish to call into question the efficacy of the Rollback Argument in establishing that causal indeterminism rules out the responsibility-grounding relationship. (After my discussion of the Rollback Argument, I shall go on to show how my critique can be *generalized* to various other plausible ways of defending the third premise.) Imagine a causally deterministic world W1 in which everything goes as it is supposed to in the sequence issuing in a given human choice and action. That is, suppose causal determinism obtains and Jones chooses (for his own reasons, in the

---

[19] Perhaps it is not surprising that the Rollback Argument and The Conseqence Argument have a common structure in so far as it might be argued that, at a deep level, they both rely on the notion of luck. Indeed, Saul Smilansky has argued that the problem with the fact that, under causal determinism, all our choices and behaviour would be the consequences of the past and laws of nature, is precisely the problem of luck. See Saul Smilansky, *Free Will and Illusion* (Oxford, Clarendon Press, 2000), 284–5; and 'Compatibilism: The Argument from Shallowness' (2003) 115 *Philosophical Studies* 257–82. In simple form, Smilansky's argument here is as follows. If causal determinism is true, then all our deliberations, choices, and behaviour are the result of casually deterministic sequences that began well before we were even born (or had any sense of the relevant options and the values that might be brought to bear on them, and so forth). Since we are not responsible for initiating these sequences, and since our decisions and behaviour are the necessary results of them, we are not 'ultimately' in control of our deliberations and actions in the sense relevant to robust moral responsibility and ethical desert. If causal determinism is true, then it can be seen from a more objective or expansive perspective that what we choose to do, and in fact do, are purely a matter of 'luck': what we choose may be 'up to us' in a superficial sense, but what we choose is causally determined by our values and background dispositions, which are causally determined by our previous experiences, and so forth.

'ordinary way') at t2 to raise his hand at t3, and Jones does in fact raise his hand at t3. Imagine, further, that whatever is required for the responsibility-grounding relationship between prior states of Jones and his choice at t2 to raise his hand at t3 is present; that is, let us say that the *requisite glue* that connects the agent's prior states with his choice at t2 is present. I am not sure exactly what this glue consists of; that is, I am not exactly sure what is required for the responsibility-grounding relationship. But we can suppose that in W1 it—whatever it is—obtains in the case of Jones's choice at t2 to raise his hand at t3.[20] (Of course, it is not enough for Jones's freely choosing at t2 to raise his hand at t3 that this glue be present; but at least we cannot say in W1 that Jones does not freely choose at t2 to raise his hand at t3 because the glue is absent.)

Now imagine another possible world W2 in which everything is the same as W1 in respect of the way the causal sequence that actually leads to Jones's choice at t2 to raise his hand at t3 (everything, that is, apart from causal determination). In W2, as in W1, Jones chooses for his own reasons, in the 'normal way', at t2 to raise his hand at t3 (and Jones does indeed raise his hand at t3). In general, whatever exactly it is that makes it the case that the responsibility-grounding relationship is present in the actual sequence flowing through Jones to his choice at t2 (and action at t3)—everything, that is, apart from causal determination—is also present in the sequence that takes place in W2.

But now we add that there is a genuinely random machine in W2, but not in W1. Let us say that Jones begins his deliberations at t1 about whether to raise his hand; his last moment of deliberation is t1.9 and he makes his choice at t2. The random machine 'operates' in W2 between times t1 and t2. (By 'operating' I simply mean that the machine goes through a series of internal states culminating in either M1 or some other state at t1.9.) For our purposes, we can focus on state M1. That is, if the machine is in state M1 at t1.9, there are two possibilities, each with a 50 per cent objective probability attached to it. The first possibility is that the machine does nothing—it 'goes to sleep', as it were, and does not trigger any causal interaction with the world (including Jones). The second possibility is that it will initiate a causal sequence that would *preempt* Jones's choice at t2 to raise his hand at t3. That is, on the second possibility, the machine would trigger a causal sequence that would terminate in (say) a direct electronic stimulation of Jones's brain sufficient to ensure that Jones chooses at t2 to *refrain* from raising his hand at t3. It is assumed that the process involving the machine in W2 is genuinely random (whatever is required for genuine randomness). Further, let us suppose that, as things actually go in W2, the machine's state at t1.9 is indeed M1, and, further, the

---

[20] Note that this case, as I have presented it in the text, will not move a committed *sceptic* about whether the glue can be present *ever*—even in a causally deterministic context. Rather, my primary target in this paper is a theorist who thinks that *indeterminism* poses a special problem for moral responsibility in virtue of posing a special challenge to the possibility of the glue's obtaining. That is, there are philosophers who do not worry about the choice and action's being an 'outflowing' of the agent under causal determinism, but *do* worry that *indeterminism* would call into question the contention that the relevant behaviour is an outflowing of the agent. It is to such philosophers that I address my argument.

machine simply 'goes to sleep' and never triggers any causal interference in the sequence flowing through Jones to his choice at t2 to raise his hand at t3. (That is, the machine is in M1 at t1.9 and the first possibility is realized—no intervention in the causal sequence flowing through Jones.)

In both worlds W1 and W2, Jones chooses and does exactly the same thing (type-identical choices and actions), as a result of relevantly similar processes. More specifically, we have assumed that the causal process linking Jones's prior states and his choice is the same in relevant respects in both worlds; thus, if the requisite glue connecting the prior states with the choice obtains in W1, it also obtains in W2. Presumably, the mere existence and operation of the machine in W2 should not in any way threaten these claims about the responsibility-grounding relationship. How could the mere existence of such a machine affect the responsibility-grounding relationship, given that the machine does not causally interact with the sequence flowing through Jones and issuing in the choice at t2?[21] Indeed, it should be intuitively obvious that the mere existence and operation of the machine in W2 is *irrelevant* to whatever it is that makes it the case that the responsibility-grounding relationship obtains in the sequence flowing through Jones.

Notice, however, that W2 is causally indeterministic during the relevant interval. Indeed, W2 is causally indeterministic in the relationship between the prior states of Jones at t1.9 and Jones's choice at t2; after all, Jones's actual deliberations could have been pre-empted by a causal sequence that was not in fact triggered in W2. So we could obviously run the Rollback Argument with respect to W2: if we were to roll back the universe to t1.9 and allow it to go forward a thousand times, then in say 467 'replays' Jones will choose at t2 to raise his hand at t3, and in say 533 replays Jones will be caused to choose at t2 to refrain from raising his hand at t3. The Rollback Argument clearly 'applies' to the conditions of W2, as W2 is explicitly an indeterministic world.

But the key point is that intuitively it is obvious, as I claimed above, that the mere existence and operation of the machine in W2 cannot in itself show that the requisite glue is not present in W2—it cannot show that whatever underwrites the responsibility-grounding relationship is missing. (Recall, again, that the machine's 'operating' refers simply to its going through a sequence of internal states; it does *not* imply the triggering of its capacity to initiate a preempting sequence.) Perhaps my point could be put as follows. Whatever underlies the

---

[21] As Randy Clarke has reminded me, there are tricky issues here. Is the machine's not being triggered a cause? If so, it might not be exactly right to say that the machine does not causally interact with Jones. Still, I here rely on some intuitive notion—difficult to specify—according to which the machine does not actually causally interact with the sequence flowing through Jones. Note that similar worries come up in the context of evaluating the classical Frankfurt cases (and variations on them). For example, the untriggered machine in W2 is parallel to resting Black in the original Frankfurt case. In both cases my contention is that they (resting Black and the untriggered machine) are (in some sense) *not* part of the actual sequence flowing through Jones and thus are *irrelevant* to Jones's moral responsibility. In the context of a defense of their version of a Frankfurt case, Alfred Mele and David Robb discuss such issues: Alfred R. Mele and David Robb, 'Bbs, Maagnets and Seesaws: The Metaphysics of Frankfurt-style Cases' in Widerker and McKenna (eds.), *Moral Responsibility and Alternative Possibilities: Essays on the Importance of Alternative Possibilities* 127–38.

responsibility-grounding relationship—whatever constitutes the relevant glue that binds together the prior states of the agent with his choice—is a matter that is *intrinsic* (in some sense) to the relevant causal sequence. It is a matter of *the way the prior states of the agent lead to the choice in question*, and this cannot be affected by the mere presence of something (such as the random machine in W2) that *plays no role in the causal sequence flowing through the agent*. And if this is correct, then the *mere fact* that the Rollback Argument can successfully be run cannot *in itself* show that the responsibility-grounding relationship is not present. After all, in W2 the responsibility-grounding relationship is indeed present; it is present to the same extent that it is present in W1. Yet we can run the Rollback Argument relative to the conditions present in W2.

In a nutshell, then, my argument is as follows. Let us suppose, what is not implausible, that we have some sufficiently determinate intuitive notion of 'the way a causal sequence goes', where this notion abstracts away from whether causal determinism obtains or not. We now suppose that in W1 the actual causal sequence goes in the 'normal' way typically thought to ground attributions of moral responsibility, apart from considerations pertinent to causal determination. Now add to W1 that causal determinism does in fact obtain. It should be widely accepted that, whatever the requisite glue (the responsibility-grounding relationship) is, it obtains in W1. Given that the responsibility-grounding relationship is present in W1 and intuitively cannot be expunged simply because of the existence of the genuinely random machine in W2, it also is present in W2. But we can successfully run the Rollback Argument with respect to the conditions present in W2; indeed, the mere existence of the machine in W2 makes it the case that the relationship between Jones's prior mental states and his choice at t2 is *indeterministic* (even though 'the way the causal sequences go' in W1 and W2, as defined above, is the same). Thus, the mere fact of the application of the Rollback Argument does *not* show what it is intended to show, namely that the responsibility-grounding relationship is absent. So the application of the Rollback Argument cannot be the reason why causal indeterminism threatens the responsibility grounding relationship—threatens to make us come unglued, as it were.

Above I noted that the Consequence Argument and the Rollback Argument have the same abstract structure. It is interesting to notice, also, that the responses to the two threats—the threat from causal determinism and the threat from causal indeterminism—employ similar ingredients.

To explain. Recall that Frankfurt pointed to an important intuition about Black (the counterfactual intervener) in the Frankfurt-cases. As Frankfurt noted, Black did not play any role in how the actual sequence unfolds; as he put it, we could 'subtract' Black from the story and everything would proceed in just the same way. Frankfurt then said, as quoted above, 'When a fact is in this way irrelevant to the problem of accounting for a person's action it seems quite gratuitous to assign it any weight in the assessment of his moral responsibility.' Black is a merely counterfactual intervener; he is, as I described him above, an 'untriggered ensurer'. Frankfurt's point, and it is an intuitively compelling point, is that a merely *counterfactual* intervener—an individual or device not actually triggered to have any effect on the causal sequence in question—is *irrelevant* to an agent's moral responsibility.[22]

Now the random machine in W2 is also a *merely counterfactual* intervener. The random machine in W2 is an *untriggered pre-empter*. But both untriggered ensurers (such as Black) and untriggered pre-empters (like the random machine) are equally *untriggered*; that is, they are *merely counterfactual* interveners, and, as such, they are plausibly thought to be *irrelevant* to attributions of moral responsibility. Thus, just as the Consequence Argument and the Rollback Argument are structurally similar at an abstract level, so are the responses to the threats they underwrite from causal determinism and causal indeterminism; both responses rely on a basic intuition, first formulated by Harry Frankfurt, to the effect that mere counterfactual interventions are irrelevant to ascriptions of moral responsibility. As I would put it, moral responsibility is a matter of what happens in the *actual sequence*.

The Rollback Argument is only one way of seeking to support premise 3 of the Dilemma of Determinism; I have contended that the possibility of certain sorts of untriggered pre-empters shows that the Rollback Argument cannot in fact support premise 3. As far as I can tell, various of the most salient ways of attempting to support premise 3 fall prey to the possibility of untriggered pre-empters (and the associated argumentation). I turn in the next section to a brief defence of this claim.

4.4 Other Arguments for the Third Premise

There is a large literature surrounding the 'luck problem'. This problem is, after all, why some compatibilists have claimed that not only is moral responsibility compatible with causal determinism, but moral responsiblility *requires* causal determinism.[23] This is not, however, the direction I would take, because it would imply that our moral responsibility would, as it were, hang on a thread—just as much as it would, if moral responsibility were deemed *incompatible* with causal determinism. Rather, I have sketched (in the previous section) a response to the problem of luck—or at least a response to one version of it. I think the response can be

---

[22] Classical Frankfurt cases involve untriggered ensurers (such as Black). I have presented what I have somewhat immodestly called, 'Fischer-type cases', in which (say) Jones would be instantaneously annihilated, if he were about to choose to do otherwise. These somewhat chiliastic cases involve what might be dubbed 'untriggered destroyers'. Untriggered destroyers are equally as efficacious as untriggered ensurers in challenging the notion that moral responsibility requires alternative possibilities. See John Martin Fischer, 'Responsibliity and Agent-Causation' in Widerker and McKenna (eds.), *Moral Responsibility and Alternative Possibilities: Essays on the Importance of Alternative Possibilities* 235–50; reprinted in John Martin Fischer, *My Way: Essays on Moral Responsibility* (New York, Oxford University Press, 2006) 143–58.

[23] Peter Van Inwagen calls the argument that moral responsibility *requires* causal determinism the '*Mind* argument' because many of the important presentations and discussions of this argument appeared in *Mind*. See, for example, R. E. Hobart (whose given name was Dickenson S. Miller), 'Free Will as Involving Determinism and Inconceivable Without It' (1934) 58 *Mind* 1–27. Van Inwagen's discussion of the *Mind* argument is in Van Inwagen, *An Essay on Free Will*, 126–52.

generalized to various other versions of the argument from luck, and it will be helpful simply to consider a few other versions here.[24]

Ishtiyaque Haji has presented what might be called the 'ensurance formulation' of the Luck Problem. The basic idea here is that indeterminism implies that an agent cannot ensure or guarantee, just prior to his choice and action, that she would make that choice and act in that way. Given this lack of capacity to ensure, it allegedly follows that the agent is not in control of her behaviour—some factor *outside the agent* makes the difference, in the end. That is, given that (under indeterminism) the agent herself cannot guarantee the outcome; it is only when some external factor is added that the outcome ensues, and thus it is not *the agent* who makes the difference; *she* does not control the outcome. It is, then, a matter of luck whether the agent chooses and acts as she does, rather than choosing and acting in some other way.

Robert Kane also takes it that causal indeterminism entails that agents lack the capacity to ensure or guarantee the behaviour in question prior to it. Kane calls this capacity, 'antecedent determining control', and he says that ' . . . the ability to be in, or bring about, conditions such that one can guarantee or determine which of a set of outcomes is going to occur *before* it occurs, whether the outcomes are one's own actions, the actions of others, or events in the world generally' is something the libertarian agent cannot have.[25] Kane however goes on to say:

No doubt, such 'antecedent determining control' as we might call it, is valuable in many circumstances; and we cannot help but value it by virtue of an evolutionary imperative to seek security and get control of our surroundings. But it does not follow that because you cannot determine which of a set of outcomes occurs *before* it occurs, you lack control over which of them occurs, *when* it occurs. When the conditions of plural voluntary control are satisfied, agents exercise control over their future lives *then and there* in a manner that is not antecedently determined by their pasts.[26]

Kane's view, then, is that we do not need antecedent determining control—the capacity to ensure our choices and actions prior to them—in order to exercise a kind of control that would render us morally responsible. My argument in the previous section can be construed as a defence of Kane's position here. Clearly, there is no reason to suppose that in W1 Jones lacks the power prior to t2 to ensure or guarantee his choice at t2 (and action at t3); but in W2 he does lack this power. Thus, in W1 he has antecedent determining control, whereas in W2 he lacks it. But, as argued above, the requisite glue—whatever it is about the relationship between Jones's prior states and his choice at t2 that underwrites the view that he is in control—is present in W2, just as much as in W1. Thus, the mere fact that an

---

[24] In thinking about the various versions of the Luck Objection, I have benefited greatly from reading unpublished material by Christopher Franklin, in particular a draft of ch. 4 of his PhD dissertation, 'Farewell to the Luck (and *Mind*) Argument' (Department of Philosophy, University of California, Riverside).

[25] Robert Kane, *The Significance of Free Will* (New York, Oxford University Press, 1996) 144. See also Robert Kane, 'Responsibility, Luck, and Chance: Reflections on Free Will and Determinism' (1999) 96 *Journal of Philosophy* 217–40.

[26] Robert Kane, *The Significance of Free Will* 144.

agent lacks antecedent control does not in itself show that he is not in control in the sense required for moral responsibility.

Alfred Mele has developed another version of the Luck Problem.[27] In Christopher Franklin's terminology, this is the 'Explanatory Formulation'. As many philosophers have pointed out, it appears that if causal indeterminism obtains, then we cannot give a *contrastive explanation* of an agent's choices and behaviour. (I do not include Mele himself in this group, although he gives a version of what might be called the Explanatory Formulation.) That is, it seems that we cannot give an explanation of why (say) Jones chooses as he actually does *rather than making a different choice*. Given the putative lack of availability of contrastive explanations under indeterminism, many have concluded that agents in causally indeterministic worlds cannot have the sort of control required for moral responsibility; after all, if we cannot even in principle explain why the agent chooses X rather than Y, it does not seem that it is genuinely *up to the agent* whether to choose X rather than Y.

Mele brings out this sort of worry by considering a goddess Diana who is creating agents who satisfy event-causal libertarianism. She worries, however, about the luck problem:

Her worry, more specifically, is that if the difference between the actual world, in which one of her agents judges it best to A straightaway and then, at t, decides accordingly, and any possible world with the same past up to t and the same laws of nature and he makes an alternative decision while the judgment persists is just a matter of luck, then he does not freely make that decision in that possible world, W. Diana suspects that his making that alternative decision rather than deciding in accordance with his best judgment—that is, the difference between W and the actual world—is just a matter of bad luck, or, more precisely, of worse luck in W for the agent than in the actual world. After all, because the worlds do not diverge before the agent decides, there is no difference in them to account for the difference in decisions.[28]

Mele's contention is the cross-world difference between the actual world and W is just a matter of luck, and thus the existence of luck poses a significant challenge to the agent's being free in either world. Just as Robert Kane seeks to address the problem of the lack of antecedent determining control, Alfred Mele goes on to address the problem posed by the apparent fact that the relevant cross-world differences are just a matter of luck.[29]

Again, the argument of the previous section shows that the mere lack of the availability of a contrastive explanation (of the relevant kind) does *not* in itself show that an agent does not possess the sort of control required for moral responsibility. After all, in W1, but not in W2, it is presumably possible to provide a contrastive explanation of Jones's behaviour by reference to states 'internal to Jones' in the relevant sense—his motivational states and whatever particular states realize or

---

[27] Alfred Mele, *Free Will and Luck* (New York, Oxford University Press, 2006). Here again I am indebted to ch. 4 of Christopher Franklin's dissertation.
[28] Alfred Mele, *Free Will and Luck* 8.
[29] John Martin Fischer, 'Review of Alfred Mele's, *Free Will and Luck*' (2008) 117 *Mind* 187–91; also, see Kane, 'Responsibility, Luck, and Chance: Reflections on Free Will and Determinism'.

constitute them. Similarly, in W2 (but not W1) Mele could run his argument from cross-world differences. But, as we have seen above, the responsibility-underwriting glue is present in W2, just as much as in W1. I thus conclude that the mere fact that the relevant cross-world differences are purely a matter of luck does not in itself show that the agent's behaviour results from luck in a sense that would rule out his moral responsibility.

Consider the pair of worlds, W2 and W2*. As we know, in W2 the state of the machine at t1.9 is M1 and the machine 'goes to sleep'. Imagine that in W2* the state of the machine at t1.9 is M1, and it swings into action (the other possible result of being in M1). Nothing else is different about W2 and W2* (up to 1.9). Thus, Mele's argument applies: if it is sound, one could conclude that the difference between W2 and W2* is just a matter of luck. But (as above) if the responsibility-underwriting glue is present is W1, it is present in W2. So, although Mele's argument applies to W2, the glue is nevertheless present in W2. Mele's argument thus cannot in itself show that there is an insuperable problem (pertaining to luck) with causal indeterminism.

4.5 Conclusion: An Approach to the Problem of Luck

A traditional conception of moral responsibility has it that in order to be morally responsible, I must make a difference to the unfolding world by selecting one from various genuinely open alternatives. There are two crucial ideas here. First, I must have alternative possibilities; I must select from among genuinely available options. Second, the path taken must be an outflowing of me—it is *I* who makes a difference by selecting the path. The two key premises of the Dilemma of Determinism call into question each of these ideas. The second premise states that if causal determinism is true, then I do not have alternative possibilities, and thus I do not select from among more than one genuinely available path into the future. The third premise states that if causal determinism is false, then it is not *I* who makes the selection.

It is interesting that the Consequence Argument (underwriting the second premise) and the Rollback Argument (supporting the second premise) have a similar abstract structure; in both cases we are essentially asked to imagine that the universe is rolled back. In the case of the Consequence Argument, we note that no matter how many times we allow it to go forward, it will take the same path. In the case of the Rollback Argument, we note that the universe will *not* take the same path every time we allow it to go forward. Both observations can be disturbing.

I have sketched replies to both premises (both horns of the dilemma). First, I have argued that moral responsibility does not require alternative possibilities. Here the Frankfurt cases are helpful; they are cases involving counterfactual interveners (such as Black) who are *untriggered insurers*. Second, I have argued that indeterminism does not by itself entail the absence of the control required for moral responsibility. Here the genuinely random machine is helpful; it is a counterfactual intervener that is an *untriggered pre-empter*. In both cases it is crucial that we have a *counterfactual* intervener—an individual or device that is poised to intervene but remains dormant. I have contended that the basic intuition elicited by Harry Frankfurt in his famous challenge to the second premise—that moral responsibility ascriptions are based on the actual sequence and thus mere untriggered devices or causally dormant individuals (counterfactual interveners) are not relevant to ascriptions of moral responsibility—also plays a crucial role in addressing the problem of luck. This basic intuition, after all, drives my argument that employs the genuinely random machine.[30]

Think of it this way. Assume (for simplicity's sake) that the counterfactual intervener, Black, in the Frankfurt example makes it the case that Jones cannot do otherwise. (More carefully, it would be Black together with other factors that would have this implication, but put this aside for now.[31]) Frankfurt's point is that in so far as Black's presence is irrelevant to Jones's moral responsibility, it follows that the fact that Jones cannot do otherwise is similarly irrelevant to Jones's moral responsibility. Now assume (again, for simplicity's sake) that the genuinely random machine makes it the case that the Rollback Argument applies in my example above (and that Jones lacks antecedent determining control, and so forth). In so far as the mere presence of the machine is irrelevant to Jones's moral responsibility in the example, it follows that the fact that the Rollback Argument applies (and that Jones lacks antecedent determining control, and so forth) is irrelevant to Jones's moral responsibility.

In previous work, I have urged that we focus like a laser on the 'actual sequence' that issues in the behaviour in question. In developing an account of the control that grounds moral responsibility, I have suggested that we prescind from counterfactual interveners, such as Black, and attend to the properties of Jones as manifested in the unfolding actual sequence of events. In identifying those properties, we pretend that Black does not exist; as it were, we put him in a box. (As Frankfurt put it, if we were to 'subtract Black away', the flow of the actual sequence would be exactly the same.) Similarly, in evaluating the relationship between indeterminism and control-grounding moral responsibility, we need to have a laser-like focus on the actual sequence. And when we 'subtract away' the genuinely random machine, we see clearly that the sequence issuing in the relevant behaviour in W2 is exactly the same as in W1. Thus, in so far as the presence of the machine is irrelevant to Jones's responsibility, Jones has exactly the same glue in W2 as in W1.

---

[30] I think that Frankfurt is basically correct that mere counterfactual interveners or untriggered factors are irrelevant to ascriptions of moral responsibility. But this simple, important point needs to be interpreted and perhaps even qualified in light of cases such as 'Shark', which are discussed in John Martin Fischer and Mark Ravizza, *Responsibility and Control: A Theory of Moral Responsibility* (New York, Cambridge University Press, 1998) 127–8. See also the fascinating (as yet unpublished) paper by Carolina Sartorio, 'Actuality and Responsibility' (Department of Philosophy, University of Arizona). Although Sartorio rejects the simple formulation of the point—that counterfactual interveners are irrelevant to ascriptions of moral responsibility—she does formulate and defend an alternative way of capturing the basic insight. I believe that the basic insight (articulated suitably) is all that is needed for my purposes in the text.

[31] For further discussion, see John Martin Fischer, 'The Frankfurt Cases: The Moral of the Stories' (2010) 119 *Philosophical Review* 315–36.

I have then sketched an approach to the problem of luck. I do not call it a 'solution' to this problem, in part because of a natural conservatism about such claims, but also in part because I do not give an *account* of the responsibility-underwriting glue. Also, I have not argued that *no* articulation of the problem of luck can avoid the sort of strategy of response I have sketched.[32] It should be noted that my suggestion does not vindicate indeterminism in a context in which the agent has 'dual control'—that is, moral-responsibility conferring control in both the actual sequence and the alternative sequence. In so far as certain libertarians adopt the dual-control picture, my solution does not in itself provide a full defence against the problem of luck that could be invoked by all libertarians.[33] Instead, I have had the more modest project of arguing that *if* the requisite glue is indeed present in a context of causal determination, then it is *also* present in certain contexts under causal indeterminism. It would seem to follow that causal indeterminism *per se* does not rule it out that we are glued together as morally responsible agents. That is, the mere possibility of running the Rollback Argument *in itself* does not threaten moral responsibility.[34]

In defending a compatibilist account of moral responsibility (elsewhere), I have contended (as any compatibilist must) that not all causally deterministic sequences

---

[32] If one is not convinced by my strategy that *no* articulation of the problem of luck will be forceful, then perhaps one can see my paper as forcing the proponent of (say) the Rollback Argument (or various of the other arguments) to be clearer and more precise as to how exactly to formulate the worry. For example, a proponent of the Rollback Argument might seek to place some constraints on the admissible scenarios that would rule out cases such as W2, where the 'randomness' is installed in a place that is 'external to and independent of' the agent. Both Anthony Brueckner and Tim O'Connor have suggested the possibility of such strategies to me. Brueckner notes that in my example of W2 the randomness is 'injected' into the machine (but not the brain), and O'Connor wonders whether we can put aside 'environmental wildcards' (such as meteorites, lightning bolts, and, presumably, causal sequences initiated by distill and causally independently-triggered machines), in evaluating agency. Recall, however, that the machine's randomness makes it the case that the connection between Jones's prior mental states and his choice is indeterministic in W2, so there is indeterminacy in just the 'right spot' in W2. That is, by 'injecting randomness' into the machine in W2, we *thereby* inject indeterminacy into the relevant place in the sequence flowing through Jones. Perhaps one could reformulate the Rollback Argument to require that we only consider scenarios where the indeterminacy in the agent is somehow 'non-derivative'—it is, as it were, injected *directly* in the agent. I do not here argue that one could not seek to reformulate the Rollback Argument in this way, although I am somewhat sceptical about this possibility; here I am simply claiming that it is important to note that it *must* be reformulated. Further, such a reformulation, if it can be given, should help us to see more precisely what the real basis of the Rollback Argument is.

[33] I am indebted to Dana Nelkin and David Brink for this point.

[34] I believe that this is an important *first step* in a more complete defence of the coherence of indeterministic control. Just as with the Frankfurt examples, I would counsel patience and at least a *two-step* argument. Here, the first step is to show that the mere applicability of the Rollback Argument does not in itself establish that the control required for moral responsibility is absent. One can then offer *additional* argumentation for the particular model of indeterministic control one wishes to defend. Thus, even though the first step of the argument cannot in itself establish an indeterministic 'dual-control' model, this step is entirely consistent with going on to defend just such a model. From the fact that my Random Machine Argument does not provide a model in which there is dual-control, if does *not* follow that it could not be an important first step in the defense of such a model. (Again, the dialectical situation here is analogous to the two-step approach I have defended with respect to the Frankfurt Examples. This is yet another way in which the two dialectical contexts exhibit a striking symmetry.)

are created equal.³⁵ That is, I have contended that it is not the case that all causally deterministic sequences crowd out moral responsibility; causal determination *qua* causal determination does not rule out moral responsibility. Similarly, here I have argued that not all contexts of causal indeterminism threaten the responsibility-underwriting relationship; thus, causal indeterminism *qua* causal indeterminism does not rule out moral responsibility.³⁶

In so far as we have good responses to both of the horns of the Dilemma of Determinism, moral responsibility does not hang on a thread. This is a good thing, since it does not seem to be a comfortable position to have to proclaim from our armchairs that we know that causal determinism is false. And it would appear to be equally uncomfortable to feel the need to pronounce, from our philosophical LazyBoys, that causal determinism is true. It is thus a good thing that we have found plausible and 'independent' arguments for the *resilience* of moral responsibility to the truth or falsity of causal determinism—arguments that are not simply statements of the desirability of such resilience. In the end, then, moral responsibility is neither inconceivable nor puzzling.

---

³⁵ See, for example, John Martin Fischer, *The Metaphysics of Free Will: An Essay on Control*; John Martin Fischer and Mark Ravizza, *Responsibility and Control: A Theory of Moral Responsibility* (New York, Cambridge University Press, 1998); John Martin Fischer, *My Way: Essays on Moral Responsibility*; and John Martin Fischer et. al., *Four Views on Free Will*.

³⁶ Indeed, my account of moral responsibility is consistent with either causal determination or indeterminism: John Martin Fischer and Mark Ravizza, *Responsibility and Control: A Theory of Moral Responsibility* 253.

5

Neuroscience and Criminal Responsibility: Proving 'Can't Help Himself' as a Narrow Bar to Criminal Liability

Henry T. Greely*

I have been working on issues of law and neuroscience for about eight years. During that time, those working in the field have devoted most of their attention to how neuroscience might affect defences to criminal liability. Professor Robert Sapolsky, neuroscientist and primatologist, has argued that neuroscience both should and, ultimately, will make the criminal justice system disappear. Neuroscience will prove, he urges, that criminal defendants could not help but act as they did. As a result, we will not be able to hold them morally accountable for their actions and we will abandon the criminal justice system.[1]

Although I believe Sapolsky's vision played an important role in the creation of the Law and Neuroscience Project, financed by the John D. and Catharine T. MacArthur Foundation, it has generally not been well received by legal scholars, judges, or legal philosophers. They point out that the criminal law does not currently require that a criminal acted of his own 'free will', nor is retribution for moral wrongdoing the only reason for the criminal justice system. Specific deterrence of future bad actions by any one defendant, general deterrence of bad actions by others in the population, prevention of future crimes by incapacitating convicted criminals, and even (at least in theory) rehabilitation are all justifications for applying a criminal justice sanction against a defendant, even one who 'can't help himself'.

I have avoided that topic the way I might avoid a drunken and belligerent brother-in-law at a family reunion: by keeping a very careful distance. I have never written about the topic, other than to note its existence and interest to others. Part

---

* Deane F. and Kate Edelman Johnson Professor of Law; Professor, by courtesy, of Genetics; Stanford University. The author would like to thank his research assistant, Mark Hernandez, as well as the many participants in the University College London colloquium in July 2009 at which he first presented these thoughts. He particularly wants to single out Professor Stephen Morse for his tireless, and perhaps Sisyphisean, efforts to educate him about criminal responsibility.

[1] Robert M. Sapolsky, 'The Frontal Cortex and the Criminal Justice System' in Semir Zeki and Oliver Goodenough (eds.), *Law and the Brain* (Oxford, Oxford University Press, 2006) 227.

of my reluctance has been a matter of background. I am not a scholar of criminal law, a field that been cultivated deeply and broadly for centuries. Part of it has been a matter of taste. I recognize that intricate discussions of criminal responsibility can fascinate reasonable (and sometimes brilliant) people, but they do not fascinate me. But part of my reluctance has been that I have largely been convinced by those who are sceptical about a large role for neuroscience in criminal responsibility, mainly for reasons described by Stephen Morse, among others.[2]

And yet. Something in this area has been nagging at me. Sapolsky sometimes illustrates his argument with the example of Tourette Syndrome. This condition afflicts somewhere between 0.1 and 1 per cent of the population, usually in a mild form. Those with the syndrome are subject to physical tics and at least one vocal tic. About 10 per cent of people with Tourette Syndrome have a kind of vocal tic called coprolalia. These people will frequently begin using crude, offensive, or obscene language for no apparent reason. As Sapolsky points out, a century ago people exhibiting coprolalia as a result of Tourette Syndrome would have been arrested and taken to jail; five or six centuries ago, they might have burnt at the stake or, at the least, been subject to exorcism to drive off the demons that possessed them. Today, we recognize that people exhibiting coprolalia as a result of Tourette Syndrome are sick, and we excuse their behaviour. Furthermore, we believe we *should* excuse that behaviour. Surely Sapolsky is right about Tourette Syndrome; how can he be wrong about the rest?

This paper, and the talk that gave rise to it, are my effort to answer that question. That question is complicated, but I believe one good way to look at it is to consider how one would go about proving that someone 'can't help himself' from committing criminal actions. It is very hard to prove with satisfying certainty that an individual could not stop himself from committing a criminal act. This difficulty may, in part, explain why the criminal law and its scholars reject Sapolsky's argument so strongly. Discussion of these issues of proof may ultimately help us understand better how neuroscience may, and may not, raise new issues of responsibility for criminal law.

This paper will largely ignore the legal and philosophical arguments about whether, how, and to what extent a claim of 'can't help himself' does or should serve to limit a defendant's criminal responsibility. Instead, it will assume that there is, or should be, such a doctrine and ask 'How could such a claim be proven?' It will first examine that question as a collective matter, looking across lots of individuals. It will then turn to a claim made about, and arising out of the special circumstances of, one individual. Finally, it will propose a narrow opportunity for a criminal defendant to prove that he 'can't help himself' and thereby avoid criminal responsibility.

---

[2] See, e.g., Stephen J. Morse, 'Moral and Legal Responsibility and the New Neuroscience' in Judy Illes (ed.), *Neuroethics: Defining the Issues in Theory, Practice, and Policy* (Oxford, Oxford University Press, 2005).

5.1 General Proof of 'Can't Help Himself'

Some behaviours clearly are, at least to some extent, beyond our control. Breathing is a behaviour. We can make ourselves breathe faster or slower, deeper or shallower, but healthy people cannot voluntarily stop breathing for more than a few minutes. We cannot hold our breaths long enough to pass out, let alone to die. Our body eventually *forces* us to take a breath.[3]

How would we prove that someone could not control his behaviour—that he 'can't help himself?'—when that behaviour is criminal? In general, how does biomedical science prove that anything is a sufficient cause or explanation for a particular behaviour? Because of the focus of my own work, I am most interested in genetic and neuroscientific evidence for uncontrollable behaviour, which I discuss below as 'physical' evidence. I will also discuss briefly 'behavioural' evidence of mental health disorders.

5.1.1 'Physical' evidence—of genes and brains

In the late nineteenth century, medicine urgently confronted questions of physical proof, as the combination of good microscopes and Pasteur's 'germ theory of disease' launched a search for microbes that 'caused' various diseases. The problem, of course, was that there are a lot of microbes, and many are found in healthy people as well as sick ones.

The German scientist Robert Koch discovered the microbial sources of anthrax and of tuberculosis and contributed to identifying the pathogen that causes cholera. In 1890 he published what came to be called 'Koch's Postulates'.

To establish that a microbe is the cause of a disease:

1) The microorganism must be found in abundance in all organisms suffering from the disease, but should not be found in healthy animals.

2) The microorganism must be isolated from a diseased organism and grown in pure culture.

3) The cultured microorganism should cause disease when introduced into a healthy organism.

4) The microorganism must be reisolated from the inoculated, diseased experimental host and identified as being identical to the original specific causative agent.

---

[3] Of course, in science as in law, almost every statement has exceptions. Some people, who have suffered damage to their brain stems, cannot take a breath, either automatically or voluntarily. To survive, they need to help from a mechanical ventilator. In another, rarer, situation, people can breath *only* through an active, voluntary decision. If they fall asleep or just forget to breath, they can become unconscious and, eventually, die. This rare condition, generally of unknown cause, is known as congenital central hypoventilation syndrome or primary alveolar hypoventilation—and is sometimes called, more poetically, Ondine's Curse.

Koch initially proposed his Postulates as 'the' test for disease causation: if the Postulates were met, the microbe caused the disease; if they were not, it did not. It quickly became clear that Koch's Postulates were too stringent to be required for proof that a microbe caused a disease. Some pathogens are found in healthy people who have subclinical or asymptomatic infections. Similarly, some pathogens, when injected into a person, may not cause disease, either because they do not become established as an infection or because they cause subclinical or asymptomatic infections. And some pathogens, notably viruses, cannot be grown in pure culture. Compliance with Koch's Postulates is not necessary to prove that a pathogen causes a disease, but it is sufficient—anything that satisfies them can be said to 'cause' the disease.

Something similar to Koch's Postulates has been used in genetics. In searching for 'disease genes' (or, more properly, 'disease alleles', since different variants of a particular gene are called alleles), researchers typically looked at genes in families that had a high rate of a particular disease thought to be genetic. Paralleling Koch's first postulate, they would search for genetic variations that were found in every family member with the disease but in no family members without the disease. The second postulate, growth in pure culture, was irrelevant in genetics. The third, causing the disease by 'injecting' the pathogen, was not ethically, or medically, feasible in humans, but some human disease alleles have been successfully transferred into the genomes of laboratory animals and have sometimes (but not always) caused symptoms in that species that are identical or similar to those it causes in humans. (Stanley Falkow published 'Molecular Koch's Postulates' in 1988, but those, like Koch's, dealt with infectious diseases, not genetic ones, and referred to genes in pathogens that caused the pathogen to be virulent.[4])

Some genetic diseases fit nicely into the equivalent of Koch's Postulates. Huntington disease, for example, is a fatal neurodegenerative condition, usually diagnosed in middle age. It is caused by the patient having more than thirty-six copies of the nucleoside bases 'CAG' (cytosine-adenine-guanine) repeating themselves, one after another, in a particular part of either of their two copies of the Huntington gene. No one with thirty-six or fewer copies of this repeat has ever been known to have Huntington disease; no one with thirty-seven or more repeats who has lived a long life is known to have avoided the disease. If you do not have more than thirty-six repeats, you cannot have Huntington disease; if you have more than thirty-six repeats, the only way to avoid it is to die first from something else.

Most genetic diseases are not so perfectly determinative. Mutations in several genes (Presenilin 1, Presenilin 2, and the Amyloid Beta Precursor Protein gene) appear to lead inevitably to early-onset Alzheimer disease, but the vast majority of people with Alzheimer disease do not have these mutations. These alleles are sufficient but not necessary causes of the disease.

Another gene, APOE, is also associated with Alzheimer disease, but not as strongly. This gene comes in three common alleles—APOE 2, APOE 3, and

---

[4] S. Falkow, 'Molecular Koch's Postulates Applied to Microbial Pathogenicity' (1988) 10 *Rev. Infect. Dis.* S274–S276.

APOE 4. (APOE 1, the first allele discovered, was easy to find because it causes very serious blood problems, but it is also quite rare.) All humans have two copies of the APOE gene (one from each parent). People who carry two copies of APOE 4 are at very high risk, about 85 per cent, of developing Alzheimer disease in their old age, but some very elderly people with two copies of that allele show no signs of the disease. People with one copy of APOE 4 and one copy of APOE 3 have two to three times the average person's lifetime risk of being diagnosed with Alzheimer disease, but their risk is still under 50 per cent. Around half of all patients with Alzheimer disease carry at least one copy of APOE 4, but that means the other half do not. APOE 4 increases the risk of Alzheimer disease but is neither sufficient nor necessary for the disease.

Does any of this have any connection to the idea that a particular criminal 'can't help himself?' Are there any situations where we can say, with confidence, that everyone with condition X will perform (criminal) behaviour Y? If so, that is some strong (though not, of course, perfect) evidence that a man with condition X 'can't help himself' from doing Y. I think so. There is at least one report of a genetic variation being sufficient (though certainly not necessary) to cause criminal behaviour.

Monoamine oxidases are proteins that catalyze chemical reactions in the body by oxidizing a class of molecules called monoamines, leading to the removal from these molecules of an amine group (a nitrogen atom bound to two hydrogen atoms). Humans have two versions, monoamine oxidase A (MAO-A) and monoamine oxidase B (MAO-B), each encoded by a different gene. Both proteins help break down certain neurotransmitters, used to move messages between neurons. MAO-A breaks down serotonin, epinephrine, and norepinephrin; MAO-B breaks down phenethylamine. Both compounds break down dopamine. MAO-A is also found in the liver, where it helps metabolize monoamines contained in food. The two MAO genes are found next to each other on the X chromosome. The X chromosome is one of two sex chromosomes in humans. Normally, women have two X chromosomes; men have one X chromosome and one Y chromosome.

In 1993, H. G. Brunner and colleagues published two reports[5] about a syndrome (now called Brunner Syndrome) they found in a large family in the Netherlands. Five males in the family shared a series of behaviours: mild mental retardation along with aggressive and sometimes violent behaviour, including arson, attempted rape, and exhibitionism. This particular combination of symptoms had not previously been described.

The researchers determined that all five affected men had inherited a mutated version of the MAO-A gene, one in which the production of the MAO-A protein was terminated early, leaving the affected men with no functioning MAO-A protein. The normal males in the family did not have this mutation. Women in

---

[5] H. G. Brunner et al., 'Abnormal Behavior Associated with a Point Mutation in the Structural Gene for Monoamine Oxidase A' (1993) 263 *Science* 578. See also H. G. Brunner et al., 'X-linked Borderline Mental Retardation with Prominent Behavioral Disturbance: Phenotype, Genetic Localization, and Evidence for Disturbed Monoamine Metabolism' (1993) 52 *Am. J. Hum. Genet.* 1032.

the family were unaffected. Because this gene is on the X chromosome, the women had two copies of the gene (as they have two X chromosomes) and, even if one of those copies were the mutant version, the other copy presumably made enough MAO-A for normal development and behaviour.

This non-functional allele of the MAO-A gene appears, therefore, to meet a genetic version of Koch's Postulates: everyone with the behaviour has the mutated allele, no one without the behaviour has the mutated allele. In addition, this association is scientifically plausible. The MAO-A protein is crucial to the metabolism of certain neurotransmitters. It is not surprising that the lack of functional MAO-A would cause behavioural changes. Since criminal behaviour is one of the behaviours caused by this mutant allele of the MAO-A gene, this allele could be called 'a criminal allele'.

Of course, we cannot be absolutely certain of that. This particular syndrome, and this particular mutation, have been identified in only five people in the entire world—all from that single Dutch family. There may be other people with the mutated allele who have normal behaviour (or, at least, not *this* abnormal behaviour). It is also possible that other males in this family, who lack the mutated allele, have similar impulses but are able to control them. Yet it appears, so far, that people with Brunner Syndrome 'can't help themselves'.

I know of no other extremely powerful 'criminal allele'. Other genes or alleles are known to increase the chance that a person will engage in criminal activity. Perhaps the most important, if rarely expressly noted, is a gene called SRY. This gene (its full name is Sex Determining Region Y) is found on the Y chromosome and is necessary to turn a human embryo into a male. About 80 per cent of those convicted of serious crimes in the US are male, as are about 90 per cent of those convicted of serious violent crimes. Without inquiring into the combination of direct genetic versus cultural influences that make men more likely to be criminals, it is clear that possession of a functional SRY gene (being male) is strongly associated with criminal behaviour. But this association would fail a strong version of Koch's Postulates. Most men do not commit (or, at least, are not convicted for committing) serious crimes, and many serious crimes are committed by women. Like APOE 4 and Alzheimer disease, SRY is correlated with the likelihood of committing a crime, but it is neither sufficient nor necessary for criminal behaviour.

Another aspect of the MAO-A gene provides a second example of a genetic variation predisposing to crime. In humans the MAO-A gene has two common variations in its promoter region. One of the variants leads to the production of more MAO-A protein; the other leads to the production of less. In a widely discussed article, Avsalom Caspi and colleagues found an association between these two promoter variations and the criminal records of young men.[6] The research used a large, long-term study of about 1,000 people in Dunedin, New Zealand, who had been carefully followed from their births in the early 1970s.

---

[6] Avsalom Caspi et al., 'Role of Genotype in the Cycle of Violence in Maltreated Children' (2002) 297 *Science* 851.

Among the roughly 450 men in the study, about 35 per cent had the low activity variant. The researchers found a strong association between having the low activity variant and antisocial behaviour—but only in men who had been abused as children. The study has become famous as one of the most prominent examples of a gene/environment interaction, but the underlying finding is that a particular genetic variation, when combined with environmental conditions of moderate or severe childhood abuse, is associated with two to three times higher risk of subsequent conviction of violent crime.

We have seen one example of a strong (but vanishingly rare) genetic association with some criminal behaviours and two examples of weaker (but more common) associations. Are there any similar examples from non-genetic neuroscience?

I know of no examples of particular brain variations (in this context, abnormalities) powerfully associated with criminal behaviours. There are, of course, many behaviours or abilities strongly associated with particular kinds of brain damage. Vision, speech, movement, sensation, memory formation, and many other abilities can be disrupted or eliminated by damage to specific areas of the brain. Thus, the lack of a functional Broca's area (by stroke, tumour, or surgery) makes it impossible for adults to speak. (Children who suffer this damage early are sometimes able to adapt to the damage and to speak using different brain areas.) But there seems to be no equivalent of Brunner Syndrome, no particular brain region that, when damaged, leads inevitably to criminal behaviour. That does not mean that such a strong association may not be discovered; neuroscience is exploding with new findings as a result of new tools, both physical and statistical.

On the other hand, at least one form of brain damage does seem to be associated with some criminal (or antisocial) behaviour. Frontal lobe damage, whether by trauma or disease, often causes people to lose their inhibitions. Phineas Gage, the famous railroad worker who had a long metal bar fly through his skull, is one example; so are many people with frontotemporal dementia (FTD). As far as I know, there are no broad statistics on the percentage of people with FTD who commit criminal acts, but at least one small study found that fifteen out of thirty patients with FTD had committed misdemeanours.[7] Thirteen had stolen items (mainly by shoplifting) and ten had physically threatened other people. (This study is in German; the conclusions described above are in the English language abstract.) But this is not a powerful association. Many people with FTD lapse into lethargy rather than become disinhibited. Many disinhibited FTD patients do not seem to commit crimes. Many people with FTD 'can' help themselves, or, at least, do 'help themselves' by not committing crimes. And certainly the vast majority of people who commit criminal behaviour have no known frontal lobe damage.

So far we have looked for physical characteristics, in the genome or in the brain, that correlate strongly with criminal behaviour. We have found one genetic variation that is very strongly linked to criminal behaviour, but is very rare. And we have

---

[7] J. Diehl et al., 'Misdemeanor in Frontotemporal Dementia' (2006) 74 *Fortschr. Neurol. Psychiatr.* 203.

seen several genetic or brain-based characteristics that substantially increase the likelihood of criminal behaviour but by no means make it certain. Remember, though, that this situation is dynamic. As genetics and neuroscience, separately and, increasingly, together, work to understand human behaviour, new associations may be discovered or previously weak associations may become much stronger. It is also possible, though less remarked, that currently strong associations will seem weaker—the discovery of even one man with a non-functional MAO-A gene who did not have Brunner Syndrome would vastly weaken that association.

5.1.2 Behavioural evidence—of psychiatric syndromes

This paper was sparked by Tourette Syndrome as an example of 'can't help himself', but there is no good 'physical' evidence of Tourette Syndrome. It is thought to have strong genetic associations—about half of the children of Tourette parents have the syndrome—but no particular genetic variations have been confirmed as being associated with the condition. Mutations in one gene, called SLITRK 1, have been found in a small number of people with Tourette Syndrome, but the association is, thus far, uncertain. It is also thought to have environmental causes, though what those are is, so far, unclear.

Physicians diagnose Tourette Syndrome when a person has one or more physical tics and at least one vocal tic. In about 10 per cent of the cases, the vocal tics include coprolalia. Obviously, a great many people who do not have Tourette Syndrome use crude or offensive language and sometimes for no reason apparent to the observer. If, however, that kind of language is used regularly, without a reasonable motivation, by someone with physical tics characteristic of Tourette Syndrome, we may assume, as a first conclusion, that the person is using foul language because of Tourette Syndrome, that he 'can't help himself'.

Of course, this might not be true. The person might be faking the physical tics and intentionally using shocking language. He might have physical tics for some known reason that it is not associated with Tourette Syndrome and be swearing intentionally. Or he might have Tourette Syndrome, complete with physical and vocal tics, but not coprolalia, and may speaking that way for some purpose. But if those possibilities were explored, and dismissed, we could conclude that he really had no ability to control his speech, even though we could find no genetic, brain-based, biochemical, or other physical explanation for the disorder.

This psychiatric diagnosis is based on comparing the behaviour of the patient to generalizations drawn from observations of many people by psychiatrists and other physicians over many years. The most widely used collection of psychiatric diagnoses (at least in the US) is found in the Diagnostic and Statistical Manual of Mental Disorders, which currently available in a 2000 'text revision' of its fourth edition, published in 1994 (referred to as DSM-IV-TR). The next edition, DSM V, is in preparation and scheduled for publication in 2013. Outside the US, the mental health section of the International Statistical Classification of Diseases and Related Health Problems is used more often.

The DSM-IV-TR lists 297 disorders. Some of the diagnoses *require* criminal behaviour. Kleptomania, for example, has five diagnostic criteria:

1) recurrent failure to resist impulses to steal objects that are not needed for personal use or for their monetary value;
2) increasing sense of tension immediately before committing the theft;
3) pleasure, gratification, or relief at the time of committing the theft;
4) the stealing is not committed to express anger or vengeance and is not in response to a delusion or hallucination;
5) the stealing is not better accounted for by Conduct Disorder, a Manic Episode, or Antisocial Personality Disorder.[8]

It is not clear to me how many psychiatric diagnoses should be viewed as providing very strong evidence that the person 'can't help himself'. One problem is that the diagnoses are based mainly on behaviours. Behaviours can be mimicked or lied about much more easily than can genotypes or neuroimages of brain abnormalities.

Beyond that is the question of how often the criminal or antisocial impulse actually *is* resisted. The criteria for kleptomania include 'Recurrent failure to resist impulses to steal...'. The diagnostic criteria for paedophilia, to use another example, state, in part:

1) Over a period of at least six months, there are recurrent, intense sexually arousing fantasies, sexual urges, or behaviours involving sexual activity with a prepubescent child or children (generally age thirteen years or younger).
2) The person has acted on these urges, or the sexual urges or fantasies cause marked distress or interpersonal difficulty.[9]

Note that, to meet the definition, the patient may have acted on those urges, which would always seem to require criminal behaviour, or the urges may have caused the patient 'marked distress or interpersonal difficulty', which does not necessarily require criminal behaviour.

Without more, a listing in the DSM-IV-TR does not imply a conclusion that the behaviours *cannot* be resisted. One would like to know how many people with the diagnosis have, in fact, successfully resisted their criminal impulses, or, in even more (and probably unobtainable) detail, what percentage of criminal impulses each subject has resisted. Only if the syndrome came with evidence that everyone (or almost everyone) with the syndrome *could not* resist the impulse to criminal behaviour would it provide the kind of very strong evidence we are seeking, akin to Brunner Syndrome. In the absence of that evidence, these diagnoses would seem to be evidence of predisposing but not compelling causes.

---

[8] Amer. Psychiatric Ass'n, *Diagnostic and Statistical Manual of Mental Disorders DSM-IV-TR* (Washington, DC, Amer. Psychiatric Ass'n, 2000) Diagnosis 312.32.
[9] Ibid. at Diagnosis 302.2.

Note that Tourette Syndrome, which started my inquiry, does not necessarily meet that standard for strong evidence of 'can't help himself'. The DSM-IV-TR diagnostic criteria for this syndrome state:

1) Both multiple motor and one or more vocal tics have been present at some time during the illness, although not necessarily concurrently. (A *tic* is a sudden, rapid, recurrent, non-rhythmic, stereotyped motor movement or vocalization.)
2) The tics occur many times a day (usually in bouts) nearly every day or intermittently throughout a period of more than one year, and during this period there was never a tic-free period of more than three consecutive months.
3) The onset is before age eighteen years.
4) The disturbance is not due to the direct physiological effects of a substance (e.g. stimulants) or a general medical condition (e.g. Huntington's disease or postviral encephalitis).[10]

In Sapolsky's example of coprolalia with Tourette Syndrome, one assumes that the urge is irresistible, but something beyond meeting the DSM criteria (such strong evidence that anyone who meets the diagnostic criteria cannot control his impulses) would be needed to be confident that the person's coprolalia is truly uncontrollable. Clinical experience, either with a particular patient or with all patients with the diagnosis, may (or may not) provide that assurance.

5.2 Individualized Proof of 'Can't Help Himself'

Thus far, this paper has discussed situations where a broad set of evidence, from genetic, neuroimaging, or psychiatric studies, provides some general evidence, strong or weak, about whether a defendant 'can't help himself'. But more often a defendant will come into court (or before the prosecutors) without that kind of evidence. He may have some evidence of an abnormality in his genes or his brain, but will be able to produce little to no evidence about others with the same abnormality and their behaviour. Without that kind of evidence, he cannot make the strong showing that everyone with condition X performs behaviour Y, with the implication that the behaviour is, in those circumstances, unavoidable.

If we fully understood the determinants of behaviour, in the genome or in the brain, we might be able to say with confidence that a particular abnormality would, with a high degree of certainty, compel a particular behaviour even if that abnormality had never, or rarely, been seen. We are very far from having that degree of understanding of human behaviour. Yet defence counsel are bringing evidence of genetic and brain abnormalities into court, in efforts to win acquittals, or convictions of lesser offences, at trial, or perhaps to obtain milder sentences.

[10] Ibid. at Diagnosis 307.23.

Occasionally, those efforts are successful. A CT scan of John Hinckley's brain was introduced into evidence in his 1982 trial for the attempted assassination of President Ronald Reagan in order to support his claim that he was mentally ill. The jury ended up finding Hinckley not guilty by reason of insanity, although the role of the CT scan in the jury's verdict is unclear.[11] In a 1992 case in New York, Herbert Weinstein was charged with second degree murder in the killing of his wife.[12] The judge's decision to allow admission into evidence in a murder trial of a 'positron emission tomography' (PET) scan showing that Weinstein had an arachnoid cyst inside his skull led the prosecution to allow Weinstein to plead guilty to the lesser offence of manslaughter. And in 2004, sixty-five-year-old Peter Chiesa was tried on two counts of first degree murder for shooting two neighbours to death.[13] The trial court allowed the admission of a 'single proton emission computed tomography' (SPECT) scan in his case to show that he had diminished control. The jury convicted Chiesa on two counts of second degree murder, rejecting the first degree murder charges. Of course, in all three of these cases there was other evidence, sometimes, as with Hinckley, quite strong, that the defendant was not 'normal' or that his behaviour had deteriorated before the crime.

Although scholars are only now trying to collect these kinds of cases and quantify the outcomes, it is widely believed that these cases are exceptions. Usually the evidence is either not admitted or, if admitted, does not sway the jury or judge. Part of the reason is that no one can say 'every person with this genetic variation or with this kind of brain damage has committed this kind of crime'. In fact, in one case the prosecution pointed out that the defence expert had found that a neuroimaging scan of his own brain was 'disturbingly similar' to that of a serial killer he had studied. With our present knowledge, defence counsel might be able to make a case that, based on neuroimaging of a defendant's brain, the defendant is unusually likely to have some particular cognitive or emotional deficits, but seems unlikely to prove the strong point that those deficits would make it impossible for him to have avoided committing the criminal acts.

Defendants may be able to make useful arguments from the lesser evidence they can present, but are there any situations where they might be able to make the strong scientific or medical case, that this particular individual had an abnormality that made it impossible for *him* to resist the crime? There may be such a case, reported both in the scientific literature[14] and in the popular press.[15]

In 2000, a previously normal forty-year-old school teacher in Virginia, in a stable, two-year-old second marriage, began to collect an expanding library of pornography, including, for the first time, child pornography. He also solicited

---

[11] See Lincoln Caplan, *The Insanity Defense and the Trial of John W. Hinckley, Jr.* (Boston, Godine, 1984).
[12] *People v Weinstein* 156 Misc 2d 34, 591 NYS 2d 715 (Supreme Ct, NY Co 1993).
[13] M. Lasden, 'Mr. Chiesa's Brain' (2004) *Calif. Lawyer* (Nov.) 27.
[14] J. M. Burns and R. H. Swerdlow, 'Right Orbitofrontal Tumor with Pedophilia Symptom and Constructional Aprxia Sign' (2003) 60 *Archives of Neurology* 437.
[15] 'Doctors Say Pedophile Lost Urge After Tumor Removed' (2003) *USA Today*, 28 July, available at <http://www.usatoday.com/news/health/2003-07-28-pedophile-tumor_x.htm>.

prostitution, for the first time. Eventually he made advances to his twelve-year-old stepdaughter, who told her mother. The man was removed from the home, diagnosed with paedophilia, put on a 'chemical castration' regimen, and found guilty of child molestation. The judge ordered him either to complete a twelve-step residential programme for sexual addiction or to go to prison. He chose the programme, but was unable to complete it. He solicited sex from clients and staff of the programme and was expelled from it. He was therefore ordered back to court to be sentenced to prison.

The evening before his sentencing, he came to the emergency room at the University of Virginia hospital, complaining of a headache. He also talked about suicide, and about possibly raping his landlady. The next day he was given a neurological exam, during which he solicited female medical staff for sex. He also showed other mental and physical symptoms. The doctors ordered a magnetic resonance imaging (MRI) scan of his skull, which revealed a tumour the size of a chicken egg in his right orbitofrontal lobe.

In December 2000 the tumour, of a type called a hemangiopericytoma, was removed. His physical problems improved almost immediately and he passed the sexual addiction course. In July 2001, he was viewed as no longer a threat to his stepdaughter and was allowed to return home. By October, however, he was suffering from headaches again and began again to collect pornography. Another MRI was performed, showing that his tumour had returned. It was removed a second time in February 2002 and, again, his physical and sexual problems seemed to resolve. The published record ends there. Through a personal communication, I learned that, at least through 2006, the man had lived a legally uneventful life. He had tried, unsuccessfully, to resume his marriage after the second operation, but eventually separated and moved to another state, where he lived alone. His tumour did not grow back, though he did show some signs of subtle frontal lobe dysfunction. After 2006 my source lost contact with him.

The man always seemed to know that what he was doing was wrong, and, at least in the early stages, tried to conceal it, but he said 'the pleasure principle overrode' his efforts at restraint. One of his doctors said 'This guy, he knew what he was doing was wrong, but he thought there wasn't anything wrong with him, and he didn't stop.'[16]

It is well known that damage to the orbitofrontal lobes can trigger bad behaviour. There are some reports of patients with tumours committing crimes, apparently because of the tumours. But there are no reports, other than this one, of this kind of tumour leading irresistibly to this kind of behaviour. Should we believe it?

This may be the rare individual case report where we can believe it. The reason is that this subject was his own control, twice. Each time he had the tumour, he exhibited similar, inappropriate behaviour. Each time the tumour was removed, the behaviour disappeared. Now, it is possible that he could have controlled his behaviour when he had the tumour and only chose to control it after the operation,

---

[16] Ibid.

in an effort to create a false impression of lack of control. This seems highly unlikely, particularly as he presumably did not know after his conviction that he had a brain tumour that he could use as an excuse for his behaviour. His doctors noted his strong desire to avoid prison, but he still failed the sexual addiction programme—miserably. When he had condition X (the tumour), he had behaviour Y (lack of sexual control). When condition X was removed, behaviour Y disappeared. It is unclear how many, if any, similarly clear cases we can expect, but this shows that they are not impossible.

I know of no other defendant who would makes so strong a case. It is highly unlikely that there was evidence in the *Hinckley*, *Weinstein*, or *Chiesa* cases about how many people in society had the brain abnormalities identified in those defendants—undoubtedly, no one knew, or knows. But it seems certain that there was no evidence that every person with those brain abnormalities committed the criminal acts charged in those cases. Neither was there evidence in their cases that those defendants had committed similar cases at other times when they had the same brain abnormalities—that, for any of the defendants, his condition X always led to (criminal) behaviour Y. They could only, at best, say 'before we had condition X, we did not commit behaviour Y; after we had condition X, we did'.

The 'paedophile tumour' did not, of course, prove that condition X *always* led to behaviour Y in the defendant. Still, the ability to remove the tumour, see the behaviour disappear, and then observe the tumour grow back and the behaviour reappear certainly is much stronger proof than the other defendants could offer. It looks more like the fulfilment of a variant of one of Koch's postulates—causing the disease by introducing 'the same pathogen', though, in this case, in the same patient a second time rather than in a different individual.

5.3 What Does It All Mean?

To return to where this paper started, I do not think we should hold a Tourette Syndrome patient with coprolalia criminally liable for uttering obscene words as a result of his disease. I would not hold someone with Brunner Syndrome criminally liable for actions associated with that syndrome. And, at least in retrospect, I would not have held the Virginia school teacher criminally liable of child sexual abuse. In each case I would avoid criminal liability because the defendant could produce scientific or medical evidence showing with reasonable certainty (or some other standard involving very high likelihood) that he 'can't help himself'.

I am not a scholar of the criminal law and I am agnostic about the right doctrinal method for avoiding criminal liability in these cases (and therefore I have avoided using the term 'defence'). Issues of mental capability can enter into the criminal law through many paths. Some, like legal insanity, are affirmative defences, but others could be used to negate necessary elements of the crime. One might argue that a mental problem prevented a defendant from having the *mens rea* necessary to be found guilty, either generally or under a special mental requirement applicable to a particular crime, such as the need in some cases for premeditation as part of the highest murder charge. In other situations there may be question of the *actus reus*, whether the defendant 'acted' at all. Motions that are made during a seizure or during unconsciousness are generally not considered 'actions' and so proof of 'automatism' can prevent a defendant from being found guilty.

I suspect my 'can't help himself' doctrine would best be viewed as part of an insanity defence. It might, in the US, be a return to the irresistible impulse test or the Model Penal Code's formulation of the insanity defence, which absolves a defendant if 'as a result of mental disease or defect he lacks substantial capacity... to conform his conduct to the requirements of the law'.[17] This test, which had been widely adopted in the US between the Code's 1962 promulgation and the 1980s, was then widely abandoned in a reaction against the insanity defence after John Hinckley was found not guilty by reason of insanity. (In jurisdictions that retain the full Model Penal Code definition or the older 'irresistible impulse' test, a broader version of my proposed defence probably already exists.)

But, whether as part of an insanity defence or another existing criminal law doctrine, or as a new doctrine, the key aspect is the need for a high level of scientific or medical proof that the defendant could not have acted otherwise. In the best case that proof would be based on large-scale studies of physical attributes that correlated overwhelmingly with criminal behaviours. Evidence of psychiatric syndromes would be a slightly weaker form of admissible proof, more convincing the greater the statistical linkage among the various symptoms. And, in very rare cases, such as the Virginia school teacher, evidence derived solely from the defendant might be able to provide sufficient proof.

This high level of proof serves two functions. First, it diminishes our concern that a defendant is faking the symptoms of insanity, or whatever mental claim he is making. A savvy defendant might be able to fake the appropriate behaviour for an insanity defence; he cannot, without great difficulty, fake his genetic test or MRI results. The psychiatric syndromes are more troubling, but, depending on the number and nature of the symptoms used to diagnose these syndromes, they may also be harder to falsify convincingly.

Secondly, though, it avoids the fear that this kind of defence will make the criminal justice system disappear. That fear could be based on a scientifically attractive determinism—the idea that all of our behaviours are determined by the workings of our brains (mainly the firings of neurons), which in turn are determined by other physical phenomenon (largely the earlier firings of neurons). Determinists in this sense may or may not believe in free will, depending on whether they are, in the language of the free will philosophers, 'compatibilists' or 'incompatibilists'. But, whether or not they believe the criminal had something we should call 'free will', either kind of philosopher will believe that he 'can't help himself'.[18]

---

[17] American Law Institute, *Model Penal Code* 4.01 (Final Draft, Philadelphia, PA 1962).
[18] For a useful introduction to philosophical arguments about free will, see Thomas Pink, *Free Will: A Very Short Introduction* (Oxford, Oxford University Press, 2004). For more extensive argument, with

The test I propose, however, requires more than a conclusion that the defendant 'can't help himself'. It requires specific proof that ties some characteristics of that defendant, some Condition X (genetic, brain-based, or behavioural) that correlates extremely strongly with the criminal Behaviour Y, so strongly that we can say, not just that, in general, the defendant 'can't help himself', but that anyone with those characteristics, including the defendant, 'can't help himself'. Requiring such strong scientific or medical proof might condemn some defendants who really couldn't help themselves, but it will at least protect from a guilty verdict those with the strongest and clearest cases.

Such a doctrine is likely not to be very important, as a practical matter. For one thing, at least now and in the foreseeable future, it cannot be successfully proven very often. We know of very, very few associations between particular characteristics and criminal behaviour for which we have good proof that a defendant 'can't help himself'. Brunner Syndrome appears to be one. Some of the psychiatric syndromes may also qualify, although that may depend on just how good the proof is that everyone diagnosed with the syndrome is irresistibly compelled to the criminal behaviour. With coprolalia as part of Tourette Syndrome, the proof seems strong. With DMS IV-TR diagnoses like paedophilia or kleptomania, it is not clear how strong that evidence is.

And there is an interesting problem for even (or perhaps especially) the strongest cases of solely individual proof. Look more closely at the Virginia school teacher. It really was the second experience—the regrowth of the tumour at the same time as the resurgence of the drive to criminal behaviour—that makes this a convincing case. But, at that point, the defendant may have some obligation to take action or be held liable. Just as someone with a relevant medical problem might be held liable for driving when he knows the medical problem is resurgent, if the Virginia school teacher had committed another sexual offence during the second growth of the tumour, we might have held him liable for not taking appropriate action in light of his returning compulsion to criminal acts. It is possible that by the time the proof for a 'single case' defence of 'can't help himself' is strong enough to be sufficient, the defendant has started to incur liability for not taking what he should know is appropriate protective action.

Of course, as Nicole Vincent pointed out during discussion of this paper at the colloquium, the revolutions in the biological sciences, particularly neuroscience and human genetics, *might* vastly expand the number of cases where strong evidence can be submitted to prove that a particular defendant *had* to do the criminal act. Given the complexity of both of those sciences, I think this is unlikely for the foreseeable future and uncertain even in the distant future. To get close to perfect correlations between Conditions X and Behaviours Y may well never been broadly possible. But, even if it does become much more common, the practical consequences should be limited.

---

particular relevance to criminal, or moral, responsibility, see John M. Fischer, *My Way: Essays on Moral Responsibility* (New York, Oxford University Press, 2006).

This doctrine will not be very important in criminal prosecutions because those defendants who successfully invoke it will be free from criminal liability, but need not, and presumably will not, be free to reoffend. Just as defendants found not guilty by reason of insanity are usually forcibly confined in psychiatric institutions until the day (which often never comes) when they can be safely released, so the defendant who avoids criminal liability because of Brunner Syndrome is likely to be confined or otherwise restricted in ways that prevent him from being a danger to others.

This is an important way in which Sapolsky's example of Tourette Syndrome leads us astray. Coprolalia as part of Tourette Syndrome is not really dangerous. At worst, it inflicts annoyance on bystanders and might, possibly, prompt a retaliatory disturbance of the peace, but it does not directly maim or kill anyone. When that kind of defence is applied to truly dangerous conditions, a successful defence does not, and should not, free the defendant. In fact, depending on the condition and what, if anything, can be done about it, the constraints could be more intrusive and longer lasting than a criminal sentence.

So what is the point of my proposed doctrine? To avoid applying criminal guilt to a defendant who 'can't help himself'. I believe this is consistent with the general thrust of at least the Anglo-American criminal law. We do not (generally) find people guilty for the results of their unconscious or uncontrollable motions. We do not find them guilty when they could not understand or appreciate that what they did was wrong. We allow a defence of duress in some situations, where threats to the defendant overrode his will. It seems to me entirely consistently with our general approach to criminal liability, and the folk psychology we apply to it, to hold that someone who can offer specific and very convincing proof that he couldn't help himself should not be found guilty. That is not a very tangible benefit from a new doctrine that would likely be a controversial. On the other hand, if the proof requirement is kept strict, the costs are low.

5.4 Conclusion

I hope readers find this paper useful. For me, it has successfully scratched the itch caused by Sapolsky's discussion of Tourette Syndrome. The idea that a defendant with very strong scientific or medical proof that he 'can't help himself' should not be held criminally liable explains, for me, why the Tourette Syndrome patient with coprolalia should not be convicted of a crime and usefully extends that precedent to at least a few other cases, like Brunner Syndrome. And, having satisfied myself, I now intend to go back to ignoring largely legal and philosophical discussions of criminal liability, where I find myself truly 'a stranger in a strange land'.

I would point out, however, that, to the extent science produces convincing evidence strongly associating a Condition X with a (criminal) Behaviour Y, the important social consequences are likely *not* to revolve around guilt in criminal trials. Instead, criminal law, or society more generally, *will* use those associations to predict criminal behaviour and to prevent or 'treat' it, through criminal sentences or otherwise.[19] This is the most important likely effect of such research findings. We will need to be careful, both to examine the evidence for such associations searchingly and sceptically and, if they are reliable, to make sure we use those associations in ways consistent with the fundamental human rights.

---

[19] N. Rose, 'Screen and Intervene: Governing Risky Minds' *Hist. Hum. Sciences* (Feb. 2010). See also H. Greely, 'Neuroscience and Criminal Justice: Not Responsibility but Treatment' (2008) 56 *Kan. L. Rev.* 1103.

6

Madness, Badness, and Neuroimaging-Based Responsibility Assessments

*Nicole A. Vincent**

Suppose that neuroimaging scans could reveal people's mental capacities, and that a scan of Conrad's brain—a man with a history of violence who has been convicted of murder and diagnosed as a psychopath—revealed that he lacks the capacity to empathize. Precisely what, if anything, would such evidence tell us about Conrad's responsibility; would it *excuse* him (even if only partially), or would it *condemn* him?

In my view, such evidence would at least partially excuse Conrad since it would show that he lacks a mental capacity which is arguably required for moral agency—i.e. on my account such evidence would testify for the *defence*.[1] However, on Heidi Maibom's account, such evidence would condemn Conrad even further since she thinks that it would graphically show (via images of his brain) that he is a bad person all the way down to the core—i.e. on her account such evidence would testify for the *prosecution*.[2] So who is right, Maibom or me—would such evidence excuse or would it condemn? Marga Reimer thinks that neither of these two positions is exclusively right, claiming instead that such evidence *both* excuses *and* condemns. Reimer writes that 'both defence and prosecuting attorneys might make use of the relevant brain images. Indeed, they might appeal to *the very same* images, but characterize them in radically different ways—brains that are "damaged" vs. brains that are "evil".'[3]

If Reimer was right then such evidence would be a lot less useful to the defence than I have suggested. However, I will argue that once we clearly distinguish condemnation of people *for who they are* from condemnation of people *for what they do*, and once we realize that each of these two kinds of condemnation plays a role at a *different* stage in a criminal trial, we will see that at the guilt determination

---

* Postdoctoral researcher in the Philosophy Department at Delft University of Technology in the Netherlands.

[1] Nicole A. Vincent, 'Responsibility, Dysfunction and Capacity' (2008) 1 *Neuroethics* 199–204. See also Nicole A. Vincent, 'Neuroimaging and Responsibility Assessments' (2009) *Neuroethics* 10.1007/s12152-008-9030-8.

[2] Heidi L. Maibom, 'The Mad, the Bad, and the Psychopath' (2008) 1 *Neuroethics* 167–84.

[3] Marga Reimer, 'Psychopathy Without (the Language of) Disorder' (2008) 1 *Neuroethics* 185–98, p. 196, emphasis added.

stage such evidence clearly favours the defence. In what follows, after offering an account of how a future neuroscience might one day play a role in helping us to assess people's responsibility, I will then explain Reimer's and Maibom's concerns and relate them to this account, and I will end by rejecting Reimer's claim that such evidence *both* excuses and condemns and Maibom's claim that such evidence condemns *rather than* excuses.

6.1 Neuroscience and Responsibility Assessments

The point of this section is to explain how a future neuroscience might play a role in helping the criminal law to assess people's responsibility. I will begin by describing a 'murder mystery' example; I will then explain how an assessment of responsibility would normally be conducted in such a case; and finally, I will explain what neuroscience might one day add to the process of assessing responsibility in cases like this. Since these ideas are discussed at length elsewhere,[4] and the main point of the present paper is not to develop these ideas but to defend them from a novel objection, I will therefore keep this discussion brief, although references will be provided to sections of that other paper where greater detail and citations to relevant literature can be found.

6.1.1 Murder mystery

Imagine that you stumble upon Smith's dead body while strolling through a forest. It seems that Smith died of a gunshot wound to the head, and that his body was then hastily concealed beneath the bush where you found him. What processes might we engage in to determine who is responsible for Smith's death?

6.1.2 Assessing responsibility without the help of neuroscience[5]

Our first question will probably be 'Who dunnit?', and so in the beginning our inquiry will involve such things as: gathering up witnesses and suspects; perhaps conducting a line-up where eye-witnesses can identify prime suspects; the police might then interrogate various people; in court, both witnesses and suspects might eventually be questioned; and physical evidence such as fingerprints, spent bullet cartridges, DNA samples, and so on might also be collected, examined, and presented in court. Many epistemic barriers stand in the way of answering the *who dunnit* question, but once these pieces of the puzzle are put together we may discover that Jones is the one who shot Smith dead. Very roughly, in criminal trials the *actus reus* inquiry addresses the *who dunnit* question.

---

[4] Nicole A. Vincent, 'On the Relevance of Neuroscience to Criminal Responsibility' (2010) 4 *Criminal Law and Philosophy* 77–98.
[5] I discuss this in detail Ibid. 80–5.

But to establish that Jones is responsible for Smith's death, we need to show a lot more than just that she 'dunnit'. Rather, given causal indeterminacy—i.e. the fact that any outcome is a result of many causal contributions—we must also show that her causal contributions were of particular significance, and the way that the criminal law does this is by conducting a *mens rea* inquiry to determine whether the party who committed the prohibited act was at fault, guilty, or culpable. To see how this might work, consider a sample of defences that might be offered by Jones to relieve herself of responsibility for Smith's death despite her obviously pertinent causal contribution. For instance, Jones might point out that she shot Smith in *self-defence* when he ambushed her on her stroll through the forest, and as long as what she did is viewed as a reasonable response—e.g. not an unwarranted use of extreme force—then this should suffice to establish that Smith was at fault for his own death (because he should not have attacked her), and thus Jones's causal contribution would not translate into a finding that she was responsible for his death. Alternatively, suppose that Jones was out hunting in a well-known, sign-posted, and cordoned-off area of the woods that was only supposed to have other hunters in it, all of whom wore brightly coloured clothing and who knew each others' locations, and that there was simply no way of her knowing (nor any reason to suspect) that Smith—a thrill-seeking prankster who liked to frustrate hunters by hiding in bushes and scaring away their game—was hiding in those bushes. In such a case it is again likely that Smith's rather than Jones's actions will be viewed as the salient causes of his own demise, and hence that he would again be picked out as the person who is responsible for his own death, because Jones acted *reasonably* (her actions were not unduly risky) whereas Smith did not (his actions were too risky).[6] Furthermore, to the extent that *intentions* affect culpability, given that Jones did not even realize nor had any reason to suspect that Smith was in that bush, she certainly did not shoot him intentionally (and *ex hypothesi* she was not negligent), and so she cannot be culpable on account of having culpable intentions either. Finally, Jones's culpability may also be reduced by citing psychological considerations—for instance, that she is insane; that she was drugged, dazed and confused; or that she is a child. These and other *mens rea* defences offer reasons (excuses or justifications) to suppose that in the final analysis the party in question should not be blamed for how they acted, and hence that they are not responsible for the outcome despite their causal involvement.

However, what is specific to how the cited *psychological* defences work in the context of the *mens rea* inquiry,[7] is that culpability is reduced by showing that the

---

[6] My point is not that we would necessarily make this substantive evaluation about the risks involved, but rather that *if* we made that evaluation *then* Smith would be deemed responsible.

[7] Not all psychological defences address the *mens rea* inquiry. For instance, the 'automatism' defence—e.g. that the actions were performed while sleep-walking—defeats an accusation of responsibility, by appeal to the idea that true actions (unlike mere body movements) must be voluntary—i.e. the automatism defence addresses the *actus reus* inquiry. For discussion of violence related to sleep disorders, see Mark R. Pressman, 'Disorders of Arousal From Sleep and Violent Behavior: The Role of Physical Contact and Proximity' (2007) 30 *SLEEP* 1039–47.

accused person lacked a mental capacity that is necessary for acting in the prescribed way, and schematically the law's reasoning here looks something like this:

$$\text{mental capacity} \rightarrow \text{can \& can't} \rightarrow \text{duties} \rightarrow \text{culpability}$$

From right to left this schematic reads: one is culpable when one violates one's duties;[8] one's duties depend at least in part on what one can and cannot do;[9] what one can and cannot do depends at least in part on one's mental capacities;[10] and so the scope and degree of one's culpability varies among other things in proportion with one's mental capacities. Put a different way, one is culpable when one acts contrary to how one ought to act, and one important determinant of how we ought to act—or perhaps a consideration that *undermines* claims about how we ought to act[11]—is what we *can* and *cannot* do. But since some of the things that we can and cannot do relate to our mental capacities, a person who non-culpably[12] lacks a mental capacity that is needed to act in some prescribed manner may therefore cite this incapacity or reduced capacity as an excuse for failing to act as prescribed. The cited psychological defences therefore reveal the law's *capacitarianism*—i.e. the underlying idea that among other things *responsibility tracks capacity*.

In law, as in morality, responsibility is partly a matter of whether the party in question is culpable—i.e. whether they acted contrary to how they should have acted—and partly a matter of their causal involvement, and this is why we must conduct a *mens rea* inquiry as well as an *actus reus* inquiry.

---

[8] I take this to be an analytic truth, though I use the term 'duty' in a broad sense that is meant to capture the many different things all of which we ought to do.

[9] There are two ways of conceiving the relationship between capacity, duty and blame. On one account, capacity plays a *positive* role by *generating* duties: we ought to do what we have most reason to do, and what we can and cannot do (presumably along with a range of many other things) generates the reasons that we have to do various things. On this first account, if I cannot save a child from drowning—perhaps because I do not know that they are drowning, or because I cannot swim, or because I do not have a rope to throw to them—then in the final analysis it is simply not true that I ought to save them (unless I am responsible for the fact that I cannot do this); thus the reason why I would not be blameworthy for not saving them is because I did not, in the final analysis, actually have that saving duty. On another account though, capacity plays a *negative* role by *regulating* duties: regardless of the source of our duties, on this second view our *in*capacities can justify departures from those duties. On this latter account, the three cited considerations do not outweigh the saving duty, but rather they justify departing from it—i.e. the reason why I would not be blameworthy is because although I did have the saving duty, my incapacity provided an excuse for departing from it. Two advantages of the second view are that only it has an explicit place for excuses (which occupy an important place in legal discourse), and it arguably also more adequately captures the rich structure of practical reasoning in which some considerations *discount, undermine* and *invalidate* (rather than just *outweigh*) other considerations. Nevertheless, I suspect that both views of the relationship between capacity, duty and blame generate the same conclusions about when a person is to blame and when they are not to blame, and since I find it simpler to explain ideas within the conceptual framework of the first account, in what follows I will therefore reason within that framework.

[10] Though we arguably have greater control over our *internal* constraints than over *external* constraints, this does not negate the fact that internal constraints can prevent us from being able to do various things.

[11] See note 9, above, for an explanation of the significance of this qualification.

[12] The 'non-culpably' clause is needed because (as I hinted in note 9, above) the exculpatory value of incapacities for which we are responsible is usually discounted.

6.1.3 How a future neuroscience might fit in[13]

A future neuroscience—i.e. one that overcomes the limitations of current science and technology—might fit into the above picture in the following ways.

Firstly, neuroscience might help us to gather evidence for the *actus reus* inquiry. For instance, improved versions of *lie-detection* techniques—perhaps refined far descendants of the current No Lie MRI or Cephos Corp technologies—might help us to figure out whose testimony to trust and whose testimony to treat with caution. Alternatively, rather than asking people to tell us what happened and then using technology to sieve out the lies, perhaps witnesses might be *compelled* to tell the truth (or at least what they *take* to be the truth) through hypnosis, through the use of refined truth drugs, through the use of techniques like the one being developed by F. Andrew Kozel and Mark George that uses transcranial magnetic stimulation to temporarily disable those parts of the brain that are used in intentional deception, or maybe even mind-reading techniques might one day be developed that extract information directly from people's brains. *Brain fingerprinting*, like the MERMER-based technique (Memory and Encoding Related Multifaceted Electroencephalographic Response), might also help us to identify key witnesses who were present at the scene of the crime, and who might therefore need to be questioned more thoroughly. And finally, perhaps some techniques might also be developed to determine whether a particular person's behaviour qualifies as *genuine actions*—for instance, by uncovering that they cannot reliably form or act upon intentions.

However, neuroscience might also be useful in the *mens rea* inquiry. As I suggested above, among other things responsibility tracks capacity. But since some of our capacities are mental, to the extent that our mental capacities are implemented within our brains,[14] a brain scan that conclusively shows that I lack an important mental capacity—i.e. one that is presupposed by the claim about how I ought to have acted—could therefore at least partially reduce my culpability.[15] This account of how neuroscience might one day fit into the *mens rea* inquiry might be captured by adding just one element to the left-hand side of the previous schematic representation:

brain scan → mental capacity → can & can't → duties → culpability

From right to left this schematic reads: one is culpable when one violates one's duties; one's duties depend at least in part on what one can and cannot do; what one can and cannot do depends at least in part on one's mental capacities; and brain scans might one day reveal people's mental capacities. Thus, for instance, if a brain scan conclusively showed that I lack the capacity for rational deliberation, and this mental capacity was thought to be necessary for moral agency, then that might

---

[13] I discuss this in detail in note 4, above, 85–93.
[14] Proponents of the *extended mind hypothesis* and *substance dualists* might object to this.
[15] Note 1, above.

excuse me for acting culpably. Likewise, if a brain scan showed that I lack another important agency-relevant mental capacity—for instance, an affective rather than a cognitive one, perhaps like Conrad's inability to empathize—then that too might at least partially excuse me. Thus, if a neuroimaging scan of Conrad's brain conclusively showed that he lacks the mental capacity to empathize, and if that capacity was thought to be necessary (e.g.) for someone to truly understand that it is wrong to kill others, then Conrad's reduced capacity for empathy might at least partially excuse him.[16] After all, we would now surmise that the reason why Conrad killed is because he did not truly understand that killing is wrong, and people who do not truly understand the reasons that there are to not do certain things cannot be expected in the strongest sense of this term to refrain from doing those things, and so they cannot be fully blamed for doing those things.

6.1.4 Summary

Nothing that I said above should be taken as recommending that any of these (and perhaps other) techniques and technologies are ready for 'prime time' in courtrooms. In fact, as I have argued elsewhere,[17] there are many scientific, technological, procedural, and moral reasons to consider before introducing this or similar technology into courtrooms. However, these considerations do not affect my present rather minimal claim that there *are* ways in which a future neuroscience might help us to assess people's responsibility. Firstly, in much the same way as testimonial evidence and forensic techniques currently help us to answer the 'Who dunnit?' question, so too neuroscience may one day provide further pieces of the puzzle for the *actus reus* inquiry. Secondly, just as psychology today provides the law with information about people's mental capacities, so too one day neuroscience may also provide information about people's mental capacities which will help to inform legal determinations of culpability in the *mens rea* inquiry.

6.2 Two Concerns

In the previous section I explained how a future neuroscience might one day play a role in helping the criminal law to assess people's responsibility; on my account it might do this by providing pieces of the puzzle that inform the *actus reus* and *mens rea* inquiries. An important detail of my account of how neuroscience might fit into the *mens rea* inquiry was the capacitarian idea that responsibility tracks capacity—i.e. that if someone has significant deficits in certain mental capacities, then to the extent that possession of those mental capacities is presupposed by claims about what that person ought to do or not do, the degree of their culpability for what they actually do should be reduced to take account of those deficits.

---

[16] This would be an instance of extending the application of the *cognitive* requirement of *Daniel M'Naghten's Case* 8 ER 718, [1843] UKHL J16 to include *affective* incapacities.
[17] I discuss this in detail in note 4, above, 93–5.

6.2.1 Heidi Maibom

However, Heidi Maibom has argued for a picture that looks radically different to the capacitarian one that I endorsed above. On her account, people like Conrad suffer from:

> [a] *moral disorder* [in the sense that] in a world with no moral and no social values, [they] would present with relatively few problems. To put it colloquially, [a person like Conrad] is *more bad than mad* [and so] the question of whether [people like Conrad] should be regarded as legally responsible boils down to whether we should excuse those whose mental disorder primarily consists in them *being bad*. That, I hope, is counterintuitive.[18]

Without repeating the details of Maibom's argument, her account might be schematically represented like this:

$$\text{brain scan} \rightarrow \text{mental capacity} \rightarrow \text{mad or bad} \rightarrow \text{culpability}$$

From left to right this schematic reads: brain scans may one day reveal people's mental capacities; once we know their mental capacities we will be able to tell whether they are mad or bad; and once we know whether they are mad or bad we will then be able to tell whether to excuse or to condemn them (i.e. excuse the mad, but condemn the bad). Given this picture, on Maibom's account the scan of Conrad's brain would show that he has no room for feelings of empathy—that this is a person who is clearly bad all the way down to his core[19]—and if anything people like that should be scorned and condemned rather than excused! Put another way, if Conrad's history of callous and mean behaviour had not already convinced us that he is a bad man, then the scans of his brain that show that he lacks even the basic capacity to feel empathy for others should cement in our minds the fact that he is a bad man, and people like that should be condemned rather than excused.

Contrary to what I said above, Maibom believes that such evidence of reduced capacity condemns rather than excuses.

6.2.2 Marga Reimer

Maibom's comments may leave the reader perplexed, wondering which of the two interpretations is right—i.e. does such evidence condemn (i.e. what Maibom said) or does it excuse (i.e. what I said). Unfortunately, what Marga Reimer says on the topic offers little relief from this feeling of confusion, because what she thinks is that such evidence would do both of these things—i.e. that it would *both* condemn *and* excuse—and so she thinks that whether such evidence does one thing or the other will ultimately depend on the language that we use to characterize Conrad's mental economy (i.e. what capacities he has, to what degree he has them, and what capacities he lacks).

---

[18] Note 2, above.   [19] Note 2, above, 177.

On Reimer's account, if we characterize Conrad's situation as involving a *deficit*, disorder or illness of some sort—e.g. a *lack* or a *deficiency* in his affective capacities—then we will tend to view such evidence as a reason to excuse him.[20] After all, Conrad will be seen as a victim of a mental disease that expresses itself in his killing other people—'It's not him,... it's *his disease* that makes him do those terrible things' we might explain. But if we characterize Conrad's situation using different terminology—for instance, if we say that his emotional coldness is an evolutionary adaptation that allows a few non-cooperators like him to exist in a society populated otherwise by cooperators—then we may instead view such evidence as a reason to condemn him as a self-centred, callous, free-rider who only looks out for number one.[21] However, the really troubling thing about what Reimer says is that she does not seem to think that there is any objective fact about whether one characterization is better than the other, or at least this is how I interpret her suggestion that whether we see such evidence as grounds for excuse or as grounds for condemnation will hinge 'in effect, [on] a *gestalt shift*. The image formerly seen and characterized as a duck, will now be seen and characterized as a rabbit'.[22] This interpretation of her position is further reinforced by her claim that '*both* defence and prosecuting attorneys might make use of the relevant brain images. Indeed, they might appeal to *the very same images*, but *characterize* them in radically different ways—brains that are "damaged" vs brains that are "evil"'.[23]

Contrary to what I said above, and indeed contrary to what Maibom said too, Reimer thinks that this sort of evidence *both* excuses *and* condemns at the same time (though not necessarily to the same degree).

6.2.3 The significance of what Maibom and Reimer say

On the capacitarian account responsibility tracks capacity, and in 6.1 I argued that since neuroscience might one day help us to assess people's mental capacities, it might therefore help us to assess their responsibility. However, if what Reimer and Maibom say is right, then the degree of a person's responsibility will not necessarily track the degree to which they possesses certain mental capacities—on Maibom's account sometimes a reduction in capacity will result in an *increase rather than a decrease* in culpability; and on Reimer's account some reductions in capacity will *both reduce and increase* a person's culpability—and so measurements of mental capacities will not necessarily translate into the results that I suggested for their responsibility.

6.3 My Response

What Maibom and Reimer say seems to pose a challenge to capacitarianism, and this in turn poses a challenge to what I said earlier about how neuroscience might

---

[20] Note 3, above, 185.   [21] Ibid.
[22] Ibid., 190, emphasis added.
[23] Ibid., 196, emphasis added.

one day help the law to assess responsibility. However, this section will explain why what Maibom and Reimer say does not in fact challenge capacitarianism. My discussion will begin by drawing attention to two features of their analyses that I find unhelpful; I will then introduce two conceptual tools; and finally those tools will then be employed to defend my account of how neuroscience might one day help the law to assess people's responsibility, and to explain where Reimer and Maibom seem to have taken the wrong turn in their analyses.

6.3.1 Two unhelpful features of Reimer's and Maibom's analyses

A feature of Reimer's analysis that I find unhelpful is her tacit assumption that the things that she says pose a *particular* problem for a neuroscientific approach to responsibility assessment, because I believe that the sorts of worries that she raises are in fact neither new nor specific to the neuroscientific approach.

For instance, reflections about what might be said of someone like Susan Wolf's fictitious JoJo character, might also lead us to be torn about whether to call JoJo 'mad' or to call him 'bad'. To recap, 'JoJo is the favorite son of Jo the First, an evil and sadistic dictator [who] is given a special education and is allowed to accompany his father and observe his daily routine[, i]n light of [which] little JoJo takes his father as a role model and develops values very much like Dad's'[24]—i.e. JoJo becomes an evil and sadistic dictator just like his dad. In that paper, as well as in a later one,[25] Wolf notes that one can indeed be torn between two apparently competing assessments of JoJo—on the one hand one wants to condemn him, but on the other hand one cannot help but to at least partially excuse him on account of his terribly unfortunate upbringing (after all, few people could say with a clear conscience that if they had been brought up as JoJo had then they would not have turned out just like him). But the thing to notice here is that we do not need to be presented with any scientific or *neuro*scientific evidence of what JoJo's brain looks like to feel torn about whether to call him 'bad' and condemn him, or to call him 'mad' and to excuse him (even if only partially), because descriptions of his behaviour and stories about his upbringing will suffice perfectly well to leave us perplexed.

A similar dilemma is also encountered in Gary Watson's reflections about what we might say about someone like Robert Alton Harris, the man who was executed in San Quentin's gas chamber on 21 April 1992, after spending nearly fourteen years on death row for the cold-blooded murder of John Mayeski and Michael Baker on 5 July 1978. While the brutality and callousness of Harris's actions and his patent lack of any recognisably human feeling or remorse incline one to condemn the man, the tragic story of how a sensitive little Robert was moulded into the hard,

---

[24] Susan Wolf, 'Sanity and the Metaphysics of Responsibility' in Ferdinand Schoeman (ed.), *Responsibility, Character and the Emotions: New Essays in Moral Psychology* (Cambridge, Cambridge University Press, 1987) 46–62 (page cited 53).
[25] Susan Wolf, 'The Reason View' in Laura Waddell Ekstrom (ed.), *Agency and Responsibility: Essays on the Metaphysics of Freedom* (Boulder, Westview Press, 2001) 205–26.

mean and nasty Harris by denials of love and experiences of brutality even from his own mother, inclines one to at least partially excuse the man.[26]

My point is that in both cases—i.e. JoJo and Harris—good old-fashioned behavioural evidence and a story about their upbringing will quite effectively leave us feeling perplexed about whether to condemn or to excuse them, and the significance of this is that if there is indeed a problem of the sort that Reimer identifies, then this will not just be a problem for the high-tech neuroscientific approach to responsibility assessments that I endorsed above, but it will rather be a problem that affects the good-old-fashioned behavioural approach to responsibility assessment as well. Exactly the same sorts of dilemmas are encountered irrespective of whether we are presented with neuroscientific evidence or just behavioural and historic evidence, and so the fact that neuroimaging evidence might leave us troubled in this way is not a reason to suppose that the high-tech neuroscientific approach to responsibility assessment suffers from a unique flaw that does not afflict other approaches to assessing responsibility as well.

On the other hand, a feature of Maibom's analysis that I find unhelpful is her use of the *mad–bad* dichotomy in the context of discussing responsibility. Throughout her discussion, Maibom revisits the question of whether such neuroscientific evidence would establish that people like Conrad are mad and thus that their responsibility is diminished (i.e. reduced culpability), or whether it would establish that they are bad and thus that their responsibility is not diminished (i.e. maybe even greater culpability). Maibom of course thinks that such evidence would exclusively show that such people are bad, hence her insistence that such evidence would condemn. However, what is implicit in her use of this *mad–bad* dichotomy—and this is the feature to which I now draw attention—is her assumption that once we establish whether a person is mad or whether they are bad, this will then tell us something about that person's responsibility. Put another way, the operative assumption behind Maibom's use of the *mad–bad* dichotomy is that claims about whether someone is mad or bad provide premises in support of conclusions about their culpability. But what troubles me is that on the capacitarian account this operative assumption gets things completely the wrong way around, because what we should do *first* is to determine whether a particular incapacity affects someone's status as a moral agent, and only once we have already determined whether their status as moral agents is compromised might we *then* be in a position to call wrong-doers whose agency is compromised 'mad' and to call those whose agency is not compromised 'bad'.

There are two opposite ways of viewing the grounding relationship between claims about responsibility and claims about madness and badness: on one view claims about madness and badness *ground* claims about responsibility, and on the other view claims about madness and badness are *grounded by* claims about responsibility. Evidently, Maibom thinks that the former view of this grounding

---

[26] Gary Watson, 'Responsibility and the Limits of Evil: Variations on a Strawsonian Theme' in Ferdinand Schoeman (ed.), *Responsibility, Character, and the Emotions* (Cambridge, Cambridge University Press, 1987) 256–86 (pages cited 268–75).

relationship is correct. However, given that capacitarians would endorse the latter view, it is therefore highly inappropriate for Maibom at the outset to adopt a hostile analysis of the grounding relationship between claims about responsibility and claims about madness and badness, since by doing this her argument effectively begs the question against capacitarians. Put another way, my point is not that the view which Maibom adopts is wrong, or that the view which capacitarians adopt is right, but it is rather that if adopting either view would unfairly bias our analysis against either side then an altogether different way of discussing responsibility should be adopted.

This objection to Maibom's use of the *mad–bad* dichotomy in her analysis has a counterpart in the sorts of problems that are encountered when we think through similar issues by using the *disease–person* dichotomy. As Robert Sapolsky notes,

when we consider individuals who have committed grotesquely violent, sociopathic crimes, but who have no demonstrable [brain] damage..., it seems a fatuous tautology to say that there *must* be an organic abnormality in such cases—'it is *only* an organically abnormal brain that produces [such patently] abnormal behaviour'—and that we simply lack sufficiently sensitive techniques for demonstrating it.[27]

Sapolsky's comments suggest that when we are confronted with people like this, it is at least as easy to view their vile actions as symptoms of some underlying though presently undetectable organic disease—the thought here would be something like 'No *healthy* person could be like that, right?'—as it is to view those actions as expressions of their evil personality. However, as the recent debate in response to a paper by John Z. Sadler in the journal *Philosophy, Psychiatry & Psychology* highlights,[28] it is very difficult to classify people's behaviour in a value-neutral manner as either a symptom of disease/damage/disorder or as an expression of their vice, evil or perversity of character. As Sadler notes in a concluding collective response to the various authors who commented on his paper, '[e]very misbehavior has its psychology, and if we proclaim [it] abnormal or deviant, *then* we have a...dysfunction'[29]—i.e. it is only in virtue of the pre-judgement that the behaviour in question is abnormal or deviant that we can even arrive at the subsequent judgement that what we have in front of us is a dysfunction. Potter and Zachar put this point by saying that although psychologists and psychiatrists should indeed take a serious stab at trying to specify the 'underlying pathological processes' of disorders, they also recognize that the problem with attempting to do this is going to be that 'defining impairment in terms of underlying processes *requires having some conception* of what counts as normal'[30]—i.e. yet again we are urged that an evaluative

---

[27] Robert M. Sapolsky, 'The Frontal Cortex and the Criminal Justice System' (2004) 359 *Philosophical Transactions of the Royal Society of London* 1787–96 (page cited 1793, emphasis added).
[28] John Z. Sadler, 'Vice and the Diagnostic Classification of Mental Disorders: A Philosophical Case Conference' (2008) 15 *Philosophy, Psychiatry & Psychology* 1–17.
[29] John Z. Sadler, 'Cause, Fault, Norm' (2008) 15 *Philosophy, Psychiatry & Psychology* 51–5 (page cited 52, emphasis added).
[30] Nancy Nyquist Potter and Peter Zachar, 'Vice, Mental Disorder, and the Role of Underlying Pathological Processes' (2008) 15 *Philosophy, Psychiatry & Psychology* 27–9 (page cited 28, emphasis added).

exercise is the precursor to deciding whether something is a disease or a disorder. Finally, as Stephen Morse puts it, 'values always play a role in the definition of mental disorder'.[31] My point is that whether we classify a particular behaviour in one way (as a disorder, disease, damage, or illness) or in another (as an expression of that person's character) is not a straightforward matter—it is *not* something that might, for instance, be discovered under a microscope by a pathologist in a lab, or during a health check-up with your physician—because the physical facts underdetermine our characterization.

Admittedly, there are some cases in which the *disease* interpretation suggests itself more naturally—for instance, the cases of Phineas Gage and of Burns's and Swerdlow's[32] forty-year-old patient who suddenly developed paedophilic urges due to a brain tumour that pressed on the orbitofrontal region of his brain, immediately come to mind. Both of these examples involve trauma and sudden changes that we intuitively view as 'damage' and as 'disease', and so it seems more appropriate to attribute the observed changes in such people's behaviour to the *disease* rather than to these *people*. The people, we might suppose, lie buried somewhere deep beneath the diseases, heavy under their weight and unable to move—or maybe the real people no longer even exist because they were destroyed by the damage or by the disease—which is why it seems inappropriate to attribute such behaviour to those people. However, what should we say about cases in which *congenital* brain differences result in similar behaviour to that exhibited by Burns's and Swerdlow's patient? Should the behaviour of such 'natural-born paedophiles'—i.e. people who have *always* had brains that function in a similar way to Burns's and Swerdlow's patient's brain when it was compressed by the tumour—be attributed to a disease even though their brains may have always been like that, or should their behaviour in such cases be attributed to them? Alternatively, how about people who *develop* certain brain features *in early life* because of their terrible upbringing—perhaps people like Robert Alton Harris and JoJo—that predispose them later in life to committing atrocities; is their behaviour due to a disease of their mind, or is it an expression of their evil selves?

In these latter two cases we are in danger of falling prey to the *gestalt* switch that Reimer describes, one moment viewing these people's actions as expressions of their rotten character, and the next viewing them as expressions of a disease of the mind (albeit a disease of the mind which they have always had). Lawrence Tancredi's solution to this problem is to suggest that since the brains of people like Harris and JoJo may in the end be little different to the brains of people like Phineas Gage and of Burns's and Swerdlow's patient, if in the former cases we would attribute the behaviour to damage and disease, then for the sake of consistency we should do likewise in the latter cases too.[33] However, the arbitrariness of the way in which

---

[31] Stephen J. Morse, 'Vice, Disorder, Conduct, and Culpability' (2008) 15 *Philosophy, Psychiatry & Psychology* 47–9 (page cited 48).

[32] Jeffrey M. Burns and Russel H. Swerdlow, 'Right Orbitogrontal Tumor with Pedophilia Symptom and Constructional Apraxia Sign' (2003) 60 *Archives of Neurology* 437–40.

[33] Laurence R. Tancredi, 'The Bad and the Mad' in *Hardwired Behavior: What Neuroscience Reveals about Morality* (New York, Cambridge University Press, 2005) 143–61.

Tancredi chooses which side is the baseline worries me—perhaps instead of taking the disease interpretation as the baseline and then claiming that all bad people should be reclassified as mad, maybe instead we should take the person interpretation as the baseline and then claim that all mad people should be reclassified as bad. My aim is not to seriously suggest that we should indeed embrace the latter suggestion, but it is rather to point out that sheer *consistency* only requires us to jump one way, but it does not yet tell us which way to jump, and so for this reason I do not find appeal to the *disease–person* dichotomy any more helpful than the *mad–bad* dichotomy. Both dichotomies presuppose contestable substantive decisions about the accused person's status as a moral agent, and hence both of these dichotomies lack the sort of impartiality that is required to settle disputes about responsibility.

To sum up my disenchantments, the worries which Reimer raises are general worries about all responsibility assessments, and hence they do not pose any specific new challenge to the high-tech approach that I endorsed above. Secondly, my objection to Maibom is that the *mad–bad* and *disease–person* dichotomies presuppose certain judgements about people's status as moral agents, and hence such dichotomies should not be used as premises in support of conclusions about responsibility since doing so begs the question against capacitarians.

6.3.2 Some conceptual tools: two kinds of culpability, and two stages of a criminal trial

I now introduce the two conceptual tools that will be used to defend my account of how neuroscience might one day help the law to make responsibility assessments.

The first conceptual tool is a distinction between two responsibility concepts—what I call 'virtue responsibility' and 'capacity responsibility'—which relate to culpability for two different kinds of things. To see this distinction, consider the following example about my two fictitious children, Jane and John:

Jane is eight years old and very well behaved. She gets up in the morning all by herself, she washes and gets dressed and even makes her own lunch, she does not fight with other kids at school, after school she does her homework, she cleans up after herself, she helps me make dinner, and she even looks out for her older brother John. However, despite the fact that Jane is such a responsible little girl, I also know that she is ultimately only a little girl—she still lacks the mental capacities that one needs to have in order to be a fully responsible person—and so I do not really blame her when she sometimes fails to do these things. Jane is a responsible person, even though she is not yet a responsible person; or put in a less ambiguous way, Jane is a responsible little girl, even though she is not yet a fully responsible person.

Now consider John, Jane's older brother. John is seventeen years old and not at all like his much younger sister—some would call John an irresponsible young man. He wags school, he will not make his own lunch even though bought lunch is much more expensive, if and when he gets to school he always gets in some kind of trouble with the teachers, his bedroom is a pigsty and we are always cleaning up after him in the rest of the house; he cannot even be trusted to look after his little sister. Nevertheless, despite the fact that John is an irresponsible young man, we all know him well enough to know that he is actually a fully responsible person—we have seen what he is capable of doing when he puts his mind to things—and that is precisely why we are so very dark on him for his constant misbehaviour. John is not a responsible person, even though he is a responsible person; or put in a less ambiguous way, John is an irresponsible young man, even though he is in fact a fully responsible person.[34]

The point of this example is to demonstrate that we can make two similar-sounding but ultimately very different kinds of responsibility assessments. Sometimes, we make assessments about what a person is like—i.e. whether they are good or bad; culpable, admirable or just normal like most other people; or what I call their *virtue responsibility*—and it is in this virtue responsibility sense that Jane is said to be a responsible little girl while John is said to be an irresponsible young man. But at other times we make assessments about what I call *capacity responsibility*—these are assessments about whether the person in question has the mental capacities that it takes to even be a moral agent or a legitimate candidate for attributions of blame for the things that they do—and it is because of our different judgements about their capacity responsibility that Jane is said to not be fully responsible for what she does whereas John is said to be responsible for what he does. This distinction is not particularly novel. In his paper *Two Faces of Responsibility* Gary Watson also argues that we should distinguish these two kinds of responsibility assessments from one another, except that he gives them different names: what I call 'virtue responsibility' he calls the *aretaic* face of responsibility, and what I call 'capacity responsibility' he calls the *accountability* face of responsibility.[35] But putting aside terminological differences, my first point is simply that if we accept the legitimacy of this distinction, then we will notice that people can be culpable on account of two very different kinds of things. Firstly, people can be culpable on account of *what they did* but, secondly, they can also be culpable on account of *who they are*, and as the Jane and John example demonstrates, saying that someone is culpable in one sense does not yet commit us to saying anything about whether they are culpable in the other sense since these two judgements can come apart.[36]

The second conceptual tool that I now introduce is the observation that these different assessments of responsibility and culpability play a role at different stages in the criminal trial. At the *guilt determination* stage, where the focus is on determining whether the accused party is responsible for some outcome or state of affairs—i.e. where the focus is on assessing the sort of responsibility that was the topic of discussion in 6.1 above—what matters is the degree of the accused party's *capacity responsibility*. After all, the whole point of the insanity defence (and the other psychological defences that were cited in the main text of that section) is to establish that the party in question does not satisfy all of the criteria of a

---

[34] Nicole A. Vincent, 'Responsibility: Distinguishing Virtue from Capacity' (2009) 3 *Polish Journal of Philosophy* 111–26 (pages cited 115–6).
[35] Gary Watson, 'Two Faces of Responsibility' in *Agency and Answerability* (Oxford, Clarendon Press, 2004) 260–88.
[36] Note 34, above, 124–5.

fully fledged moral agent, in virtue of which their responsibility for the said outcome or state of affairs should be reduced. However, once we progress to the *sentencing* stage, as we ponder how that person should now be treated on account of what they have done and on account of the degree of their responsibility for it, we may take into account mitigating and aggravating factors, and one such factor is precisely whether that person is responsible or irresponsible in the *virtue responsibility* sense. After all, we may choose to be lenient on a person who has always been responsible in the virtue responsibility sense and for whom this was their first transgression, but a repeat offender who has a history of irresponsibility (yet again in the virtue responsibility sense of the term) may be given a harsher sentence. Thus, while assessments of people's *capacity responsibility* and the related culpability for *what they did* play a role during the *guilt determination* stage, at the *sentencing* stage those things are taken for granted and our focus shifts to assessments of people's *virtue responsibility* and to culpability for *who they are*.

With these two conceptual tools in place—i.e. the distinction between the two kinds of culpability assessments, and the different stages of the criminal trial at which each assessment plays a role—I will now defend my earlier claims about how a future neuroscience might help the law to assess responsibility, and I will also comment on where I think Reimer's and Maibom's analyses went wrong.

6.3.3 A defence of my position

I said earlier that Maibom's and Reimer's concerns seem to cast doubt on the capacitarian claim that responsibility tracks capacity, and that with the capacitarian idea shaken up, it no longer seems to make sense for me to suggest that responsibility might one day be assessed in part by using neuroscience to measure people's mental capacities. However, I believe that when the two points made immediately above are taken into account, we will see that Maibom's and Reimer's concerns do not in fact cast doubt on the capacitarian claim that responsibility tracks capacity.

Maibom's comments were meant to cast doubt on the capacitarian claim that responsibility tracks capacity by showing that sometimes (e.g. in Conrad's case) a reduction in capacity would result in an increase rather than a decrease in culpability. The idea here was meant to be that if sometimes a reduction in capacity will result in an increase rather than a decrease in culpability, then sometimes responsibility would *track* capacity but on other occasions responsibility would *inversely track* capacity. However, on closer inspection we should notice that on the *reverse tracking* occasions a different sense of responsibility is in play to the sense that is in play on the *normal tracking* occasions. Specifically, when Maibom insists that Conrad's reduced capacity to empathize would be a reason to condemn him, what she means is that Conrad is culpable on account of *who he is*—i.e. the *virtue responsibility* sense is in play here. However, in asking whether Conrad's reduced capacity to empathize should excuse him at the guilt determination stage, our interest is in his culpability for *what he does*—i.e. our interest is in his *capacity responsibility* not in his virtue responsibility—and so the fact that his culpability on account of *who he is* goes up rather than down is hardly a counter-example to the capacitarian claim that his responsibility for *what he did* would go down. Put another way, Maibom might indeed be right that on some occasions a reduction in capacity will result in an increase rather than a decrease in culpability; however the increase will be in culpability for *who that person is* (at the sentencing stage) and not for *what that person did* (at the guilt determination stage), and so this is not a counter-example to the capacitarian claim that responsibility *for what a person does* tracks capacity.

Turning now to Reimer, her comments were meant to cast doubt on the capacitarian claim that responsibility tracks capacity by showing that some reductions in capacity will both reduce and increase a person's culpability, rather than only reducing their culpability. But, using Conrad as the example, once we notice that we would excuse him *for what he did* whereas we would condemn him *for who he is*—as well as the fact that the excuse would occur at the guilt determination stage, whereas the condemnation would show up at the sentencing stage—there is again little reason to view Reimer's observation that *both* excuse *and* condemnation are on order as a counter-example to the capacitarian claim that responsibility tracks capacity, since the claim that responsibility tracks capacity only applies to responsibility for what a person does—i.e. it only applies to our assessments of responsibility at the guilt determination stage. Put a different way, the sense in which Conrad's responsibility would be increased (i.e. we would condemn him on account of *who he is*) is completely different to the sense in which his responsibility would be decreased (i.e. we would excuse him, even if only partially, for *what he did*). Furthermore, the increase would happen at a different time (i.e. during sentencing) than the decrease (i.e. during guilt determination), and so there is no need (and in fact no possibility even) to offset the excuse against the condemnation.

Thus, in addition to the problems that I discussed in 6.3.1 above, Reimer's and Maibom's analyses also seem to have gone wrong in two further ways. Firstly, they both failed to distinguish the two very different sorts of things on account of which someone might be judged culpable—i.e. that people can be culpable on account of *what they do* or on account of *who they are*; and, secondly, they both failed to give due recognition to the fact that while the excuse would play a role at the *guilt determination* stage, the condemnation would play a role at the *sentencing* stage of the criminal trial. Taken together, these two problems lead to the following three shortcomings: firstly, both of them failed to notice that there is *no genuine conflict* between the excuse and the condemnation interpretation of the neuroscientific evidence;[37] secondly, this oversight led Maibom to think that she must choose between *either* saying that psychopaths should be condemned *or* saying that they should be excused (and she chose the former); and thirdly, this oversight also led Reimer to think that the exculpatory effects of reduced capacities *would be reduced* by their condemnatory effects.

I now conclude that neither Maibom's nor Reimer's comments pose a problem for the capacitarian claim that responsibility tracks capacity, and I see this as a

---

[37] Putting the same point, though in the terminology of the *mad–bad* dichotomy, Morse also insists that '[a]ctions can be *just* mad, *just* bad, or mad *and* bad': note 31, above, 47, emphasis added.

defence of my earlier account of how neuroscience might one day play a role in helping the law to assess people's responsibility.

6.4 Conclusion

This paper has advanced two points. First, I argued that although Reimer is right in claiming that some evidence of reduced capacity *both* condemns and excuses, she is wrong in thinking that such evidence condemns people *for the same thing* as the thing for which it excuses them. On my account, what such evidence *condemns* people for is *for who they are*, whereas it *excuses* them *for what they do*. This observation is important because it helps us to notice that such evidence would support the *defence* at the *guilt determination* stage (i.e. when our interest is in assessing the degree of a person's responsibility for what they have done), whereas it would support the *prosecution* only at the *sentencing* stage (i.e. when we are interested in assessments of their character as either a mitigating or an aggravating factor).

Secondly, I also rejected Maibom's claim that some evidence of reduced capacity condemns *rather than* excuses. Like Reimer, Maibom too fails to notice that such evidence condemns people *for different things* from the things for which it excuses them, and that it does this *at different stages* of the criminal process. But unlike Reimer, who was driven by these oversights to conclude that both of these assessments must therefore be valid, Maibom instead *took sides* and insisted that only one of them is right and that the other is mistaken. However, I have argued that there is *no need to choose* between *either* condemning *or* excusing people with incapacities like Conrad's, because while the condemnation pertains to *who they are* and is relevant to the *sentencing* stage, the *exculpation* pertains to *what they did* and applies to the *guilt determination* stage of a criminal trial.

A positive consequence of these two largely negative arguments is that at the guilt determination stage a finding that conclusively shows that Conrad lacks such a capacity may indeed reduce the degree of his responsibility for what he did—i.e. such evidence would not testify for both the defence and for the prosecution at the guilt determination stage, but only for the defence—however admittedly at the sentencing stage such evidence may indeed condemn him on account of who he is.

7

Brain Images as Evidence in the Criminal Law

*Adina L. Roskies and Walter Sinnott-Armstrong**

Brain images are appearing more and more often in courtrooms. Introduction of evidence from neuroscience has become common in capital sentencing, and a number of prominent neuroscientists have served as expert witnesses in criminal trials involving fMRI evidence. In a recent case, a motion was filed to exclude the testimony of an expert witness because it was not accompanied by fMRI and EEG evidence.[1] Although the motion was denied, it illustrates the perceived importance of neuroimaging as a forensic technique. As neuroscience teaches us more about brain function the trend for introducing neuroimages in the courtroom is likely to increase, and so the decision about whether to admit neuroscience evidence is one judges are likely to face more and more often. Whether brain images should be admitted into trials depends on how probative they are for specific legal issues and whether they are likely to mislead fact-finders in trials.

7.1 Legal Standards for Introducing Evidence into Trials

To introduce a brain image as evidence in a legal federal trial, the defence needs to meet standards for evidence (such as exhibits) as well as standards for scientific expert testimony. The testimony is needed in order to interpret the images during the trial.

The rules governing expert witnesses derive from both court decisions and statutes. The classic standard, announced in *Frye v US*,[2] requires only that the expert witnesses testimony be based on methods that are generally accepted by a relevant scientific community. To apply this standard, judges need only a

---

* Adina L. Roskies, Associate Professor, Department of Philosophy, Dartmouth College; Walter Sinnott-Armstrong, Chauncey Stillman Professor in Practical Ethics in the Department of Philosophy and the Kenan Institute for Ethics at Duke University. We would like to thank Emily Murphy, Scott Grafton and Teneille Brown for comments on earlier drafts of this paper.

[1] *United States of America v Naee, J. Williams* 2009 U.S. Dist. LEXIS 13472; 78 Fed. R. Evid. Serv. (Callaghan) 1204 (United States District Court for the District of Hawaii 2009).

[2] *Frye v United States* 54 App. D.C. 46, 293 F. 1013 No 3968 (1923).

sociological assessment of the scientific community. In many (though not all) jurisdictions, however, the Frye rule was replaced by a new rule in *Daubert v Merrell Dow*:[3]

Faced with a proffer of expert scientific testimony... the trial judge... must make a preliminary assessment of whether the testimony's underlying reasoning or methodology is scientifically valid and properly can be applied to the facts at issue. Many considerations will bear on the inquiry, including whether the theory or technique in question can be (and has been) tested, whether it has been subjected to peer review and publication, its known or potential error rate, and the existence and maintenance of standards controlling its operation, and whether it has attracted widespread acceptance within a relevant scientific community.

This *Daubert* standard requires the judge to determine what is 'valid' and 'proper' (a normative scientific judgment, not just a sociological one). The *Daubert* decision influenced the Federal Rules of Evidence:

FRE 702: If scientific, technical, or other specialized knowledge will assist the trier of fact to understand the evidence or to determine a fact in issue, a witness qualified as an expert by knowledge, skill, experience, training, or education, may testify thereto in the form of an opinion or otherwise, if (1) the testimony is based upon sufficient facts or data, (2) the testimony is the product of reliable principles and methods, and (3) the witness has applied the principles and methods reliably to the facts of the case.

Most states follow the Federal Rules in formulating related standards for admitting expert witnesses.

Even if neuroscientists are admitted as expert witnesses, it is a separate question whether they should be allowed to use brain images as exhibits during their testimony. The following rules govern the introduction of brain images as evidence in the form of exhibits during trials:

FRE 401: 'relevant evidence' means evidence having any tendency to make the existence of any fact that is of consequence to the determination of the action more probable or less probable than it would be without the evidence.

FRE 403: Although relevant, evidence may be excluded if its probative value is substantially outweighed by the danger of unfair prejudice, confusion of the issues, or misleading the jury, or by considerations of undue delay, waste of time, or needless presentation of cumulative evidence.

Again, most states in the US follow something like these federal rules. Although the issue of whether neuroscientists would meet the *Frye*, *Daubert* and FRE 702 standards is both fascinating and important, we will focus for simplicity on whether brain images pass muster under FRE 401 and FRE 403.

To apply FRE 403 to brain images in criminal trials, courts must answer three central questions: 1) How probative for criminal responsibility is the brain image? 2) How dangerous (prejudicial, confusing, misleading, needless) is the brain image? 3) Does its danger substantially outweigh its probative value? Here we assume that the reader is familiar with the basics of fMRI.

---

[3] *Daubert v Merell Dow Pharmaceuticals* 509 U.S. 579 (1993).

Although brain images could be used for many different purposes within the legal system—such as for detection of lies, pain, or consciousness or for decisions about parole or involuntary commitment—here we will focus on the potential use of brain images to reduce responsibility of the defendant during a criminal trial, such as to deny an element of the crime during the guilt phase or to argue for an insanity defence. Many of the points we mention could also apply to uses of brain imaging in other legal contexts, but some considerations may differ. For example, the burden of proof differs in criminal and civil cases, the costs of errors can differ, and so on. In addition, the status of the background scientific knowledge varies. For these reasons, we reiterate that there is unlikely to be a one-size-fits-all analysis for the uses of neuroimaging in the courts.

7.2 Are Brain Scans Dangerous as Legal Evidence?

The balancing rule of FRE 403 allows evidence to be excluded if the danger of prejudice outweighs its probative value. Here we first review the research relevant to assessment of dangerousness.[4] After that, we will turn to the more difficult issue of probative value.

The potential dangers are listed in FRE 403: '[1] unfair *prejudice*, [2] *confusion* of the issues, or [3] *misleading* the jury, or [4] by considerations of undue delay, waste of time, or *needless* presentation of cumulative evidence.' (our emphasis) It is not completely clear what these italicized terms mean.

Allen et al.[5] explain them in this way: evidence is unfairly *prejudicial* to the extent that it leads the jury to decide a case on an improper basis (such as dislike or pity for the defendant). Evidence is *confusing* to the extent that it focuses the jury's attention on collateral or inessential issues (such as how the defendant's brain damage occurred). Evidence is *misleading* to the extent that it leads the jury to draw a mistaken inference (such as that the defendant felt an irresistible impulse). Evidence can cause undue delay, waste of time, or *needless* presentation of cumulative evidence if other evidence is easier to present and reliable enough.

Brain images might be needless if they do not add additional information that could not also be obtained by behavioural measures. To the extent that neurological information could add to the total evidential picture, brain images could still be needless if they do not add additional information that could be given just as well by tables or graphs. However, a brain image would not be needless if jurors might understand it better than a table or a graph.[6] It is also hard to see why brain images would be prejudicial to the defendant or to the prosecution any more than neuroscientific evidence presented in other forms. Hence, the most likely dangers are that brain images confuse juries (by making them focus on the spatial location

---

[4] See also Sinnott-Armstrong et al. (2008).
[5] Allen et al. (2006) 138–41.
[6] Cf. Feigenson (2006) 236, on substantive *versus* demonstrative evidence, where the latter 'is offered merely to illustrate testimony' so it needs to meet less strict standards.

rather than function or behaviour) and mislead juries (by making them infer that certain data and explanations are more revealing and relevant than they really are).

Do brain scans create these dangers? The answer is not clear at this time, since no studies directly assess these issues for juries. However, the few studies there are do suggest that lay people are apt to be prejudiced by brain images as evidence. Here we summarize the findings of a few studies. These are more thoroughly discussed elsewhere.[7]

7.2.1 The power of photographs

Brain images are not photographs of brain activity. However, it is possible that the lay person takes them to be very like photographs.[8] Both photographs and neuroimages are created mechanistically, are counterfactually dependent on the state of the objects being imaged, and visually display properties of what is being imaged. These and other qualities of photographs are responsible for the 'transparency' of photography[9] as well as its evidential value. What is less apparent to the lay person are the many ways in which the two imaging techniques differ, and the ways in which these differences affect the interpretability of the image. The old adage 'seeing is believing' seems to apply here, even if what one sees need not provide additional reason to believe. If people are subject to this confusion, their misapprehension of brain images as photographs could incline people to overestimate the reliability of inferences based on brain images, and attribute to them an unwarranted epistemic status as sources of evidence.[10]

The degree to which this is a worry in the legal context is not clear. There is some independent evidence that photographs do unduly influence juror's deliberation, so that the mere admission of photographic evidence, even when it does not substantially add to the case, affects jurors' decisions. Bright and Goodman-Delahunty[11] had subjects read and issue verdicts in fictional criminal cases. Some of their subjects saw no photographs, but other subjects saw either gruesome photographs or neutral photographs, such as photographs of scratches on the door that had been jimmied open. Bright & Goodman-Delahunty found that the conviction rate with neutral photographs (38%) was almost as high as with gruesome photographs (41%) and much higher than with no photographs (8.8%), even though the neutral photographs added no new information that would have justified conviction. This result suggests that photographs as a form of presentation might influence jurors more than the content of the photographs, such as whether the photographs were gruesome or neutral. These photographs in this study had nothing to do with the brain, but this study raises suspicions about whether brain images might have similar effects if they are confused with photographs.

---

[7] Sinnott-Armstrong et al. (2008).
[8] This analogy is with traditional film photography. With the rise of digital techniques photography no longer has quite the evidential import it used to enjoy.
[9] Walton (1984).   [10] Roskies (2007).   [11] Bright and Goodman-Delahunty (2006).

7.2.2 The potential biasing effects of adverting to brain data

Beyond their resemblance to photographs, brain images could mislead because of a bias for people to privilege information that they deem to be a product of hard sciences. Weisberg et al.[12] tested the effects of adverting to information about the brain without accompanying visual images. They had subjects assess the quality of explanations with and without the addition of nonsensical or irrelevant brain information. Subjects were given either good or bad psychological explanations of cognitive phenomena. Some explanations were purely psychological, while others included some mention of neuroscientific terms or data, even though the neuroscience did not add to the quality of the explanation. Weisberg et al. found that bad explanations were still rated more 'satisfying' by both novices and neuroscience college students when accompanied by irrelevant brain information (means: novices = 0.16 and neuroscience students = 0.2) than without (respectively –0.73 and –1.1). Reassuringly, this effect was not seen with experts. These findings suggest that irrelevant and nonsensical neuroscientific information may confuse and mislead unsophisticated consumers. The implication is that the mere mention of brain evidence or mechanism adds to people's confidence in an explanation, when of course it should not.

What is the source of this difference in the credence given to these explanations? It is possible that the mere addition of more information or the mere suggestion that a psychological phenomenon can be explained in reductionistic terms makes an explanation seem better. The addition of brain images as exhibits to accompany expert testimony could potentially bias jurors by making them more confident in the testimony than the testimony warrants.

Although this study raises concerns about the potential biasing effects of any neuroscientific information, including neuroimages, some caution should be exercised in assessing the implications of this study for the law. First, the scenarios in this study introduced many kinds of neuroscientific information, not only brain images. In addition, since these scenarios only mentioned but did not display brain images (or other images), these scenarios do not provide any evidence that images themselves are biasing or misleading. Whether the introduction of brain images would increase or lessen or have no impact on this effect remains to be seen. Second, although it is clear that novices and even those with some neuroscientific education are affected by the presence of irrelevant information from neuroscience, these results do not show that accurate and useful neuroscience confuses or misleads. At most, it suggests that the layperson, even if not neuroscientifically naïve, is prone to be irrationally swayed by neuroscience evidence, and is unable to discriminate relevant from irrelevant evidence. Third, as experts were unaffected by the irrelevant information (and in fact rated those explanations as worse), this bias should not affect expert testimony, even if it does affect jury understanding of that testimony. Finally, this study did not have control conditions using relevant brain

---

[12] Weisberg et al. (2008).

data to support good arguments. Hence, it cannot show that brain data is always confusing or misleading or that brain data never increases accuracy.

7.2.3 The potential biasing effects of brain images

Two studies to date have directly explored the effect of brain images on judgements about explanation. McCabe and Castel[13] presented subjects with articles containing poor arguments, such as 'Watching TV helps with math ability because both activate the temporal lobe'. These articles included either brain images, bar graphs, or neither. McCabe and Castel found that subjects rated the articles as making more sense when accompanied by brain images (2.9) than when accompanied by only a bar graph illustrating the same point (2.7) or by neither (2.7). This effect was not large, and it does not involve legal cases, but it suggests that the form in which data is presented can affect people's assessment of arguments: brain images seem slightly more compelling than brain data. Whether or not this points to a biasing effect is unclear, since images do contain more information than do bar graphs, which focus on a (presumably relevant) subset of the information contained in the image. This study was not placed in a legal context, so it is silent on the matter of whether the image vs graph would affect legal decisions.

A second study explored the effect of structural brain images on NGRI (not guilty by reason of insanity) verdicts. Gurley and Marcus[14] asked subjects to read about a violent crime and then decide whether the defendant should be found NGRI. Some subjects read about expert testimony that the defendant had a psychosis, whereas others were told of testimony that the defendant had psychopathy. Some subjects read expert testimony about traumatic brain injury, but others did not. Some subjects were shown brain images suggesting damage in the frontal lobes, whereas others were not shown any brain images. Gurley and Marcus found that the percentage of subjects who found the defendant NGRI after reading expert testimony on mental disorder (psychopathy/psychosis) was higher when accompanied by a neural image (19% psychopathy / 37% psychosis), by testimony about traumatic brain injury (27% / 43% respectively), or by both (44% / 50% respectively) than when subjects received neither (11% / 22% respectively). Thus, the introduction of testimony about traumatic brain injury as well as images of brain damage increased the NGRI rate from 11% to 44% in the case of psychopathy. That is a big effect, so neural images and neuroscience do seem to affect legal decisions. Nonetheless, the fact that the images affected the judgements are not an indication that they were misleading. Structural brain scans can provide relevant evidence of malfunction. In addition, nothing in the Gurley and Marcus study shows which rate of NGRI is correct. Although brain images and testimony increased the rate of NGRI verdicts, that would be a *good* result if the defendant really deserved an NGRI verdict. Thus, the fact that they

---

[13] McCake and Castel (2007).   [14] Gurley and Marcus (2008).

were shown to have an effect might be reassuring: brain scans might inform and increase accuracy instead of being misleading. Moreover, because these images were structural images that showed dramatic holes in the brain, rather than functional images with altered patterns of functioning, it is difficult to extrapolate from these findings to claims about the potential misleadingness of functional brain images.

7.2.4 Summary of the potential prejudicial impact of brain images

The above studies may seem to indicate that brain data, and in particular, brain images, can and do affect people's assessment of arguments and even deliberation in legal cases. But although these studies are suggestive, none conclusively shows that brain images are misleading.

A second problem with concluding that brain scans are dangerous is that the above studies test only certain kinds of people in certain kinds of circumstances. Lab subjects differ from real jurors in important ways: real jurors are not all college students.[15] Real jurors hear more details of each case. Real jurors know that real lives are affected. Real jurors hear both sides of argument as well as cross-examination. Real jurors deliberate and know that they are accountable to other jurors in so far as they will probably be asked to give reasons for their beliefs and decisions. Because of such differences, we cannot quickly draw conclusions about jurors from studies done with lab subjects. These problems might be overcome with better studies.

In addition, the studies discussed above point to a potential bias that is relevant to the law, but they do not go very far in addressing the causes of that bias, or possible ways of mitigating it. Why do brain data and in particular brain images have these effects? To what extent are they dependent upon the format used to display the data, or on the implied reductionism? How can bias be minimized without ignoring the actual evidential value some data from brain scans may have? Roskies,[16] for instance, has postulated that photographs have a number of features that render them 'inferentially close' to that which they are about, so that they are apt to be interpreted correctly. In contrast, brain images may seem inferentially close to brain activity, but in fact they are not. This gap between the apparent and actual inferential distance could cause people to place more stock in the evidential value of brain images than is warranted. If this is so, are there ways to instruct or educate jurors that will eliminate this effect? Studies are underway that more directly explore the effect of brain images on judgements made in more realistic legal contexts. They may provide more insight into some of these questions. Results from these studies may also suggest ways in which potentially misleading effects of brain images might be counteracted.

---

[15] Some studies (Elliott and Robinson (1991), discussed in Bornstein (1999)) have found that decisions by student and non-student mock jurors did not differ significantly.
[16] Roskies (2008).

7.3 Are Brain Scans *Probative* of Criminal Responsibility?

The previous section discussed the possibility that brain images might mislead jurors, and concluded that whether and in what ways they do requires further study. However, the possible prejudicial effects of imaging evidence are moot if images as evidence do not have sufficient probative value. What is probative value? Although the Federal Rules of Evidence do not explicitly define probative value, a standard textbook identifies probative value as degree of relevance:

> Remember that evidence is 'relevant' if it has 'any' tendency to make the fact of consequence more or less probable; probative value measures the strength of the effect on the probabilities, even if only in general terms like 'highly,' 'somewhat,' or 'minimally' probative.[17]

This account makes probative value equivalent to a relative conditional probability. We doubt that the issue is this simple, because values enter into the equation in ways that we will see. Still, a good starting point for assessing the probative value of any evidence is to ask how much the evidence increases the probability of some fact that matters.

To apply this standard to brain images, we need to consider the precise nature of the information that is presented in the image and also which fact the image is supposed to be evidence for. This topic is immensely complex; here we can only run through five main obstacles that might arise when trying to use brain images as evidence of facts that are relevant to criminal responsibility.

7.3.1 What typical brain imaging studies show

The aims of experimental research differ from the aims of the law. While the law is usually concerned with the behaviour of a particular individual and the causes underlying that behaviour, science is concerned with the causes of behaviour in general. Both the research aims of scientists and scientific constraints to which they are subject influence the potential for neuroscientific data to be useful in the courtroom.

The vast majority of knowledge we have about patterns of brain functioning in normal humans derives from experiments conducted by scientists aimed at determining general principles underlying normal cognition. Thus, in most cases neuroscientists are more concerned with describing invariants across the population, rather than the differences within populations.

In addition, the BOLD signal recorded by fMRI is a noisy signal, with the changes of interest corresponding to cognitive activity accounting for only a few per cent of the baseline signal, and often far less. Because of the low signal-to-noise of the technique, data must be averaged over many trials and usually many different subjects in order for statistically significant results to be obtained. As a consequence

---

[17] Allen et al. (2006) 135.

of these factors, most scientific studies are on groups of people recruited to subject pools by university labs, and data is averaged across this population to draw conclusions about patterns of brain activity that correlate with task performance. The fact that most knowledge we have about brain activity patterns derives from these group studies has a number of consequences for what we can infer from them.

First, we know from anatomical studies that there is significant structural variability among individual brains. In group studies anatomical variability is dealt with by warping algorithms that enable different brains to be warped into a common stereotactic space so that data from different subjects can be aligned. Individual structural scans are usually aligned to a common atlas, such as the Talairach Atlas.[18] It bears mention that even this atlas is far from representative, being derived from just three brains. It thus is not representative of a normal population, but serves mainly as a common anatomical standard. Although the warping algorithms used are generally thought to be fairly good for registering the large-scale anatomy, there are theoretical difficulties with determining error rates for such algorithms, since there is no functionally significant standard for comparison.[19] In addition, problems arise in registering grossly structurally abnormal brains, such as brains with large lesions.[20] Since it is reported that a large percentage of people in prison have weathered some brain insult or injury, this could be a significant limitation in interpreting results of functional scans of the criminal population. The extent to which the criminal population is structurally grossly abnormal remains to be determined, but experimental research is underway.

In contrast to earlier techniques for functional neuroimaging, fMRI is entirely noninvasive and does not employ radioactive isotopes. Because of this a single subject can be scanned for long periods of time, or even in multiple sessions, making it possible to acquire sufficient data from a single subject to draw conclusions about individual brain activity patterns in some experimental paradigms. Comparison of single-subject studies in the same task shows distinct similarities in patterns of brain activity that corresponds well to earlier results obtained with PET data. However, these studies also underline the variability in functional profiles across subjects. The variability is present even when anatomical differences are minimized with warping algorithms. As will be discussed in more detail below, the residual functional-anatomical variability poses significant problems for the use of functional brain data in legal contexts. One obvious difficulty lies in assessing the relative degree to which observed deviations are due to functional as opposed to anatomical differences.

Because of these differences and the practical and theoretical constraints and limitations of experimental science, the conclusions reached by scientific studies are often not focused on the issues nor reported in the manner that would be of most use to the legal profession. These differences in scope and focus put limits on the probativeness of brain images in many legal contexts.

---

[18] Talairach and Tournoux (1988). [19] Brett et al. (2002); Gholipour et al. (2007).
[20] Crinion et al. (2007); Kim et al. (2008).

7.3.2 Assessing abnormality

Perhaps the most frequent use of brain images is to convince the jury that a defendant is abnormal in some way that removes or reduces criminal responsibility. Brain images, whether structural or functional, may be introduced to show abnormality. However, the notion of normality is difficult to assess with brain images. Most neuroscientific studies using fMRI report group averages, but individual functional profiles can vary so much that it is not unusual for most individuals to differ from the group average. This point is made graphically in this diagram from Miller et al.,[21] which shows activation patterns for individuals during an episodic retrieval task:

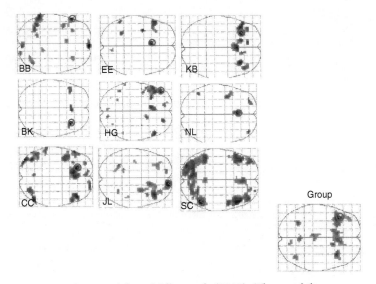

Figure 7.1 Reprinted from Miller et al. (2002). The variability among individuals and group data for an fMRI study on episodic retrieval.

In this study, subjects SC, JL, and BK look very different from the group average, and subjects EE and NL seem fairly different as well. Thus, five of nine subjects seem 'abnormal,' in that they vary significantly from the mean. Nonetheless, all the subjects recruited for this study are 'normal' adults who chose to participate in a scientific study. None exhibit functional deficits in the task being investigated. It is likely that this group of subjects is representative of the population. Thus, mere deviation from the mean has no bearing on assessments of abnormality in any sense that would be relevant to aims of a criminal defence. In cases like these, significant deviation from the mean does not show much at all with respect to the issue of criminal responsibility.

---

[21] Miller et al. (2002).

Even assuming structural differences can be negated, determining the source of individual differences in functional anatomy is still problematic. There is evidence that the different patterns of activation seen in normal subjects, such as in the Miller et al.[22] study are not a statistical glitch: these patterns of activation are fairly stable over time for a given individual, even when that individual varies far from the group average.[23] Miller et al. hypothesize (but do not show) that the different patterns found in their study reflect differences in cognitive strategy, which would be consistent with far less individual variability in basic functional anatomy than a first glance at the data might suggest: had subjects used the same cognitive strategy for the retrieval task, patterns of activation might have been more consistent. The data are also consistent with considerable variability in functional architecture, even when people use the same strategy. This question will not be resolved until better behavioural methods for assessing cognitive strategies are developed and brought to bear upon this sort of paradigm.

It is worth noting that the degree of individual functional variability is likely to depend heavily on the nature of the functional task being investigated. More basic processes are likely to be implemented in more canalized ways; more complex and higher-level tasks that depend more on learning and cognitive control may be more variable in the population, because of brain plasticity and because performance of that task may lend itself to multiple strategies for execution, such as is likely with episodic retrieval. To some extent, different people may simply process the same information differently, even when none is 'abnormal' in any way that would be relevant to criminal responsibility. A lower level of activity in a certain area might indicate in some people that the task was easier for them, so they did it more quickly; whereas in others reduced activity might show that they did not consider some of the factors relevant to the task. The range of individual variability in functional architecture poses a real problem for using functional scans to determine abnormalities in individuals that would be relevant to legal issues.

The individual variability in functional scans depends both upon individual variability in underlying brain structure, on variability in the functional architecture, and on variability in strategy and performance. Determining how much of the variability to attribute to each of these aspects is a difficult but scientifically tractable question that researchers are currently addressing. Moreover, the amount of variability in the 'normal' population is likely to differ for different brain regions and for different kinds of tasks. Assessment of the degree and distribution of structural and of functional variability for a variety of tasks will be necessary in order to have a relevant baseline against which conclusions relevant to criminal responsibility can be drawn. The data will have to show a statistically significant difference between populations, one that outstrips differences that can be attributed to individual differences in the normal population.

---

[22] Miller et al. (2002). [23] Ibid.

7.3.3 Inferences from group to individual

The difficulties in assessing abnormality can be extended to any attempt to draw conclusions about the status of an individual relative to the general population with respect to the neural basis of some legally relevant aspect of behaviour. In general, the aim of introducing an individual's data is to make an argument that he or she does or does not fall into some group. Ideally, to do that, one would have data on both members and non-members of the group in question. With this kind of information, one can say with some confidence whether an individual is more likely to belong to one population rather than another.

This raises a question: what are the appropriate comparison classes to use in this assessment? The answer is not self-evident. Is the appropriate comparison matched for age, IQ, gender, race, or socioeconomic status? Does one compare criminals and non-criminals? Violent offenders and non-violent subjects? Typical scientific experiments are not designed with these factors in mind. The populations studied in the lab tend not to be as representative of the general population or populations of interest as would be desired to reach legal conclusions.

Assuming one could identify appropriate comparison classes, how probative a scan from an individual would be would depend first upon whether appropriate comparison classes existed for the populations at issue. Second, without good evidence that a specific type of damage had effects that warranted mitigation of responsibility, merely showing an abnormality (i.e. deviation from the norm) has no immediate relevance for judgements of culpability. For example, we would want to know whether the abnormal activity was a cause of culpable behaviour, whether the behaviour was unavoidable, etc. What percentage of the population that did not engage in illegal behaviours shared that pattern of activation?

The foregoing does not mean that abnormality can never be determined using imaging, or that that abnormality cannot have mitigating value. Indeed, some structural damage is so extensive that when coupled with behavioural evidence and/or personal history, the data pointing to mitigation seems overwhelming. In these cases, however, we already understand that extensive damage to or death of brain tissue is a cause of severe cognitive deficits. Understanding how more subtle differences in brain activity affect cognitive and behavioural capacities is currently in its infancy. For brain images effectively to address the issue of abnormality in the context of the legal system, it will be necessary to develop a better understanding of the functional anatomy governing complex behaviour and a much more extensive database of brain data than we currently possess, both of the normal and the criminal populations.

7.3.4 Other problems attending the use of functional scans for legal culpability

7.3.4.1 Temporal and situational factors

Determining abnormal functional patterns may be a first step to identifying a criminal mind, but there are other problems with drawing inferences between patterns seen in a functional scan and criminal behaviour. First, scans would typically be carried out well after the criminal act, and in an experimental context that differs radically from the one in which criminal behaviour takes place. Even if there are functional abnormalities at work in producing criminal behaviour, it is not clear whether they would be seen in situations temporally and contextually far removed from the one in which behaviour actually occurred (although this temporal gap might be less of a problem when using structural as opposed to functional scans). Limitations due to these factors reduce the likelihood of finding any information relevant to legal purposes, both in terms of establishing baselines for criminal and normal populations, and in terms of gathering relevant evidence from individuals charged with crimes. Crucial questions thus involve whether there are standing differences between criminal and non-criminal populations, and whether particular tasks can be found that can be exploited in the scanner to probe for relevant differences.

7.3.4.2 Experimental design

In many cases, determining relevant baselines for functional scans will require performing functional scans during some kind of relevant behavioural task. It may be very difficult or even impossible to design an experimental task that accurately produces the condition of interest in a criminal case in order to establish relevant baselines. For example, arguments that certain emotion-eliciting stimuli were the cause of a crime of passion would be very difficult to test, for it is difficult to see how to elicit emotions of the necessary intensity in a laboratory setting or to test whether the right kind and strength of emotion has successfully been elicited. Similar caveats apply to the difficulty of studying lie detection experimentally.

7.3.4.3 Base rates and false alarms

Even if we can identify certain functional patterns as abnormal, we still need to determine which individual defendants display that abnormal pattern. The problem here is that imaging, like most means of assessment, is not perfect, and functional abnormalities that remove criminal responsibility are likely to be rare. Most people by far are responsible for what they do. When the base rate is low in this way, even a fairly high specificity of a test (that is, a low rate of false alarms) will yield a high number of false alarms. To illustrate this, consider a population of 10,000 with a 1 per cent base rate of a functional abnormality that causes deadly behaviour. (Luckily, this is an overestimate.) That means that 100 people in the population have that functional abnormality and 9,900 do not. If an fMRI test for this functional abnormality has 95 per cent specificity, then it will still test positive in 5 per cent of the 9,900, which is 495 false positives. Thus, even if the fMRI test is 100 per cent sensitive, a positive test still has a predictive value of only 100/595, which is less than 17 per cent.

This low predictive value of a single positive test for a functional abnormality can be improved by additional tests, but only if the defence agrees to them. If the defence gets a positive result that they think supports their claim that the defendant is not fully responsible, then they might be unwilling to subject the defendant to further testing. That will make it difficult in practice to improve upon the low positive predictive value unless courts require multiple tests, which would raise constitutional issues of privacy, search and seizure, and double jeopardy.

Of course, the positive predictive value will be greater if the base rate is greater, so defenders of fMRI evidence will point out that the base rate for criminal defendants could be much higher than the base rate in the population as a whole. Bayesian calculations will then yield a higher predictive value for the higher base rate. However, since we do not know what the base rate is for criminal defendants with functional abnormalities of this sort, we cannot say whether that is true. It is hard to see how we could guess the base rate, especially since we cannot assume the guilt of defendants in criminal trials. That makes it hard to overcome this problem of false alarms given low base rates.

7.3.4.4 Causation

Intuitively, criminal responsibility may be mitigated by a brain dysfunction that causes a person to behave criminally—that is, that the person committed the crime because of a 'bad brain'. However, functional brain scans reveal only correlations between patterns of brain activity, so we need additional evidence before we can conclude that any functional abnormality played a causal role in the production of criminal behaviour. Even if the defendant has a functional abnormality that is correlated with violent crime, correlation does not prove causation.

Causation can be inferred in rare cases of brain abnormality. Burns and Swerdlow[24] describe a forty-year-old person with little previous use of pornography and no sexual deviance, who started to use child pornography and then molested his stepdaughter. He was required to enter a twelve-step in-patient programme where he propositioned the staff, so he had to be removed. While awaiting sentencing to prison, he experienced headaches, dysgraphia, and loss of coordination, so his brain was scanned, and an egg-sized tumour was found. The tumour was removed, at which point he lost his symptoms, including his deviant sexual desires. Ten months later, however, he started to collect pornography again, and it was discovered that the tumour had grown back. When a behaviour comes and goes with the presence of a tumour in this way, then it is reasonable to infer causation.

What is striking about this case is how unusual it is. Almost no other cases relate sexual deviance to frontal-lobe damage. Moreover, we rarely get to observe the behaviour come and go with the tumour. The sudden onset of deviant behaviour with traumatic brain injury makes it a little easier to make the appropriate connection. Of crucial importance to our interpretation is the correlation with a behavioural change. When all the evidence we have is functional brain scans, in the absence of a clear-cut history or some method of intervention, then we do not have enough evidence to infer causation.

---

[24] Burns and Swerdlow (2003).

7.3.4.5 Capacity for control

More philosophical issues arise when we realize that our brains cause all our behaviours, yet we nonetheless hold each other responsible for the vast majority of them. Thus, it is not only causation that matters, but whether the defendant had the capacity to control his behaviour. Elucidating the notion of capacity is philosophically fraught, but we will simplify here by focusing on control. In the previous example of the man who developed paedophilic tendencies, the case report suggests that the person was unable to control the behaviour caused by his tumour. However, in many cases, even if the defendant has a functional abnormality that causes violent crime, causation need not prove lack of control. The brain abnormality might create a desire that causes the defendant to act but which the defendant could stop himself from acting on.

To see the point, compare thrill-seekers who jump out of aeroplanes (with parachutes), jump off bridges (with bungee cords), ski down triple black diamond slopes or drive race cars. There is some evidence that thrill seekers have a common neural abnormality (one of us, the thrill-seeker, prefers the term 'variant' to 'abnormality' here). Suppose that is true. They still have the capacity to control their desires. They can prevent themselves from acting on their impulses, and can choose when to act upon them.

Analogously, when neural abnormalities create desires to do illegal acts, people with those abnormalities still might be able to stop themselves. All of us have some desires to do acts that we know we ought not to do, and we stop ourselves. We are excused only when we cannot stop ourselves. But brain scans alone cannot reveal an inability to control oneself. To show that, we would need to know a great deal more about the nature of control systems, and how they interact with systems involved in motivation and desire.

7.3.4.6 The double-edged sword: future behaviour

Even if the defence can prove a functional abnormality in a particular defendant, what matters to law is not brain function but behaviour, and abnormal brain function need not make abnormal behaviour likely. While this may not be important for establishing relevance, since the test there is whether the evidence bears upon the likelihood of the crime, it may be of great import in sentencing, or in other circumstances in which assessing future dangerousness of the defendant is operative.

Although numerous studies reveal that brain damage, particularly in the frontal lobe, can be associated with increased aggressive or antisocial behaviour, the overall prevalence of actual violent crime in these studies is small. For example, in a study of 144 British soldiers with penetrating head injury, five (3 per cent) committed 'crimes and misdemeanours' after the injury.[25] All five had probable injury to the

---

[25] Lishman (1968).

frontal lobe. Similar results (5 per cent) were found in a Finnish study of World War II veterans.[26] More recently, in a large study (including brain scans) of Vietnam veterans with head injury, 14 per cent of subjects with injury to the frontal lobe engaged in fights or damaged property, compared with about 4 per cent of controls without head injury. Increased aggression was more likely to be present if there was evidence of damage to the medial or orbital areas of the frontal cortex.[27]

These and many other studies establish that damage to the frontal lobes, particularly the medial or orbitofrontal cortex, can lead to abnormal executive function, particularly to deficits in control that could increase the chances for future impulsive behaviour or aggression. However, these studies show only an increase above baseline, but not a high probability of violent behaviour. No study to date has reliably demonstrated a characteristic pattern of frontal lobe dysfunction based on behavioural measures or brain scanning that is predictive of a loss of control or the emergence of violent crime that is applicable to an individual case. All the worries noted in the previous section about Bayesian calculations apply here too. Thus, when the issue is prediction, it is not clear what the relevance of brain imaging will be for the law. To be sure, there is always the possibility that some signature pattern will be discovered that is highly predictive of some kind of criminal behaviour or that predictions based on behavioural evidence will become more reliable when neural evidence is added; but then a problem will likely be finding a large enough sample size to be confident of the results.

In any case, if one were to find evidence sufficiently strong to mitigate responsibility in a crime, the very same data that supports reduced culpability could be used to argue for increased, uncontrollable dangerousness. The argument for longer sentences to protect the public, despite a less punitive rationale for the sentence, might more than offset the benefit to the defendant.

7.3.5 Summary of probative value

There are a number of barriers for establishing the probative value of brain imaging in a criminal trial. Chief among them is the statistical problems attending inferences about the relevance of individual data, due to noise, low base rates, and the lack of appropriate comparison classes. None of these problems shows that fMRI brain images cannot ever be relevant to issues of criminal responsibility. Indeed, we can easily imagine circumstances where they would become relevant as markers of some disability. At present, however, the probative value of brain images for criminal responsibility seems minimal on Allen et al.'s scale of highly/somewhat/minimally probative. Their probative value might be increased for specific uses, but the problems that we listed would have to be overcome.

---

[26] Virkkunen et al. (1976).
[27] Grafman et al. (1996).

7.4 Conclusions

7.4.1 Legal conclusions

Because of the scientific uncertainties, our legal conclusion has to be conditional: If brain images are as confusing and misleading in trial contexts as they seem to be in reported experiments, and if they lack much probative value because they cannot overcome the problems listed above, then their moderate dangers 'substantially outweigh' their minimal probative value, so brain images fail the balancing test of FRE 403 and should not be admitted into trials. Although we have not emphasized these issues so far, we hope that our discussion has made it clear that what counts as evidence and whether evidence is strong enough for beliefs and decisions to be justified depends not just on pure probabilities but also on values, including costs of errors. It is also worth recalling that brain scans still might be admissible in some situations, such as capital sentencing, where the defence is strongly favoured and allowed to admit almost anything. This is where neural evidence is most often admitted now, and this practice seems to rest on the traditional value judgement that it is particularly abhorrent to find someone guilty who is innocent or who has not been given every reasonable chance to defend himself or herself.

7.4.2 Considerations for a future legally sensitive neuroscience

Gathering and assessing legal evidence from neuroscience is a task that requires the involvement of people from different fields, including neuroscience, the law, and psychology. We argued above that current brain scans are only minimally probative of criminal responsibility. This might change, of course, as techniques and our knowledge of brain function improve. Advances on the horizon include pattern classifiers and new ways to base individual predictions on group data given knowledge of individual differences. It seems that by far the most important start in doing so would be to develop a far more sophisticated means of exploring and cataloguing individual variability on different cognitive tasks, and determining the sources of that variability, be they anatomical, functional, or cognitive/strategic. In addition, we need to elucidate better philosophically and biologically relevant notions of control, and to strive to understand how control systems in the brain interact with those that motivate and produce behaviour. On the sociological side of things, further studies of the effects of images on deliberation in realistic cases will do much to shed light on the issue of prejudice/dangerousness. The balancing equation codified in FRE 403 will be one that will require frequent reassessment as the state of our knowledge of these diverse factors changes.

7.5 References

Allen, R. J., Kuhns, R. B., Swift, E., and Schwartz, D. S. (2006) *Evidence: Texts, Problems, and Cases*, 4th edn (New York, Aspen).

Bornstein, B. H. (1999) 'The Ecological Validity of Jury Simulations: Is the Jury Still Out?' 23(1) *Law and Human Behavior* 75–91.

Bright, D. A. and Goodman-Delahunty, J. (2006) 'Gruesome evidence and emotion: Anger, blame, and jury decision-making' 30 *Law Hum. Behav.* 183–202.

Burns, J. M. and Swerdlow, R. H. (2003) 'Right orbitofrontal tumor with pedophilia symptom and constructional apraxia sign' 60 *Arch. Neurol.* 437–40.

Elliott, R. and Robinson, R. J. (1991) 'Death Penalty Attitudes and the Tendency to Convict or Acquit: Some Data' 15(4) *Law and Human Behavior* 389–404.

Feigenson, N. (2006) 'Brain imaging and courtroom evidence: on the admissibility and persuasiveness of fMRI' 2 *International Journal of Law in Context* 233–55.

Grafman, J., Schwab, K., Warden, D., Pridgen, A., Brown, H. R., and Salazar, A. M. (1996) 'Frontal lobe injuries, violence, and aggression: A report of the Vietnam head injury study' 46 *Neurology* 1231–8.

Gurley, J. R. and Marcus, D. K. (2008) 'The effects of neuroimaging and brain injury on insanity defenses' 26 *Behavioral Sciences and the Law* 85–97.

Lishman, W. A. (1969) 'Brain damage in relation to psychiatric disability after head injury' 114 *Br J Psychiatry* 373–410.

McCabe, D. P. and Castel, A. D. (2007) 'Seeing is believing: The effect of brain images on judgments of scientific reasoning' 107 *Cognition* 343–52.

Miller, M. B., Van Horn, J. D., Wolford, G. L., Handy, T. C., Valsangkar-Smyth, M., Inati, S., et al. (2002) 'Extensive Individual Differences in Brain Activations Associated with Episodic Retrieval are Reliable Over Time' 14(8) *Journal of Cognitive Neuroscience* 1200–14.

Roskies, A. L. (2007) 'Are brain images like photographs?' 74 *Philosophy of Science* 860–72.

Roskies, A. L. (2008) 'Neuroimaging and Inferential Distance' 1 *Neuroethics* 19–30.

Virkkunen, M., Nuutila, A., and Huusko, S. (1976) 'Effect of brain injury on social adaptability' 33 *Acta Psychiatr. Scand.* 168–72.

Sinnott-Armstrong, W., Roskies, A. L., Brown, T., and Murphy, E. (2008) 'Brain images as legal evidence' (2008) 5(3) *Episteme: A Journal of Social Epistemology* 359–73.

Talairach, J. and Tournoux, P. (1988) *Co-Planar Stereotaxic Atlas of the Human Brain* (M. Rayport, Trans.) (New York, Thieme Medical Publishers, Inc.).

Walton, K. (1984) 'Transparent pictures: on the nature of photographic realism' 11 *Critical Inquiry* 246–76.

Weisberg, D. S., Keil, F. C., Goodstein, J., Rawson, E., and Gray, J. R. (2008) 'The Seductive allure of neuroscience explanations' 20 *Journal of Cognitive Neuroscience* 470–7.

8

The Neural Correlates of Third-Party Punishment

Joshua W. Buckholtz*, Christopher L. Asplund**, Paul E. Dux***, David H. Zald[†], John C. Gore^{††}, Owen D. Jones[#], and René Marois^{##}

8.1 Summary

Legal decision-making in criminal contexts includes two essential functions performed by impartial 'third parties': assessing responsibility and determining an appropriate punishment. To explore the neural underpinnings of these processes, we scanned subjects with fMRI while they determined the appropriate punishment for crimes that varied in perpetrator responsibility and crime severity. Activity within regions linked to affective processing (amygdala, medial prefrontal, and posterior cingulate cortex) predicted punishment magnitude for a range of criminal scenarios. By contrast, activity in right dorsolateral prefrontal cortex distinguished between scenarios on the basis of criminal responsibility, suggesting that it plays a key role in third-party punishment. The same prefrontal region has previously been shown to be involved in punishing unfair economic behaviour in two-party interactions, raising the possibility that the cognitive processes supporting third-party

---

* PhD candidate in Neuroscience at the Vanderbilt Brain Institute and Department of Psychology, Vanderbilt University.
** Department of Psychology, Neuroscience Graduate Program of Vanderbilt University.
*** Department of Psychology of Vanderbilt University.
[†] Department of Psychology, Center for Integrative and Cognitive Neurosciences of Vanderbilt University.
^{††} Institute of Imaging Science, Center for Integrative and Cognitive Neurosciences, Departments of Radiology and Radiological Sciences and Biomedical Engineering of Vanderbilt University.
[#] New York Alumni Chancellor's Chair in Law and Professor of Biology at Vanderbilt University, Director of the MacArthur Foundation Law and Neuroscience Project.
^{##} Associate Professor of Psychology at Vanderbilt University. Scientific correspondence should be addressed to: René Marois, 530 Wilson Hall, Department of Psychology, Vanderbilt University, 111 21st Ave S., Nashville, TN 37203; phone: 615-322-1779; email: rene.marois@vanderbilt.edu; or Owen Jones, Vanderbilt University Law School, 131 21st Ave. South, Nashville, TN 37203–1181; tel: (615) 322–7191; email: owen.jones@vanderbilt.edu.

legal decision-making and second-party economic norm enforcement may be supported by a common neural mechanism in human prefrontal cortex.[1]

8.2 Introduction

Though rare in the rest of the animal kingdom, large-scale cooperation among genetically unrelated individuals is the rule, rather than the exception, in *homo sapiens*.[2] Ultra-sociality and cooperation in humans is made possible by our ability to establish social norms—widely shared sentiments about appropriate behaviours that foster both social peace and economic prosperity.[3] In turn, norm compliance relies not only on the economic self-interest often served by cooperation and fair exchange, but also on the credible threat of unwelcome consequences for defection.[4] Social order therefore depends on punishment—which modern societies administer through a system of state-empowered enforcers, guided by state-governed, impartial, third-party decision-makers, who are not directly affected by the norm violation and have no personal stake in the execution of its enforcement.

The role of legal decision-makers is two-fold: determining responsibility and assigning an appropriate punishment. In determining responsibility, a legal decision-maker must assess whether the accused has committed a wrongful act and, if so, whether he did it with one of several culpable states of mind (so-called '*mens rea*').[5] For many of the most recognizable crimes, the defendant must have engaged in the proscribed conduct with intent in order to merit punishment. Moreover, in sentencing an individual for whom criminal responsibility has been determined, a legal decision-maker must choose a punishment that fits the crime. This sentence must ordinarily be such that the combined nature and extent of punishment is proportional to the combined harmfulness of the offence and blameworthiness of the offender.[6]

Despite its critical utility in facilitating prosocial behaviour and maintaining social order, little is known about the origins of, and neural mechanisms underlying, our ability to make third-party legal decisions.[7] The cognitive ability to make

---

[1] This article originally appeared in (2008) 60 *Neuron* 930–40. Please refer to the original publication for all supplementary data mentioned in this paper: <http://download.cell.com/neuron/mmcs/journals/0896-6273/PIIS0896662730800889 1.mmcl.pdf>.

[2] Joseph Henrich, 'The cultural and genetic evolution of human cooperation' in Peter Hammerstein (ed.), *Genetic and Cultural Evolution of Cooperation* (Cambridge, MA, MIT Press in Cooperation with Dahlem University Press, 2003), 445.

[3] Ernst Fehr and Urs Fischbacher, 'Social norms and human cooperation' (2004a) 8 *Trends in cognitive sciences* 185. Manfred Spitzer, Urs Fischbacher, Bärbel Herrnberger, Georg Grön, and Ernst Fehr, 'The neural signature of social norm compliance' (2007) 56 *Neuron* 185.

[4] Spitzer et al. (2007), note 2, above.

[5] Paul H. Robinson, 'Mens rea' in Joshua Dressler (ed.), *Encyclopedia of Crime & Justice* (New York, Macmillan Reference USA, 2002) 995.

[6] Nita A. Farahany and James E. Coleman Jr., 'Genetics and responsibility: To know the criminal from the crime' (2006) 69 *Law and Contemporary Problems* 115; Wayne R. LaFave, *Criminal Law*, 4th edn (St Paul, Thomson/West, 2003).

[7] Brent Garland (ed.), *Neuroscience and the Law: Brain, Mind and the Scales of Justice* (New York, Dana Press, 2004); Brent Garland and Paul W. Glimcher, 'Cognitive neuroscience and the law' (2006) 16 *Current opinion in neurobiology* 130; Semir Zeki and Oliver Goodenough, 'Law and the brain: introduction' (2004) 359 *Philosophical transactions of the Royal Society of London* 1661.

social norm-related judgments likely arose from the demands of social living faced by our hominid ancestors.[8] These demands may have promoted the emergence of mechanisms for assessing fairness in interpersonal exchanges and enacting personal retaliations against individuals who behaved unfairly (second-party punishment).[9] Recent work has greatly advanced our understanding of how the brain evaluates fairness and makes decisions based on the cooperative status and intentions of others during two-party economic exchanges.[10] Notably, these studies have elucidated the neural dynamics that underlie human altruistic punishment, in which the victim of a social norm transgression, typically unfairness in an economic exchange, punishes the transgressor at some significant additional cost to himself. These findings have specifically highlighted the importance of reward and emotion-related processes in fuelling cooperative behaviour.[11] However, how, or even whether, neural models of economic exchange in dyadic interactions apply to impartial, third-party legal decision-making is currently unknown.[12] Furthermore, the importance of uncovering neural mechanisms underlying third-party punishment is underscored by the proposal that the development of stable social norms in human societies specifically required the evolution of third-party sanction systems.[13]

Given that, in great measure, criminal law strives towards the stabilization and codification of social norms, including moral norms, in legal rules of conduct, moral decision-making is inherently embedded into the legal decision-making process.[14] The relevance of moral decision-making to an investigation of legal reasoning is highlighted by experimental findings which suggest that individuals punish according to so-called 'just deserts' motives; i.e. in proportion to the moral

---

[8] Note 1, above; Peter J. Richerson, Robert T. Boyd, and Joseph Henrich, 'Cultural Evolution of Human Cooperation' in Peter Hammerstein (ed.), *Genetic and Cultural Evolution of Cooperation* (Cambridge, MA, MIT Press, 2003) 357.

[9] Note 2, above.

[10] Dominique J.-F. de Quervain, Urs Fischbacher, Valerie Treyer, Melanie Schellhammer, Ulrich Schnyder, Alfred Buck, and Ernst Fehr, 'The neural basis of altruistic punishment' (2004) 305 *Science* 1254; Mauricio R. Delgado, Robert H. Frank, and Elizabeth A. Phelps, 'Perceptions of moral character modulate the neural systems of reward during the trust game' (2005) 8 *Nature Neuroscience* 1611; Brooks King-Casas, Damon Tomlin, Cedric Anen, Colin F. Camerer, Steven R. Quartz, and P. Read Montague, 'Getting to know you: reputation and trust in a two-person economic exchange' (2005) 308 *Science* 78; Daria Knoch, Alvaro Pascual-Leone, Kaspar Meyer, Valerie Treyer, and Ernst Fehr, 'Diminishing reciprocal fairness by disrupting the right prefrontal cortex' (2006) 314 *Science* 829; Alan G. Sanfey, James K. Rilling, Jessica A. Aronson, Leigh E. Nystrom, and Jonathan D. Cohen, 'The neural basis of economic decision-making in the Ultimatum Game' (2003) 300 *Science* 1755; Tania Singer, Stefan J. Kiebel, Joel S. Winston, Raymond J. Dolan, and Chris D. Frith, 'Brain responses to the acquired moral status of faces' (2004) 41 *Neuron* 653; Tania Singer, Ben Seymour, John P. O'Doherty, Klaas E. Stephan, Raymond J. Dolan, and Chris D. Frith, 'Empathic neural responses are modulated by the perceived fairness of others' (2006) 439 *Nature* 466. Spitzer et al. (2007), note 2, above.

[11] Ben Seymour, Tania Singer, and Ray Dolan, 'The neurobiology of punishment' (2007) 8 *Nature Reviews* 300.

[12] Fehr and Fischbacher (2004a), note 2, above.

[13] Jonathan Bendor and Piotr Swistak, 'The Evolution of Norms' (2001) 106 *American Journal of Sociology* 1493.

[14] Paul Robinson and John M. Darley, *Justice, Liability and Blame* (Boulder, Westview Press, 1995).

wrongfulness of an offender's actions.[15] As such, the seminal work of Greene and others—which has demonstrated distinct contributions of emotion-related and cognitive control-related brain regions to moral decision-making—is germane to the study of legal decision-making.[16] However, despite the conceptual overlap between moral and legal reasoning, the latter process is not entirely reducible to the former.[17] Indeed, whereas determining blameworthiness may in many cases fall under the rubric of moral decision-making, the distinctive core and distinguishing feature of legal decision-making is the computation and implementation of a punishment that is appropriate both to the relative moral blameworthiness of an accused criminal offender, and to the relative severity of that criminal offence.[18] The present study is focused on elucidating the neural mechanisms underlying this third-party, legal decision-making process.

In this study, we used event-related fMRI to reveal the neural circuitry supporting third-party decision-making about criminal responsibility and punishment. Given that these two legally distinct judgments are rendered on the basis of differing information and considerations,[19] we were particularly interested in determining whether these two decision-making processes may rely on at least partly distinct neural systems. To address this issue, we scanned sixteen participants while they determined the appropriate punishment for actions committed by the protagonist (named 'John') in a series of fifty written scenarios. Each of these scenarios belonged to one of three categories: Responsibility (R), Diminished-Responsibility (DR) and No-Crime (NC). Scenarios in the Responsibility set

---

[15] Adam L. Alter, Julia Kernochan, and John M. Darley, 'Transgression wrongfulness outweighs its harmfulness as a determinant of sentence severity' (2007) 31 *Law and Human Behavior* 319; Kevin M. Carlsmith, John M. Darley, and Paul H. Robinson, 'Why do we punish? Deterrence and just deserts as motives for punishment' (2002) 83 *Journal of Personality and Social Psychology* 284; John M. Darley and Thane S. Pittman, 'The psychology of compensatory and retributive justice' (2003) 7 *Personality and Social Psychology Review* 324.

[16] Joshua D. Greene, Leigh E. Nystrom, Andrew D. Engell, John M. Darley, and Jonathan D. Cohen, 'The neural bases of cognitive conflict and control in moral judgment' (2004) 44 *Neuron* 389; Joshua D. Greene, R. Brian Sommerville, Leigh E. Nystrom, John M. Darley, and Jonathan D. Cohen, 'An fMRI investigation of emotional engagement in moral judgment' (2001) 293 *Science* 2105; Hauke R. Heekeren, Isabell Wartenburger, Helge Schmidt, Kristin Prehn, Hans-Peter Schwintowski, and Arno Villringer, 'Influence of bodily harm on neural correlates of semantic and moral decision-making' (2005) 24 *NeuroImage* 887; Hauke R Heekeren, Isabell Wartenburger, Helge Schmidt, Hans-Peter Schwintowski, and Arno Villringer, 'An fMRI study of simple ethical decision-making' (2003) 14 *Neuroreport* 1215; Jorge Moll, Ricardo de Oliveira-Souza, Ivanei E. Bramati, and Jordan Grafman, 'Functional networks in emotional moral and nonmoral social judgments' (2002a) 16 *NeuroImage* 696; Jorge Moll, Ricardo de Oliveira-Souza, Paul J. Eslinger, Ivanei E. Bramati, Janaína Mourão-Miranda, Pedro Angelo Andreiuolo, and Luiz Pessoa, 'The neural correlates of moral sensitivity: a functional magnetic resonance imaging investigation of basic and moral emotions' (2002b) 22 *The Journal of Neuroscience* 2730.

[17] H. L. A. Hart, 'Positivism and the Separation of Law and Morals' (1958) 71 *Harvard Law Review* 593; Oliver Wendell Holmes, Jr, 'Lecture II: The Criminal Law' in *The Common Law* (New York, Dover Publications, 1991) 39; Richard A. Posner, 'The problematics of moral and legal theory' (1998) 111 *Harvard Law Review* 1637; Paul H. Robinson, *Structure and Function in Criminal Law* (Oxford, Oxford University Press, 1997); Robinson and Darley (1995), note 13, above.

[18] Robinson (1997), note 16, above. Robinson and Darley (1995), note 13, above.

[19] Wayne R. LaFave, Jerold H. Israel, Nancy J. King, and Orin S Kerr, *Criminal Procedure*, Vol. 6, 3rd edn (St Paul, West Group, 2007).

(N=20) described John intentionally committing a criminal action ranging from simple theft to rape and murder. The Diminished-Responsibility set (N=20) included actions of comparable gravity to those described in the Responsibility set but also contained mitigating circumstances that may have excused or justified the otherwise criminal behaviour of the protagonist by calling his blameworthiness into question. The No-Crime set (N=10) depicted John engaged in non-criminal actions that were otherwise structured similarly to the Responsibility and Diminished-Responsibility scenarios (scenarios available as Supplementary Methods). Participants rated each scenario on a scale from 0–9, according to how much punishment they thought John deserved, with '0' indicating no punishment and '9' indicating extreme punishment. Two groups of fifty scenarios (equated for word length between conditions and between groups) were constructed and their presentation counterbalanced across the sixteen participants. The Responsibility set of group 2 consisted of group 1 Diminished-Responsibility scenarios for which the mitigating circumstances had been removed, while the Diminished-Responsibility set of group 2 consisted of group 1 Responsibility scenarios with mitigating circumstances added. Thus, each criminal scenario (e.g. depicting theft, assault, or murder) in the Responsibility and Diminished-Responsibility condition was created by modifying identical 'stem' stories, with salient details such as magnitude of harm matched between conditions.

8.3 Results

8.3.1 Behavioural data

Behavioural data showed a significant effect of scenario category on punishment ratings ($F(1,15) = 358.61, p < 0.001$) (Figure 1), with higher mean ratings for the Responsibility (Mean = 5.50, S.E. = 0.22) than for the Diminished-Responsibility scenarios (Mean = 1.45, S.E. = 0.21) ($p < 0.001$, paired t-test), indicating that assessed punishment was strongly modulated by the protagonist's criminal responsibility. By the same token, the fact that the mean punishment rating for the Diminished-Responsibility condition was greater than 0 suggests that some participants still attributed some blameworthiness to the protagonist despite the extenuating circumstances. To examine the subjective emotional experience elicited by the scenarios, all participants completed post-scan ratings of emotional arousal for each scenario. These ratings also demonstrated an effect of condition ($F(1,15) = 94.61, p < 0.001$) (Figure 1), with greater mean arousal scores for the Responsibility (Mean = 4.83, S.E. = 0.41) compared to Diminished-Responsibility scenarios (Mean = 3.48, S.E. = 0.35) ($p < 0.001$, paired t-test). Additionally, we found a significant interaction between rating type (punishment vs. arousal) and condition (Responsibility vs. Diminished-Responsibility) ($F(1,15) = 68.8, p < 0.001$) such that, while the punishment and arousal ratings were not significantly different for the Responsibility scenarios ($p > 0.05$, paired t-test), punishment ratings were significantly lower than the arousal ratings for the Diminished-Responsibility

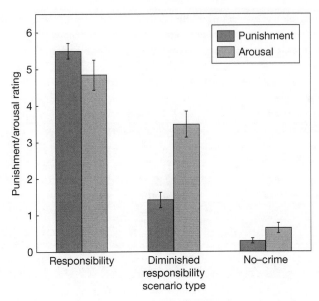

Figure 1 Punishment and arousal ratings for each scenario type. While punishment and arousal scores were similar in the Responsibility condition, punishment scores were significantly lower than arousal scores in the Diminished-Responsibility condition.

scenarios (p < 0.001, paired *t*-test) (Figure 1). Lastly, we found a main effect of scenario condition on reaction times (RTs) ($F(1,15) = 21.87$, $p < 0.001$), such that RTs were shortest for the No-Crime condition and longest for the Diminished-Responsibility condition (mean, S.E. for: Responsibility = 12.69s, 0.46; Diminished-Responsibility = 13.76s, 0.46; No-Crime = 11.12s, 0.44) (all paired comparisons p < 0.01).

8.3.2 fMRI Data: Criminal Responsibility

To identify brain regions that were sensitive to information about criminal responsibility, we contrasted brain activity between Responsibility and Diminished-Responsibility scenarios. The resulting statistical parametric map (SPM) revealed an area of activation in the right dorsolateral prefrontal cortex (rDLPFC, Brodmann Area 46, peak at Talaraich coordinates 39, 37, 22 [x,y,z]; Figure 2a) that was significantly more activated in the Responsibility than in the Diminished-Responsibility condition. Time course analyses of peak activation differences confirmed that there was greater rDLPFC activity in Responsibility compared to Diminished-Responsibility or No-Crime conditions (L>NL, p = 0.002; L>NC, p = 0.0004; paired *t*-tests; see Figure 2b) and no difference between the Diminished-Responsibility and No-Crime conditions (p = 0.19). No effect of condition was found in the

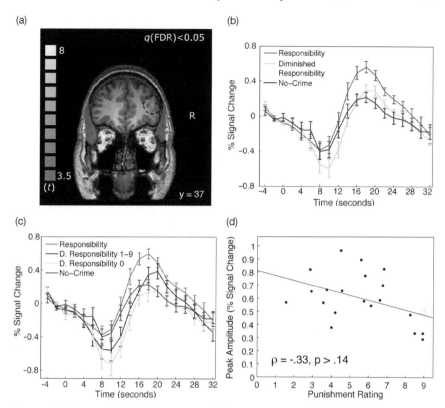

Figure 2 Relationship between responsibility assessment and right DLPFC activity. (a) SPM displaying the right DLPFC VOI (rendered on a single subject T1-weighted image), based on the contrast of BOLD activity in the Responsibility condition compared to the Diminished-Responsibility condition, $t(15) > 3.5$, $q < 0.05$, random effects analysis. R = Right Hemisphere. (b) BOLD activity time courses in right DLPFC for the Responsibility, Diminished-Responsibility and No-Crime conditions. BOLD peak amplitude was significantly greater in the Responsibility condition compared to both the Diminished-Responsibility and No-Crime conditions ($p = 0.002$, $p = .0004$, respectively). Peak was defined as the single TR with maximal signal change from baseline within the first 13 volumes after scenario presentation onset. t-tests were performed on these peak volumes, which were defined separately for each condition and each subject. (c) BOLD activity time courses in right DLPFC for Responsibility, 'non-punished' Diminished-Responsibility (Diminished-Responsibility 0), 'punished' Responsibility (Diminished-Responsibility 1–9) and No-Crime scenarios. BOLD peak amplitude was significantly greater in 'punished' compared to 'non-punished' Diminished-Responsibility scenarios ($p = 0.04$), while no difference was observed between 'non-punished' Diminished-Responsibility and No-Crime scenarios ($p = 0.98$). (d) Relationship between BOLD peak amplitude in right DLPFC and punishment ratings in the Responsibility condition. These two variables were not significantly correlated ($p > 0.15$).

Editor's note: In the original publication, the image (a) appears in colour. In black and white, dashed circles were added.

left DLPFC (p > 0.2 for all paired comparisons; see Methods), and the right DLPFC was significantly more engaged than the left DLPFC in the Responsibility condition (p = 0.04, paired *t*-test), suggesting that punishment-related prefrontal activation is confined to the right hemisphere. Bilateral anterior intraparietal sulcus (aIPS) demonstrated a pattern of responsibility-related activity that was similar to rDLPFC (Table 1, Supplementary Figure 1, Supplementary Results), whereas the temporo-parietal junction (TPJ) showed the reverse pattern, with more activity for the Diminished-Responsibility than the Responsibility condition (Table 1, Figure 3, below).

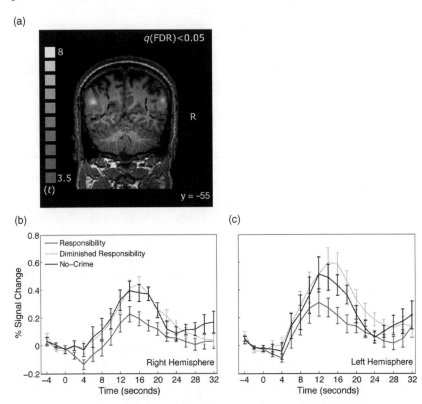

Figure 3 Relationship between responsibility assessment and bilateral temporo-parietal junction (TPJ) activity. (a) SPM displaying the right and left TPJ VOIs (rendered on a single subject T1-weighted image), based on the contrast of BOLD activity in the Diminished-Responsibility condition compared to the Responsibility condition, $t(15) > 3.5$, $q < 0.05$; random effects analysis. R = Right Hemisphere. BOLD activity time courses in right (b) and left (c) TPJ for the Responsibility, Diminished-Responsibility and No-Crime conditions. BOLD peak amplitude was significantly greater in the Diminished-Responsibility condition compared to the Responsibility and conditions for right (p = 0.0005) and left (p = 0.001) TPJ. Peak was defined as the single TR with maximal signal change from baseline within the first 13 volumes after scenario presentation onset. *t*-tests were performed on these peak volumes, which were defined separately for each condition and each subject.

Editor's note: In the original publication, the image (a) appears in colour. In black and white, dashed circles were added.

The greater rDLPFC activation in the Responsibility condition did not simply result from longer time-on-task: response times (RTs) to Responsibility scenarios were shorter than Diminished-Responsibility scenarios ($p = 0.005$, paired t-test), and the effect of condition on rDLPFC activity was still significant when response time was used as a covariate in an analysis of covariance (ANCOVA, $F(1,37) = 10.15$, $p = 0.003$) or when response times were equated between conditions (see Methods; R>DR, $p = 0.006$; R>NC, $p = 0.002$; Supplementary Figure 2). In addition, rDLPFC activity was not correlated with reaction time ($p = 0.09$ in Responsibility scenarios, $p = 0.12$ in Diminished-Responsibility scenarios). We also assessed whether the activity pattern in rDLPFC might have been driven by between-condition differences in emotional arousal rather than by differences in criminal responsibility. To this end, we performed a peak activation difference analysis between the Responsibility and Diminished-Responsibility conditions after equating their mean arousal ratings (Responsibility = 3.62, Diminished-Responsibility = 3.50; $p > 0.10$, paired t-test; see Methods). The results still revealed greater rDLPFC activity in the Responsibility compared to the Diminished-Responsibility condition even in the absence of arousal differences ($p = 0.0005$, paired t-test).

If rDLPFC is involved in the decision-making process to punish blameworthy behaviour, then this brain region should be more activated during Diminished-Responsibility scenarios in which subjects still decided to punish (punishment ratings of 1 or greater) compared to Diminished-Responsibility scenarios in which they did not (punishment rating of 0). Consistent with this hypothesis, rDLPFC activity was higher in 'punished' Diminished-Responsibility trials than in 'non-punished' Diminished-Responsibility trials ($p = 0.04$, paired t-test, Figure 2). In turn, rDLPFC activity during 'non-punished' Diminished-Responsibility trials was not greater than in No-Crime trials. ($p = 0.98$, Figure 2). These results, as well as those for aIPS (Supplementary Results, Supplementary Figure 1), strongly support the notion that prefrontal and parietal activity is modulated by a punishment-related decisional process.

In addition to the peak activation differences, the timecourse of rDLPFC activity revealed an early deactivation (negative percent signal change from baseline) around eight seconds post-stimulus onset. Importantly, this early deactivation ('dip') does not account for the peak activation results outlined above: the activation differences between conditions at the dip do not predict corresponding activation differences at the peak (correlation of subjects' activity differences between the Responsibility and Diminished-Responsibility conditions at the dip and at the peak: $\rho = -0.19$, $p = 0.49$; Supplementary Figure 3; see Methods). Furthermore, rDLPFC activity during 'non-punished' Diminished-Responsibility and No-Crime trials strongly differed at the dip ($p = 0.008$) but not at the peak ($p = 0.97$), indicating that peak activation differences are not simply carry-over effects from differences during the dip.

8.3.3 fMRI Data: Punishment Magnitude

The finding that rDLPFC activity was higher when subjects decided to punish, in either Responsibility scenarios or in 'punished' Diminished-Responsibility trials, raised the possibility that this brain region might track the amount of assessed punishment for a given criminal scenario. However, rDLPFC signal amplitude was not linearly correlated with punishment ratings ($\rho = -0.33$, p = 0.15; Figure 2d) in the Responsibility condition. This finding suggests that the magnitude of punishment is not simply coded by a linear increase in rDLPFC activity.

Although rDLPFC activity was not proportional to punishment amount, a linear relationship between peak BOLD amplitude and punishment magnitude was found in a set of brain regions that have been extensively linked to social and affective processing. To isolate such effects, we compared Responsibility scenarios with high punishment ratings to those with low ratings (median split by scenario across subjects; see Methods). The resulting SPM revealed activation in the right amygdala (peak Talairach coordinates 29, −7, −13; Figure 4; Supplementary Figure 5) as well as in other brain regions commonly associated with social and affective processing,[20] including the posterior cingulate, temporal pole, dorsomedial and ventromedial prefrontal cortex, and inferior frontal gyrus (Supplementary Table 2; Supplementary Figure 4, Supplementary Figure 5). The association between amygdala activity and punishment magnitude was further demonstrated by a strong correlation between amygdala BOLD signal and punishment ratings across Responsibility scenarios ($\rho = 0.70$, $p = 0.001$; Figure 4). However, punishment rating was not the only variable that correlated with amygdala function, as participants' arousal ratings yielded a similar correlation with amygdala activity ($\rho = 0.67$, p = 0.001), and punishment and arousal ratings were themselves highly correlated ($\rho = 0.98$, p = 0.000001). Correlations between peak BOLD signal and punishment ratings (and between peak BOLD signal and arousal ratings) also held for a number of the other affective regions, including ventromedial prefrontal cortex and posterior cingulate cortex (Supplementary Table 2; Supplementary Figures 4 and 5), indicating that the relationship between affective processing and punishment involved a distributed neural circuit.

Although the correlation between amygdala activity and punishment scores could be interpreted as evidence for a role of emotional arousal in the assignment of deserved punishment, it is also possible that such activity simply reflected subjects' emotional reaction to the graphical content of the scenarios rather than its involvement in the decision-making process *per se*. To avoid the potential arousal confound inherent to an examination of criminal scenarios that differ in graphic

---

[20] Joseph E. LeDoux, 'Emotion circuits in the brain' (2000) 23 *Annual Review of Neuroscience* 155; Elizabeth A. Phelps, 'Emotion and cognition: insights from studies of the human amygdala' (2006) 57 *Annual Review of Psychology* 27; Mary L. Phillips, Wayne C. Drevets, Scott L. Rauch, and Richard Lane, 'Neurobiology of emotion perception I: The neural basis of normal emotion perception' (2003) 54 *Biological Psychiatry* 504; Joseph L. Price, 'Free will versus survival: brain systems that underlie intrinsic constraints on behavior' (2005) 493 *The Journal of Comparative Neurology* 132.

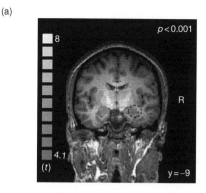
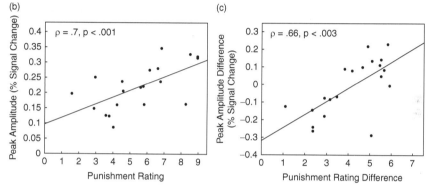

Figure 4 **Relationship between punishment and right amygdala activity.** (a) SPM displaying the right amygdala VOI (rendered on a single-subject T1-weighted image), based on the contrast of BOLD activity between high and low punishment (computed from the median split for Responsibility scenarios), thresholded at $t(15) > 4.1, p < 0.001$ (uncorrected) for visualization. This amygdala activation survives correction for multiple comparisons, q(FDR) < 0.05; random-effects analysis. R = Right Hemisphere. (b) Relationship between BOLD peak amplitude in the right amygdala and punishment ratings in the Responsibility condition. These two variables were significantly positively correlated (p = 0.001). (c) Relationship between condition differences in right amygdala BOLD peak amplitude (Responsibility minus Diminished-Responsibility) and condition differences in punishment score (Responsibility minus Diminished-Responsibility); these two variables are significantly correlated (p = 0.001).

Editor's note: In the original publication, the image (a) appears in colour. In black and white, dashed circles were added.

content (as was the case for our comparison of high vs low punishment scores within the Responsibility condition), we examined the relationship between punishment ratings and amygdala activity after controlling for the possible confounding effect of graphic arousal. Because Responsibility and Diminished-Responsibility scenarios were equated for graphic content and differed only by the presence of mitigating circumstances (see Methods), the potentially confounding contribution of graphic arousal to amygdala activity in the Responsibility scenarios can be controlled for by subtracting amygdala activity in the Diminished-Responsibility scenarios from that in the corresponding Responsibility scenarios. If amygdala activity appertains to punishment magnitude rather than, or in addition to, emotional arousal related to the graphic content of the scenarios, it should still track punishment ratings even after subtracting out graphic content differences in the scenarios. To this end, we created, for each pair of Responsibility and Diminished-Responsibility scenarios, punishment rating difference scores (Responsibility minus Diminished-Responsibility) and assessed whether these scores were correlated with the corresponding difference scores for peak amygdala BOLD signal. That correlation was significant ($\rho = 0.62$, p = 0.001; Figure 4), indicating that the magnitude of amygdala BOLD signal difference between Responsibility and Diminished-Responsibility conditions for a given scenario predicted a corresponding change in punishment rating for that scenario. Similar correlations were found in posterior cingulate and ventromedial prefrontal cortex (Supplementary Table 2). These findings suggest that activity within brain regions previously implicated in social and affective processing reflect third-party decisions about how much to punish, even after controlling for the potentially confounding arousal associated with the 'graphic' content of the criminal scenarios.

8.4 Discussion

The present findings suggest that the two fundamental components of third-party legal decision-making, determining responsibility and assigning an appropriate punishment magnitude, are not supported by a single neural system. In particular, the results reveal a key role for the right dorsolateral prefrontal cortex in third-party punishment. This brain region appears to be involved in deciding whether or not to punish based on an assessment of criminal responsibility. The only other brain region demonstrating a comparable pattern of responsibility-related activity (R>DR, R>NC, DR=NC) to rDLPFC was the anterior intraparietal sulcus (Supplementary Table 1, Supplementary Figure 1, Supplementary Results). This parietal region has been associated with a number of diverse cognitive functions including general response selection[21] and quantitative numerical comparisons,[22] which may hint at a role for this area in associating a specific action (punishment outcome) with a given scenario.

Our results also implicate neural substrates for social and affective processing (including amygdala, medial prefrontal cortex and posterior cingulate cortex) in third-party punishment, albeit in ways distinct from the rDLPFC. Specifically, while prefrontal activity was linked to a categorical aspect of legal decision-making

---

[21] Silke M. Göbel, Heidi Johansen-Berg, Tim Behrens, and Matthew F. S. Rushworth, 'Response-selection-related parietal activation during number comparison' (2004) 16 *Journal of Cognitive Neuroscience* 1536.

[22] Stanislas Dehaene, Manuela Piazza, Philippe Pinel, and Laurent Cohen, 'Three parietal circuits for number processing' (2003) 20 *Cognitive Neuropsychology* 487; Stanislas Dehaene, Elizabeth Spelke, Philippe Pinel, Ritta Stanescu, and Susanna Tsivkin, 'Sources of mathematical thinking: behavioral and brain-imaging evidence' (1999) 284 *Science* 970; Lisa Feigenson, Stanislas Dehaene, and Elizabeth Spelke, 'Core systems of number' (2004) 8 *Trends in Cognitive Sciences* 307.

(deciding whether or not to punish on the basis of criminal responsibility), the magnitude of assigned punishments for criminal transgressions parametrically modulated activity in affective brain regions, even after controlling for the potentially confounding arousal-related activity associated with the graphic content of the criminal scenarios. Our findings suggest that a set of brain regions (e.g. amygdala, medial prefrontal cortex, and posterior cingulate) consistently linked to social and emotional processing[23] is associated with the amount of assigned punishment during legal decision-making. As such, these results accord well with prior work pointing to social and emotional influences on economic decision-making and moral reasoning,[24] and provide preliminary neuroscientific support for a proposed role of emotions in legal decision-making.[25] Our data concur with behavioural studies that have proposed a link between affect and punishment motivation in both second- and third-party contexts, and are consistent with the hypothesis that third-party sanctions are fuelled by negative emotions towards norm violators.[26] However, it must be acknowledged that the present conclusions rest exclusively on correlational data. Thus, additional research will be required to determine confidently the contributions of socio-affective brain regions to third-party punishment in the absence of any graphic arousal confound. In particular, it will be important in future experiments fully to dissociate the factors of crime severity and arousal by employing task conditions that manipulate arousal without affecting crime severity. Furthermore, future research should also focus on determining how these affective

---

[23] Ralph Adolphs, 'Neural systems for recognizing emotion' (2002) 12 *Current Opinion in Neurobiology* 169; David M. Amodio and Chris D. Frith, 'Meeting of minds: the medial frontal cortex and social cognition' (2006) 7 *Nature Reviews* 268; Lisa Feldman Barrett, Batja Mesquita, Kevin Ochsner, and James J. Gross, 'The experience of emotion' (2007) 58 *Annual Review of Psychology* 373; Matthew D Lieberman, 'Social cognitive neuroscience: a review of core processes' (2007) 58 *Annual Review of Psychology* 259; Phelps (2006), note 19, above; Phillips, et al. (2003), note 19, above; David H. Zald, 'The human amygdala and the emotional evaluation of sensory stimuli' (2003) 41 *Brain Research Reviews* 88.

[24] Benedetto De Martino, Dharshan Kumaran, Ben Seymour, and Raymond J. Dolan, 'Frames, biases, and rational decision-making in the human brain' (2006) 313 *Science* 684; Delgado, et al. (2005), note 9, above; Michael Koenigs and Daniel Tranel, 'Irrational economic decision-making after ventromedial prefrontal damage: evidence from the Ultimatum Game' (2007) 27 *Journal of Neuroscience* 951; Joshua Greene and Jonathan Haidt, 'How (and where) does moral judgment work?' (2002) 6 *Trends in Cognitive Sciences* 517. Greene, et al. (2004), note 15, above; Greene, et al. (2001), note 15, above. Jonathan Haidt, 'The emotional dog and its rational tail: a social intuitionist approach to moral judgment' (2001) 108 *Psychological Review* 814; Heekeren, et al. (2003), note 15, above; Michael Koenigs, Liane Young, Ralph Adolphs, Daniel Tranel, Fiery Cushman, Marc Hauser, and Antonio Damasio, 'Damage to the prefrontal cortex increases utilitarian moral judgements' (2007) 446 *Nature* 908. Moll, et al. (2002b), note 15, above; Jorge Moll, Roland Zahn, Ricardo de Oliveira-Souza, Frank Krueger, and Jordan Grafman, 'Opinion: the neural basis of human moral cognition' (2005) 6 *Nature Reviews* 799.

[25] David J. Arkush, 'Situating emotion: a critical realist view of emotion and nonconscious cognitive processes for the law' (2008) Available at SSRN (<http://ssrn.com/abstract=1003562>); Terry A. Maroney, 'Law and emotion: a proposed taxonomy of an emerging field' (2006) 30 *Law and Human Behavior* 119.

[26] Darley and Pittman (2003), note 14, above; Fehr and Fischbacher (2004), note 2, above; Ernst Fehr and Urs Fischbacher, 'Third-party punishment and social norms' (2004b) 25 *Evolution and Human Behavior* 63; Seymour, et al. (2007), note 10, above.

brain regions interact with dorsolateral prefrontal cortex during third-party punishment decisions.

An additional concern in interpreting our findings, or any others based on simulated judgments, is whether they are relevant to real-world decision-making. After all, the punishment decisions made by our participants did not have direct, real-world consequences for real criminal defendants. Thus, it remains to be seen if our findings, generated by examining brain activation patterns during 'hypothetical' judgments, will generalize to circumstances in which 'real' punishments are made. However, there is some evidence suggesting that the hypothetical judgments made by our subjects may be a good proxy measure for real-world legal judgments. For example, post-scan debriefing of our subjects indicated that their punishment assessments were implicitly legal, with lower numbers corresponding to low prison sentences and higher numbers corresponding to high prison sentences (see Supplementary Table 3). Thus, participants appeared to adopt an internal punishment scale based on incarceration duration—a legal metric—when making their judgments, even in the absence of explicit instructions to do so. Further, we found that participants' decisions about punishment amount for each of the crimes depicted in the Responsibility scenarios were strongly correlated with the recommended prison sentences for those crimes, according to the benchmark sentencing guidelines of North Carolina, a model state penal code ($\rho = 0.8$, p<.0001; Supplementary Figure 6; see Methods). Thus, although our subjects were not literally applying a criminal statute to an accused individual, these data suggest that subjects' punishment decisions were consistent with statutory legal reasoning. However, despite these suggestions, further empirical studies are required to confirm our supposition that neuroimaging studies of simulated third-party legal decision-making can be valid models for understanding the neural basis of real-world legal reasoning.

8.5 Relative Contributions of Temporo-Parietal Junction (TPJ) and rDPLFC to Third-Party Punishment Decisions

The neural mechanisms of third-party punishment are undoubtedly complex, involving a dynamic regional interplay unfolding in a temporally specific manner. In particular, the decision to punish a person for his blameworthy act is generally preceded by an evaluation of that person's intention in committing that act.[27] Such an evaluation ought therefore to activate brain regions that underlie the attribution of goals, desires, and beliefs to others, referred to as theory of mind (TOM).[28] One

---

[27] Alter, et al. (2007), note 14, above. Carlsmith, et al. (2002), note 14, above; Darley and Pittman (2003), note 14, above; John M. Darley and Thomas R. Schultz, 'Moral rules: their content and acquisition' (1990) 41 *Annual Review of Psychology* 525; Robinson and Darley (1995), note 13, above; Paul H. Robinson, Robert Kurzban, and Owen D. Jones, 'Shared intuitions of justice' (2007) 60 *Vanderbilt Law Review* 1633; Thomas R. Shultz, Kevin Wright and Michael Schleifer, 'Assignment of moral responsibility and punishment' (1986) 57 *Child Development* 177.

[28] Helen L. Gallagher and Christopher D. Frith, 'Functional imaging of "theory of mind"' (2003) 7 *Trends in Cognitive Science* 77.

such region, the TPJ—a key node in the distributed TOM network[29]—might be predicted to serve this function during legal decision-making given recent evidence of its role in attributing mental beliefs in moral judgements[30] and its involvement in dyadic economic exchange games.[31] Given this context, it is noteworthy that the TPJ was activated in all of our conditions (Figure 3). Furthermore, TPJ came online during the period when rDLPFC was deactivated (see Figure 2b), a result that is consistent with the suggestion that temporo-parietal cortex and dorsolateral prefrontal cortex operate within largely distinct and at times functionally opposed networks.[32] Given this proposed antagonistic response pattern in the TPJ and DLPFC, we speculate that the early rDLPFC deactivation may reflect a perspective-taking based evaluation of the beliefs and intentions of the scenarios' protagonist, which is followed by a robust rDLPFC activation as subjects go on to make a decision to punish based on assessed responsibility and blameworthiness. However, the conclusion that rDLPFC's biphasic timecourse reflects an initial socio-evaluative process followed by a decisional process must be viewed as tentative because the present experiment did not constrain the temporal sequences of evaluative and decisional processes involved in this task.

8.6 Moral versus Legal Decision-Making

The results of the present neuroimaging study underscore the conceptual relationship between moral and legal decision-making. Indeed, the general involvement of both the prefrontal cortex and affective brain regions in legal reasoning is reminiscent of their roles in moral judgement.[33] Specifically, moral decision-making studies have indicated that regions of lateral prefrontal cortex and inferior parietal lobe may be preferentially involved in impersonal moral judgements whereas socio-affective areas (e.g. amygdala, medial prefrontal cortex, and posterior cingulate cortex) may be primarily engaged during personal moral decision-making.[34] Thus, both legal and

---

[29] Jean Decety and Claus Lamm, 'The role of the right temporoparietal junction in social interaction: how low-level computational processes contribute to meta-cognition' (2007) 13 *The Neuroscientist* 580; Gallagher and Frith (2003), note 27, above; Rebecca and Nancy Kanwisher, 'People thinking about thinking people: the role of the temporo-parietal junction in "theory of mind"' (2003) 19 *Neuroimage* 1835; Birgit A. Völlm, Alexander N.W. Taylor, Paul Richardson, Rhiannon Corcoran, John Stirling, Shane McKie, John F.W. Deakin, and Rebecca Elliott, 'Neuronal correlates of theory of mind and empathy: a functional magnetic resonance imaging study in a nonverbal task' (2006) 29 *Neuroimage* 90.
[30] Liane Young, Fiery Cushman, Marc Hauser, and Rebecca Saxe, 'The neural basis of the interaction between theory of mind and moral judgment' (2007) 104 *Proceedings of The National Academy of Sciences of the USA* 8235.
[31] James K. Rilling, Alan G. Sanfey, Jessica A. Aronson, Leigh E. Nystrom, and Jonathan D. Cohen, 'The neural correlates of theory of mind within interpersonal interactions' (2004) 22 *Neuroimage* 1694.
[32] Michael D. Fox, Abraham Z. Snyder, Justin L. Vincent, Maurizio Corbetta, David C. Van Essen, and Marcus E. Raichle, 'The human brain is intrinsically organized into dynamic, anticorrelated functional networks' (2005) 102 *Proceedings of The National Academy of Sciences of the USA* 9673.
[33] Greene et al. (2004); Greene et al. (2001), note 15, above.
[34] Greene et al. (2004); Greene et al. (2001), note 15, above.

moral decision-making may rely on 'cold' deliberate computations supported by the prefrontal cortex and 'hot' emotional processes represented in socio-affective brain networks, although the extent to which these two decision-making processes rely on the same brain circuitry remains to be determined.

While these findings serve to highlight an important conceptual overlap between moral reasoning and legal reasoning in criminal contexts, they do not imply that third-party punishment decisions are reducible to moral judgement. Indeed, while legal decision-making may in most (but not all) criminal cases have an essential moral component, there are crucial distinctions between morality and law.[35] Perhaps the most critical distinguishing feature of legal decision-making, compared to moral decision-making, is the action of punishment—intrinsic to the former and secondary to the latter.[36] Although our participants likely engaged in the process of evaluating the moral blameworthiness of the scenarios' protagonist, our study was designed to investigate the neural substrates of a fundamental legal decision—assigning punishment for a crime—that is not a defining characteristic of moral judgement. Indeed, while moral decision-making studies to date have focused on assessing brain function during decisions about the moral rightness or wrongness of actions depicted in written scenarios, they have not specifically addressed the issue of punishment.[37]

8.7 Neural Convergence of Second-Party and Third-Party Punishment Systems

The prefrontal cortex area activated in the present third-party legal decision-making study corresponds well to an area that is involved in the implementation of norm enforcement behaviour in two-party economic exchanges (peak Talairach coordinates of 39, 37, 22 [x,y,z];[38] vs 39, 38, 18 [x,y,z] for the present study), raising the possibility that rDLPFC serves a function common to both third-party legal and second-party economic decision-making. In this respect, it is noteworthy that this

---

[35] Hart (1958); Holmes Jr, (1991); Posner (1998), note 16, above.
[36] Robinson (1997), note 16 above.
[37] Jana Schaich Borg, Catherine Hynes, John Van Horn, Scott Grafton, and Walter Sinnott-Armstrong, 'Consequences, action, and intention as factors in moral judgments: an FMRI investigation' (2006) 18 *Journal of Cognitive Neuroscience* 803; Greene et al. (2004); Greene, et al. (2001), note 15, above. Heekeren, et al. (2005); Heekeren et al. (2003), note 15, above; Gayannée Kédia, Sylvie Berthoz, Michele Wessa, Denis Hilton, and Jean-Luc Martinot, 'An agent harms a victim: a functional magnetic resonance imaging study on specific moral emotions' (2008) 20 *Journal of Cognitive Neuroscience* 1788; Qian Luo, Marina Nakic, Thalia Wheatley, Rebecca Richell, Alex Martin, and R. James R. Blair, 'The neural basis of implicit moral attitude—an IAT study using event-related fMRI' (2006) 30 *NeuroImage* 1449; Moll et al. (2002a, 2002b), note 15, above; Jorge Moll, Paul J. Eslinger, and Ricardo de Oliveira-Souza, 'Frontopolar and anterior temporal cortex activation in a moral judgment task: preliminary functional MRI results in normal subjects' (2001) 59 *Arq. Neuropsiquiatr.* 657; Young et al. (2007), note 29 above. Liane Young and Rebecca Saxe, 'The neural basis of belief encoding and integration in moral judgment' (2008), 40 *NeuroImage* 1912.
[38] Knoch et al. (2006); Sanfey, et al. (2003), note 9, above.

region of rDLPFC is recruited when participants decide whether or not to punish a partner by rejecting an unfair economic deal proposed by that partner;[39] this result is analogous to our finding that rDLPFC is activated by the decision to punish the perpetrator of a criminal act. Furthermore, while disruptive magnetic stimulation of this region impairs the ability to punish economic norm violations in dyadic exchanges,[40] this manipulation has no effect on norm enforcement behaviour when the unfair economic exchanges are randomly generated by a computer instead of a human agent.[41] This result accords well with our finding that rDLPFC was much less activated when the scenario protagonist was not criminally responsible for his behaviour, and supports the notion that this prefrontal cortex area is primarily recruited when punishment can be assigned to a responsible agent.[42] Finally, we still observed greater rDLPFC activity in the Responsibility condition (compared to Diminished-Responsibility scenarios) when we restricted our analysis to scenarios that only contained physical harms ($p < 0.005$, paired t-test), suggesting that the overlap of rDLPFC activity between studies of economic decision-making and the present examination of legal decision-making is not solely driven by scenarios describing economic transgressions.

The parallels between these previous findings and our current results lead us to suggest that the right DLPFC is strongly activated by the decision to punish norm violations based on an evaluation of the blameworthiness of the transgressor. This proposed function of rDLPFC appears to apply equally to situations where the motive for punishment is unfair behaviour in a dyadic economic exchange or when responding to the violation of an institutionalized social norm in a disinterested third-party context. Of course, confirmation of this hypothesis will require further experimental evidence that legal and economic decision-making (and perhaps moral decision-making as well) rely on the same neural substrates. That said, this apparent overlap illustrates an important point: that the brain regions identified in our study are not specifically devoted to legal decision-making. Rather, a more parsimonious explanation is that third-party punishment decisions draw on elementary and domain-general computations supported by the rDLPFC. In particular, on the basis of the convergence between neural circuitry mediating second-party norm enforcement and impartial third-party punishment, we conjecture that our modern legal system may have evolved by building on pre-existing cognitive mechanisms that support fairness-related behaviours in dyadic interactions. Though speculative and subject to experimental confirmation, this hypothesis is nevertheless consistent with the relatively recent development of state-administered law enforcement institutions, compared to the much longer existence of human

---

[39] Sanfey et al. (2003), note 9, above.
[40] Knoch et al. (2006), note 9 above; Mascha van't Wout, René S. Kahn, Alan G. Sanfey, and André Aleman, 'Repetitive transcranial magnetic stimulation over the right dorsolateral prefrontal cortex affects strategic decision-making' (2005) 16 *Neuroreport* 1849.
[41] Knoch et al. (2006), note 9 above.
[42] Ibid.

cooperation;[43] for thousands of years before the advent of state-implemented norm compliance, humans relied on personal sanctions to enforce social norms.[44]

8.8 Experimental Procedures

8.8.1 Subjects

Sixteen right-handed individuals (eight males, aged eighteen to forty-two) with normal or corrected-to-normal vision participated for financial compensation. The Vanderbilt University Institutional Review Board approved the experimental protocol, and informed consent was obtained from each subject after they were briefed on the nature and possible consequences of the study. A brief psychological survey was also administered to exclude individuals who may react adversely to the content of the criminal scenarios. Exclusion criteria included history of psychiatric illness, being the victim of or having witnessed a violent crime (including sexual abuse), and having experienced any trauma involving injury or threat of injury to the subject or a close friend/family member.

8.8.2 Paradigm

In this experiment, subjects participated in a simulated third-party legal decision-making task in which they determined the appropriate level of punishment for the actions of a fictional protagonist described in short written scenarios. The principal goal of our study was to isolate the neural processes associated with the two fundamental processes of legal decision-making: deciding whether or not an accused individual is culpable for a given criminal act, and determining the appropriate punishment for that act (a parametric process based on the ordinal severity of a crime). Correspondingly, our design manipulated responsibility in a dichotomous fashion and crime severity in a continuous fashion. Each participant viewed fifty scenarios (some inspired by prior behavioural studies of relative blameworthiness[45]) depicting the actions of the protagonist named 'John'. The fifty scenarios were subdivided into three sets (complete scenario list available as Supplementary Methods). In the Responsibility set (N = 20), the scenarios described John intentionally committing a criminal action ranging from simple theft to rape and murder. The Diminished-Responsibility set (N = 20) included similar actions comparable in gravity to those in the Responsibility set, but contained circumstances that would often legally excuse or justify the otherwise

---

[43] Richerson et al. (2003), note 7, above.
[44] Ernst Fehr, Urs Fischbacher, and Simon Gächter, 'Strong reciprocity, human cooperation and the enforcement of social norms' (2002) 13 *Human Nature* 1; Ernst Fehr and Simon Gächter, 'Altruistic punishment in humans' (2002) 415 *Nature* 137.
[45] Robinson and Darley (1995), note 13, above; Paul H. Robinson and Robert Kurzban, 'Concordance and conflict in intuitions of justice' (2007) 91 *University of Minnesota Law Review* 1829.

criminal behaviour of the protagonist. The No-Crime set (n=10) depicted John engaged in non-criminal actions that were otherwise structured similarly to the Responsibility and Diminished-Responsibility scenarios. The No-Crime scenarios were included to assist in interpreting activity differences between Responsibility and Diminished-Responsibility scenarios (see e.g. Figure 2).

Two groups of fifty scenarios were constructed and their presentation counterbalanced across the sixteen participants (eight subjects received group 1 scenarios, and eight others received group 2 scenarios) and across gender (equal numbers of men and women received scenarios from each group). The Responsibility set of group 2 consisted of group 1 Diminished-Responsibility scenarios from which the mitigating circumstances had been excised, while the Diminished-Responsibility set of group 2 consisted of group 1 Responsibility scenarios with mitigating circumstances added. As a result, the Responsibility and Diminished-Responsibility scenarios were counterbalanced across subjects, and differed only by the presence of mitigating circumstances. Thus, exactly the same scenario premises were used in constructing the Responsibility and Non-Responsibility conditions. Finally, the No-Crime set was identical in both groups of scenarios, and all scenario sets were equated for word length.

Participants rated each scenario on a scale from 0–9, according to how much punishment they thought John deserved, with '0' indicating no punishment and '9' indicating extreme punishment. Punishment was defined for participants as 'deserved penalty'. Participants were asked to consider each scenario (and thus, each 'John') independently of the others and were encouraged to use the full scale (0–9) for their ratings. In the scanner but prior to the functional scans, subjects were shown five practice scenarios that were designed to span the punishment scale. Scenarios were presented as white text (Times New Roman) on a black background (14.2 degrees [width] x 9.9 [height] degrees of visual angle). Below each scenario an instruction reminded participants of the task instructions: 'How much punishment do you think John deserves, on a scale from 0 to 9 where 0 = No punishment and 9 = Extreme punishment. By punishment, we mean deserved penalty.' Participants were instructed to make a response as soon as they had reached their decision.

Each trial began with the presentation of a scenario, which remained on screen until participants made a button press response, or up to a maximum of thirty seconds. Participants then viewed a small white fixation square (0.25 degrees of visual angle) for twelve to fourteen seconds (as stimulus onset was synched to scan acquisition [TR = 2s], while stimulus offset was synched to subject response), which was followed by a larger fixation square (0.49 degrees of visual angle) for two seconds prior to the presentation of the next scenario. Ten scenarios (four Responsibility, four Diminished-Responsibility, and two No-Crime)—selected randomly without replacement from the fifty scenarios—were presented in each of the five fMRI runs. Scenario identity and condition order were randomized for each run. The duration of each fMRI run was variable, with a maximum length of 7.33 minutes. The experiment was programmed in Matlab (Mathworks, Natick, MA) using the Psychophysics Toolbox extension and was presented using a Pentium IV PC.[46]

Following the scanning session, participants rated the same scenarios along scales of emotional arousal and valence. They first rated each of the fifty scenarios (presented in random order on a computer screen outside the scanner) on the basis of how emotionally aroused they felt following its presentation (0 = calm, 9 = extremely excited). They then rated each of the scenarios, presented again in random order, on the basis of how positive or negative they felt following its presentation (0 = extremely positive, 9 = extremely negative). In these sessions, subjects rated the same scenarios they viewed in the scanner. The valence data were highly correlated with arousal ratings, and multiple regression analysis demonstrated that they did not account for any additional variance in punishment ratings that is unaccounted for by the arousal data. Therefore, the valence data are not further discussed in this manuscript.

8.8.3 Internal scale questionnaire

In a post-scan debriefing, participants were questioned about the internal scale of punishment they used during the scan. Specifically, participants were asked 'what kind of punishment did you imagine?' for punishment scores of 1, 3, 5, 8 and 9. There was strong agreement among participants about their internal scale of justice. While low punishment scores (1, 3) were generally associated with financial or social penalties, greater punishment scores (5, 8) included incarceration time, with higher scores associated with longer jail times and, at the extreme (9), life imprisonment or state execution.

8.9 Relationship between Punishment Ratings and Legal Statutes

To investigate the relationship between punishment ratings for Responsibility scenarios obtained in the present experiment and an existing, statutorily prescribed punishment for each of the crimes depicted in these scenarios, we coded each Responsibility scenario using the criminal law and criminal procedure statutes of the state of North Carolina. Among those states that have a sentencing statute, North Carolina's is widely considered to be both comprehensive and exemplary.[47]

For each responsibility scenario, we determined the crime(s) (such as larceny, involuntary manslaughter, or murder) with which John might reasonably be charged under the criminal code of North Carolina (2005 General Statutes of North Carolina, Chapter 14). We then determined, for each crime, the authorized

---

[46] David H. Brainard, 'The psychophysics toolbox' (1997) 10 *Spatial Vision* 433; Denis G. Pelli, 'The videotoolbox software for visual psychophysics: transforming numbers into movies' (1997) 10 *Spatial Vision* 437.
[47] Lisa Stansky, 'Breaking up prison gridlock' (1996) 82 *ABA Journal* 70; Ronald F. Wright, 'Counting the cost of sentencing in North Carolina, 1980–2000' in Michael Tonry (ed.), *Crime and Justice: A Review of Research* (Chicago, University of Chicago Press, 2002) 39.

presumptive sentencing range (such as fifty-eight to seventy-three months in prison), assuming no aggravating or mitigating factors that could, under the statute, increase or decrease the authorized sentencing range (2005 General Statutes of North Carolina, Chapter 15A, Article 81). We then calculated and assigned to each scenario the mean for this range, in months. As the distribution of sentence values was highly right-skewed, we log-transformed (natural log) to create a normal distribution of sentence values (we verified that non-transformed data produced similar correlations as transformed data). For scenarios with multiple crimes, the averages for each respective crime were summed (whether this summed value or simply the mean value for the most severe crime depicted in a given scenario was used in the correlation analysis did not significantly affect the results). Where the upper limit of the sentencing range was life in prison, it was coded as twenty-nine years (which has been estimated as the average time likely to be served by lifers newly admitted in 1997).[48] Similarly, where the upper limit of the sentencing range was death, it was also quantified as life in prison (twenty-nine years). The log-transformed mean sentences for each of the twenty scenarios were then correlated with the group-averaged punishment ratings for these scenarios.

8.10 Statistical Analysis

Mean punishment and arousal scores and reaction times were calculated for each subject for each condition (Responsibility, Diminished-Responsibility, and No-Crime) and entered into a repeated-measures Analysis of Variance (ANOVA) using SPSS 15 (SPSS Inc. Chicago, IL) to determine main effects and interactions. Data from sixteen subjects were used for all analyses. Punishment, arousal scores, and reaction times were compared between conditions and post-hoc tests were performed using Fisher's Least Significant Difference (LSD) measure using an alpha level of 0.05. Two-tailed tests were used in all cases. For correlational analyses, data from Responsibility scenarios ($N = 20$) were averaged across all ($N = 16$) subjects. Examination of scatterplots for the correlation of rDLPFC signal and punishment suggested the presence of outliers. As non-parametric correlations tend to be more robust to outliers, we used Spearman's ρ to measure correlations between fMRI signal, behavioural measures, and recommended sentences. All correlations that were significant using Spearman's ρ were also significant ($p < 0.05$) when we employed Pearson's r.

8.10.1 fMRI data acquisition

High resolution 2D and 3D anatomical images were acquired with conventional parameters on a 3T Philips Achieva scanner at the Vanderbilt University Institute of Imaging Science. The visual display was presented on an LCD panel and back-projected onto a screen positioned at the front of the magnet bore. Subjects lay

---

[48] Marc Mauer, Ryan King, and Malcolm C. Young, *The Meaning of 'Life': Long Prison Sentences in Context* (Washington, DC, Sentencing Project, 2004).

supine in the scanner and viewed the display on a mirror positioned above them. Stimulus presentation was synchronized to fMRI volume acquisition. Manual responses were recorded using two five-button keypads (one for each hand; Rowland Institute of Science, Cambridge, MA). Functional (T2* weighted) images were acquired using a gradient-echo echoplanar imaging (EPI) pulse sequence with the following parameters: TR 2000 ms, TE 25 ms, flip angle 70°, FOV 220x220mm, 128x128 matrix with 34 axial slices (3 mm, 0.3 mm gap) oriented parallel to the gyrus rectus. These image parameters produced good T2* signal across the brain except in ventromedial frontal cortex, where some signal dropout was evident in all subjects (Brodmann area 11).

Each of the sixteen participants performed five fMRI runs, except for two participants who could only complete four runs due to technical malfunctions.

8.10.2 fMRI data preprocessing

Image analysis was performed using Brain Voyager QX 1.4 (Brain Innovation, Maastricht, the Netherlands) with custom Matlab software (MathWorks, Natick, MA).

Prior to random effects analysis, images were preprocessed using 3D motion correction, slice timing correction, linear trend removal, and spatial smoothing with a 6mm Gaussian kernel (full width at half maximum). Subjects' functional data were coregistered with their T1-weighted anatomical volumes and transformed into standardized Talairach space.

8.10.3 Responsibility analysis

This analysis was performed to isolate brain regions that were sensitive to responsibility during punishment assessment. Signal values for each fMRI run were transformed into Z-scores representing a change from the signal mean for that run and corrected for serial autocorrelations. Design matrices for each run were constructed by convolving a model hemodynamic response function (double gamma, consisting of a positive γ function and a small, negative γ function reflecting the BOLD undershoot—SPM2, <http://www.fil.ion.ucl.ac.uk/spm>) with regressors specifying volumes acquired during the entire trial (stimulus onset to stimulus offset) for a given condition. These were entered into a general linear model with separate regressors created for each condition per subject (random effects analysis). We then contrasted the beta-weights of regressors using a t-test between conditions to create a statistical parametric map (SPM) showing voxels that demonstrated significantly increased activation in the Responsibility condition compared to the Diminished-Responsibility condition. Predictors for the No-Crime condition were weighted with a zero (i.e. not explicitly modelled). We applied a False-Discovery Rate (FDR) threshold of $q < .05$ (with $(c(V) = ln(V) + E)$) to correct for multiple comparisons. Only activations surviving this corrected threshold are reported.

Volumes of interest (VOIs) were created from the suprathreshold clusters isolated in the above SPM at the conservative FDR threshold. The boundary of these VOIs were drawn from SPMs thresholded using a less conservative implementation of FDR (q <.05, $c(V) = 1$). The signal for each trial (event) included the time course from two TRs (four seconds) before stimulus onset to thirteen TRs (twenty-six seconds) after. Each event's signal was transformed to a percent-signal change (PSC) relative to the average of the first three TRs (0–4 seconds before stimulus onset). Event-related averages (ERAs) were created by averaging these PSC-adjusted event signals; separate ERAs were created for each combination of VOI, condition, and subject. These ERAs were then averaged across subjects for display purposes.

As subjects were instructed to make a response as soon as they had reached a decision about punishment amount, and in keeping with other neuroimaging studies of decision-making,[49] decision-related activity should correspond to the portion of the time course that follows subjects' response. Given that mean RTs hovered around twelve seconds (mean, S.E. for: Responsibility = 12.69s, 0.46; Diminished-Responsibility = 13.76s, 0.46; No-Crime = 11.12s, 0.44) and accounting for a hemodynamic peak rise time of about five seconds post-stimulus,[50] then peri-decision activity should occur approximately seventeen seconds after trial onset, which corresponds well with the time of peak hemodynamic response observed in rDLPFC (see Figure 2). We therefore used the peak hemodynamic response as a measure of decision-related activity. To determine condition effects on BOLD signal within a given brain region, we then contrasted each condition's activation averaged across subjects by using paired t-tests applied on these peak estimates. The peak was experimentally defined as the single volume with maximal signal change from baseline between volumes 1 and 13 (two to twenty-six seconds post stimulus onset). However, we ascertained that the same results were obtained when the peak was defined using a narrower volume range of fourteen to twenty-two seconds post-stimulus (R>DR, p = 0.00070; R>NC, p = 0.00025, DR>NC, p = 0.19), or even when using a single volume sixteen seconds post-stimulus (R>DR, p = 0.00023; R>NC, p = 0.00027, DR>NC, p = 0.84). Thus, our rDLPFC peak activation results are insensitive to the temporal width of the analysis window.

---

[49] Adam R. Aron and Russell A. Poldrack, 'Cortical and subcortical contributions to Stop signal response inhibition: role of the subthalamic nucleus' (2006) 26 *Journal of Neuroscience* 2424. Giorgio Coricelli, Hugo D Critchley, Mateus Joffily, John P. O'Doherty, Angela Sirigu, and Raymond J. Dolan, 'Regret and its avoidance: a neuroimaging study of choice behavior' (2005) 8 *Nature Neuroscience* 1255; Paul E. Dux, Jason Ivanoff, Christopher L. Asplund, and René Marois, 'Isolation of a central bottleneck of information processing with time-resolved FMRI' (2006) 52 *Neuron* 1109; Jason Ivanoff, Philip Branning, and René Marois, 'fMRI evidence for a dual process account of the speed-accuracy tradeoff in decision-making' (2008) 3 *PLoS ONE* e2635; Benjamin Rahm, Klaus Opwis, Christoph P. Kaller, Joachim Spreer, Ralf Schwarzwald, Erich Seifritz, Ulrike Halsband, and Josef M. Unterrainer, 'Tracking the subprocesses of decision-based action in the human frontal lobes' (2006) 30 *Neuroimage* 656.

[50] Geoffrey M. Boynton, Stephen A. Engel, Gary H. Glover, and David J. Heeger, 'Linear systems analysis of functional magnetic resonance imaging in human V1' (1996) 16 *Journal of Neuroscience* 4207; Karl J. Friston, Peter Jezzard, and Robert Turner, 'Analysis of functional MRI time-series' (1994) 1 *Human Brain Mapping* 153. David J. Heeger and David Ress, 'What does fMRI tell us about neuronal activity?' (2002) 3 *Nature Reviews* 142.

8.10.3.1 Arousal- and reaction-time equated analyses

To determine whether activation differences between the Responsibility and Diminished-Responsibility conditions were driven by punishment assessment rather than any differences in arousal, these two conditions were compared after equating for arousal ratings. This was accomplished by deleting the six trials with the highest arousal ratings from the Responsibility condition for each subject. Time courses were extracted and peak differences were compared as above.

We also determined whether reaction time differences between the Responsibility, Diminished-Responsibility and No-Crime conditions affected the brain activation results by comparing these conditions after equating for response times. This was accomplished by deleting, for each subject, the trials with the highest reaction times for Diminished-Responsibility scenarios and the trials with the lowest reaction times for the No-Crime scenarios until the RTs across conditions (for each subject) were approximately equal ($p > 0.1$ for all paired t-tests between conditions). In addition, we compared rDLPFC activation between Responsibility and Diminished-Responsibility scenarios controlling for reaction time by performing a GLM analysis of covariance (ANCOVA) using the extracted rDLPFC BOLD signal and punishment reaction times for each Responsibility and Diminished-Responsibility scenario averaged across subjects.

8.10.3.2 Dissociation of activation peak and deactivation dip

To assess the relationship between early (~8s) deactivation in the rDLPFC time-course and later (~16s) peak activation, we calculated peak and 'dip' values for the Responsibility and Diminished-Responsibility conditions from each subject's ERA. 'Peak' and 'dip' were defined, respectively, as the volume with the maximal positive and maximal negative change from baseline. For each subject, we subtracted the Diminished-Responsibility peak value from the Responsibility peak value, and the Diminished-Responsibility dip value from the Responsibility dip value. Per-subject peak and dip difference values were then correlated via Spearman bivariate correlation in SPSS 15.

8.10.3.3 Laterality analyses

To confirm the lateral specificity of Responsibility-related activation in right DLPFC, we extracted BOLD signal from the corresponding left DLPFC volume of interest (i.e. 'x-mirrored' VOI, centered on talairach coordinate $-39, 37, 22$). We performed a two-way ANOVA with 'Condition' (Responsibility, Diminished-Responsibility and No-Crime) and 'Side' (Left and Right) as independent variables and BOLD signal as the dependent variable. Post-hoc comparisons between conditions in each hemisphere, and between hemispheres for the Responsibility condition, were performed using paired t-tests.

8.11 Punishment Rating Analysis

To identify brain regions that tracked the degree of punishment subjects assigned to a scenario, we performed a median split for punishment scores given during Responsibility scenarios. Based on the median punishment value for each scenario in the Responsibility condition across subjects, scenarios were separated into two groups, high and low. Design matrices and GLMs were constructed as above, with predictors for high and low scores for each subject specifying volumes acquired during Responsibility trials on which a high or low punishment score was given, respectively. We contrasted the beta-weights of these predictors using a t-test between high and low punishments to create an SPM showing voxels that demonstrated significantly increased activation during Responsibility trials in which subjects gave high (at or above the median) punishments compared to Responsibility trials in which subjects gave low (below the median) punishments. We applied a threshold of $q < 0.05$ False-Discovery Rate (FDR) to correct for multiple comparisons. Using a conservative implementation of the FDR correction technique ($c(V) = ln(V) + E$), we did not find significant activation differences. We report activations significant at FDR $q < 0.05$, using a less conservative implementation of FDR ($c(V) = 1$). The differences between the two implementations relate to assumptions about the independence of tests being performed on the data; both are valid for controlling multiple testing in functional imaging data.[51]

VOIs were created as described for the Responsibility analysis. The extracted peak activation values were used for a correlation analysis between punishment rating and BOLD response. Specifically, for each of the twenty Responsibility scenarios, the peak amplitude of the group-averaged ERA was computed, and the resulting value was correlated with the corresponding group-averaged punishment rating for that scenario. These peak values were also used in the between-condition difference score analyses.

8.12 Acknowledgements

This research was supported by grants from the John D. and Catherine T. MacArthur Foundation Law and Neuroscience Project, the Vanderbilt University Central Discovery Grant Program, the Vanderbilt Law and Human Behavior Program, and the Cecil D. Branstetter Litigation and Dispute Resolution Program of Vanderbilt University. The authors wish to thank Martha Presley for providing valuable background research and Jeffrey Schall, Nita Farahany, Terry Maroney, Michael Treadway, Eyal Aharoni, Terry Chorvat, and Walter Sinott-Armstrong for useful comments.

---

[51] Christopher R. Genovese, Nicole A. Lazar, and Thomas Nichols, 'Thresholding of statistical maps in functional neuroimaging using the false discovery rate' (2002) 15 *NeuroImage* 870.

9

Law, Neuroscience, and Criminal Culpability

*Lisa Claydon**

This paper examines the relationship between explanations of human behaviour drawn from neuroscientific imaging and the approach adopted by the criminal law to establishing the legal framework for evaluating guilt or innocence. It questions whether purely scientific approaches provide a sufficient understanding of what it means to be guilty or innocent of a crime. It will consider what pictures of brain states may tell us about culpability and concludes by considering whether the present philosophical basis of the criminal law is sufficiently robust to withstand some claims that are being made about what neuroscience tells us about brain states.

One of the most disturbing truths about neuroscience is that it has the capacity to change profoundly the manner in which human identity is perceived. There is also concern that neuroscientific evidence may be harnessed and used in evidence, or used to determine disposals or treatment of offenders, in a manner that may stretch the science. This is certainly true of the areas of research that relate to human mental or behavioural disorders.[1] Neuroscience and related technologies seem to proffer new explanations of who we are and what we might become. The ability to understand and perhaps alter or modify behaviour brings with it responsibilities. Susan Greenfield talks of the importance of human identity: our identity *is* 'our brains, or rather our minds'.[2] This paper examines how the use of imagery of the brain may or may not shed light on some of the key questions of responsibility that underpin criminal law judgments.

Speculation as to what the implications of advances in neuroscience may tell us about human culpability and how this could play out in courtrooms has been the subject of wider discussion in the US than in the UK. In January 2009, in an article in the science section of the *Wall Street Journal*,[3] Robert Lee Holtz speculated on the possibility of being able to analyse a criminal defendant's case using neuroscientific

---

* Principal Lecturer in Law, Department of Law, Bristol Law School, University of the West of England.
[1] See Ilinia Singh and Nikolas Rose, 'Biomarkers in Psychiatry' (2009) 460 *Nature*, 202.
[2] Susan Greenfield in *Id: The Quest for Meaning in the 21st Century* (London, Sceptre, 2008) x. Greenfield talks about 'the dynamism, the "plasticity", of the human brain and the wonderful benefits and terrifying threats that that malleability brings.' Emphasis in original.
[3] (2009) *Wall Street Journal*, 15 January, A7.

evidence. He talked about the possibility of using data obtained from neuroscientific testing to evaluate the ability of particular defendants to recall events, or to assist the jury in determining whether defendants were sane at the time of committing the crime. He also speculated as to whether further research that helped to develop a greater understanding of the functioning of the brain could help the jury when attributing criminal responsibility.

However, so far as is known,[4] the English courts have been reluctant to distinguish the treatment of neuroscientific evidence from other types of evidence utilized in court. Perhaps the most disturbing reports of the use of neuroscientific imaging in court have come from India. The case concerned the murder of a man called Udit Bharati. The accused was his 'former fiancee', Aditi Sharma. The evidence used to convict her of the poisoning of Udit was a brain electrical oscillation signature (BEOS) test. The claims made for this test were that it provided a brain fingerprinting technique. Aditi Sharma had agreed to take the test. Further claims made in relation to the test are that it enables the person interpreting the data to distinguish normal brain activity from the stimulation of memory. Aditi was read passages of description during the test. Some of these passages described the police view of what she had done to poison Udit, others were described as 'value neutral' in terms of the test and included statements such as 'the sky is blue'.[5] The evidence produced by the BEOS test is reported as having helped convict Aditi of the killing. The *New York Times*, commenting on the case, said:

> The technologies, generally regarded as promising but unproved, have yet to be widely accepted as evidence—except in India, where in recent years judges have begun to admit brain scans. But it was only in June, in a murder case in Pune, in Maharashtra State, that a judge explicitly cited a scan as proof that the suspect's brain held 'experiential knowledge' about the crime that only the killer could possess, sentencing her to life in prison.[6]

It is to say the least debatable whether the evidence produced to convict Aditi Sharma had any probative value at all. The case raises interesting issues with regard to developing interdisciplinary understandings of the meaning to be attributed to neuroscientific evidence and how such evidence should be utilized in court. The disciplinary clashes that distinguish approaches to matters of proof in the criminal law and in other disciplines provide at least three significant impediments to the integration of neuroscientific and legal approaches to arguments about proof of legal responsibility for actions.

Firstly, the definitions of *mens rea* or other terms that define criminal liability are very specialist. The fact that the accused's conduct falls within these definitions has to be proved by the prosecution to establish to the satisfaction of a jury guilt or

---

[4] As it is only possible to look at decisions from recorded cases then most of the information on which this paper rests is drawn from the reported case law of the appeal courts.

[5] <http://reason.com/blog/2008/09/15/brain-scans-used-to-convict-wo> (accessed 5 November 2009).

[6] Anand Giridharadas, 'India's use of brain scans in courts dismays critics' (2008) *New York Times*, 15 September.

innocence. This will, where appropriate, necessitate the use of expert evidence to establish the requisite proof. However, these specialist definitions often do not accord with ordinary, scientific, or philosophical meanings of the words employed by the criminal law. These definitions are often said to be normative, serving a declaratory purpose in highlighting the types of behaviour that deserve the most serious punishment.

A second difference is the separation of acts and the circumstances in which they take place from the mental attitude that accompanies the act. Where a subjective mental element forms part of the definition of criminal culpability this introduces into the law a mild form of dualism. It has served the criminal law well to separate the prohibited circumstances of a crime from the accompanying mental attitude. Such separation enables detailed examination of issues, for example whether the description of the accused's action falls within the ambit of the behaviour proscribed by law. If it does not, then she cannot be guilty of the crime. Perhaps this can be most clearly seen in the specialist meanings given to causation in the criminal law, for example in the early case of a man trying to poison his mother. The mother died of a heart attack before she had consumed sufficient of the poison to kill her. The son could not be guilty because he could not be said to have caused the prohibited result, the unlawful killing.[7] In twenty-first-century legal reasoning there is little to suggest that the criminal law is dualist in the Cartesian sense of the word. The law merely gives a specialist interpretation to the reporting of mental states. Naturally such a manipulation of the actual circumstance surrounding a crime does have implications for determinations of guilt and innocence, which will be considered later in this paper.

A third complicating factor is the manner in which cases are handled. The criminal law requires the jury finds that the factual evidence establishes a defendant's guilt beyond reasonable doubt. A further layer of complexity exists because the roles of judge and jury are separate and clearly defined. The judge advises on the application of the law to material facts. The jury determines guilt and innocence on the facts. Then there is another complicating factor in relation to neuroscientific evidence in the form of brain images. This evidence is highly visual and claims to have scientific validity. Images are powerful and likely to affect a jury's reaction to the evidence. They need careful handling and the claims made in respect of these images will need to be measured and balanced against the other evidence at trial or on appeal.

This area is also bedevilled by the public and media interest in criminal justice and science. This may lead to distorted claims being made about what science is able to add to notions of blameworthiness. How the evidence is presented in court and how it is treated by the judge becomes a matter of primary importance. This raises a separate but linked issue that needs to be addressed. This is how different disciplines that contain quite complex systems for arguing issues and balancing and distilling the results of deliberations should proceed when seeking to work together.

[7] *White* [1910] 2 KB 124.

You could add a further question which is: why would they want to work together when it may actually be in both their interests to maintain distinct and separate approaches that are suitable to the legal or medical issues with which they deal? All of these questions will be considered in this paper but some will be explored more thoroughly than others.

9.1 Mind, Brain, and Culpability

There is no doubt that scientific approaches, particularly to the examination of the brain and the functioning and workings of consciousness, are of relevance to the criminal law. There are a number of reasons for this but perhaps the main reason is that scientific experiments are helpful in establishing how brains work. They may also assist us in establishing just how much freedom of choice we have in relation to our actions and finally enable us to dismiss the notion that all our actions are determined and allow a more nuanced scientific explanation of the causes of action. However, much scientific work will simply be to establish correlative connections, for example those that link images in the brain with diseases. There is of course a genuine scientific interest in how brains work, coupled with a genuine excitement about what science may reveal. This area of study has broad intellectual appeal and is perhaps typified by the comments of Steven Rose speaking on the *Today* programme on Radio 4 in 2009.

It's clear that we cannot understand the brain of any of us, or certainly the minds of any of us, without putting neuroscience together in a much broader context than many of us are able to think about.[8]

The serious question to be answered by those who administer and amend the criminal law is how relevant this research is to their enterprise; and how much is it a speculative scientific venture? Furthermore, how much do we need to understand the brain of any defendant when making determinations of guilt or innocence? What is neuroscience, or the philosophy that draws upon the science, likely to add to our present understanding of the criminal law?

9.2 Areas of Genuine Difference

To return to the issue of culpability it has always seemed unlikely that neuroscience will assist much in terms of the definition of *mens rea* states. Descriptions of *mens rea* are largely normative and tend to describe behaviour in terms where the behaviour is separated into elements of culpability. So in the case of intention, which could be viewed as the most blameworthy mental state, the criminal law demands that the defendant intended his action in the sense that it was his aim or

---

[8] *Today*, 7 January 2009.

purpose to bring it about. Alternatively the criminal law states that the jury may find from the evidence that the there was foresight of virtual certainty that the defendant's act would bring about the criminal result or circumstances; and, if it does so, this finding will be sufficient to establish the *mens rea* requirements of the most serious criminal law offences such as murder. The further proviso is that the jury must also find that the defendant knew that the act was virtually certain to bring about the prohibited result.[9] This second type of intention is often referred to by lawyers as oblique intention.

As was pointed out many years ago by Lord Ackner before he became a Law Lord:

There is no scientific measurement or yardstick for gauging a person's intention. Unfortunately, there is no form of meter which one can fix to an accused person, like an amp meter or something of that kind, in order to ascertain what the intention is, no X-ray machine which will produce a useful picture.[10]

Nor, is there ever likely to be such a measure as the brain snapshot would have to be taken at the precise moment of the crime. However, there is no doubt that the criminal law, in its evaluation of the mental state of a murderer for example, could be viewed by some as indulging a type of voyeuristic epiphenomenalism. This seems a harsh criticism but the criminal law is certainly accepting a form of dualism in that the intention to act is something above and beyond brain processes causing action. The definition of intention does not equate to philosophical descriptions of action that commonly refer to things that must happen as a result of our actions as being intentional. Indeed for Jennifer Hornsby the word 'intentionally' performs a central role in defining action. 'An *action* is a person's doing something *intentionally*.'[11] This definition of the word 'intentionally' in the discourse of philosophy is fairly uncontroversial but in criminal law it would be unacceptable. Arguably criminal law has its own philosophical bases and those bases are highly contested. In terms of how legal reasoning is applied, perhaps the best description is that of Alan Norrie, who considers why the normative judgement reached by the law has driven 'a wedge between legal and moral judgement'.[12] Norrie explores the issue of the legal meaning of intention in relation to a woman who swims naked. He suggests that she could argue she was entitled 'so to swim'. The law however might take a different view. He explains the bifurcation between law and morality clearly, arguing 'the law may hold the conduct is intentionally disgusting'. Norrie goes on to explain that in his view: 'the law forecloses the issue of normative disagreement

---

[9] *R v Woollin* [1998] 3 WLR 382 (HL).
[10] Extract from Ackner J's summing up to the jury in *Hyam* No. 6530, C. 72, Warwick Crown Court; tried 22–24 November 1972.
[11] Jennifer Hornsby, 'On what's intentionally done' in Stephen Shute, John Gardner, and Jeremy Horder (eds.), *Action and Value in the Criminal Law* (Oxford, Oxford University Press, 1993) 55, emphasis in original.
[12] Alan Norrie, *Law and the Beautiful Soul* (London, Glasshouse, 2005) 66.

by insisting that the oblique intention is present whatever the individual's own normative view'.[13]

At first sight this may seem similar to the claims made by scientists such as Colin Blakemore, that the need for distinguishing intentional actions from other sorts of actions is unnecessary:

> All our actions are the product of the activity of our brains. It seems to me to make no sense (in scientific terms) to try to distinguish sharply between acts that result from conscious intention and those that are pure reflexes or that are caused by disease or damage to the brain. We *feel* ourselves, usually, to be in control of our actions, but that feeling is in itself a product of the brain, whose machinery has been designed, on the basis of its functional utility, by means of natural selection.[14]

Blakemore goes on to emphasize the distinctions between medical science and legal reasoning and to underline the point that the problem arises when an attempt is made to 'expect' the two systems to use the 'same language'. He makes the further point that scientific descriptions of causal influences on behaviour have little to do with attributions of legal responsibility.[15]

Undoubtedly for most criminal lawyers attributions of legal responsibility for actions will form a significant part of their professional practice. Identifying who is responsible and establishing the basis of the responsibility is critical in much of the work undertaken in magistrates' and higher courts. Tied to this need is a related theoretical imperative: the imperative to protect an individual from the spurious imposition of blame. Thus the edges of responsibility in law need to be clearly defined. One means of doing this is to utilize a concept of individual autonomy. In the criminal law this underpins and supports the conception of the distinction between acts that are culpable and acts that are not. The legal philosophical basis of this claim was fairly succinctly expressed by Joseph Raz.[16] It has also been stated explicitly by the House of Lords in a case specifically dealing with mental responsibility for actions.[17] Lord Bingham, in his opinion, makes the following statement: 'conviction of a serious crime should depend on proof not simply that the defendant caused (by an act or omission) an injurious result to another but that his state of mind when so acting was culpable.'

---

[13] Ibid. all quotations. This discussion relates to hypothetical indecency offences and not to murder highlighting that the criminal law has more than one meaning of intention, meanings may vary from offence to offence.

[14] Colin J. Blakemore, *The Mind Machine* (London, BBC Books, 1998) 270, emphasis in original.

[15] Ibid.

[16] *The Morality of Freedom,* (Oxford, Oxford University Press, 1986) 425. Raz defines autonomy thus: 'Three main features characterise the autonomy-based doctrine of freedom. *First,* its primary concern is the promotion and protection of positive freedom which is understood as the capacity for autonomy, consisting of the availability of an adequate range of options, and of the mental abilities necessary for an autonomous life. *Second,* the state has a duty not merely to prevent the denial of freedom, but also to promote it by creating conditions of autonomy. *Third,* one may not pursue any goal by means which infringe people's autonomy unless such action is justified by the need to protect or promote the autonomy of those people or others.'

[17] *R v G and another (Appellants)* [2003] UKHL 50.

Without doubt central to the rhetoric of the criminal law are notions of justice and fairness. Thus evaluations of mental states that accompany actions provide for criminal lawyers and criminal case juries an additional element to add to the depth of reasoning concerning guilt or innocence. This is identified by Lord Bingham as culpability for the potentially criminal act or omission. Lord Bingham goes on to put a little flesh on the bones of the criminal law requirement that the defendant should be culpable in terms of the *mens rea* element of recklessness: 'It is clearly blameworthy to take an obvious and significant risk of causing injury to another. But it is not clearly blameworthy to do something involving a risk of injury to another if... one genuinely does not perceive the risk.' He explains why this is the case quite clearly: 'such a person may fairly be accused of stupidity or lack of imagination, but neither of these failings should expose him to conviction of serious crime or the risk of punishment'.[18]

Now it is debatable whether Lord Bingham's view of criminal culpability is reflected throughout the criminal law in terms of his assertion that neither 'stupidity or lack of imagination should expose him to conviction of a serious crime'.[19] But what is not debatable is that many of the words he uses, such as 'obvious and significant risk' seem to be related to blame. It is because they are so related to culpability that the meaning of seemingly ordinary words have been the subject of appeals and there is a body of case law that has determined specialist meanings for words that normally have other meanings. No scientific measurement will make clear the distinction between the stupid or unimaginative person's acts. They may illuminate damage to the faculties that permit reasoning, but sheer brute stupidity or insensitivity is unlikely to be captured. To make matters even more complex, the law within its normative framework is careful in its consideration of when excuses may enter the process of judgment. To return to the argument put by Norrie, the woman's view of her act of swimming naked is not relevant to her liability if the law views her actions as indecent and the offence has not been formulated to permit her mental view of her action to be taken into account. However, if the law were to take the approach set out in Norrie's hypothetical example, some psychological explanation of her view might mitigate any punishment meted out by a court.

Norrie continues: 'The issue of judgment is settled by the court, according to its view of what is acceptable.' Norrie argues that the admission of issues of justification and excuse may only be heard after the broader, normative, determinations of *actus reus* and *mens rea* have been considered.[20] Thus at least in relation to determinations of whether the defendant may have an excuse, the scientific evidence in court is likely to emerge once some of the main arguments as to whether this defendant could be criminally responsible have already been heard. This means that scientific evidence is likely to be considered as supporting evidence

---

[18] Ibid. para. 32 all quotations.
[19] The crime of manslaughter by gross negligence is an area where arguably Lord Bingham's argument may not run, being constrained to some extent by an analysis of whether the defendant's actions put someone at risk of serious harm—see for example *R v Yaqqob* [2005] EWCA Crim 2375.
[20] Note 12, above, 66.

that the defendant's behaviour should in some manner be excused, or alternatively, and more rarely to support an assertion by the prosecution that issues of *mens rea* are irrelevant because the defendant is insane. Scientific evidence may also be utilized to mitigate responsibility in terms of sentencing, or by the defence to argue that a defendant's responsibility is negated,[21] or that responsibility for the offence is partially removed.[22]

To conclude this section of the argument, the fundamental point being made here is that normative frameworks exist within the criminal law. Within these frameworks evidence is admitted that may determine guilt or innocence. These frameworks are complex, they are the foundations upon which criminal liability rests. These foundations reflect a view of how behaviour should be judged, which has been specifically developed by and within the criminal law. It will be useful to consider the way that neuroscientific evidence has been viewed by the appeal courts to see how it has been applied in evaluations of guilt or innocence.

9.3 The Use of Neuroscientific Evidence in Court

One of the pertinent questions this paper aims to address is what neuroscientific analysis can bring to the assessment of culpability in a criminal law court. Neuroscientific evidence has been utilized recently in the appeal courts in relation to the partial defence of diminished responsibility, sentencing and the reliability of witness testimony. There is no doubt that the use of magnetic resonance imaging (MRI) scans is increasing particularly where a plea by a defendant rests on claims of brain damage.[23] Before examining recent case law in this area it is perhaps as well to evaluate the offence definition of murder and the defence requirements for one of the partial defences to murder. This offence and potential partial defence were at issue in the first case to be considered, *Hill v R*.[24] The *mens rea* requirement for murder is that the defendant has the intention to kill or cause grievous bodily harm to his victim. The *actus reus* requirement is the unlawful killing of a person in being under the Queen's peace.[25] If both elements are established the defendant may raise the partial defence of diminished responsibility.[26] Section 2 of the Homicide Act

---

[21] In the case of the defence of insanity successful proof of insanity at the time of the act will mean a verdict of not guilty by reason of insanity (NGRI).
[22] In the case of partial defences to murder—such as the partial defence of diminished responsibility.
[23] See *R v Hanson* [2005] EWCA 1142. MRI scan evidence was submitted to allow consideration of the argument that the failure to disclose neuroscientific evidence at the time of trial had prejudiced the defendant's plea of diminished responsibility.
[24] [2008] EWCA Crim 76.
[25] Coke 3 Inst. 47.
[26] Homicide Act 1957, s. 2: '(1) Where a person kills or is a party to the killing of another, he shall not be convicted of murder if he was suffering from such abnormality of the mind (whether arising from a condition of arrested or retarded development of mind or any inherent causes or induced by disease or injury) as substantially impaired his mental responsibility for his acts and omissions in doing or being a party to the killing...'.

1957[27] defined the partial defence that reduced murder to manslaughter where the plea was successful. A manslaughter verdict permits the defendant to escape the mandatory sentence for murder and, where appropriate, a range of disposal options are available to the court. The partial defence required evidence of abnormality of the mind, which had been widely defined by case law.[28] A plea of diminished responsibility under the 1957 Act had to be based on fairly specific causes and medical evidence was generally required to establish that the abnormality has a legally accepted cause. However, medical evidence was not determinative. The whole of the plea was to be viewed by the jury against the broader background of the circumstances surrounding the death.[29]

In *Hill* neuroscientific evidence was used, at trial, to assess the effect of a major traffic accident on a defendant who was charged with murder.[30] Although there was a great deal of evidence that might have mitigated the appellant's culpability for the act—Hill was convicted of murder. Hill appealed against the verdict. The Court of Appeal considered further evidence that might support a defence of diminished responsibility at the time of the killing. The court considered electroencephalogram (EEG) evidence that had been taken in the years immediately following the accident in 1971; and Hill's mother's statement that he had become more aggressive following the accident. Fourteen years after the murder new and improved brain scan evidence was also presented to the court.

There was a clear purpose to the new neuroimaging evidence in the form of an MRI scan. The aim of using the MRI image to support the claim was that the damage had occurred to the *corpus callum*. According to the case report the 'damage includes the frontal lobes of the brain, which are those particularly relevant to (inter alia) impulse control, planning, and sequencing'.[31] However, differences of opinion as to the relevance of the neuroscientific evidence to the plea of diminished responsibility existed between prosecution and defence experts. Furthermore, there was a disagreement as to how additional neuroscientific evidence should be interpreted: 'the experienced neuropsychologist tester is of the opinion that these

---

[27] This will shortly be replaced by a new partial defence which has different requirements: Coroners and Justice Act 2009; see note 108, below.

[28] *R v Byrne* [1960] 2 QB 396 (CA): '"Abnormality of mind"... means a state of mind so different from that of ordinary human beings that the reasonable man would term it abnormal. It appears to us to be wide enough to cover the mind's activities in all its aspects, not only the perception of physical acts and matters, and the ability to form a rational judgment as to whether an act is right or wrong, but also the ability to exercise will-power to control physical acts in accordance with that rational judgment.'

[29] *R v Sanders* (1991) 93 Cr App R 245 (CA), per Lord Parker CJ: 'We are also satisfied that the judge was not called upon to go further than he did with regard to the medical evidence, that is to say beyond reminding them, as he did, that it was so to speak all one way, that it was in the purely medical sense that it did not stand alone for the jury's consideration of the appellant's state of mind... Regardless of that the jury had to bear in mind, among other matters, the manner of the killing, the contents of the will and the letters, when the last letter was written and certain admissions made by the appellant to the police in interview.'

[30] [2008] EWCA Crim 76 para. 24: 'The appellant had had a serious road accident in April 1971. He had been unconscious for a prolonged period of about 18 days thereafter.'

[31] Ibid. para. 26.

results show abilities in the low to average range without any significant cognitive impairment, Dr Fenwick disagrees; his view is that they do demonstrate significant cognitive disability'.[32] The Court of Appeal was prepared to proceed on the basis that Dr Fenwick's interpretation 'might be right'. However, when reviewing the rest of Fenwick's evidence in relation to the link between the impairment and the defendant's sleep patterns they concluded that this link could not be made out on the facts of the case.

The evidence produced by the defence team in this case, which was referred to the Court of Appeal by the Criminal Cases Review Commission (CCRC), suggested that Hill had killed following being woken from sleep, the argument being that because of his injuries his ability to control impulsive behaviour would be weakened by being suddenly aroused from sleep. This for the Court of Appeal was one assertion too far. The possibility of cognitive impairment was accepted. However, the court took the view that the other available evidence against the defendant was such that it was not possible to view the impairment as establishing the excusing conditions required for a successful plea of diminished responsibility. Indeed the court switched its attention from the argument that the defendant had killed whilst in a slightly confused state having been awoken from slow wave sleep and focused on the defendant's previous convictions for 'persistent fraud'. In the eyes of the court what emerged from this evaluation reduced the credibility of Dr Fenwick's assertion that being awoken from slow wave sleep 'with existing organic damage to the frontal lobes would significantly reduce... [Hill's] responsibility, indeed his culpability, for the killing'.[33] The main direction of Hughes LJ's reasoning[34] was to discount the elements of Dr Fenwick's argument in respect of a series of previous fraud offences. These fraud offences, in the opinion of the Court of Appeal, showed preplanning in a manner that made the evidence that the brain damage affected planning less credible.

There is in the judgment clear judicial commentary with regard to the consequences of accepting Dr Fenwick's argument that it 'was highly likely that a record for persistent fraud... would be attributable to brain damage'. If this proposition were to be accepted there could be clear policy implications for the prosecution of those who committed persistent frauds. It is not so much the neuroscientific evidence that Hughes LJ rejects as the implications being read into such evidence. He states:

To the extent that that involved the implicit assertion that a large proportion of persistent fraudsters suffer from a condition which would amount to a substantial impairment of their responsibility for their crimes, we cannot avoid saying that this is very surprising.[35]

---

[32] Ibid. all quotations.
[33] Ibid. para. 27.
[34] The judges hearing the case in the Court of Appeal were Hughes LJ who delivered the judgment, Saunders J and Sir Christopher Holland.
[35] Ibid. para. 28.

It is clear from this and the treatment of MRI scan evidence in the case of *R v McCann*[36] that the Court of Appeal sees neuroscientific evidence as having a value as one form of evidence submitted in court. However, there is a strong assertion that the court will evaluate such evidence by viewing it in the context of the existing legal framework for establishing culpability. Specific reference is made in both *McCann* and *Hill* to the importance of the jury's interpretation of all the circumstances surrounding the act. In *Hill*, referring to the need for substantial impairment of responsibility for the killing, the Court of Appeal said that such matters were always 'a broad question of fact and degree'. The court went on to say of the partial defence of diminished responsibility: 'It is not determined by the medical evidence, although there are of course some clear cases where the medical evidence is such no one could doubt the answer. It is determined at trial by the jury.'[37]

This point is further supported by the use of scientific evidence in two appeals made by Charles Hanson. The first appeal, heard by the Court of Appeal,[38] was based in part around the Crown's failure to disclose some of the expert neuroscientific evidence at Hanson's trial for the murder of his wife. The excluded evidence was that of one medical expert. Five doctors actually gave evidence at trial with regard to the claim of diminished responsibility.[39] However, the expert whose evidence was excluded had reviewed the MRI scans taken of Hanson's brain in 1994 and 1996 and had concluded that there was a possibility that in the 1996 scan there could be evidence of temporal lobe epilepsy. This evidence was not disclosed at Hanson's trial. The Court of Appeal case report quotes this expert's evidence stating: 'he could not accept that the 1996 scan was normal'.[40] All of the experts who were asked to give evidence at trial had agreed that there was evidence of abnormality of mind. According to the report: 'There was no dispute between them that the appellant suffered from abnormality of mind which impaired his responsibility for the killing to some extent. The issue was to what extent: in other words was there substantial impairment?'[41]

Before the Court of Appeal the issue revolved around the effect of the exclusion of the medical evidence that suggested that there might be evidence of temporal lobe epilepsy. The doctor whose evidence had been excluded was asked to reconsider the scans and a third scan carried out in 1996. He concluded that the MRI scans could indicate temporal lobe abnormality but that 'they need to be treated with extreme caution and cannot be taken as clear indication of temporal lobe damage'.[42] The issue of what MRI scans are or are not able to determine will be considered later in this paper.

---

[36] Information taken from a transcript of a Court of Appeal hearing App. no. 199902101/W2 heard on 28 November 2000.
[37] *Hill v R* [2008] EWCA Crim 76 para. 28.
[38] *Hanson v R* [2005] EWCA Crim 1142.
[39] Again this plea was based upon the claim as outlined in s. 2 the Homicide Act 1957. The law in force at the time of Hanson's trial.
[40] Para. 8.
[41] Para. 4.
[42] Para. 11.

Acting on behalf of Charles Hanson, Dr Peter Fenwick suggested further MRI scans and tests in view of the improved detection abilities of the new neuroimaging techniques. In the light of this new evidence he concluded that the further scans and tests 'point towards a severe brain injury in the past'. However, the scan revealed nothing of relevance to the temporal lobes. Fenwick admitted that 'no clear evidence of residual brain damage can be seen in the MRI scan'.[43] Hanson dismissed his legal team and pursued the appeal himself, arguing that the failure to disclose the evidence of possible brain damage at his trial in 1996 undermined the fairness of his conviction. The court rejected his argument saying that any original unfairness was cured by the defendant being allowed the right of appeal. His conviction was upheld.

Undeterred Hanson continued to pursue his case using the neuroscientific evidence in part to support his application for the determination by the High Court of the minimum term he should serve before the parole board could review his circumstances.[44] The issues which were of relevance to the appeal under Sch. 21 of the Criminal Justice Act 2003 were, inter alia, the issue of the following mitigating factors: 'the fact that the offender suffered from any mental disorder or mental disability which although not falling within section 2(1) of the Homicide Act 1957, lowered his degree of culpability'.[45] The case report does not give lengthy consideration to the neuroscientific evidence, merely noting that when Hanson appealed against his conviction the Court of Appeal determined that the new neuroscientific evidence 'did not affect the safety of the conviction'.[46] In the High Court it is noted that, in respect of Sch. 21, the matters listed that go to mitigation are 'only illustrative' rather than definitive.[47] What is interesting is that in the view of the High Court judge[48] the defendant's preplanning and his use of a knife had to be weighed against evidence of his mental condition. Unsurprisingly, this suggests that evidence of preplanning and the implications that it carries with it of risk to the public and the policy of public protection is likely to be at the forefront of a judge's mind when reviewing an application of this type.

9.4 How Might Neuroscientific Evidence Assist a Court in its Determinations of Guilt or Innocence in the Future?

R v McCann[49] was a case referred to the Court of Appeal by the CCRC which examined the possibility of false accusations by a witness at trial. McCann had been convicted of murder and robbery. The argument advanced was that one of the key

---

[43] Para. 12.
[44] *Re Hanson* [2006] EWHC 284 (Admin).
[45] For further details see Sch. 21 Criminal Justice Act 2003, which lists the aggravating and mitigating factors to be borne in mind.
[46] *Re Hanson* [2006] EWHC 284 (Admin) para. 10.
[47] Para. 19.
[48] Beatson J.
[49] [2009] EWCA Crim 1098.

witnesses in the case was unreliable, due to organic brain disease caused by alcoholism. In this case the argument was that the failure to reveal medical evidence that would have provided evidence to the jury of the witness's lack of credibility had assisted the Crown case in a manner which was detrimental to McCann. It was argued that where a witness was illiterate then a brain scan might provide the best means of obtaining clear evidence of Korsakoff's syndrome. Such evidence would have supported the appellant's assertion that the witness's evidence was unreliable. The statement made in this case by one of the expert witnesses is interesting. He said he would have liked evidence of a computerized tomography (CT) or MRI scan to help in the assessment of the witness's mental condition and her reliability as a witness.[50] The witness was dead by the time of the appeal, but a scan, taken while she was alive, would have provided a continuing visual representation of her brain. This raises interesting issues with regard to just how evidence from brain scans may come to be used in the future by the criminal justice system of England and Wales.

Evidence relating to brain states has been given in court for some time. Indeed many defences that involve the submission of a plea that an underlying mental condition caused the accused to behave in a particular manner would be the poorer without such evidence. However, the use of images of the brain in the courtroom is a matter of concern. Sometimes the use of brain imaging may be of great concern, as Aditi Sharma's conviction in India on the evidence of a neuroscientific test of debatable utility illustrates. Therefore it will be useful to examine and consider, before going further, the claims that have been made for modern neuroscience. Perhaps it would be a good idea to establish first a perspective on how new approaches to studying the brain led to the claims that are made on behalf of neuroimaging and its related techniques for interpreting brain states.

In his book *The Future of the Brain*, Steven Rose describes a change of approach to studying the brain:

Researchers have studied brain and behaviour from the beginning of recorded science, but until recently it was left to chemists to analyse the molecules, physiologists to observe the properties of ensembles of cells, and psychologists to interpret the behaviour of living animals. The possibility—even the hope—of putting the whole jigsaw together only began to emerge towards the end of the last century.[51]

Steven Rose describes the huge growth of activity in this area since the 1990s. He describes the linking of the 'Human Genome Project' to the use of neuroscience to map and decode areas of the mind. Rose charts the growth of the new science and looks at those who are investing 'billions of dollars' into the research—the US, Europe, and Japan. He concludes that this is now a major industry funded by 'government—including its military wing—and the pharmaceutical industry'. Accompanying this development he suggests that the domain of neuroscience, or as he refers to them 'neurosciences', is such that 'what were once disparate

---

[50] Ibid. para. 44.
[51] (Oxford, Oxford University Press, 2005) 2–3.

fields—anatomy, physiology, molecular biology, genetics and behaviour—are now all embraced within "neurobiology".' He comments that those who pursue research have ambitions that 'have reached still further into the disputed terrain between biology, psychology and philosophy'.[52] Rose is concerned about the claims made for neurotechnologies. In particular he has concerns about the creation and identification of new behavioural states and, making reference to the *American Diagnostic and Statistical Manual* to support his argument, he writes: 'The reductionist fervour within which they are being created argues that a huge variety of social and personal ills are attributed to brain malfunction, themselves a consequence of faulty genes.'[53]

Some assessment of the claims being made by Steven Rose is necessary. It is clear that a number of eminent scientists are making claims that seem to support his assertions and are of relevance to the law. Paul Churchland, writing about sociopathology and corrective policy, claimed that 'few will deny that courts are deeply unreliable at determining the many dimensions of cognitive, emotional and social competence in any defendant'.[54] Despite a reassuring degree of scepticism about the state of technology in 1995, Churchland goes on to argue that in time these technologies may produce:

> real benefit of identifying truly problematic people in this high-tech way... the specific *nature* of their neurosocial problem can thus be identified, and they will thus become a candidate for possible relief, repair or continuing modulation.[55]

One of the measures that Churchland identifies as assisting in this work is the use of positron emission tomography (PET), MRI and functional magnetic resonance imaging (fMRI) scan data. Churchland sees one of the values of such work as being the prediction and prevention of future behaviour that puts the general population at risk:

> Identifying the truly problematic offender is the first thing... But if diagnostic and treatment technologies improve as projected, then problem—specific neurological interventions may allow us to return a dangerously dysfunctional personality almost immediately to a state much closer to social and psychiatric *normal.*[56]

Examining more recent claims made on behalf of neurotechnologies, an article written by a group of German academics underlines and provides evidence for the assertions made by Steven Rose. Whilst admitting that the data obtained from brain scans will not permit such work as yet, because it is not sufficiently accurate, the authors suggest that 'brain scanning could provide biological brain-based criteria of the offenders' personality traits'. The article's authors are keen to see

---

[52] Ibid. 3 all quotations.
[53] Ibid. 6. The behaviours identified include: 'oppositional defiance disorder' and 'disruptive behaviour disorder 'attention deficit hyperactivity disorder'.
[54] Paul M. Churchland, *The Engine of Reason the Seat of the Soul* (Cambridge, MA, MIT Press, 1995), in ch. 11, entitled 'Neurotechnology and Human Life' 312.
[55] Ibid., emphasis in original.
[56] Ibid. 314, emphasis in original.

more research: 'the causal link between abnormal brain function on the one hand and striking abnormal personality traits and legal offenses on the other still remains to be explored intensively'.[57] They promote their paper as providing the opportunity to move away from using 'free and voluntary decision making' as exclusive criteria to judge criminal responsibility. They advocate looking at responsibility rather as being founded on 'brain behaviour relationships'[58] as identified by a view of the 'embedded brain'. That is 'the brain embedded within a psychosocial environment'. The claim made in the article is that 'the metaphysical problem of the mind-brain relationship becomes replaced by the question of the brain-environment relationship'.[59] Again the basis of this claim is the advances in neuroimaging that the authors see as leading to a 'paradigm shift' to the application of criteria relating to legal responsibility. The claim is that the substitution of 'psychosocial subjective criteria' with 'bio-physical and objective criteria' would achieve a better and fairer legal system.[60]

There is no doubt that much money is being invested in improving the database for studying brain function. Another recent paper, funded in part by the US Office of Naval Research and two research foundations,[61] aims to build an evidence base on which to formalize relationships between structure and function in the brain, with the aim of understanding 'how mental processes map to neural processes'.[62]

Some consideration of how brain imaging works is therefore necessary, as well as some consideration of scientific evaluations of its effectiveness. This consideration will necessarily be short and only be pursued in so far as it relates to the main arguments being made in this paper. In terms of imaging, fMRI is viewed as amongst the most effective techniques for certain purposes. It is therefore worth examining critiques of this technique to consider some of the practical constraints on the actual use of neuroimaging.

MRI and fMRI are still techniques in their relative infancy. Gregory Brown estimates that these neuroimaging techniques have been in use since 1990.[63] His research identifies the use of fMRI to study, inter alia, 'deception, the inference of intention, self-concept, phantom pain, the influence of culture on charitable donations, and socio economic deprivation'. He notes that 'investigators are using fMRI to provide information about applied problems in law, economics and marketing'.[64] The paper considers some of the methodological difficulties with

---

[57] Joachim Witzel, Martin Walters, Bernhard Bogerts, and Georg Northoff, 'Neurophilosophical Perspectives of Neuroimaging in Forensic Psychiatry—Giving Way to a Paradigm Shift?' (2008) 26 *Behav. Sci. Law* 113–30, 118.
[58] Ibid. 113.
[59] Ibid. 124.
[60] Ibid. 125.
[61] James S. McDonnell Foundation and The National Sciences Foundation.
[62] Russell A. Poldrack, Yaroslav O. Halenko, and Stephen José Hanson, 'Decoding the Large-Scale Structure of Brain Function by Classifying Mental States across individuals' (2009) 20(11) *Psychological Science* 1364–72, 1371.
[63] Gregory G. Brown, 'Functional Magnetic Resonance Imaging in Clinical Practice: Look Before You Leap' (2007) 17 *Neurophyschol. Rev.* 103–6, 103.
[64] Ibid. all quotations.

using fMRI data. Brown identifies some issues surrounding the use of blood oxygenate level dependent (BOLD) contrast responses to measure brain activity. Susan Greenfield explains what BOLD responses measure.

> MRI... measures changes in blood oxygen concentration serving brain areas that are more active... Oxygen is carried by the protein haemoglobin. MRI exploits the fact that the actual amount of oxygen present affects the magnetic properties of haemoglobin: these properties can be measured by the presence of a magnetic field... The radio signal is unique to the amount of oxygen carried by haemoglobin in the sample and therefore gives a very sensitive measure of brain activity of different regions of the brain.[65]

The problems that Brown identifies with BOLD measurement include, disruptions to blood flow and metabolism, '[i]mpaired motivation, inattentiveness, failure to understand directions, or differences in time on task'.[66] He is also concerned that further errors may occur whilst the images are processed. Each constructed image of the brain having to pass through a number of processes before the final picture of brain activity is produced. The fitting of individual brains to a common atlas space[67] that allows comparisons to be made also means that images are manipulated to fit. This makes the individual image less reliable as a source of data.[68] Other studies note that the number of manipulations of the image will seriously affect the accuracy of the final image. Swallow et al. point out that: 'If more than one resampling step is used, spatial normalisation may result in the loss of the high-frequency components of the signal and decrease the reliability and power of methods utilising it.' They go on to conclude that multiple manipulations of images may remove 'crucial components of the task related BOLD signal'.[69]

In 2009 in a paper originally entitled 'Voodoo Correlations in Social Neuroscience' scientists from Massachusetts Institute of Technology and the University of California examined fifty-five articles describing research and fMRI data collected in relation to studies of emotion, personality and social cognition.[70] The conclusions that they reach are of interest for a number of reasons. The suspicions of the researchers were aroused because of 'extremely high correlations between measures

---

[65] Susan Greenfield, *The Human Brain* (London, Weidenfeld and Nicholson, 1997) 30–1.
[66] Brown (2007) 104.
[67] Such activity places the scanned brain into a space taken from an atlas of averaged brains. This means that the new image allows those investigating the brain to see variances from average brains. This technique permits researchers to identify where the brain scanned differs from the average of other brains scanned. This obviously requires some resizing of the brain scan and therefore some loss of accuracy in the amended scanned image.
[68] Brown writes (ibid. 106): 'Beyond agreement about the importance of a general lineal model as a core element to brain mapping pathway, laboratories agree to differ on how to correct for physiological and behavioural movement, manage temporal noise, blur images to smooth spatial noise, solve the Type 1 error problem related to the massive number of statistical tests performed, warp brains of different shapes and sizes into a common atlas space, and identify regions of interest to name some salient decision points.'
[69] Khena M. Swallow, Todd S. Braver, Abraham Z. Snyder, Nicole K. Speer, and Jeffrey M. Zacks, 'Reliability of Functional Localization Using fMRI' (2003) 20 *Neuroimage* 1561–77, 1562.
[70] Edward Vul, Christine Harris, Piotr Winkielman, and Harold Pashler, 'Puzzling High Correlations in fMRI Studies of Emotion, Personality and Social Cognition, Perspectives on Psychological Science' (2009) 4(3) *Perspectives on Psychological Science* 274–90.

of individual differences relating to personality, and social cognition and measures of brain activity obtained using fMRI'. They note that the correlations found by fMRI were often stronger than correlations found in similar investigations based 'in purely behavioural studies'.[71]

To ascertain why this might be the researchers looked at a number of issues including some research into the accuracy of correlations drawn from BOLD measures. From this they concluded that measures 'computed at the voxel level will not often have reliabilities greater than about 0.7'.[72] Yet fMRI studies often reported correlations much higher than this. The researchers realized that the degree of activation recorded by the studies depended on how the measurement of brain activity was collected by the researchers.[73] The researchers therefore emailed a number of questions to those who had written the articles they had selected. They concluded from this that the methods adopted to measure brain activity explained the high levels of correlation that had stimulated their interest in the first place. The article identifies the key as being 'in the 53% of respondents' who answered their survey by stating that '"regression across subjects" was the functional constraint used to select voxels, indicating that voxels were selected because they correlated highly with the behavioural measure of interest'.[74] Why this was seen as significant is that such an approach to measurement will 'inflate observed across-subject correlations'.[75] Further, the researchers identify the correlation values, using voxels in over half the studies under review as being 'inflated to the point of being completely untrustworthy'. The researchers question whether the type of statistical calculation utilized in the papers reviewed in the study biased the results obtained.[76] The paper has a worrying conclusion, which is that 'a disturbingly large and quite prominent segment of fMRI research on emotion, personality and social cognition is using seriously defective research methods and producing a profusion of numbers which should not be believed'. Worryingly they

---

[71] Ibid. 274, both quotations.
[72] Ibid. 275.
[73] The measurement 'was always a correlation value—a correlation between each person's score on some behavioural measure and some summary statistic of their brain activation. The latter summary statistic reflects the activation or activation contrast within a certain set of voxels. In either case the critical question is, "How was this set of voxels selected?" As we have seen voxels may be selected on anatomical criteria, functional criteria or both. Within these broad options are a number of more fine grained choices' (ibid. 277). The authors note that these complex choices were not covered in the literature they studied and therefore often the basis of the calculation of the correlations remained opaque necessitating further questioning of the authors of the studies.
[74] Ibid. 278.
[75] Ibid. 279.
[76] Ibid. 284 both quotations. To clarify this point the following argument is made on the same page: 'A correlation of 0.96 (as in Sander et al., 2005), indicates that 92% of the variance in proneness to anxiety is predicted by the right cuneus response to angry speech. A relationship of such strength would be a milestone in the understanding of brain–behavior linkages and would promise potential diagnostic and therapeutic spin-offs. In contrast, suppose—and here we speak purely hypothetically—that the true correlation in this case were 0.1, accounting for 1% of the variance. The practical implications would be far less, and the scientific interest would be greatly reduced as well. A correlation of 0.1 could be mediated by a wide variety of highly indirect relationships devoid of any generality or interest.'

continue: 'we suspect the questionable analysis of methods discussed here are also widespread in the fields of cognitive neuroscience, clinical neuroscience and neurogenetics'.[77]

This leads to a conclusion that the images taken from brain scans should be viewed as one form of evidence but not taken as convincing evidence of the existence of behavioural problems. This has clear messages for those using these images to predict risk. In cases such as *McCann*, where evidence is required to support a behavioural diagnosis related to damage to areas of the brain, it is not suggested that brain images would lack explanatory force. The thrust of the argument is simply that the evidence at present does not support assertions that the philosophy and jurisprudence in relation to criminal responsibility will have to be realigned to take account of objective criteria based upon brain behaviour relationships.

What an examination of the burgeoning research in relation to the neurosciences suggests is that these issues will not go away. It also suggests that the law will need an extremely robust means of assessing the pertinence of evidence. MRI or fMRI scan evidence are highly visual and therefore may have greater impact in terms of explaining motivation or mental conditions. Additionally, neuroscientific research is often backed by governments or pharmaceutical companies and may receive press coverage that inflates its predictive powers. I will return to these issues in my conclusion when reviewing the excellent proposals put forward by the English Law Commission in its consultation paper No. 190 on 'The Admissibility of Expert Evidence in Criminal Proceedings in England and Wales'.

9.5 Voluntariness: Legal Meanings and Further Differences in the Approach of Law and Science to Meaning

As seen earlier in this paper, much that is written in this area revolves around voluntariness, agency and moral responsibility. However, a further point of practical interest is the great divide between legal and scientific views of causation; and whether this distinction poses any problems in practice. Consideration of this issue may lead to the conclusion that dialogue between lawyers and scientists will have to be carefully thought through. Before considering the treatment by the appeal courts of circumstances that should be given causal relevance, it is necessary briefly to explore another issue canvassed in this paper. The issue is the role of free will in legal determinations of guilt or innocence.

Much ink has been expended on the requirements for a voluntary act on which to found criminal responsibility. There is a philosophical ideal on which the idea is based and that ideal is largely undisputed by criminal lawyers in England and Wales. The ideal is partly based on the principle of individual autonomy. Most criminal law textbooks cover this principle and, in particular, Andrew Ashworth's

---

[77] Ibid. both quotations, 285.

book, *Principles of the Criminal Law,* addresses it in some depth. Ashworth talks of the factual element of the principle of individual autonomy being based on the view that 'individuals in general have the capacity and sufficient free will to make meaningful choices'.[78]

There is no doubt that this view is not shared by all as the basis of moral responsibility for actions. Indeed this point is addressed by Ashworth in his book, drawing on the work of Barbara Hudson who writes:

the notion of free will that is assumed in ideas of culpability... is a much stronger notion than that usually experienced by the poor or the powerless. That individuals have choices is a basic legal assumption: that circumstances constrain choices is not. Legal reasoning seems unable to appreciate that an existential view of the world as an arena for acting out free choices is a perspective of the privileged, and that the potential for self-actualization is far from apparent to those whose lives are constricted by material or ideological handicaps.[79]

Clearly the discussion of these issues, by Hudson and earlier by Witzel et al.,[80] reflect differences of opinion about what acts committed by actors have moral relevance and more pertinently the relevance of the events, be they mental or environmental, which lead actors to make choices. These matters are rightly hotly debated by lawyers and non-lawyers. However, it is also necessary to acknowledge the reality of the law as it is practised, just as it is necessary to acknowledge and understand the practices of scientific endeavour.

There is no actual requirement of voluntariness in the criminal law; there is rather a presumption of voluntariness, just as there is a presumption of sanity. The manner in which the presumption of voluntariness has been expressed in case law is as follows:

Whilst the ultimate burden rests on the Crown of proving every element essential in the crime, nevertheless in order to prove that the act was a voluntary act, the Crown is entitled to rely on the *presumption* that every man has sufficient mental capacity to be responsible for his crimes: and that if the defence wish to displace that presumption they must give some evidence from which the contrary may be reasonably inferred.[81]

Viewed in the light of this definition the view of commentators that English law requires much in the way of exercise of free will has to be viewed with some scepticism, hence the comments earlier in this paper with regard to the statements made by the House of Lords in *R v G (and another)*.[82] In a sense therefore the law recognizes that claims of involuntariness or lack of voluntariness may be made.

---

[78] 6th edn (Oxford, Oxford University Press, 2009) 23.
[79] Ibid. 24 citing B. Hudson, 'Punishing the Poor a Critique of the Dominance of Legal Reasoning in Penal Practice' in W. E. Heffermen and J. Kleinig (eds.), *From Social Justice to Criminal Justice* (Oxford, Oxford University Press, 2000) 302. In terms of the criminal law it is at least debatable whether the law views all acts as unconstrained. The defences of necessity and duress attest to the laws view that some acts are constrained and choices may not be unfettered. It has to be said the law does not tend to take 'an existential view of the world.'
[80] Note 57, above and accompanying text.
[81] *Bratty v Attorney General for Northern Ireland* [1963] AC 386, 413 per Lord Denning.
[82] See note 17, above and the accompanying discussion.

However, it does not in its assessment of facets of responsibility assume that free will in the broader contexts as discussed in attributions of moral culpability has much of a role to play. Indeed the criminal law looks at issues of culpability through the separation of attitudes of mind from events. Voluntariness, where it is an issue, floats free of either *mens rea* or *actus reus* descriptions. Lord Simon describes the volitional element as follows:

> Lastly, *actus reus* and *mens rea* are misleading terms; since (other than exceptionally) a mental state is not criminal without an accompanying act and an act is not criminal without some accompanying mental element. Both terms have, however, justified themselves by their usefulness; and I shall myself employ them in their traditional senses—namely, *actus reus* to mean such conduct as constitutes the crime if the mental element involved in definition of the crime is also present (or, more shortly, conduct prohibited by law); and *mens rea* to mean such mental element, over and above volition, as is involved in the definition of the crime.[83]

Thus the law takes its own stance in relation to culpability, which is quite separate from the stance taken by those from different disciplines. Indeed arguably it has to given the approach it adopts to causation. No one would dispute that the meaning of the word 'cause' is the basis of much philosophical argument particularly by those of a sceptical bent. These discussions have been noted by the law and by science. Indeed establishing causation and replication are central to the validation of many scientific experiments. In the criminal law the use of causation in attributing responsibility has been subject to reductionist arguments in two ways.

The actual definition in English criminal law of 'legal causation'[84] requires that a defendant's acts be a significant and operative cause of the criminal act. But pragmatism requires that the definition given to this term be something other than suggested by the words. A good example of the conundrum faced by the courts is the case of *R v Pagett*.[85] The facts of the case are complicated, but put simply Pagett used Gail Kinchen as a human shield against police officers who were trying to arrest him. He was armed with a shotgun, a police marksman fired at Pagett and Gail was killed as a result. Pagett was tried and convicted of Gail's manslaughter. Evidence at the trial stated that he had fired at the police before they had fired at him. The Court of Appeal had to consider who in law had caused Gail's death. The court made it very clear that the interpretation of the word 'causation' was a legal interpretation and the meaning adopted has a specialist meaning.

> We have no intention of embarking in this judgment on a dissertation of the nature of causation, or indeed of considering any matters other than those which are germane to the decision of the issues now before us. Problems of causation have troubled philosophers and lawyers throughout the ages; and it would be rash in the extreme for us to trespass beyond the boundaries of our immediate problem. Our comments should therefore be understood to be confined not merely to the criminal law, but to cases of homicide (and possibly also other crimes of violence to the person); and it must be emphasised that the problem of

---

[83] *DPP for Northern Ireland v Lynch* [1975] AC 653 (HL), 690.
[84] The criminal law also uses factual 'but for' causation; see note 7, above and the discussion preceding the note.
[85] *R v Pagett* (1983) 76 Crim App R 279, 287.

causation in the present case is specifically concerned with the intervention of another person (here one of the police officers) whose act was the immediate cause of the death of the victim, Gail Kinchen.

It would be naive in the extreme to suggest that neuroscience would take such an approach to establishing causal arguments relating to how the brain works. A different form of reductionism is being applied legally to that applied by the scientific community. What counts as causes here is defined and honed from a nexus of competing forces. The Court of Appeal had to find a just and fair solution. They also had to bear in mind the likely press reaction to Pagett's case. Pagett had violently assaulted Gail's mother on at least two occasions and shot and seriously injured Gail's father during the events leading to Gail's manslaughter. Clearly any decision the court made would have an effect on policy development. There was also a need to frame this policy so as to minimize any risks to public safety, to protect the operational competence of the police force, and, furthermore, they had to avoid bringing the law into disrepute. Scientists may have their own problems with policy issues, but in establishing what causation might mean this is usually a matter of statistical analysis to establish significance—not the tightrope that judges have to walk. Causation in a court of law has in respect of different offences very different meanings indeed. In some of the dangerous driving cases its significance in attributing blame has been said to require no more than a 'slight or trifling link'.[86]

9.6 When Scientific Opinion and the Law Meet, Some Past Practice, and its Implications for the Relationships between Neuroscientists and Criminal Lawyers

Differences in the meanings attributed to words present obvious difficulties to collaborative work between different disciplines. But these difficulties are not insurmountable. The legal and scientific community continue to work together in their assessment of defendants' acts and a number of decisions taken concerning the merits of prosecutions. An example of this relationship is the case of Brian Thomas, who was charged with murder following the killing of his wife during a night terror. The Crown Prosecution Service (CPS) decided not to pursue an insane automatism verdict. This decision was made following the consideration of expert psychiatric evidence. The case had already reached court so the explanation of the CPS as to their decision not to proceed is of interest. Iwan Jenkins, Chief Crown Prosecutor for CPS Dyfed Powys, said:

This has been a unique case with a unique set of circumstances.

We have a duty to keep cases under continuous review, and following expert evidence from a psychiatrist it was suggested no useful purpose would be served by Mr Thomas being

---

[86] *R v Kimsey* [1996] Crim LR 35 (CA): 'the recorder's reference to a "slight or trifling link" was permissible and a useful way of avoiding the term de minimis.'

detained and treated in a psychiatric hospital, which would be the consequence of a special verdict in this case.

Once it was raised, the CPS had a duty to review the case and decided that, guided as we must be and have been throughout this case by the views of the experts, the public interest would no longer be served by continuing to seek a special verdict of not guilty by reason of insanity.

The consequences of such a finding would have meant Mr Thomas' detention in a psychiatric hospital, but it is now clear that the psychiatrists feel that that would serve no useful purpose and the risk of reoccurrence is very, very small. We therefore have offered no further evidence and asked the jury to return a simple verdict of not guilty.[87]

The whole of the manner in which this issue is reported raises interesting questions about the intersection of culpability and science and the proper maintenance of that relationship. The CPS seem to have conflated the disposition following a verdict with the trial process which determines guilt or innocence. The CPS reasoning continues:

There has never been any dispute that Mr Thomas caused his wife's death... but the prosecution accepted that he should be found not guilty of murder or manslaughter, based on the evidence of a number of experts in the field of sleep-related disorders.

Further expert evidence led to a range of opinions on whether Mr Thomas presented any ongoing risk, and therefore whether he required further medical intervention.

... For that further treatment to be made a requirement, a special verdict of 'not guilty by reason of insanity' would be required. The law dictates that this is a verdict that cannot be determined by anyone other than a jury which is why the case had to go to court.

The death of Christine Thomas was thoroughly investigated by Dyfed Powys Police. Investigations continued after Mr Thomas was charged and involved the instruction of experts in several fields.

I must emphasise that the circumstances of this case are almost unique in the UK and there have been fewer than 50 instances recorded worldwide. It is only because of highly sophisticated tests carried out by sleep experts that Mr. Thomas' condition could be confirmed.

Our thoughts remain with the family of Brian and Christine Thomas who have remained dignified throughout this difficult time.[88]

It is accepted that continuing the trial may have meant that the jury passed a not guilty verdict because of their understandable sympathy for the accused; also that the CPS may have been concerned that there would be policy implications if the jury were to accept evidence of non-insane automatism.[89] However, the conflation of the needlessness of psychiatric detention with determination by the jury of criminal culpability obscures more than it reveals. Arguably, in this case, expert

---

[87] <http://www.cps.gov.uk/news/press_releases/156-09/>, CPS press release dated 20 November 2009 (accessed 25 November 2009).

[88] Ibid.

[89] There was some evidence which would have supported an argument that the cause of Mr Thomas's sleep disorder the night that he killed his wife was an external cause. There was a disturbance outside the camper van in which they were sleeping. This may have led the jury to conclude that the defence of non insane automatism should excuse his conduct.

evidence may have dictated the result of the trial because the experts were of the opinion that Mr Thomas should not be detained in a mental health institution.[90]

Sometimes the unquestioning acceptance of scientific evidence and statistics and deductions drawn from it present the courts with almost insurmountable difficulties. Scientific argument is traditionally seen as advancing by the testing of hypotheses and then evaluating the resultant evidence that tests those hypotheses. Commonly the scientific explanation that emerges from this type of investigation is therefore the most likely explanation, rather than an explanation that is indisputably correct. This type of proof has greater similarities with the civil burden of proof that is established on the balance of probabilities. The law seeks certainty and in the case of the criminal law proof beyond reasonable doubt. Further exploration of this issue may be achieved by looking at a series of cases where expert evidence came to be viewed as having brought about two serious and openly acknowledged miscarriages of justice.[91] The cases concern cot deaths. Professor Roy Meadow was clearly viewed by the prosecuting authorities as an excellent expert witness. Yet, subsequently, the evidence he gave to the courts of first instance in respect of the cases of Angela Cannings[92] and Sally Clark has been accepted by the Court of Appeal as flawed. The Court of Appeal quashed the convictions of the two women, who had been convicted of killing their own children.

Sally Clark had been convicted of the murder of her two sons. Harry, who died in 1998 aged eight weeks and Christopher, who died in 1996 aged eleven weeks. In *Clark* the evidence of two expert witnesses was criticized by the Court of Appeal when it quashed her conviction.[93] The first expert failed to disclose test results for one of the deceased children. The test revealed that one of the children was suffering from an infection that could have caused his death. The manner of the presentation of the statistical evidence in court was also seen as highly relevant to Clark's conviction by the jury. The Court of Appeal reviewed the manner in which Professor Meadow responded to cross-examination, at the trial, in respect of his statistical evidence with regard to the likelihood of two deaths occurring in one family. The Court of Appeal was particularly critical of the use of a misleading statistic. Meadow, giving his expert evidence, suggested that there was only 'one in 73 million chance of having two cot deaths in the same family'. This was found by the Appeal Court to be a 'gross misrepresentation'. The court also expressed concern that there was no consideration at trial as to how the statistical evidence should be submitted.

---

[90] An insanity verdict in a murder case would have resulted in the making of a hospital order with a restriction order, so the views of the experts were relevant. The sentence is mandated by the Criminal Procedure (Insanity Act) 1964, s. 5. The requirements of Art. 5 of the European Convention for the Protection of Human Rights and Fundamental Freedoms would have posed considerable difficulties in this disposal as Brian Thomas seems to have been viewed as of sound mind by the medical experts at the time of his trial.
[91] This expression is used in the non-legal sense of a miscarriage of justice. That is the conviction of people who later had their conviction quashed on the basis that the expert evidence at the trial which resulted in the conviction was flawed.
[92] *R v Cannings* [2004] EWCA Crim 1.
[93] *Clark* [2003] EWCA Crim 1020, source of quotations, para. 99.

All of this evidence was given without objection from the defence but Mr Bevan QC (who represented the appellant at trial and at the first appeal but not before us) cross-examined the doctor. He put to him figures from other research that suggested that the figure of 1 in 8,543 for a single cot death might be much too high. He then dealt with the chance of two cot deaths and Professor Meadow responded: 'This is why you take what's happened to all the children into account, and that is why you end up saying the chance of the children dying naturally in these circumstances is very, very long odds indeed one in 73 million.' He then added:

> ... it's the chance of backing that long odds outsider at the Grand National, you know; let's say it's a 80 to 1 chance, you back the winner last year, then the next year there's another horse at 80 to 1 and it is still 80 to 1 and you back it again and it wins. Now here we're in a situation that, you know, to get to these odds of 73 million you've got to back that 1 in 80 chance four years running, so yes, you might be very, very lucky because each time it's just been a 1 in 80 chance and you know, you've happened to have won it, but the chance of it happening four years running we all know is extraordinarily unlikely. So it's the same with these deaths. You have to say two unlikely events have happened and together it's very, very, very unlikely.[94]

There will always be difficulties at trial in interpreting expert evidence in a manner that is accurate and comprehensible to the jury. However, in this case the Court of Appeal clearly took the view that this explanation was completely outside of the reasonable range of inferences that could have been drawn from the statistics and as such highly misleading:

> It is unfortunate that the trial did not feature any consideration as to whether the statistical evidence should be admitted in evidence and particularly, whether its proper use would be likely to offer the jury any real assistance. Inherent in the evidence were dangers ... and the 1 in 73 million figure was wholly irrelevant.[95]

In both the cases of Brian Thomas and Sally Clark issues that are pertinent to the interpretation and use of scientific evidence in court emerge and they are as pertinent to neuroscientific evidence as they are to other sorts of evidence. Firstly, where the evidence of medical experts suggest a mental condition may have contributed to the criminal actions of a defendant or have been entirely causal of those actions. Is it appropriate that the determination of whether a trial proceeds should be made by the CPS in accordance with their interpretation of the likely sentencing disposal by the judge, basing their decision on the issue of a defendants' medical condition at the time of trial? Secondly, what should the process be to ensure that expert evidence in court is a clear and reliable statement of the scientific knowledge in the area to date? There are already measures in place that require that all relevant evidence be disclosed and there are legal protections in place to allow appeals where evidence is not produced at trial and subsequently emerges. Whether those protections are sufficient is a matter for debate but the main interest of this paper is how neuroscientific evidence, brain images should be dealt with when they are produced in court.

---

[94] Ibid.
[95] Ibid. para. 173.

In *Clark* it was the description of the statistical likelihood of two cot deaths in one family which was grossly overestimated with tragic results for Sally Clark.[96] Given the problems with neuroscientific images and the doubts expressed in respect of measures of brain activity during fMRI scans there must be real concerns about how this evidence will be presented in court.[97] Brain scans are visual images and therefore they may seem to present a picture of the accused's brain to the jury. However, as has already been argued this is most likely to be a picture of a brain displayed by means of a standardized atlas of brains to help in the identification of abnormalities. The picture will be there to back up other types of evidence. Worryingly there is something that is convincing about the visual representation of evidence, including brain images. It adds a layer of seemingly visual proof to statements that are merely the statistical assertions of an expert witness. Who will be responsible for telling the court where an image has been resampled a number of times? Who will say as Susan Greenfield writes, 'we cannot treat isolated brain regions as though they were autonomous brains; nor can we localise highly sophisticated and complex functions to a single brain area'?[98] The pressure on the trial process where such evidence exists in high profile cases will be very great indeed.

In November 2009 Leveson LJ, giving a paper at a conference on expert evidence reported by the BBC news website, remarked that television programmes, such as *CSI* tended to compound problems of getting witnesses to testify in court. 'There can be a temptation, certainly in the eyes of the public, to think there can be expert evidence to prove the essential point in a case to the extent that you don't need regular, old fashioned, normal witnesses anymore.'[99] The judge, according to the report, hinted that 'some experts were exaggerating their evidence or straying into subjects in which they did not have specialist knowledge'.[100] It is submitted that the presentation of neuroscientific evidence through the media and political treatment of the evidence will be likely to add to the difficulties in dealing with expert evidence in court.

9.7 Proposals to Address the Problems Posed by Expert Evidence

In April 2009 the Law Commission produced consultation paper No. 190 on 'The Admissibility of Expert Evidence in Criminal Proceedings in England and Wales'. This paper is subtitled 'A New Approach to the Determination of Evidentiary

---

[96] Sally Clark committed suicide following her release from prison. According to Sally Clark's family, as reported on the BBC news website, she 'never fully recovered from the effects of this appalling miscarriage of justice': <http://news.bbc.co.uk/go/pr/fr/-/1/hi/england/essex/7028071.stm> published 4 October 2007 (accessed 25 November 2009).
[97] For the earlier discussion of this issue see note 69, above and the surrounding text.
[98] Susan Greenfield, *Id: The Quest for Meaning in the 21st Century* (London, Sceptre, 2009) 223.
[99] <http://news.bbc.co.uk/go/pr/fr/-/1/hi/uk/8347410.stm> published 6 November 2009 (accessed 25 November 2009).
[100] Ibid.

Reliability'. In the paper reference is made to the problems experienced by Sally Clark and the need to thoroughly interrogate expert evidence is set out. In terms of admissibility of evidence the paper suggests that the judge should have a 'gate keeping' role. The paper then examines three questions that it sees as pertinent to the issue:

(1) Is the evidence logically relevant to a disputed matter?
(2) Would the evidence provide the jury with substantial assistance?
(3) Does the witness qualify as an expert in the field, and would he or she be able to provide an impartial opinion?[101]

The proposals aim to test both the logic and relevance of the expert evidence. They look at the expertise of the witness and evaluate whether the evidence will be 'a substantial help to the jury'. It must be impartial and demonstrably be relevant to the facts at issue in the case.[102]

Additionally they suggest a threefold test of the assessment of the reliability of expert evidence, the paper suggests:

Evidence may be reliable if:

(a) the evidence is predicated on sound principles, techniques and assumptions;
(b) those principles, techniques and assumptions have been properly applied to the facts of the case; and
(c) the evidence is supported by those principles, techniques and assumptions as applied to the facts of the case.

The party relying on the evidence must demonstrate that it is reliable:[103]

The judge would then be asked to assess the reliability of the evidence looking at the hypothesis involved, methodology and the assumptions made by the expert and how these relate to the facts of the case under consideration. Some test would have to be made as to whether the expert evidence had been properly applied and whether the expert's conclusions were both logical and sustainable.[104]

There is then a listing within the paper of the kinds of issues which are relevant to the determination of the admissibility/reliability of scientific expert evidence. These are extensive and well thought out.[105] The implementation of such controls on the admission of expert evidence would be helpful and is perhaps long overdue.

---

[101] No. 190 on 'The Admissibility of Expert Evidence in Criminal Proceedings in England and Wales' para. 6.6.
[102] Ibid. 6.7.
[103] Ibid. 6.10.
[104] Ibid. 6.11.
[105] Ibid. 6.26: '(1) In determining whether scientific (or purportedly scientific) expert evidence is sufficiently reliable to be admitted, the court shall consider the following factors and any other factors considered to be relevant: (a) whether the principles, techniques and assumptions relied on have been properly tested, and, if so, the extent to which the results of those tests demonstrate that they are sound; (b) the margin of error associated with the application of, and conclusions drawn from, the principles, techniques and assumptions; (c) whether there is a body of specialised literature relating to the field;

9.8 Conclusion

These proposed changes will not reduce the difficulty of making the assessment of just what neuroscientific imaging evidence is relevant. As yet only tentative conclusions may be drawn from images of the brain and how this relates to other types of neuroscientific evidence. It is clear from Aditi Sharma's conviction that for some people even untested and highly debatable neuroscientific evidence may be convincing. Neuroscientists will continue to push the bounds of their discipline and make arguments that will need to be examined about the implications of the findings of their research for the allocation of causal attributions of behaviour that may, or may not, be relevant to criminal culpability. Legal culpability in terms of the criminal law is clearly separated from other determinations of culpability. Case law interprets and applies the criminal law in a manner that means that words are defined in a way that is sometimes only pertinent to specific offences or defences. Such definitions may have no wider applicability.

From this we might deduce that the normative, pragmatic, philosophical basis of the criminal law presents real problems for interdisciplinary endeavour. However, such an obvious conclusion is clearly refutable. Neuroscience is beginning to develop an understanding of the meaning that may be attributed to brain states. It can already diagnose diseases that clearly have an influence on behaviour, such as epilepsy. Epilepsy has for long been associated with violent behaviour, and is a condition that can often be effectively controlled by medication. Cases such as *R v Sullivan*[106] have perplexed and troubled criminal lawyers and the medical profession, as have criminal acts related to sleep disorders. Neuroscientific research may help by clarifying the nature of certain disorders. This would enable the criminal law to reach a more informed judgment about the nature of excusing conditions. This in turn would assist judges and lawyers to pose more meaningful questions to defendants and witnesses, both expert and otherwise, in a manner that enables a clearer presentation of the facts of cases to juries. It is to be hoped this would help juries to make better-informed and fairer decisions on the facts before them.

Does this mean that lawyers will have to abandon their present philosophies and slightly dualist definitions? Well to some extent the definitions of criminal law offences and defences have been subject to major changes over the past decade that have challenged and undercut many of the previous assumptions made by the

---

(d) the extent to which the principles, techniques and assumptions have been considered by other scientists . . . and if so, the extent to which they are regarded as sound in the scientific community; (e) the expert witness's relevant qualifications, experience and publications and his or her standing in the scientific community; (f) the scientific validity of opposing views (if any) and the relevant qualifications and experience and professional standing in the scientific community of the scientists who hold those views; and (g) whether there is evidence to suggest that the expert witness has failed to act in accordance with his or her overriding duty of impartiality.'

[106] [1983] 2 All ER 675 (HL) where an epileptic who committed a violent act following a seizure was persuaded to plead guilty. The defendant wished to claim that his actions that caused the injury were involuntary—but his legal team, following directions given to the jury by the trial judge, persuaded him to plead guilty as charged to avoid an insanity verdict.

law.[107] Where it will be necessary to exercise caution is in making sure that, however law defines offences and defences, the issues to be determined by the jury do not become wholly dependent on one type of expert evidence. Legislation passed in November 2009 suggests that unfortunately the law may be moving towards more mechanistic definitions of accountability.[108] The new definition of the partial defence of diminished responsibility has abandoned the previous reference to abnormality of mind and replaced this wording with a reference to abnormality of mental function. This abnormality must be linked to a recognized medical condition. Arguably such a functionalist definition invites neuroscientific evidence and could provide the platform from which MRI or genetic marker evidence of abnormality of mental function *via* MRI scans may be produced in court. This evidence could help to determine whether the abnormality of mental function caused the defendant to kill. Whether the use of this evidence in such a manner is helpful or will advance the cause of justice is debatable. The new legislative provision does not make it clear what the role of the jury will be in making such determinations. Previously when reaching a decision as to whether there was evidence from which the jury could find that the defendant's responsibility was substantially impaired the jury were able to take into account all the surrounding evidence and the medical evidence was only one factor that was to be taken into account. It is hard to see that in terms of abnormality of mental function they could be allowed such latitude. The determination of mental function and its causal power seems largely a medical or neuroscientific issue. It is to be hoped this is not a sign that the criminal law is abandoning its previous view of defendants as having minds and simply viewing them as individuals with mental

---

[107] For example the Sexual Offences Act 2003 changed many of the existing sexual offences and created new offences. The traditional balance between the *mens rea* and *actus reus* definitions of some of the more serious sexual offences such as rape was considerably altered and revised. The balance of the two partial defences to murder has been substantially altered by the Coroners and Criminal Justice Act 2009 with significant alterations to the balance of the matters to be proved to establish the partial defences.

[108] Coroners and Justice Act 2009: '**52 Persons suffering from diminished responsibility (England and Wales)** (1) In section 2 of the Homicide Act 1957 (c. 11) (persons suffering from diminished responsibility), for subsection (1) substitute—

"(1) A person ("D") who kills or is a party to the killing of another is not to be convicted of murder if D was suffering from an abnormality of mental functioning which—
  (a) arose from a recognised medical condition,
  (b) substantially impaired D's ability to do one or more of the things mentioned in subsection (1A), and (c) provides an explanation for D's acts and omissions in doing or being a party to the killing.

(1A) Those things are—
  (a) to understand the nature of D's conduct;
  (b) to form a rational judgment;
  (c) to exercise self-control.

(1B) For the purposes of subsection (1)(c), an abnormality of mental functioning provides an explanation for D's conduct if it causes, or is a significant contributory factor in causing, D to carry out that conduct."

(2) In section 6 of the Criminal Procedure (Insanity) Act 1964 (c. 84) (evidence by prosecution of insanity or diminished responsibility), in paragraph (b) for "mind" substitute "mental functioning".'

function. If it were to be such a sign, it would be a most worrying development. We are clearly more than our mental functions; or as Steven Rose expresses it:

> the question of how we can be free if our acts and intentions are ascribed to neurons is a classic category error. When Francis Crick tried to shock his readers by claiming that they were 'nothing but a bunch of neurons' he splendidly missed the point. 'We' are a bunch of neurons and other cells. We are also, in part by possessing those neurons, humans with agency. It is precisely because we are biosocial organisms, because we have minds constituted through the evolutionary, developmental and historical interaction of our bodies and brains (the bunch of neurons) with the social and natural worlds that surround us, that we retain responsibility for our actions, that we, as humans, possess the agency to create and recreate our worlds.[109]

It is for precisely this reason that more functional reductionist explanations of behaviour reveal very little about agency. Similarly Susan Greenfield refuses to refer to the brain as the centre of identity or personality in her work; she talks persistently about the relevance of the mind and conscious states. These are the things that she views as giving us our character and our personality. Like Rose she sees us as neurons plus experiences.[110] Arguably, both see the meaning of human experience as something much more. Lawyers need to bear in mind that our humanity and social interactions form the basis of our intuitions about human agency and responsibility for actions. Judgements about responsibility reached by juries in the court room will have a different foundation from the type of judgement exercised by scientists when interpreting the results of their research. However, few dispute that neuroscience will add to our understanding of the world and that it may, in time, add to our understanding of ethical issues. But, for the foreseeable future, it is unlikely to be advisable to allow a scientific world view to replace a philosophical or legal world view. The systems that are put in place to control the admissibility and ensure the probative value of expert neuroscientific evidence in court are of huge importance. In view of the excessive claims being made by a few neuroscientists there is a need to retain a realistic view of the position to be accorded to neuroscientific evidence in the evaluation of guilt or innocence. It is to be hoped that this will not lead to a total rejection of such evidence. Legislators, lawyers, and expert witnesses in the law courts need to view neuroscience as part of the explanation of human behaviour and not the whole formula for evaluating individual human actions. It is very important that the laws controlling admissibility of evidence permit the admission of the best, most suitable, scientific evidence. However, equally important is the need to allow the jury to see a meaningful objective assessment of the probative value of such evidence. The present English Law Commission proposals on expert evidence, reviewed in this paper, encapsulate such aims. It is to be hoped that they are enacted. Such an approach should enable neuroscientific advances to inform the law in a way that clarifies the legal issues before the court. This should provide jurors and judges with more meaningful assistance when trying to understand the complex matters which underpin determinations of criminal culpability.

---

[109] Rose (2005) 305.
[110] Greenfield (2009) 292.

10

How (Some) Criminals Are Made

Theodore Y. Blumoff *

> A person can be morally responsible only for what he does; but what he does results from a great deal that he does not do; therefore he is not morally responsible for what he is and is not responsible for.**

And some kids have bad luck. Consider abused kids. It is a brute sociological fact that they tend to become abusers themselves.[1] Bad luck it is when children are the unfortunate victims of poor antecedent and constitutive conditions that conduce to harmful behaviour, *mala in se*. In this paper, I hope to explain why this is the case, at least for some children, and perhaps especially for those who enter the world with low levels of monoamine oxidase type A (MAOA), a brain enzyme responsible for inhibiting neurotransmitters associated with aggression. I will also argue that individuals of this type—genetic predisposition (G) coupled with abusive and/or neglectful early environments (E)—have suffered from those poor constitutive, antecedent, and circumstantial conditions that generate bad moral luck and, therefore, they (a) are less blameworthy than those who have not endured the combination of such conditions and (b) are particularly good candidates for therapeutic justice, that is for treatment for those who we know or *prima facie* should know suffer neurobiological deficits.

---

* Professor of Law, Mercer University, Macon, GA. The author's sincere thanks go to many individuals who have helped along the path to this manuscript, including Christof Engel, Oliver Goodenough, Morris Hoffman, Owen Jones, Karen Kovach, Harold Lewis, Stephen Morse, and Emily Murphy for useful comments and suggestions on an earlier version of this paper. He has presented portions of this paper at the Annual Scholarship Meeting of the Society for the Evolutionary Analysis of Law (Vanderbilt University, April 2009), the Annual Meeting of the Gruter Institute of Law and Behavioral Research (Squaw Valley, CA, May 2009) and a Colloquium on Law and Neuroscience, University College London (July 2009). As always, he is grateful for the ongoing support of Mercer University School of Law.

** Thomas Nagel, 'Moral Luck' in *Mortal Questions* (Cambridge, Cambridge University Press, 1979) 34 (originally published in *Proc. Arist. Soc'y* (supp vol. L, 1976)).

[1] E.g. Dorothy Lewis, 'From Abuse to Violence: Psychophysiological Consequences of Maltreatment' (1992) 31 *Journal of the Amer. Acad. Of Child and Adolescent Psychiatry* 383, at <http://www.pbs.org/wgbh/pages/frontline/shows/little/readings/lewis.html>; Terri Morritt, 'Gene May Protect Abused Kids Against Behavior Problems' *Bio-Medicine.org*, <http://news.bio-medicine.org/biology-news-2/Gene-may-protect-abused-kids-against-behavior-problems-7072-1/>.

This paper begins with a brief vignette about two boys, born on the same date but reared in very different environments. It then discusses the concept of moral luck and how it figures into tendencies towards violence. The paper next examines how genetic endowment can and does affect behaviour, for good or ill, in the appropriate environment, G x E. In the penultimate part the paper takes up the related objections that there is no such thing as moral luck and, that if one pushes too hard on the concept, one undermines free will. The final part lays out brief recommendations for dealing with certain types of individuals who engage in violent misconduct.

10.1 Imagining Moral Luck

Just suppose there was a young and devilishly innocent looking child who, unbeknownst to him, was be born with a brain enzyme deficiency. We'll call him Boy I. He was raised in a family in which being honest was an unquestioned default expectation; it was one of the signal qualities that constitute the very healthy ambience of love and accountability this child, Boy I, grew up in. Manifestly, Boy I was lucky.

This developing child could not claim with full honesty that insincerity was a quality that was entirely foreign to his intentions. More often than he would be comfortable divulging in great detail, he distorted (and still could distort) deadlines and destinations and diverse data of every kind; he was especially prone to doing so as a young boy caught crossing the boundary between acceptable and not. But for so many reasons, Boy I was lucky, very lucky, in fact. The lovingly and reasonably composed boundary-setting refrains that played constantly in his background embodied love and intellect, compassion, emotion and cognition. Those qualities worked in harmony, and assured the child-to-be-man a sense of decency, duty, fair dealing, responsibility, and so on. These choral lines were presented and, thankfully, heard within a household that provided necessary and abundant amounts of unconditional love. Because of these blessings, the mature man's consciousness of his own rationality is grounded in and so measured by a child's rich emotional and intellectual life.

It—the message that nourished him—could have done a lot worse. Through no fault of Boy I's he might have come from and into a vastly different world. One can easily envision a home in which the lessons of love and duty exist only exceptionally, so that even when the practice of civility is faintly detectable, the ambience is composed of alcoholism, drug abuse, and the debilitating tirades that accompany abuse; countless but unpredictable offers and demonstrations of violence, including sexual or physical or emotional or all of the above. Discipline might be intense but compassion was non-existent; and, as I have said, it would almost always be manifest as brief respite from incessant verbal abuse. Which version of 'rationality' among all we might envisage will emerge from a background like this? What cognitive and neuroanatomical loops might be selected and activated as an adaptation strategy?

I want to suggest two different scenarios—one mostly real, the other imagined—that may hold some clues to and sensibilities about an answer. First of BOY I: when he was seven, he was caught at a neighborhood five-and-ten cent store stealing a 50¢ fish tank toy. (Yes, a fish tank toy: a little painted ceramic building that sat on the bottom of the gold fish bowl on the top of and amidst small colourful pebbles and through which the tiny creatures sometimes swim.) Experiencing thoughts that resembled rolling claps of thunder, his pulse racing, his heart rate and body temperature elevating, his eyes casting about in a futile effort to avoid detection, and his guilt surging—all the emotions one might suspect would accompany conscious wrongdoing—he took it. Except that he wanted the toy, he would not know to this day why he took it when he knew he should not have, it was almost as if his hands moved on their own; he had little impulse control. It seemed mostly stupid—even then and certainly now. As he stepped away from the fish counter, his shoulders sagged, his heart and pulse raced, a numbed affect filled his chest, his skin conductance doubtlessly changed.[2] His desire to bolt out of the store was stopped by the long arm and stern voice of the store manager who caught him.

What's in your pocket, young man?
Nothing.
Don't tell me 'nothing' son, because I saw you stick 'something' in your pocket there. [Pointing to the right front pocket]. Give it to me, now! And then we're going to call your father.

BOY I was caught.

BOY I's sense of guilt and shame and dread has never exceeded the levels attained at that moment. Dishonesty was just not acceptable behaviour in his home, and worse by far: now he had to face his Dad, who would have to leave work in the middle of the day to retrieve him from his shame. Having to meet his Dad's disappointment was more than BOY I could stand. He broke down crying. Whatever representations and images he had mapped in his brain that concerned the subject of personal honesty and the expectations therein encompassed triggered an unreflecting but naturalized emotional response: BOY I did not tell himself to cry; he did not decide to cry; he did not weigh carefully (or even negligently) the benefits of crying or not crying. Like the snatching itself, his response was primitive: He just cried. Crying was a rational adaptive reaction rooted in his particular cultural, social and selectional background. As I have said, BOY I was lucky. He cried both because of what he had learned and in furtherance of his learning: he continued to receive and store bits of information for future associated circumstances.

Now imagine a different, less fortunate, and perhaps more primitive sketch. Suppose the child in this version, BOY II, was born the same day and in the same hospital as BOY I. Unbeknownst to him, he too was be born with a brain enzyme deficiency, but in this envisaged variation, the surrounding tone is not one of

---

[2] *See* Antonio R. Damasio, *Descartes' Error: Emotion, Reason, and the Human Brain* (New York, HarperCollins, 1994) 207–12 (surveying the neuroscientific research that supports the author's 'somatic marker hypothesis').

expected honesty enveloped in loving kindness, at least not as we understand those terms as abstractions from educated Western norms.[3] Here, in contrast, DYS-FUNCTION (writ in big letters) is the prevailing norm: assume BOY II came from a mostly single-parent, socially (if not economically) impoverished household, which blended together with violence or alcohol or drugs or toxic waste or all of the above. Models for the control of his impulses were wholly lacking. Now the seven-year-old kid is caught stealing—a toy for a fish bowl or a pack of cigarettes or some Hostess Twinkies or whatever. Instead of standing mute, overwhelmed by the emotions of an embryonic sense of accountability, though, BOY II runs when he is nabbed, and does so for a variety of reasons imprinted as representations and dispositions that merge together in a vague but familiar image of the world he lives in and of his orientation to it. He runs to escape detection, runs to escape punishment (with which he is only too familiar), runs to make off with his entitlements, runs because that is what his somatic markers signal:[4] the message—'Go, BOY II, Get outta here!'—courses through his body. The features on his face tighten, his heart rate and pulse rise also, and his skin conductance doubtless varies. He too is engaged in adaptive behaviour, and his penchant for stealing and running may continue well into the future. BOY II felt fear but he did not cry. And BOY II was doubly unlucky: not only was he abused as a child, he too suffered BOY I's deficit that prevented some key neurotransmitters from shutting down when they should have. Later, he might say 'F**k it!' when asked about his future. He did not care.

Why would he say that? Why did he not, like BOY I, stand and cry and move on to mature reactions of shame and the healthy deterrence it generally fosters? Is it lack of impulse controls? Intellectual controls? Emotional controls? Opportunities? All of the above?

Or maybe, given his enzyme deficit and the hideous environment he grew up in, it was just bad luck?

10.2 Morality and Moral Luck

10.2.1 The Kantian premise

Contemporary efforts to poke holes in the Kantian myth of the good will began in 1976 with an essay by Bernard Williams.[5] Williams made a frontal challenge on the Kantian notion that there is a form of moral value free from contingencies of all kinds. The problem Williams addressed began with the *Groundwork Principles of the Metaphysics of Morals*, where Kant articulated his goal:

---

[3] For some sense of the extent to which this imagined scenario tracks real-life circumstances: see, e.g., Jeffrey Fagan, 'Context and Culpability in Adolescent Crime' (1999) 6 *Va. J. Soc. Pol'y & L.* 507; (Note) Catherine R. Guttman, 'Listen to the Children: The Decision to Transfer Juveniles to Adult Court' (1995) 30 *Harv. C.R.-C.L. L. Rev.* 507.

[4] See text accompanying notes, below.

[5] Bernard Williams, 'Moral Luck' in *Moral Luck* (Cambridge, Cambridge University Press, 1981) 20 (originally published in *Proc. Arist. Soc'y* (supp vol. L, 1976)).

Do we think it a matter not of utmost necessity to work out for once a pure moral philosophy completely cleansed of everything that can only be empirical and appropriate to anthropology? That there must be such a philosophy is already obvious from the common Idea of duty and from the laws of morality. Every one must admit that a law has to carry with it absolute necessity if it is to be morally valid, that is, as a ground of obligation.[6]

That it was possible to construct a pure moral philosophy based upon 'pure thinking', altogether *a priori*, was deemed incontrovertible.[7] Kant contended that 'matters of morality [derived from and productive of an epistemologically a priori synthetic proposition] can easily be brought to a high degree of accuracy and precision even in the most ordinary intelligence'.[8] That he could free morality from all things anthropological followed from his conception of the 'good will', which is indifferent to ends:

it is good through its willing alone—that is, good in itself... Even if, by some special disfavor of destiny or the niggardly endowment of step-motherly nature, this will is entirely lacking in power to carry out its intentions,... even then it would still shine like a jewel for its own sake as something which has its full value in itself. Its usefulness or fruitfulness can neither add to, nor subtract from, this value.[9]

As Christine M. Korsgaard notes, Kant tried to show that 'our ethical judgments can be explained in terms of rational standards that apply directly to conduct or deliberation'.[10]

10.2.2 The challenge of moral luck

Williams was having none of it and neither was Thomas Nagel, who responded to, and in important ways disagreed with, Williams's original position. All rational people, *contra* Kant, are not capable of achieving moral or rational justification immune to luck, Williams argued, using the example of Gauguin's desertion of his family to pursue a life in art as an exemplar of decision-making in the oral domain that depends—ultimately(?)—on consequences: how things turn out.[11] Williams thus describes the Kant's 'attempt to escape luck... [as] so intimate to our notion of morality... that its failure may rather make us consider whether we should not give up that notion altogether'.

In some important ways, Nagel's understanding of moral luck went beyond Williams's, but it too was a head-on challenge to the Kantian notion of the good will. The problem with Kant's position, as both point out albeit in different

---

[6] H. J. Paton, trans., *Kant's Groundwork of the Metaphysics of Moral* (New York, Harper Torchbook, 1873) 55.
[7] See note 49, below.
[8] Kant, *Groundwork,* note 6, 57.
[9] Ibid. 60.
[10] Christine M. Korsgaard, 'Skepticism About Practical Reason' (1986) 83 *J. Phil.* at 5.
[11] Not surprisingly, many commentator's disagree with Williams in the particulars at least. See, e.g., Don S. Levi, 'What's Luck Got to Do with It?' in Daniel Statman (ed.), *Moral Luck* (Albany, SUNY Press, 1993) at 109, 115 (noting that the importance of Gauguin's leaving his family 'fit[] his action into the pattern of self-deception that may become visible only after all his history is known to us').

contexts, is this: whatever conduct one undertakes, whether one fulfills one's goals or fails to fulfill them—that conduct and its moral context are *in fact* affected by factors outside the actor's control, and still we find them guilty despite the lack of control over what are often important condition. Nagel describes the problematic phenomenon this way: whenever a substantial part of the reasons for and outcome of all that one seeks to accomplish is influenced by factors beyond one's control, yet we continue to assess blameworthiness based on the actor's effects, we are dealing with moral luck.

Although Nagel describes four different instantiations of moral luck, two are most important in the present context: constitutive luck, 'the kind of person you are', a category that includes one's 'inclinations, capacities, and temperament' and including heredity and environment; and 'circumstantial luck', that is 'the kind of problems and situations one faces'.[12] The main points here are two: the gifts (or curses) one is born with and the kinds of moral tests one faces matter in terms of who the individual becomes and what outcomes he seeks. Margaret Urban Walker makes the salient point:

Moral luck is part of a picture of *impure agency*: agency situated within the causal order in such ways as to be variably conditioned by and conditioning parts of that order, without being able to draw for *all* purposes a *unitary* boundary to its exercise at either end, nor always for *particular* purposes a *sharp* one.[13]

Walker agrees with Nagel that our entanglement with the world, from an external perspective, causes the problems of moral luck: 'The truth of moral luck that the rational, responsive moral agent is expected to grasp is that *responsibilities outrun control*, although not in a single or simple way.'[14]

That individual responsibilities outrun our ability to control events is given; we could not sustain a moral order without this recognition. But when we make juridical judgments we are judging the individual himself: 'We are judging *him*, rather than his existence or characteristics.'[15] It is the case that we do not judge ourselves as simply cogs in the world's machine: 'We are unable to view ourselves simply as portions of the world, and from inside we have a rough idea of the boundary between what is us and what is not... We apply the same essentially internal conception of self to others', in moral judgement.[16]

Now to the basic question: How should we approach the judgement of individuals who we know or should know suffer neuropsychological deficits for which they are innocent, who were reared in abusive environments, and because of both unlucky conditions they are substantially more likely to engage in violent

---

[12] Thomas Nagel, *Moral Luck* in *Mortal Questions* (Cambridge University Press, 1979). at 28. The other two types of 'moral luck' include antecedent luck, 'the stripped-down acts of the will itself', and consequential or resultant luck 'in the way one's actions and projects turn out'. Ibid. Although antecedent luck could be important in this context, as Nagel notes, in fact antecedent circumstances are largely determined by constitutive and circumstantial luck. See ibid. at 35.

[13] Margaret Urban Walker, 'Moral Luck and the Virtues of Impure Agency' in Statman note 11, above, 243.

[14] Ibid. at 241.

[15] Nagel note 12, above, 36 (emphasis added).

[16] Ibid. at 37.

unacceptable behaviour? This problem is made all the more acute by the existence of fundamental attribution error; the nearly universal tendency, when a naive observer (such as a juror) is ascribing causation to an event, to over-ascribe a harmful outcome to the individual's disposition, and underestimate situational influences on behaviour.[17] The bottom line is that we tend to demonize the actor, distance ourselves from any responsibility therefor, and ultimately abandon him.[18] This seems particularly unfair when the individual whom we judge bears little or no responsibility for the deficits he suffers.

10.3 Genetic/Environment Interactions (GxE)

How is it that a minor deficit in a brain enzyme no one outside a small circle of neuroscientists has ever heard of can significantly affect one's predisposition to violent action? This part and the next survey the operation of selectional forces generally, and the effects of MAOA specifically. The answer lies in the open-endedness of the forces of selection.

10.3.1 Open-ended selection

The process of selection, operating on all creatures, entails the 'continual adaptive matching or fitting of elements in one physical domain [generally the prefrontal lobe of the cerebral cortex] to novelty occurring in elements of another [the world around us]'.[19] Natural selection operates at the level of genomes—changing or mutating our genetic alphabet—and it does so on the basis of 'recognition' rather than 'information'. The receptive/adaptive resources produced by natural selection, when operating within normal ranges of effectiveness, stand ready to absorb information from the outside and respond automatically thereto. We are designed to be that way. And these processes are always on duty, usually for the better, sometimes for the worse.

One way to illustrate the distinction between 'recognition' and other forms of information exchange is by reference to a familiar physical process, our immune system, which, like all of our systems, operates on the basis of *selection* by *recognition*. When an antigen or pathogen or almost any form of invasive trauma occurs, our immune system moves into action. The system is a spectacular biochemical product of natural selection working over time on our genome, in which various decentralized resources are recruited to attack the invaders. The system immediately identifies, and thereafter stores and remembers, products in the body that are not us. The system's magic lies in it ability to *recognize* immediately invading objects

---

[17] Theodore Y. Blumoff, 'The Problems with Blaming' in Michael Freeman and Oliver R. Goodenough (eds.), *Law, Mind and Brain* (London, Ashgate Publishing, 2009) 155–60.
[18] Garth Williams, 'Blame and Responsibility' (2003) 6 *Ethical Theory and Practice* 427.
[19] Gerald M. Edelman, *Bright Air, Brilliant Fire: On the Matter of the Mind* (New York, Basic Books, 1992) 74. Unless otherwise noted, the material in this paragraph and the next are from Edelman.

that are foreign. When 'non-self' invaders appear, a systemic biochemical process springs into action as lymphocytes recognize and bind to the molecular non-selves, targeting them for removal and destruction.[20] The majesty and mystery of the process is that the encroaching outsiders do not advertently pass information to the immune system about their novel qualities; rather, our immune system recognizes non-self antigens without an obvious information exchange from the invaders to the responders. (Within the system, communication among and between neurotransmitters and hormones and various groupings regulated in the brain is instantaneous: neurons fire within nanoseconds.) In other words, the immune system exists within us and arrives at problems ready to recognize and react to foreign invaders.[21] This selection system's repertoire is open-ended and pervasive—that is, it responds in an open way to any novel non-self invader. Selection operates within every sphere of our beings.

Despite our persistent primitive intuitions, within this selected system there is no single neurobiological control centre.[22] We do not work that way. Rather, our capacities are dispersed among various neuronal groupings and sub-cortical mechanisms in a process of diffusion that often is the progenitor for specific capacities that generally vary along familiar standard distributions.[23] This occurs because, like our immune system, to survive we must continually select and develop *based on our actual—and highly varied—experiences* so that on any day in question our cognition, our perception, and the processes we bring to our choices necessarily vary, usually along a standard distribution.[24] To choose anything or nothing is to make a choice in this system because the neural resources we bring to such decisions are called upon in every case in which a decision is required. Two important points both inform and follow from this biological fact: first, our hard-wiring—our genotype—was formed at a time when simply surviving and passing on one's genes to progeny were all that life required and probably as much as one could perform. It accounts for roughly 99.9 per cent of our shared genetic heritage.

---

[20] For an explanation of the operation of lymphocytes, see Bruce Alberts et al., *Molecular Operation of the Cell*, 4th edn (New York, Garland, 2002), at <http://www.ncbi.nlm.nih.gov/books/bv.fcgi?rid=mboc4.section.4422>.

[21] Edelman, *Bright Air*, note 19, above, 75–9; see also ibid. at 78 (noting that the immune system is a 'recognizing system [that] *first* generates a diverse population of antibody molecules and then selects *ex post facto* those that fit or match. It does this continually and, for the most part, adaptively.') Edelman won the 1972 Nobel Prize in medicine for this discovery.

[22] As one researcher points out, even within a single system, such as the visual system, dimensions like colour, motion, location, and object identification are processed in different areas of the brain. Adina L. Roskies, 'The Binding Problem' (1999) 24 *Neuron* 7.

[23] See Antonio Damasio, *The Feeling of What Happens: Body and Emotion in the Making of Consciousness* (New York, Harcourt 1999) 99; Edelman, note 19, above, 28–9; John R. Searle, *Mind, Language and Society* (New York, Basic Books, 1998) 90; Andrew E. Lelling, 'Eliminative Materialism, Neuroscience, and the Criminal Law' (1993) 141 *U. Pa. L. Rev.* 1471. This is not to say that there is not also domain specificity; there is. It is to say that along with specificity there is dispersal so that systems operate together to bring about perception. See, e.g., Semir Zeki, 'The Visual Image in Mind and Brain' in *The Scientific American Book of the Brain* (New York, Scientific American, 1999) 17–28.

[24] Joseph LeDoux, *Synaptic Self: How Our Brains Become Who We Are* (New York, Penguin, 2002) at 74–9.

Second, and at the level of phenotype, including that 0.01 per cent of difference or roughly 3,000,000 base pairs, each individual makes the choices he then *can* effect, which depends on the actual circumstances and experiences he encounters and to which he must adapt.

The take-home point here is so closely aligned with our intuitions—our folk psychology—that it should not need emphasis: the *actual circumstances* of our lives—including those that effect poor constitutive and circumstantial luck—affect both our morphology and individual development; these are indefeasible facts of human existence. Our common aphorisms—'chip-off-the-old-block', 'like father like son', 'the acorn never falls far from the tree'—illuminate our intuitions. Comparing our brains to our immune system thus 'shows that genetic evolution does not invariably lead to the kind of modularity that excludes open-ended processes. Instead, it can create processes that are themselves evolutionary and, therefore, capable of providing new solutions to new problems.'[25] Antonio Damasio states the crucial point:

> as we develop from infancy to adulthood, the design of the brain circuitries that represent our evolving body and its interaction with the world seem to depend on the activities in which the organism engages, and on the action of the innate bioregulatory circuitries, *as the latter react to such activities.*[26]

We *must* adapt.

10.3.2 Child abuse

The ever-attending GxE interaction is especially burdensome on abused children. A particularly pertinent example of this phenomenon, both as an exemplar of selection in general but, more so, as a specific genetically victimized class of young white men, was teased out of a multitude of data by a team of neuroscientists led by Avshalom Caspi and Terri Moffitt. The team tested the hypothesis that 'childhood maltreatment predisposes most strongly to adult violence among children whose MAOA is insufficient to constrain maltreatment-induced changes to neurotransmitter system'.[27] (MAOA, monoamine oxidase, is a brain enzyme located on the X gene that come in one of two forms, A and B. They are responsible, with other neurochemicals, for the necessary degradation of neurotransmitters after neurons have fired, thereby stopping the transmission of the previously successful neurochemical signal and permitting the next information

---

[25] David Sloan Wilson, *Darwin's Cathedral: Evolution, Religion and the Nature of Society* (Chicago, U. Chicago Press, 2002) at 30–1.
[26] Antonio R. Damasio, *Descartes' Error: Emotion, Reason, and the Human Brain* (New York, HarperCollins, 1994) at 111 (emphasis in the original).
[27] Avshalom Caspi et al., 'Role of Genotype in the Cycle of Violence in Maltreated Children' (2002) 297 *Science* 851, at <http://www.sciencemag.org/cgi/content/full/297/5582/851>. On the apparent limitation of this phenomenon to white males, see Cathy S. Widom and Linda M. Brzustowicz, MAOA and the 'Cycle of Violence: Child Abuse and Neglect, MAOA Genotype, and Risk for Violent and Antisocial Behavior' (2006) 60 *Biol. Psychiatry* 684.

processing neural signal to get through to the next neuron and so on.[28]) Researchers have known for years that the enzyme was strongly associated with aggressive behaviour.[29] Studying longitudinally more than a thousand Caucasian children, over half of whom were boys at various cohorts, the research team discovered that boys who suffered deficits in a key neurochemical *and* who grew up in an abusive environment were substantially more likely to engage in violent, anti-social behaviour that led to unwelcome interactions with the criminal justice system than were boys raised either with the deficit but in a reasonably healthy environment, or who were raised in a healthy environment without the MAOA deficit. Put otherwise, what the team found was that children raised in abusive environments differ significantly in the likelihood that they will engage in violent behaviour, 'depending upon whether or not their genotype conferred high or low levels of MAOA expression'.[30] Interestingly, the deficit in type A monoamine oxidase does not alone conduce to later violence; it is the combination of neurochemical anomaly *plus* abuse in the young child's environment that produce unwanted developmental patterns—GxE.

The central findings of the initial Caspi studies are summarized below:

For adolescent conduct disorder..., maltreated males (including probable and severe cases) with the low-MAOA activity genotype were more likely than nonmaltreated males with this genotype to develop conduct disorder by a significant odds ratio (OR) of 2.8 [95% confidence interval (CI): 1.42 to 5.74]. In contrast, among males with high MAOA activity, maltreatment did not confer significant risk for conduct disorder (OR = 1.54, 95% CI: 0.89 to 2.68). For adult violent conviction..., maltreated males with the low-MAOA activity genotype were more likely than nonmaltreated males with this genotype to be convicted of a violent crime by a significant odds ratio of 9.8 (95% CI: 3.10 to 31.15). In contrast, among males with high MAOA activity, maltreatment did not confer significant risk for violent conviction (OR = 1.63, 95% CI = 0.72 to 3.68). For self-reported disposition toward violence... and informant-reports of antisocial personality disorder symptoms..., males with the low-MAOA activity genotype who were maltreated in childhood had significantly elevated antisocial scores relative to their low-MAOA counterparts who were not maltreated. In contrast, males with high MAOA activity did not have elevated antisocial scores, even when they had experienced childhood maltreatment.[31]

---

[28] See 'Monoamine oxidase' in (2007) *Sci-Tech Encyclopedia*, 22 May, at <http://www.answers.com/topic/monoamine-oxidase>. MAOA is necessary to break down three important neurotransmitters that trigger or inhibit the transmission of nerve impulses. One, norepinephrine, raises blood pressure and increases alertness as part of the body's fight or flight response. The others, serotonin and dopamine, are involved in regulating mood and alertness; imbalances of these neurotransmitters are often found in patients with psychiatric disorders. All are also implicated in aggressive behaviour. Sarah Richardson, 'A Violence in the Blood' (1993) *Discover*, 1 October, at <http://discovermagazine.com/1993/oct/aviolenceinthebl293>.

[29] H. G. Brunner et al., 'Abnormal Behavior Associated With a Point Mutation in the Structural Gene for Monoamine Oxidase A' (1993) 262 *Science* 578.

[30] 'Gene-environment Interactions and the Brain' (2006) *Brain Ethics*, 25 July, at <http://brainethics.wordpress.com/2006/07/25/gene-environment-interactions-and-the-brain/>.

[31] Caspi et al., note 27, above, page cites unavailable on line (emphasis added, references to graphic figures omitted). The team surveyed 1,037 individuals (52 per cent male) at various age cohorts: 3, 5, 7, 9, 11, 13, 15, 18, and 21 and was virtually intact (96 per cent) at age twenty-six years. They used a

Note the crucial distinction: only those males with low-MAOA activity who *also* suffered early childhood abuse are at significantly higher risk (1); they are 9.8 times more likely to be convicted of violent crimes than males from any of the other three groupings. Those who suffered the deficit but were not abused (2), those who do not suffer the deficit and are abused (3), and those who do not suffer the deficiencies are not abused (4)—all fared the same.

In subsequent work, the researchers confirmed their earlier findings, expanded upon them, and mostly advanced the basic notion that it is the environment that tends to produce the (mal)adaption in light of individual differences, and not a genetic deficit alone that causes the poor behaviour.[32] This is consistent with, in fact it is an example of, the primary realization that selection is an open-ended phenomenon:

Heterogeneity of response characterizes all known environmental risk factors for psychopathology, including even the most overwhelming of traumas. Such response heterogeneity is associated with pre-existing individual differences in temperament, personality, cognition and autonomic physiology, all of which are known to be under genetic influence. The hypothesis of genetic moderation implies that differences between individuals, originating in the DNA sequence, bring about differences between individuals in their resilience or vulnerability to the environmental causes of many pathological conditions of the mind and body.[33]

Although much research is necessary before these interactions are fully understood, Caspi and Moffitt's research clearly indicates that the 'gene–environment interaction approach assumes [and strongly suggests] that environmental pathogens cause disorder'.[34] Importantly, it is *we* who, in many instances, control the environment.

---

variety of standard measures to determine aggressivity, and studied four groupings: 1. Low MAOA Abusive; 2. Low MAO A Non-abusive; 3. High MAOA Abusive; 4. High MAOA Non-abusive.

[32] The study has now been replicated fourteen times; twelve of those subsequent studies confirmed the original Caspi findings. See, e.g., Kevin M. Beaver et al., 'Gene-Environment Interplay and Delinquent Involvement' (2009) 24 *J. Adolescent Research* 147 (abstract), at <http://jar.sagepub.com/cgi/content/abstract/24/2/147>; Debra L. Foley et al., 'Childhood adversity, monoamine oxidase A genotype, and risk for conduct disorder' (2004) 61 *Arch. Gen. Psychiatry* 738; M. Skondras, M. Markianos, A. Botsis, E. Bistolaki, and G. Christodoulou, 'Platelet Monoamine Oxidase Activity and Psychometric Correlates in Male Violent Offenders Imprisoned for Homicide or Other Violent Acts' (2004) 254 *European Archives of Psychiatry and Clinical Neuroscience*, 380 (finding additional evidence implicating low activity of MAOA as a risk factor for violent behaviour); Jack P. Shonkoff et al., 'Neuroscience, Molecular Biology, and the Childhood Roots of Health Disparities: Building a New Framework for Health Promotion and Disease Prevention' (2009) 303 *JAMA* 2252, abstract at <http://jama.ama-assn.org/cgi/content/abstract/301/21/2252>; full text at <http://jama.ama-assn.org/cgi/content/full/301/21/2252>; Joan Arehart-Treichel, 'Gene Variant, Family Factors Can Raise Conduct Disorder Risk' (2004) 39 *Psychiatric News* 38, at <http://pn.psychiatryonline.org/cgi/content/full/39/17/38>.

[33] Avshlon Caspi and Terri E. Moffitt, 'Gene-environment Interaction Research and Neuroscience: A New Partnership?' (2006) 7 *Nature Reviews Neuroscience* 583 (quoted in *Brain Ethics*, at <http://brainethics.wordpress.com/2006/07/25/gene-environment-interactions-and-the-brain/>).

[34] Ibid.

10.3.3 MAOA, the environment, and causation

Although there are many different approaches to questions concerning the conditions required for holding one morally responsible, in general it is fair to say virtually all such approaches hold individuals morally responsible provided, first, they have the wherewithal to know that they will be held responsible for their actions and the capacity to control their conduct.[35] And that combination of capacity and opportunity underscores our intuitions about moral certainty. To be held responsible for one's actions requires, second, a kind of moral education for a child that begins with a certain kind of moral discipline. This includes learning the idea of individual agency (the self as a source of effect on the world), learning praise and blame, and learning what being 'held' responsible means, that is being a target sensitive to the values of praise or blame and the appropriate moral psychological stances that follow that social knowledge.[36] This requires opportunity, repetitive efforts at teaching and learning moral indignation.[37] It also requires capacity. Deficiencies in both opportunity and capacity place significant obstacles to appropriate social behaviour. Note, then, that within the basic account of the conditions for attributing moral responsibility, there is little room for children who are the products of both a violent, abusive childhood *and* bad moral constitutive luck.

Here it is worth looking more carefully at the crucial distinction within the MAOA data. 'For adult violent conviction..., *maltreated males* with the low-MAOA activity genotype were more likely than *nonmaltreated males with this genotype* to be convicted of a violent crime by a significant odds ratio of 9.8 (95% CI: 3.10 to 31.15).' Ten-fold. Thus for both groups with low MAOA, the abusive and non-abusive, the difference is accounted for solely by virtue of their significantly abusive and/or neglectful early childhood experiences. Caspi and others are right to speak in descriptive terms: associations and correlations. But the question for present purposes is how one moves from those objective facts to the normative, from data to policy? That 'is' does not lead directly to 'ought' means at most that we have to address the 'is' to determine the 'ought'. As Hume pointed out, 'nature will always maintain her rights, and prevail in the end over any abstract reasoning whatsoever'.[38] And when it comes to policy, it is we who fill whatever gap separates the two nature, G, and the environment, E.[39]

---

[35] See, e.g., Peter Strawson, 'Freedom and Resentment' in Gary Watson (ed.), *Free Will* (Oxford, Oxford University Press, 1982).

[36] See, e.g., John Martin Fischer and Mark Ravizza, *Responsibility and Control: A Theory of Moral Responsibility* (Cambridge, Cambridge University Press, 1998) 208–10.

[37] Ibid. Of course, Aristotle made this point millennia ago in his *Nicomachean Ethics* 1104b11–13.

[38] David Hume, *An Enquiry Concerning Human Understanding* (ed. L. A. Selby-Bigge and P. H. Nidditch, 3rd edn (Oxford, Oxford University Press, 1975)) at 41. In what reads like a direct challenge to Kant, (but was probably directed at Samuel Clarke, see John Rawls, *Lectures on the History of Moral Philosophy*, ed. B. Herman (Cambridge, MA, Harvard University Press, 2000) 70–3), Hume disputes the distinction between reasoning and experience, arguing that the former gives form to the latter. Reasoning is the process of making sense of our experience and *not* the 'result of our intellectual faculties, which, by considering *a priori* the nature of things... [somehow] examin[es] the effect, that must follow from their operations'. Hume, above, at 43–5, note 1.

[39] See Morris B. Hoffman, 'Evolutionary Jurisprudence: The End of the Naturalistic Fallacy and the Beginning of Natural Reform', in this volume.

One of the primary reasons neuroscience and genetics should inform law and policy lies in the ability to make use of new data that challenges our otherwise fairly healthy folk psychology. Science either affirms our folk wisdom, which it does most of the time, or it debunks it. Science has played that role in our culture at least since Copernicus and later Galileo. The question, then, is how much data do we need to conclude that, with respect to this class of people at least, we ought to seek incapacitation, as necessary, and medication, where it exists. And there is some hope that a pharmaceutical fix may lie in the future.[40]

10.3.4 The importance of GxE

Studies on MAOA, it turns out, are only one small example of the larger take home point. Allan Gibbard articulates an important basic point: 'The genetic plan for a human being will be full of contingency plans: full of schemes that in effect say "If A then do X, whereas if B then do Y."'[41]

Given a difference in how two people act, it is perfectly biological to say something like this: the two people's genetic plans [their genotypes] are the same in relevant respects. They've encountered, though, different cues as to their circumstances. The cues the two have encountered differ in ways for which the single genetic plan they share makes provision. The plan they share is to respond one way given one set of cues and another way given the other. The cues in question may be immediate ones, or they may be cues that came years ago in childhood and have affected the development of psychic mechanisms or the setting of parameters for them.[42]

Raise a child in an abusive environment who, through no fault of his own, suffers an important neurochemical deficit, and the likelihood is significantly increased that he or she will become a violent adult.[43] But we knew that all along, just as we have known that the absence of touch and warmth in the early stages of

---

[40] A team of Brazilian researchers have recently published findings suggesting that MAOA subjects given methylphenidate, a chemical that increases dopamine levels in the brain (see NIDA InfoFacts, 'Stimulant ADHD Medications—Methylphenidate and Amphetamines' (revised, 2009), at <http://www.nida.nih.gov/infofacts/ADHD.html>), are offered some relief from aggressivity. A. P. Guimarães et al., 'Maoa Is Associated with Methylphenidate Improvement of Oppositional Symptoms in Boys with Attention Deficit Hyperactivity Disorder' (2009) 12 *Intern'l. J. Neuropsychopharmacology* 709. Of course, the road from there to a genuine 'fix' is long, arduous, and (above all) uncertain.

[41] Allan Gibbard, 'Genetic Plans, Genetic Differences, and Violence: Some Chief Possibilities' in David Wasserman and Robert Wachbroit (eds.), *Genetics and Criminal Behavior* (Cambridge, Cambridge University Press, 2001) at 173.

[42] Ibid. at 174. For example, there is rich data in the neuroscience literature indicating that early maternal rejection, physical or psychological abuse, and exposure to environmental toxins conduce to violent behaviour as the child matures. See, e.g., Gary W. Evans and Elyse Kantrowitz, 'Socioeconomic Status and Health: The Potential Role of Environmental Risk Exposure' (2002) 23 *Ann. Rev. Pub. Health* 303; Dan Orzech, 'Chemical Kids—Environmental Toxins and Child Development' (March/Apr 2007) 7 *Social Work Today* 37.

[43] See also Martin H. Teicher, 'Scars that Won't Heal: The Neurobiology of Child Abuse' (2002) 286 *Sci. Amer.* 68.

development conduces to more violence,[44] that child abuse generally can affect concentrations of certain cerebrospinal fluids,[45] and that exposure to environmental toxins can produce deviant behaviour[46]—We have known all this for a very long time. Yet too often we refuse to move.[47]

10.4 Two Objections

There are commentators who insist that there is no such thing as moral luck, and others who argue that even if there is, it should make no difference in our assessments of blameworthiness. As to those who insist on both, perhaps there is no such thing as moral luck: for them the belief in a dense transcendent morality is real, but it is, in the end, a tautology. If one begins convinced that there is a Platonic/Kantian sphere of life in which each of us is and can be deemed worthy, then the problem of moral luck disappears; it cannot exist in a sphere that, *a priori*, foreswears all matters pertaining to anthropology.[48] And then what are we left with? That is the question.

But one should ask what it is that animates an idea inferred intuitively from a wholly formal ideal of rationality and autonomy,[49] one which alone determines the self-constructed categorical goodness of the will. What one finds is a conception that produces a particular and unreal view of humans as moral beings. The worth of the Kantian conception is that it 'provide[s] a shelter against luck, one realm of

---

[44] Tiffany Field, 'Violence and Touch Deprivation in Adolescents' (2002) 37 *Adolescence* 735.

[45] Randall A. Ruppel et al., 'Excitatory Amino Acid Concentrations in Ventricular Cerebrospinal Fluid After Traumatic Brain Injury in Infants and Children: The Role of Child Abuse' (2001) 138 *J. Pediatrics* 1.

[46] Eric Taylor, 'The Roots and Role of Violence in Development' in Pamela Taylor (ed.), *Violence in Society* (London, Royal College of Physicians, 1993) at 49–52.

[47] Also see Avshalom Caspi et al., 'Neighborhood Deprivation Affects Children's Mental Health: Environmental Risk Identified in a Genetic Design' (2000) 11 *Psychol. Sci.* 338; Catlin M. Jones, 'Genetic and Environmental Influences on Criminal Behavior', at <http://www.personalityresearch.org/papers/jones.html>; Dorothy Lewis, 'From Abuse to Violence: Psychophysiological Consequences of Maltreatment' (1992) *J. Amer. Acad. Child and Adolescent Psychiatry*, at <http://www.pbs.org/wgbh/pages/frontline/shows/little/readings/lewis.html>.

[48] E.g., Michael J. Zimmerman, 'Luck and Moral Responsibility' in Statman, note 11, above, at 230.

[49] As best I can strip a remarkably sophisticated argument based on a pure form and exercise of thought without unduly compromising the idea, the synonymity of 'autonomy' and 'rationality' seems to come to this: the first categorical imperative demands that a rule be 'universalizable'; that is, that the rule or maxim must apply at all times to all rational persons: 'I ought never to act except in such a way *that I can also will that my maxim should become a universal law.*' At the same time, to be autonomous means that one participates in moral reflection by prescribing 'universal laws [which are, therefore,] free in respect of all laws of nature, and obeying *only* those which he makes himself. If a maxim is itself a universal requirement derived from rational intuition, and if to qualify as a "law" means it must hold, not merely for men, but for all *rational beings as such*, not merely subject to contingent conditions and exceptions, but *with absolute necessity*', then the distinction between rules that govern others and rules that govern oneself collapses. All rational individuals are not only autonomous but, by definition, can legislate rationally only identical universal rules. In this way, rationality and autonomy are, in the end, coextensive: prescribing and (in theory) living by the only formally conceivable universal laws. And that view of the matter is, ultimately, immune to deep epistemological inspection.

value (indeed, of *supreme* value) that is defended against contingency'.⁵⁰ And therein lies the problem: inasmuch as this treasured ontological feature is inured to luck and (ultimately) closed to justification, it cannot serve as the only basis for justice.⁵¹ Kant made this point in his jurisprudential writings, when he separated the offices of law and morality.

In so far as our legal judgments are rooted too firmly and exclusively in our formal commitment to a discourse of subjective culpability (the will), they are under-analysed and reflect a particular, unforgiving vision of the interface between moral philosophy and jurisprudence. What matters with respect to the Kantian will is the agent's intention, not its effect on the outside world. Thus, moral conduct cannot be coerced. In contrast, the office of criminal law, with its heavy reliance on deterrence, exists not to protect freedom or morality; by hypothesis that simply cannot be done. Rather, it protects the *possibility* of freedom.⁵² Justice mediates between autonomous individuals, having 'regard only to the external and practical relation of one person to another, ... [to the] relation of [one person's] free action to the freedom of *action* of the other..., [and] in the reciprocal relation of voluntary action'.⁵³ In this context, 'relation' governs and presupposes individual free wills acting in conjunction with one another such that 'only the *form* of the transaction is taken into account'.⁵⁴ Whereas morality and a totally subjectivist view of culpability look to motivation alone, a legal theory must examine the external relationship—including its constituents and consequences—of one's action on another's freedom and, especially, on whether one's actions interfere with the freedom of another.⁵⁵ We are thus left to our own sensibilities about what does and, beyond that, what should work in the world we occupy. I offer only tentative answers.

The neuroscientific basis of moral luck thus presents no threat to free will. Although free will always be a normative construction, that some of us are freer than others is undoubtedly true, where a free action includes not only capacity but opportunity to act. Here it is worth a minute to look at the arguments of Professor Stephen Morse, who maintains that because our concept of free will is a social construction, nothing the brain sciences show us will undermine our ability to choose, and we live in a world where free choice is presumed to be available for almost everyone. In a fairly recent article, Morse asserts that although the hard empirical disciplines might provide potentially relevant knowledge about human behavioural capacities, it 'must fall silent about the ultimate criteria the law adopts'.

---

⁵⁰ Bernard Williams, 'Moral Luck: A Postscript' in *Making Sense of Humanity and other Philosophical Papers, 1982–1993* (Cambridge, Cambridge University Press, 1995) (emphasis added) 241.
⁵¹ See Susan K. Houser, 'Metaethics and the Overlapping Consensus' (1993) 54 *Ohio St. L. J.* 1139 (noting that intuitionism entails circularity precisely *because* it purports justify itself–that is, it insulates itself from external justification) at 1147.
⁵² Bruce Chapman, 'Agency and Contingency: The Case of Criminal Attempts' (1988) 38 *U. Toronto L.J.* 355 at 364.
⁵³ Immanuel Kant, 'Introduction to the Science of Right' in *The Philosophy of Law* (trans. by W. Hastie; Edinburgh, Clark, 1887) at 44–5.
⁵⁴ Ibid. at 45.
⁵⁵ Chapman, note 52, above, at 365.

This is the case because '[n]o general finding from any other discipline entails any general legal conclusion about legal responsibility unless it *conclusively undermines the possibility of responsibility at all*, in which case it is the basis for an external rather than an internal critique'.[56] Because all legal standards are normative, he concludes, '[n]o normative differences are logically entailed by behavioral differences unless, counter-factually to reality, the behavioral differences precisely track the normative differences'.[57]

Morse elaborates on the distinction between critiques internal and external to law. As to the former, he hypothesizes two relatively narrow classes of case, one dealing with neuroscientific evidence of something like automatism and the other evidence of brain damage; in both cases such evidence might assist a fact-finder. Although in both cases the evidence does track a specific legal standard, neither class is likely to change law beyond the odd case in which the evidence was relevant. So far so good. He then moves to the external critique of legal standards. He rightly rejects the likelihood that neuroscience will demonstrate that we are all automatons, that intentional brain states are themselves nonexistent. Morse aligns this position with eliminative materialism, the position that our common-sense psychology is radically wrong, 'that mental notions like belief or sensation could simply be abandoned in favor of a more accurate physiological account'.[58] He then gives his full embrace to compatibilism, rejecting incompatibilism, which he describes as the idea that no one 'can be genuinely responsible because neuroscience and other disciplines conclusively demonstrate that all our actions are mechanistically determined and determinism (or universal causation or some such) is incompatible with ultimate responsibility'.[59] From Morse's perspective, those who challenge compatibilism are stuck: to sustain their position, he argues, they are required to abandon the notion of responsibility *for anyone at any time* and accept that determinism 'is not selective or partial'.[60]

Here Professor Morse joins hands with Professor Michael Moore, who thoroughly rejects the common-sense factually incontrovertible idea that determinism can exist in degrees.[61] Strongly taking issue with a psychologist who argues that the feeble-minded lack the freedom of action and choice of a psychopath who lacks the freedom of action and choice that an '"average, reasonable" or "prudent" abstract standard man of the law', Moore writes:

[T]o speak of being partly determined or partly free makes as much sense as to speak of being partly pregnant. To be sure, we can make comparative judgments that one cause is

---

[56] Stephen J. Morse, 'Brain Overclaim Syndrome and Criminal Responsibility: A Diagnostic Note' (2006) 3 *Ohio St. J. Crim. L.* 397 (emphasis added), at 400.
[57] Ibid.
[58] *Eliminative Materialism*, Stan. Encyc. Phlos., at <http://plato.stanford.edu/entries/materialism-eliminative/> (citing W. V. O. Quine for this idea).
[59] Morse, note 56, above, at 402.
[60] Ibid. (emphasis added).
[61] Michael S. Moore, 'Causation and the Excuses' (1985) 73 *Cal. L. Rev.* 1091, at 1114–6.

more important than another in producing behavior. Indeed, there is quite a body of literature on the criteria we use in determining which conditions are more causally relevant than others in various contexts. But none of this literature can make sense of the quite different comparative judgment about the relative importance of all causes on the one hand, and of freedom on the other. For the degree determinist, it has to be sensible to ask: how much causation was there? The problem is that such a question seems to make no sense at all.[62]

Once again, we have an either–or situation; like Morse, one cannot abandon the idea of free will unless doing so *'conclusively undermines the possibility of responsibility at all'.*

Moore's argument is in at least one sense less defensible than Morse's. Moore's analogy to pregnancy, perhaps meant only as a light-hearted aside, is nonetheless absurd on its face. Likening being partially determined and partially free to being 'partly pregnant' attempts to compare a capacity for free action, a metaphysical construct that even Kant ultimately could not fully warrant,[63] to a demonstrable physical condition. The analogy is thus utterly flawed. We do not ask the jury to determine whether or not A is pregnant. If the question should arise as an issue in a real case, its answer is determinate; it is neither formal nor abstract. Absent extraordinary conditions, she either is or is not pregnant; there are ample testing regimens and (with time) visual proof available beyond any doubt by anyone that A is or is not pregnant. But to suggest that the average non-psychopathological individual enjoys the same freedom of action as the feeble-minded is a trivially simple formalism. Caspi and Moffitt's work with MAOA makes this point emphatically. Moore's dis-analogy fails to distinguish between legal responsibility—freedom of choice—and a factual condition—pregnant or not. Moreover, legally, we demand a finding. For example, in a proper case we might ask, 'Was D under duress or not under duress at time t?' We assign a burden of proof to this question, and the need to make such an assignment, standing alone, suggests that people may be uncertain about how much control D had in fact.

Second, Morse and Moore continually use the term 'causation' interchangeably with the term 'determinism', but doing so ignores the fact that when freedom of action issues actually arise in a proper case, the fact-finder is required to make a probabilistic determination about the mental state of the defendant: based on the evidence presented, did *this* actor, beyond a reasonable doubt (or some lesser probabilistic standard)—pick one: a) have the requisite capacity to form an intention; b) show the amount of resistance that a reasonable person would show; c) suffer diminished capacity, lose control of his emotions; and so on? These questions *assume* some amount of what Moore has dismissed as 'degree determinism'. The

---

[62] Ibid. at 1115.
[63] In his monumental effort to defend free will, Kant ultimately defends freedom normatively in the face of physical necessity: '[T]o argue away freed is as impossible for the most abstruse philosophy as it is for the most ordinary human reason. Reason must therefore *suppose* that no genuine contradiction is to be found between the freedom and the natural necessity ascribed to the very same human actions': H. J. Paton (ed.), *Kant's Groundwork of the Metaphysics of Morals* (New York, Barnes & Noble, 1873) at 116.

major point is that these legal issues are not binary. Although it is true that each juror, in the end, exercises a single 'yes' or 'no' vote on the question, the dynamics of the jury room in reaching one decision or another—argument, struggle, negotiation over each charge brought and the ultimate decisions reached—suggest a system that understands the nuance of partial determinism.

But the most important point, an elaboration on the last point, is one that Moore and Morse seem to overlook. When issues of responsibility arise in the course of ordinary litigation we have some reason to question the actual intellectual and/or volitional capacity of the individual who faces incarceration or even death. By dismissing the idea that different individuals raised in different environments have different levels of capacity, and that the law might take cognizance of that undeniable fact in one way or another, they beg the essential question that the new findings from neuroscience deliver. On virtually every measure one can think of, our capacities, including the capacity to make socially acceptable choices, are distributed among us in standard fashion.[64]

The law can take account of this fact, not in its substantive domain, but in its punishment system, by taking seriously—rather than paying lip service to—the exhortation to consider rehabilitative and therapeutic measures. As previously noted, the Model Penal Code's division of *mens rea* into four categories constitutes implicit (but only) partial recognition of the way in which our control functions actually operate—on a continuum.[65] That courts understand the inapplicability of a binary approach is necessary because the drafters take the firm position generally that 'questions about determinism and free will' have no place in its description of substantive crimes.[66] Rather, the Code's comments state that courts should take degree determinism into account in the sentencing phase of trial.[67]

Finally, their arguments are, in effect, extended tautologies. If determinism is not 'selective or partial', as Morse puts it, or if partial determinism 'makes as much sense as to speak of being partly pregnant', as Moore has it, then one begins the argument with an understanding of free will that rejects anything other than full-throated compatibilism in all but the most psychopathologically defective individuals, who are treated as if they have *no* free will. In their telling, the only possible conclusion is that uncompromising compatibilism is, and partial determinism never is, consistent with free will, so defined.[68] Neither seems open to evidence, which is a surprising conclusion given that both are also psychologists who practice

---

[64] See, e.g., Robert Plomin, *Nature and Nurture: An Introduction to Human Behavioural Genetics* (Pacific Grove, CA, Brooks/Cole, 1990).
[65] See Model Penal Code, section 2.02 (ALI 1962 Draft).
[66] Ibid. cmt to § 2.01, at 215 (1985) (requiring a voluntary act).
[67] See, e.g., cmt 2 to §2.09 at 6 (Tent. Draft No. 10, 1960).
[68] Professor Morse's conclusions follow because he assumes, as most of us do, that 'free will is a social construction'. Its details are often socially constructed and so forever open to criticism and change. But that truism raises a different, more fundamental question, namely, is there a core to the idea of free will that is, in fact, a part of human nature? This is not the place to answer such a difficult question but it is worth noting that the belief in individual free will is universal, that it plainly has adaptive qualities, and that it is functional as well as ideal.

in a discipline that seems focused mostly on those problems that neurologists and neuroscientists are working to explain empirically.

And this suggests that there seems to be some sort of odd, implicit dualism at work wherein one can concede material causes all the way down *and* assume that there is a sharp causal break between our brain processes and our belief in a free will. How can those apparently inconsistent views be reconciled? I am not certain I can answer this question but I do think I can at least lay out some of the possibilities. Either one really does not (or should not) believe in this epistemic break but remains unwilling to refine all the processing necessary to bring our punishment practices in line with the reality of a causal scheme that exists on a continuum from less to more capable of both possessing and of exercising our choice-deliberative capacities. Or, perhaps, one does not believe in the break but is convinced that we, as a society, have already achieved as much as we can to help those less fortunate than ourselves; we are content with the settled understanding. (After all, belief in free will goes back at least to Hebrew Scriptures.[69]) Or, one could believe there is a divide but has to concede that, epistemologically, this is the best we can do by way of articulation and so we continue to move 'as if' we can justify our belief in free will. Or, finally, we are driven by ideology or belief of one sort or another to trust in extra-causal forces in the world. In the alternative, one can concede that, in fact, some individuals have more freedom of choice than others and refine our punishment practice to account for that reality.

10.5 Modest Proposals

In the past, I have suggested that the judicial system should, in general, adopt a policy that moves forward initially on two fronts (understanding that the content of these categories is, for now, undetermined).[70] The two overriding criteria are capacity for meaningful cognition and volitional control, on one axis, and opportunity for moral education, on the other. The content of both criteria will be determined in actual cases, in part, by constitutive, circumstantial and antecedent luck. How broadly applicable this paradigmatic view of public assessment should be *is* the next question.

The MAOA studies and the facts of constitutive and circumstantial moral luck make a strong case for the impact of GxE on violent behaviour. We know that early exposure to abuse, neglect and a host of toxins can also effect subsequent conduct. That being the case, two goals come immediately to mind, one directed at legislatures, the second at legislatures and the courts. In the long term, we should use this information as a lever to push policy makers to invest far more heavily than

---

[69] 'Behold, I set before you this day a blessing and a curse' (Deut. xi. 26). Why is it stated this way, since it has likewise been said, 'See, I have set before thee this day life and good, death and evil' (Deut. xxx.15)? See Abraham Cohen, *Everyman's Talmud: The Major Teachings of the Rabbinic Sages* (New York, Schocken Books, 1995) at 93.

[70] See, e.g., 'The Problems with Blaming', note 17, above, at 171–2.

they do now in eliminating, or at least minimizing, early childhood abuse and neglect of all forms, including victimization by toxic wastes. The second goal is therapeutic/restorative, that is using the tools of rehabilitation, where they are available, to mitigate the effects of neurotransmitter deficits as part of treatment for those who commit crime along with some conferencing approach. The two goals, at least in this instance, are interrelated.

Early intervention in abusive family dynamics is obviously a worthwhile aim; no one doubts the motivation for the goal. But it is a risky goal as well. Space prohibits all but the most cursory look at this issue but 'primary prevention', that is, preventing abuse before it occurs, is a goal of all advocates for the reduction of child abuse of all types.[71] Studies that confirm the GxE implications of early childhood abuse merely affirm what those advocates have already known. But the potential problem is equally obvious: how do we balance that need against the intrusiveness of the tactics meant to realize the goal? The policy issue raises sensitive questions about early childhood testing, institutional data separations, which raises serious constitutional issues in every liberal democracy. We can envision the entire parade of horribles that might be raised in opposition to this early intervention, by which I mean some system that requires at least: 1) substantial, documentable reason to believe that intervention is necessary, 2) testing based on observed and memorialized observations, and 3) creating a permanent reviewable record testing.[72] Suffice it to say for now that my educated intuition suggests that intervention is necessary to improve outcomes over time but clearly not sufficient, and it is insufficient in part because such a measure could trigger our most paternalistic drives—not a pretty picture!

That said, my goals are appropriately modest. MAOA deficits have been identified with aggressive tendencies for nearly a generation,[73] and research continues to find promising therapies are in the offing.[74] If those researches lead to a sufficiently promising pharmaceutical therapy, that finding could play a significant role in bringing those with unwanted aggressive tendencies under positive rehabilitative controls. Chemical therapy alone may not be sufficient to undo the years of habituation that can bring a kid to the dock; some kind of moral education, including conferencing, restitution, and prolonged talk therapy may be required. But that seems a small price to pay when the cost of housing an individual in the penitentiary, where his tendencies are very likely to be exacerbated, can run

---

[71] E.g., Lesa Bethea, 'Primary Prevention of Child Abuse' (1999) *Amer. Family Physician*, 15 March, at <http://www.aafp.org/afp/990315ap/1577.html>; Janet Carter, 'Domestic Violence, Child Abuse, and Youth Violence: Strategies for Prevention and Early Intervention' 14 March 2005, at <http://www.mincava.umn.edu/link/documents/fvpf2/fvpf2.shtml>.

[72] I thank Cristoph Engel for raising this issue at a recent conference and talking me through the fairly recent Swedish experience.

[73] The history of MAOA research is described in 'Violence in the Blood', note 3, above.

[74] See 'Research from University Federal do Rio Grande do Sul in Area of Attention Deficit Hyperactivity Disorder Therapy Described' (2009) *NewRx*, 25 May (describing efforts to reduce symptoms of oppositional disorder), at <http://www.newsrx.com/health-articles/1526666.html>.

between $30,000 and $50,000 per year.[75] There is some empirical data, moreover, suggesting that the type of approach being advocated here has a positive influence on recidivism rates.[76]

There are many confounds in many recidivism studies, and not least the types of defendants who are willing to participate, but that should not deter efforts to use the data from genetics and neuroscience as levers to push public policy when the science justifies it. MAOA research presents just such a case. We know, conclusively, that repeated early childhood exposure to poisonous environments of all kinds disrupts healthy development. We can do something about this, if only we have the will to do so.

---

[75] These costs vary by state and prisoner. For example, five years ago there were estimates that elderly prisoners in American prisons can cost up to $70,000 per year: 'Elderly Inmates Swell Prison, Driving Up Costs' (2004) *USAToday.com*, 28 February, at <http://www.usatoday.com/news/nation/2004-02-28-elderly-inmates_x.htm>. An estimate of the average cost per prisoner for housing was roughly $24,000 per prisoner as of June 2007: 'Too Many Prisoners: States Should Stop Warehousing Nonviolent Offenders' (2008) *Washington Post*, 11 July, at <http://www.washingtonpost.com/wp-dyn/content/article/2008/07/10/AR9780199599844.html>.

[76] The most complete and appropriately moderated study I have seen is by Gwen Robinson and Joanna Shapland, 'Reducing Recidivism: A Task for Restorative Justice?' (2008) 48 *Brit. J. Criminol.* 337. Also see Teresa W. Carns, 'Therapeutic Justice in Action: An Evaluation of Three Therapeutic Courts', Alaska Justice Forum, Winter 2005, at <http://justice.uaa.alaska.edu/FORUM/22/1spring2005/b1_therapeuticcourts.html>; and Research Summary, 'Restorative Justice and Recidivism' (2003) 8 *Public Safety Canada*, January, at <http://www.publicsafety.gc.ca/res/cor/sum/cprs200301_1-eng.aspx>.

11

Neuroscience and Penal Law: Ineffectiveness of the Penal Systems and Flawed Perception of the Under-Evaluation of Behaviour Constituting Crime. The Particular Case of Crimes Regarding Intangible Goods

David Terracina*

11.1 Introduction. The Crisis of the Penal Systems

Contemporary penal systems have long suffered a profound crisis of ineffectiveness.[1] Judicial statistics show a crime rate that is constantly on the rise, whether blood crimes or crimes of a patrimonial nature. Also in constant ascent is the so-called '*cifra oscura*' (dark number, *Dunkelziffer*), a concept derived from criminology that refers to the divergence between the number of crimes effectively committed in a specific place at a specific time, and those brought to the attention of the judicial authorities.[2]

Obviously, the chronic ineffectiveness of the penal systems cannot depend on merely one factor. Otherwise, that single factor would have been identified long ago, and appropriate measures would have been put in place. Moreover, the universality of the above-mentioned ineffectiveness, involving in a more or less pervasive manner the near totality of the penal systems, leads to the probable conclusion that some of the causes are not linked to the peculiarities or characteristics of one single system, but are structural.

In addition, it is unlikely on the one hand that all factors determining a crisis of such proportions can be known; while, on the other hand, it is highly likely that the mechanisms of the factors known are not completely understood.

Hence, dealing with a combination of heterogeneous factors, both exogenous and endogenous, there can be no single solution that is able to restore the likes of

---

* Assistant Professor in Criminal Law at the University of Rome Tor Vergata.
[1] Enzo Musco, *L'illusione penalistica* (Milan, Giuffré, 2004) 117 et seq.
[2] Gabrio Forti, *L'immane concretezza. Metamorfosi del crimine e controllo penale* (Milan, Raffaello Cortina Editore, 2000) 64.

the penal systems by itself. The considerations above are so obvious that they do not need any justification of a scientific nature and that they could be acquired as tacit assumptions without fear of being contradicted. However, it would seem that there is just one possible remedy for our systems, appealing systematically nearly exclusively to penal sanctions.

All that is not surprising, and is easily explained. Appealing to penal sanctions is, in fact, the simplest and cheapest solution of all those available, and the one that pays the most at election and consensus times.

It is the simplest solution because the modern legislator is long from having developed a science of legislation that also makes use of ideas from criminology, from sociology, from psychology, etc., and through which efficient choices can be made regarding criminal policy. He has simply limited himself, over time, to using new cases in point of crimes, riding an emotional wave, each time linked to particular criminal phenomena, without worrying in the least about the possibilities of success of the adopted measures.

It is the cheapest solution because a reasoned science of legislation would surely require using vast resources to finance research and experimental activities.

Finally, it is certainly also the most profitable solution in terms of consensus, for different reasons. First, appealing to penal sanctions provides the fastest response the legislator can offer to the almost constant requests for security and protection by society members. Other alternative or supporting measures to the penal sanctions would in fact require long-term experimentation; and the effects, if any, would be noticeable too late, maybe to the benefit of a new political majority. Second, as already mentioned before, penal sanctions back up the most malevolent and irrational instincts of security that are present in a more or in a less significant manner in each of us.

11.2 A Role for Neuroscience

In this situation of absolute confusion, it may seem strange but cognitive neuroscience could provide useful instruments to comprehend some of the factors responsible for the ineffectiveness of the penal systems.

Let us look at how this is possible. First, there is no doubt about the fact that the bigger the conscience and the perception of the underevaluation and the unlawfulness of certain behaviours, the more likely that the laws sanctioning those behaviours will actually be respected by the majority of society members. The existing relationship between consensus and respect for the law has been demonstrated by penal doctrine and by criminology. Therefore, we can accept it as a given, without elaborating on it any further.

The flawed perception of the social and juridical under-evaluation of certain behaviours can depend on more diverse factors, which are more or less known. The tendency, however, is to call on factors of a cultural nature, like, for example, happens in the Anglo-Saxon systems where an effective excuse or mitigating circumstance exists, so-called *cultural defences*.

The Particular Case of Crimes Regarding Intangible Goods

Neuroscience teaches us, nonetheless, that lack of perception of the under-evaluation of certain behaviours can depend not only on 'cultural' factors, but also partly on the structure and the functioning of our brain. This would seem to be the case, for example, in relation to behaviour with as object intangible goods. It is believed that such behaviours lack an emotional component or that such a component is present only minimally.[3] Contrarily, the under-evaluation of some behaviours like those related to existing taboos about the theft of material goods, would already be present in our sense of justice based on emotion and intuition, even before being formally accepted into any legal body. Legal norms condemning aggressive behaviour towards other people's material goods would result. Or yet, already strongly internalized by those for whom the precept was developed, an emotional response would be elicited that pushes towards general acceptance of the prohibition. Differently, behaviour with as object immaterial goods, which are protected by intellectual property laws that are essentially just and balanced, would not find corresponding mental processes associated with sense of justice. The consequences are obvious to everyone.

Two possible reasons can be identified for a similar emotional *deficit*:

1) The *deficit* could depend exclusively on a problem of education, experience, socialization, or lack of fear of punishment.
2) Contrarily, mental differences could come from fundamental differences in the actual perception of goods as objects of behaviour, which would determine an emotional involvement in relation to material goods, but which would not be able to produce the same result in relation to immaterial goods.

It is exactly this last aspect that seems to require further examination. Goodenough and Prehn believe it is quite possible that the existing cognitive differences between material and immaterial goods at some point in the recognition of the moral dilemma represent so-called 'cognitive origins'. One could thus imagine that the original structure of property works better with material and tangible goods than with immaterial goods. Putting it simply, it is possible that the human brain is not equipped to perceive immaterial goods as objects suitable for normative judgements that are backed up by emotions; that it is hence not structured to completely understand relationships with as object immaterial goods on an emotional level. Contrary to what Bentham affirms, one could then say that the fundamental components of property would precede any formal institution and that the fundamental principles would be encoded in the human brain.[4]

---

[3] Oliver R. Goodenough and Kristin Prehn, 'A neuroscientific approach to normative judgment in law and justice' in Semir Zeki and Oliver Goodenough (eds.), *Law & the brain* (Oxford, Oxford University Press, 2006) 77 et seq.

[4] Jeffrey Evans Stake, 'The property "instinct"' in Semir Zeki and Oliver Goodenough (eds.), *Law & the brain* (Oxford, Oxford University Press, 2006) 185 et seq.

11.3 Origins of Property Law

Considering the above, it is interesting to look at the origins of property law, albeit briefly, to understand the fundamental differences between tangible and intangible goods. The explanations generally given about how property law was established are traditionally centered on necessity. At a certain point in the history of man, rules came into being to regulate the competition of various subjects for the use of limited consumable resources. The justification of property law is thus found in the scarcity and limitation of particular goods, of which the enjoyment is strictly reserved for the owner of the right, while the rest of society members are excluded. It is considered more advantageous for rivals to regulate the ownership of goods than to fight for them.

With just as much certainty we can assume that property law originates in relation to tangible goods, such as land and its fruits, the fruits of hunting, tools, kitchenware, etc. Consequently, there is a clear close connection between the owner of the right and the good. Proof of this can also be found in the protection that has always been recognized by juridical dispositions likening actual possession of property, and that finds its justification in the direct relationship between the subject and the object of possession. For this reason, the presumption about other people's rights is in reality substituted with the presumption that comes from the fact.

Significant in that sense is Roman law, where possession was seen as the actual domination of a thing.[5] Let us also think about the modalities of acquisition and transfer of property in ancient Rome. There, a sale was not valid immediately, like a consensual contract would be that is completed with the simple agreement of buyer and seller. Initially, the Romans referred to the *mancipatio*, which is the immediate exchange of goods against price. Buyer and seller participated in a kind of ceremony, with the goods physically present, assisted by five witnesses, and the buyer had to touch the goods of sale pronouncing a solemn formula.

For ages, if not for millennia, property law has thus had as object exclusively material, tangible goods. Even better, one could say that property law was conceived exactly with such goods in mind, and in function of them. Hence, it is impossible that this would not have influenced the functioning of our brain, and, consequently, of our mind.[6]

Modern biology has, in fact, clarified in unmistakable terms how the form, the functioning, and the behavioural stimuli of the brain are the product of interactions between genes and environment.[7] Our brain is thus also the fruit of the forces of

---

[5] Vincenzo Arangio-Ruiz, *Instituzioni di diritto romano* (Naples, Jovene, 1989) 269 et seq.
[6] Antonio R. Damasio, *Descartes' Error: Emotion, Reason and the Human Brain* (New York, Putnam, 1994).
[7] Erin Ann O'Hara, 'How neuroscience might advance the law' in Semir Zeki and Oliver Goodenough (eds.), *Law & the brain* (Oxford, Oxford University Press, 2006) 21 et seq.

evolution, and its structure influences our thoughts, our sensations, and our actions.

The environment, the brain, and human behaviour can therefore be regarded as closely related to each other. Just as human makeup, law and cultural institutions are subject to the evolution process.[8] There would hence be a continuous exchange between the environment that affects us by making us adopt norms to handle the difficulties it faces us with, and the adopted norms that interact with our brain processes and determine our behaviour. The latter, in turn, influences the environment, thus completing the circle.

Furthermore, it is necessary to keep in mind that all theories about human behaviour are, ultimately, theories about the brain, which is a complex phenomenon that has to be considered in relation to the body and to biology in general. In that case, Hobbes's old paradigm that viewed the law as a purely cultural construction to repress people's innate aggression, has to give way to a more profound understanding of law as a value in adaptation, and to a greater focus on its institutional role in expressing the evolution of social behaviour.[9]

Therefore, let us apply the scheme put forward by Chorvat and McCabe that also pertains to property law: 1) the environment provides limited resources; 2) man, to face that limitation of resources and avoid conflicts that can undermine peaceful cohabitation, puts in place rules to regulate property and the possession of such resources; 3) those rules influence the behaviour of the same subjects that have adopted them; 4) that behaviour affects the environment and the resources it makes available to man.

Completely different is the reasoning related to intangible goods which, contrarily to tangible goods, are not generally competitive resources. For example, when one listens to a song or reads a book, the use of the good at hand does not exclude other people's enjoyment of it. Extending this reasoning to illegal behaviour, one could likewise say that whoever downloads a file protected by copyright from the Internet, or yet, whoever illegally copies an intellectual work, may not have a complete understanding of how incorrect that behaviour is, exactly because they are neither consuming the good as such nor preventing other people's enjoyment or use of it.

The focus on the evolutionistic aspect of our brain and of our way of thinking consequently assumes a fundamental importance for law, and could provide some interesting answers. Nevertheless, the legislator often ignores information from many behavioural disciplines such as sociology, religion, philosophy, etc.[10] If used well, however, even evolutionistic analysis could help in the field of law, especially in understanding why some human behaviours are continuously repeated even though they are prohibited by law. In short, it could help to comprehend some of

---

[8] Terrence Chorvat and Kevin McCabe, 'The brain and the law' in Semir Zeki and Oliver Goodenough (eds.), *Law & the brain* (Oxford, Oxford University Press, 2006) 113 et seq.

[9] Morris B. Hoffman, 'The neuroeconomic path of the law' in Semir Zeki and Oliver Goodenough (eds.), *Law & the brain* (Oxford, Oxford University Press, 2006) 3 et seq.

[10] Owen D. Jones, 'Law, evolution and the brain: application and open questions' in Semir Zeki and Oliver Goodenough (eds.), *Law & the brain* (Oxford, Oxford University Press, 2006) 57 et seq.

the factors mentioned earlier that determine the ineffectiveness of the penal systems.

11.4 New Starting Points for a Lazy Legislator

All the above could lead to different contrasting strategies, increasing the effectiveness and efficiency of the system.

Even though neuroscience admits that, to this day, it has no idea of how our brain controls our mind,[11] one thing seems sure: if people are unable to comprehend and use laws as premises upon which to base their own choices, the law will not be able to influence human behaviour. This view is based on the idea that people generally use the law as a higher premise in the syllogism that determines many human actions. Hence the law presumes that each subject is moved by a practical rationality, albeit minimal, but considered sufficient by most conventional and social constructions of which the law is a part. The history of evolution explains why human beings need rules like those provided by the law. If one wanted to follow pure mechanistic theory, human beings would find it impossible to stop seeing themselves as rational beings. Certainly, like the other causes, neuroscience is unable to define what degree and capacity of control are necessary to recognize the full responsibility of an individual. This remains a normative, moral, and juridical question. However, that problem will be dealt with elsewhere.

11.5 Neuroethics

It does appear interesting to take a closer look at some ideas from disciplines related to neuroscience, for example, 'neuroethics'. This field holds that the ability to understand other people's emotions by considering them in relation to our own experience is of fundamental importance, both for social interaction and for spontaneous moral reaction that should prevent us from doing wrong.[12] In fact, the absence of empathy characterizes some psychological pathologies in which the wrong done to others neither provokes remorse nor sense of guilt. That could open the field to additional ideas of great interest, also in relation to the problems dealt with in this paper. The spontaneous question arises what happens if the absence of empathy regarding certain behaviours is not caused by psychological pathologies, but depends directly on our brain structure. That concept immediately leads us to behaviour with as object immaterial goods in relation to which there is usually no direct relationship with the owner of the infringed right and no direct relevant personal experience.

---

[11] Jonathan A. Fugelsang and Kevin N. Dumbar, 'A cognitive neuroscience framework for understanding causal reasoning and the law' in Semir Zeki and Oliver Goodenough (eds.), *Law & the brain* (Oxford, Oxford University Press, 2006) 157 et seq.

[12] Laura Boella, *Neuroetica* (Milan, Raffaello Cortina Editore, 2008) XVI.

For the same reasons, the ideas of neuroethics about the concept personal/impersonal are of great interest. A problem is considered 'personal' when it directly connects two human beings and makes the acting subject immediately responsible for the suffering caused to the victim. A problem is considered 'impersonal' when there is no direct relationship among subjects. In fact, *imaging* would demonstrate that when there is a personal relationship, the cerebral areas associated with emotion and social cognition are more active.[13] That means that behaviour with as object intangibile goods presents a double difficulty: on the one hand, there is no direct contact between subject and good, and, on the other hand, there is no relationship among subjects.

Fundamental once again is knowing how the brain works in order to look for, using Boella's words, a *moral before the moral*. The latter term refers to the 'neurobiological predisposition for moral capacity' that involves a precise awareness of the complex relationship between scientific-natural behaviour and the perspective of experience in its globality:

It is necessary to admit to the existence of a tension between a functional-objective level (with which the language and methodologies of science correspond); and a multilayered level, going from variable moments throughout the evolution of the species, cultural and social history, and symbolic creations of the various populations that inhabit the planet; to individuals' personal development derived from this multitude of elements in many ways.[14]

All this to question and possibly review the class of responsibility. The brain:

consults the information acquired through the senses and stored in memory as far as it is relevant to decisions regarding potential actions and strategies, and thus permits one to decide how and in what context to relate to the surrounding environment. That this happens automatically and involuntarily simply means that the trigger for how much in us is aware and voluntary is unaware and involuntary.

11.6 Copyright

The lack of perception of the underevaluation of certain behaviours related to immaterial goods could therefore also be explained on the basis of a different perception that our brain has of such goods compared to material goods. The matter of intellectual property appears rather significant in that sense. Many people, even those culturally and socially evolved, do not consider illegally copying works protected by *copyright* as legally important. They may not even consider it as morally reprehensible. The question is then why the same people that would feel terribly guilty even taking something as simple as someone else's pencil show no scruples in copying programs, songs, and movies.

A first simplistic explanation could be found in the secrecy that characterises most criminal behaviours infringing *copyright*, together with the considerable

---

[13] Ibid. 67 et seq.   [14] Ibid. 50 et seq.

economic gain those behaviours can guarantee. There is no doubt that similar reasons are an important component in the motivational process leading to the realization of crimes against intellectual property. The chances of being caught and convicted are also very limited. Nevertheless, if we look at previous judicial statistics following the adoption of drastic measures against the phenomenon of so-called 'intellectual piracy', we notice an exponential increase of such violations. Obviously, deterrence alone is not enough.

Hence it is necessary to ask if contrasting strategies against the phenomenon of intellectual piracy that only entail a penal sanction can have any effect without preparing instruments that increase people's emotional involvement regarding criminal behaviour.

The proof that this question is completely rhetorical is its obvious answer.

11.7 Conclusions

The ideas derived from cognitive neuroscience force us to reconsider some of penal law's age-old certainties, and especially to reconsider its role in the legal system. Goodenough and Prehn again highlight that new knowledge; and the progress made in understanding the brain, its functions, and its functioning permit us and even force us to reconsider traditional theories on normative judgement.

Hence, the importance of emotion in the decision process can no longer be ignored: the images of a fast and unconscious 'low process' and that of a slow and conscious 'high process' are contrasted. Thus it follows that judgements can come to be in two different ways: one intuitive, fast, and automatic; and the other slower, controlled by rules. Contrary to the former, the 'high process' is aware of the rules that its 'low' counterpart is sometimes inclined to violate, and intervenes to correct or substitute erroneous judgements.

The quality of normative intervention is thus fundamental. When our brain tries to solve social problems, it has access to many sources of information. In addition to their genetic nature, people have access to a great variety of cultural knowledge that evolves together with them. As highlighted by Goodenough and Prehn, such knowledge can be implicit, like the unconscious creation of social models coming from childhood; or explicit, like morals, customs, or law itself. It is highly probable that the normative judgement expected from man is the fruit of a complex and composite process; and not of a unitary process, as was believed for centuries.

Both cognitive and affective aspects are present in normative judgement. However, if there are many processes and sources of information involving normative processes, as was ascertained by cognitive neuroscience, that poses questions about what modulation of strategies for criminal policy is most suitable to fight a certain criminal phenomenon.

Nonetheless, the problems that cognitive neuroscience deals with today have already been partly touched upon by traditional doctrine. The distinction between natural and artificial crimes refers exactly to the higher or lower degree to which people agree to criminalize certain behaviours. That has forced penal doctrine to review some fundamental concepts and principles, such as the consciousness of illegality in malice and l'*ignorantia legis non excusat* (i.e. not knowing the law does not excuse illegal behaviour). Therefore, there is no doubt that the punishment of homicide, the natural crime *par excellence*, presents the highest degree of agreement. In addition, except in particular cases of cerebral dysfunction, the normative judgement of the crime of homicide can involve all the above-mentioned components to the highest level. This is not the case, as we have seen, in relation to crimes committed with as object immaterial goods, for example those against intellectual property. The consequence surely cannot be a lack of responsibility of the subjects committing crimes with as object intangible goods. As said before, responsibility is a normative concept, an elaboration of a social nature.

What role can the law thus play in relation to similar behaviour? Certainly it can act to expand the rational component of normative judgement. The disapproval a disposition expresses in relation to a specific behaviour is surely able to guide the rational component of the normative judgement that society members are requested to express. Cognitive neuroscience has shown that normative judgement results from a combination of various factors, with the cultural factors that the law itself is a part of playing a big role. Also wanting to give credit to more deterministic views, reason ultimately retains the power to *veto* choices set forth and determined by the brain.

Considering the ideas coming from neuroscience, a policy that really wants to promote compliance to laws regarding intellectual property and that wants to have a real possibility of success, should not limit itself to reinforcing the existing penal disciplinary apparatus, but should definitely aim at involving society members emotionally, so that the emotional systems also are implicated in making normative judgements. In other words, neuroscience suggests that normative judgement is the fruit of a combination of factors. In relation to certain behaviours the emotional component results are strongly flawed, if not completely absent. It seems quite obvious that it is necessary to strengthen the other components before being able to reproach a subject for having a certain behaviour.

12

Neuroscience and Emotional Harm in Tort Law: Rethinking the American Approach to Free-Standing Emotional Distress Claims

Betsy J. Grey*

American tort law traditionally distinguishes between 'physical' and 'emotional' harm for purposes of liability, with emotional harm treated as a second-class citizen. The customary view is that physical injury is more entitled to compensation because it is considered a more trustworthy harm and perhaps more important. The current draft of the Restatement of the Law (Third) of Torts (a compendium of views of preeminent legal scholars, jurists and lawyers) maintains this view. Even the name of the Restatement project itself—'Liability for Physical and Emotional Harm'—emphasizes this distinction. And comments to the Restatement explain the reasoning behind the distinction by noting that 'emotional distress is less objectively verifiable than physical harm and therefore easier for an individual to feign, to exaggerate or to engage in self-deception about the existence or extent of the harm'.[1] Advances in neuroscience suggest that this concern over verification may no longer be valid, and that the phenomena we call 'emotional' harm has a physiological basis. Because of these early scientific advances, this may be an appropriate time to re-examine our assumptions with regard to tort recovery for emotional harm.

Emotional harm is a legal,[2] not a scientific, concept, but one that relates closely to cognitive or mental disorders identified by the medical community. The medical

---

* Professor of Law and Faculty Fellow, Center for Law, Science & Innovation at the Sandra Day O'Connor College of Law at Arizona State University. The author would like to thank her students Lyn Gulley and Natalie Greaves for their research assistance and Oscar Gray, Gary Marchant, Linda Demaine, Ron Korn, Bob Dauber, Robert Bartels, Carissa Hessick, David Kaye, and Joel Nomkin for their valuable comments on drafts of this Article and discussions of its subject matter.

[1] Restatement (Third) of Torts, Scope Note at 2 (Tentative Draft No. 5, 2007).
[2] Emotional harm is a legal concept that is often defined in terms of what it is not. For example, Professor Daniel Dobbs defines emotional harm as 'the antithesis of happiness or enjoyment of life which everyone pursues'. Daniel B. Dobbs, *The Law of Torts* § 302 (Minnesota, West Pub., 2002). This includes 'distress, emotional harm, anxiety, diminished enjoyment, losses of autonomy, and similar intangible harms.' Ibid. Section 4 of the Restatement (Third) of Torts distinguishes emotional harm from physical harm. It defines 'physical harm' as 'physical impairment of the human body or of

field recognizes the physiological aspect of mental disorders. The *Diagnostic and Statistical Manual of Mental Disorders* ('DSM-IV'), the leading guide to clinical practice in psychiatry, puts it most succinctly:

> Although this volume is titled the Diagnostic and Statistical Manual of Mental Disorders, the term mental disorder unfortunately implies a distinction between 'mental' disorders and 'physical' disorders that is a reductionistic anachronism of mind/body dualism. A compelling literature documents that there is much 'physical' in 'mental' disorders and much 'mental' in 'physical' disorders. The problem raised by the term 'mental' disorders has been much clearer than its solution, and unfortunately, the term persists in the title of DSM-IV because we have not found an appropriate substitute.[3]

The DSM-IV further explains that the terms 'mental disorder' and 'general medical condition', used throughout the manual, are not intended to 'to imply that there is any fundamental distinction between mental disorders and general medical conditions, that mental disorders are unrelated to physical or biological factors or processes, or that general medical conditions are unrelated to behavioural or psychosocial factors or processes'.[4]

Studies in neuroscience have begun to document the physiological basis of cognitive disorders. Research focused on the areas of emotion, memory, and anxiety has begun to reveal the primary structural and chemical systems of the brain that are affected when an individual experiences trauma. Focusing on the anxiety disorder of Post Traumatic Stress Disorder (PTSD) in particular, neuroscience studies have begun to document and analyse changes in the neural mechanisms, discovering common patterns that identify a biological basis for the harm suffered when an individual is exposed to extreme stress. In the future, documentation of these changes could provide a verifiable physiological change to serve as a basis for an emotional distress claim. In addition, as we continue to discover the physiological origins of emotional harm, the distinction between emotional and physical harm may become outmoded. This would argue in favour of including emotional harm under the rubric of bodily or personal injury, instead of treating it as a separate (and separable) harm.

The approach in English law to tort recovery for emotional harm accommodates this unitary view. Although the English approach developed long before these advances in neuroscience occurred, English courts have consistently treated mental

---

real property or tangible personal property'. In turn, it defines 'physical impairment of the human body' as 'physical injury, illness, disease, and death'. Section 4, comment b, goes on to state that this 'definition of physical harm is meant to preserve the ordinary distinction between physical harm and emotional disturbance'. The Restatement refers to emotional harm as 'emotional disturbance' which is 'distinct from bodily harm and means harm to a person's emotional tranquility'. § 4, cmt a. This harm must be 'severe' for legal recognition: 'It eliminates claims for routine, everyday distress that is a part of life in a modern society, thereby avoiding a multitude of claims that might otherwise be brought. A seriousness threshold also assists in ensuring that claims are genuine, as the circumstances can better be assessed by a court and jury as to whether emotional disturbance would genuinely be suffered.' § 46, cmt i.

[3] American Psychiatric Association, *Diagnostic and Statistical Manual of Mental Disorder, Fourth Edition Text Revision (DSM-IV-TR)*, 4th edn (Washington, DC, American Psychiatric Press, 2000).
[4] Ibid. xxxv.

illnesses as an aspect of bodily harm. Letting medicine rather than law assume the lead in cases of emotional harm, English law requires the plaintiff to show a recognizable or diagnosable psychiatric illness before the plaintiff can proceed with the tort claim. English decisions express concerns similar to those cited by American courts, such as floodgates and the need for bright lines, but do not draw a sharp distinction between the physical and emotional. Instead, the decisions treat psychiatric illness as a part of personal injury.

At least one American court has already adopted the view that a psychiatric disorder constitutes bodily harm. In *Allen v Bloomfield Hills School District*,[5] the plaintiff, Charles Allen, was a train operator who tried to stop in time to avoid hitting a school bus that had manoeuvered around a lowered gate at a railroad grade crossing. Unable to stop the train traveling at 65 mph, he witnessed it smash into the school bus presumably full of young children. Though he was later informed that there were no children on the bus at the time of the accident, Allen was subsequently diagnosed with PTSD stemming from the accident. When Allen attempted to recover damages for this injury, the trial court denied the claim, concluding that the plaintiff had not suffered 'bodily injury' within the meaning of the motor vehicle exception to the governmental immunity statute.[6]

The Court of Appeals reversed, permitting Allen to argue at trial that he had suffered 'bodily injury' because the accident caused physical damage to his body as evidenced by a positron emission tomography (PET) scan of his brain.[7] Relying on expert testimony presented by the plaintiff, the court found that plaintiff had presented 'objective medical evidence that a mental or emotional trauma can indeed result in physical changes to the brain',[8] and the showing was consistent with dictionary definitions of 'bodily injury'.[9] One expert testified that the PET scan depicted 'decreases in frontal and subcortical activity consistent with depression and post traumatic stress disorder'[10] and that 'the abnormalities in Mr. Allen's brain as depicted on the... PET scan are quite pronounced and are clearly different in brain pattern from any normal controls' and 'consistent with an injury to Mr. Allen's brain'.[11] A second expert opined that PTSD 'causes significant changes in brain chemistry, brain function, and brain structure. The brain becomes

---

[5] 281 Mich App 49 (Mich App 2008).
[6] Mich. Comp. Laws Ann., § 691.1407(1) (2007) ('Except as otherwise provided in this act, a governmental agency is immune from tort liability if the governmental agency is engaged in the exercise or discharge of a governmental function.') This broad immunity is limited by certain exceptions, including bodily injury: ibid. § 691.1405 ('Governmental agencies shall be liable for bodily injury and property damage resulting from the negligent operation by any officer, agent, or employee of the governmental agency, of a motor vehicle of which the governmental agency is owner.').
[7] *Allen* (note 5).
[8] Ibid.
[9] The term 'bodily injury' was not defined in the immunity exception statute. § 691.1405. The court relied on precedent from a 2008 Michigan Supreme Court case to construe the meaning of the term. *Allen* (note 5) (citing *Wesche v Mecosta County Road Com'n* 480 Mich 75, 84, 746 NW2d 847, 853 (2008) ('"[B]odily injury" simply means a physical or corporeal injury to the body.'). In doing so, the *Allen* court decided that there was no requirement of 'direct trauma' in the plain wording definition of the phrase that would remove psychiatric injury from its reach: ibid. 59.
[10] Ibid. 56.   [11] Ibid. 57.

'rewired' to over-respond to circumstances that are similar to the traumatic experience'.[12] Relying on these two medical doctors' opinions, the court found that Allen had provided sufficient evidence to show that Allen had suffered an injury to the brain to overcome a motion for summary judgment.[13] As the court stated:

> [A]s a matter of medicine and law, there should be no difference medically or legally between an *objectively* demonstrated brain injury, whether the medical diagnosis is a closed head injury, PTSD, Alzheimer's, brain tumor, epilepsy, etc. A brain injury is a 'bodily injury.' If there were adequate evidence of a brain injury to meet the requisite evidentiary standards, i.e., objective medical proof of the injury, summary disposition was improper.[14]

Thus, the court allowed the plaintiff to argue, by using brain scans to show long-term changes in brain function, that the PTSD he suffered was a bodily injury to the brain. The *Allen* court's analysis may signal the beginning in the US of a rethinking of the 'physical'/'emotional' distinction—a distinction that currently dominates the jurisprudence.

This paper proceeds as follows. Part I reviews the current American common law, as well as the draft Restatement (Third) proposals, concerning emotional distress claims. Mental distress claims have never been given equal status with other tort claims and this section discusses why this value judgement has occurred. Using studies of PTSD as an example, Part II explores advances in neuroscience that have begun to shed light on the biological basis of the harm suffered when an individual is exposed to extreme stress. These advances underline the shrinking scientific distinction between physical and emotional harm. Part III examines English law and its threshold requirement of showing a diagnosable psychiatric illness. Drawing on these different lines of thinking, Part IV concludes that we should rethink the American approach to emotional distress claims. In general, it proposes that we change our approach to account for advances in neuroscience, moving toward a more unified view of bodily injury. Two potential legal applications are advanced in this paper: 1) that science can provide empirical evidence of what it means to suffer emotional distress, thus helping to validate a claim that has always been subject to greater scrutiny; and 2) that this evidence may allow us to move away from the sharp distinction between how physical and emotional injuries are conceptualized, viewing both as valid types of harm with physiological origins.

12.1 American Common Law View of Emotional Harm

America's common law treatment of emotional harm began with a presumption of no duty and gradually evolved to a system that allows some sharply circumscribed recovery in certain limited situations. Judicially imposed constraints, reflecting both pragmatic and normative concerns, shaped the doctrine as it cautiously expanded, creating a patchwork of liability rules that may seem illogical to the untutored eye.

---

[12] Ibid. 58. [13] Ibid. 60. [14] Ibid. 59–60.

Traditionally, American courts have been most receptive to tort claims where the plaintiff has suffered some physical harm to the person or to property. Sometimes emotional harm is recognized as a compensable injury in those cases as parasitic damages—damages that are recoverable as pain and suffering for some other tort.[15]

Common law also traditionally recognized several forms of emotional harm claims for trespassory torts like assault (distress from fear of bodily touching without consent), false imprisonment (distress from being detained without consent), as well as libel and slander (distress from plaintiff's good name being put in doubt by false statements), without a showing of related physical injury. This is based on the likelihood that most reasonable people would likely suffer emotional distress in those circumstances.[16]

Separate jurisprudence has developed for the stand-alone claim of emotional distress, which does not require a showing that the defendant committed another tort or inflicted physical harm. American common law recognizes two types of stand-alone claims for emotional harm or distress: those caused intentionally and those caused negligently. Satisfying the intent or *scienter* requirement for the former tort (intentional infliction of emotional distress or IIED) demands proof of conduct that was extreme and outrageous, as well as a showing that the actor purposefully caused the severe emotional disturbance. Traditional examples of this tort include intentional or reckless interference with a dead body,[17] abusive conduct by a person with authority over the plaintiff, or repeated and harassing behaviour.[18] Section 45 of the Restatement (Third) explains that the intentional infliction tort 'originated as a catchall to permit recovery when an existing tort claim was unavailable but in the narrow instances in which an actor's conduct exceeded all permissible bounds of a civilized society'.[19] The focus is on the extreme conduct, and courts presume that severe emotional distress will follow. Thus, the plaintiff generally is not required to use medical testimony to show either the severity of the distress or its cause.[20]

This paper concentrates on the second type of stand-alone claim, the claim of negligent infliction of emotional distress, which focuses squarely on the harm suffered rather than the extremity of defendant's conduct.[21]

12.1.1 Primary claim of negligent infliction of emotional distress

Negligent infliction of emotional distress (NIED) developed as an outgrowth of the IIED claim, and has always been the more controversial of the two stand-alone

---

[15] Daniel B. Dobbs, *The Law of Torts* § 302 (Minnesota, West Pub., 2002)
[16] Ibid.
[17] *Travelers Insurance Co. v Smith* 991 SW2d 591 (Ark 1999) (failure to obtain prompt autopsy).
[18] *George v Jordan Marsh Co.* 268 NE2d 915 (Mass 1971) (debt collection).
[19] *Restatement* (note 1 § 45 cmt. a).
[20] Dobbs (note 15) § 303.
[21] This paper focuses on the stand-alone emotional harm claim and does not take a position on the parasitic claim of pain and suffering. Further study may show, however, that the arguments advanced here may be equally applicable from an evidentiary and theoretical viewpoint to a claim of pain and suffering.

emotional distress claims. The general rule at common law is that a negligent actor is not responsible for injury that causes only emotional harm. Gradually exceptions have developed to allow recovery in certain circumstances, but the tort has never gained equal status to the tort claims for negligently inflicted personal injury or property damage. The trustworthiness of the claim has always been a primary concern. In addition, courts expressed concerns that the claim was trivial or might give rise to the proverbial floodgates for claims for money damages.

In the face of these concerns, courts refused to permit recovery for NIED based on foreseeability of the harm—which is the touchstone for negligence recovery for physical or property damages. Rather, courts developed a series of tests to limit NIED recovery. At first, courts required a physical injury to precede the emotional harm claim following the traditional view of emotional harm as a parasitic claim to a personal injury claim.[22] This test began to erode as courts dropped the requirement of a physical injury in favour of a showing of 'physical impact' resulting from the defendant's actions, with or without a showing of physical injury.[23]

Most states have now abandoned the impact rule,[24] but have substituted other limiting rules, still addressing concerns over the untrustworthiness and triviality of the claim. Thus, to demonstrate the genuineness or validity of the claim, some states have required the plaintiff to show evidence of an objective physical manifestation of the shock or fright that has occurred.[25] Other states have required that the emotional injury must be medically diagnosable as an emotional disorder.[26] Some states have abolished the requirement of objective physical manifestation altogether[27] or in particular circumstances.[28] Some states limit recovery to instances in which the plaintiff suffered emotional harm by being placed in the zone of danger of physical injury by the defendant's negligence.[29] Some states combine these limiting rules in various ways.[30]

In general, courts test the validity of the claim with both a subjective test ('Did the plaintiff in fact suffer severe emotional distress?') as well as an objective test ('Would most people suffer severe emotional distress under these

---

[22] *Mitchell v Rochester Railway Co.* 45 NE 354 (NY 1896).
[23] Some states still require a showing of impact. *Steel Technologies, Inc. v Congleton* 234 SW3d 920 (Ky 2007); *Atlantic Coast Airlines v Cook* 857 NE2d 989 (Ind 2006).
[24] *Battalla v State* 176 NE2d 729 (NY 1961).
[25] *Brueckner v Norwich University* 730 A2d 1086 (Vt 1999). What constitutes physical manifestation can be debatable. For example, depending on the court, nightmares may or may not be considered a physical manifestation. *Marzolf v Stone* 136 Wash 2d 122, 960 P2d 424 (1998)(requiring 'objective symptomology'; 'nightmares, sleep disorders, intrusive memories, fear and anger may be sufficient ... [if] they ... constitute a diagnosable emotional disorder'.)
[26] *Paz v Brush Engineered Materials, Inc.* 949 So2d 1 (Miss 2007); *Hegel v McMahon* 960 P2d 424 (Wash 1998).
[27] *Molien v Kaiser Foundation Hospitals* 616 P2d 813 (Ca 1980).
[28] *Clark v Estate of Rice* 653 NW 2d 166 (Iowa 2002); *Doe Parents No. 1 v State Dept. of Educ.* 58 P3d 545 (Haw. 2002).
[29] *AALAR Ltd. v Francis* 716 So2d 1141 (Ala 1998).
[30] *Willis v Gami Golden Glades, LLC.* 967 So2d 846 (Fla 2007) (plaintiff may show either impact or physical manifestation of injury); *Catron v Lewis* 712 NW2d 245 (Ne. 2006) (plaintiff must show zone of danger if he cannot show impact or physical injury).

circumstances?').[31] This dual test helps the courts avoid holding a defendant responsible for a plaintiff's abnormal difficulties or maladaptive attitudes toward adjusting to distress, but it does not mean that the amount of damages is limited to what a normal person would suffer. Instead, once the threshold is reached, the defendant can be responsible for damages to the extremely sensitive plaintiff.[32] This is a variation on the usual damages rule in tort law, the 'thin-skulled plaintiff' rule, which requires a defendant to take a plaintiff as he finds him—whether he be thin-skinned or thick-skinned, rich or poor, young or old—and pay replacement damages if found responsible for the injury.

Section 46 of the Restatement (Third) of Torts draws on these judicially developed tests to limit recovery for emotional harm caused by a negligent actor to two situations. The Restatement provides that a person whose negligent conduct causes 'serious emotional disturbance' to another is liable if the defendant's negligence either 1) places the plaintiff in 'immediate danger of bodily harm,' or 2) the negligence occurs in the context of special relationships in which negligent conduct is especially likely to cause serious emotional disturbance.[33] The Restatement views these two situations 'as two lines of exceptions to the general rule that an actor is not liable for negligent conduct that causes only emotional harm'.[34]

The first category is a direct descendant of the 'impact' and 'zone of danger' rules.[35] The second category addresses liability even though the actor's negligent conduct only creates a risk of 'emotional' and not 'bodily' harm.[36] This exception to the no-duty rule occurs when an actor undertakes certain specific obligations or gets involved in certain specific relationships.[37] The Restatement does not identify every activity or relationship that would give rise to liability for emotional distress, but indicates that foreseeability of harm is not sufficient to impose liability.[38] Rather, the Restatement suggests that 'policy issues' surrounding these relationships and situations should be examined 'to determine whether, as a category, they merit inclusion among the exceptions to the general rule of no liability'.[39]

With regard to the overarching requirement of 'serious emotional disturbance', the Restatement rejects the common law test of physical consequences or physical manifestation.[40] Although it recognizes that the physical manifestation test served as a proxy to show that the emotional harm is genuine and non-trivial, the Restatement notes that these same concerns can be met by requiring evidence

---

[31] Dobbs (note 15) § 313.
[32] *Steinhauser v Hertz Corp.* 421 F2d 1169 (2d Cir 1970); Fowler v Harper, Fleming James, Jr. & Oscar S. Gray, *The Law of Torts*, § 18.4 note 51, 3rd edn (Boston, Aspen Pub., 1996); Dobbs (note 15) § 313.
[33] *Restatement* (note 1) § 46.
[34] Ibid. § 46 cmt b.
[35] Ibid. This would include the emotional fright suffered in a doomed aircraft, but would not entail the 'exposure only' cases of toxic substances with a long latency period before manifestation of a physical injury: ibid. cmt. c.
[36] Ibid. cmt. d.
[37] Ibid. Examples of these include the negligent mishandling of a corpse by a hospital or funeral home, and the negligent transmission of a telegram that erroneously informs the plaintiff about the death of a loved one. This category would also encompass cases in which the plaintiff consumes food contaminated with a 'repulsive foreign object': ibid.
[38] Ibid. cmts. d, f.   [39] Ibid. cmt. f.   [40] Ibid. cmt. g.

that a reasonable person would suffer serious harm under similar circumstances (a descendant of the 'objective' test) as well as 'credible evidence' that the person has suffered serious harm (a descendant of the 'subjective' test).[41] The Restatement does not indicate what constitutes 'credible evidence'.[42]

12.1.2 Bystander claims

A second line of cases developed in negligence, called the 'bystander claim', which occurs when emotional harm results from witnessing injury to another. The classic example is when a mother witnesses a car strike her child. Although not challenging the mother's fright and shock,[43] courts originally denied these claims altogether.[44] Courts then opened the door slightly to this claim by allowing the mother to recover if she were found to be within the zone of danger.[45] An influential California case later dropped the zone of danger requirement in favour of a form of foreseeability test, using as guidelines the plaintiff's physical proximity to the accident, the contemporaneous witnessing of the accident, and the relational proximity of the plaintiff and the victim.[46] In a later case, California tightened these guidelines into rules, abandoning a general foreseeability test.[47] Most states now allow some form of bystander recovery for emotional distress when witnessing an injury to someone else, but only in certain limited circumstances.[48]

The Restatement (Third) adopts similar rules to curtail recovery by the bystander. It allows recovery for 'serious emotional disturbance' if the plaintiff 1) 'perceives…contemporaneously' the bodily harm to a third person; and 2) is a 'close family member' of the injured third person. This claim is derivative to the claim of the injured third person.[49] By definition, it applies as 'pure emotional harm' without any 'bodily harm'.[50] Recognizing that the requirements of contemporaneous perception and close familial relationship are in some sense arbitrary, they 'are required for emotional harm because of its ubiquity'.[51] The other alternative would be a 'rule of no liability'.[52]

12.1.3 Instrumentalist and normative concerns

The value judgement that physical harms are more entitled to compensation than emotional harms pervades tort jurisprudence. Aside from questions about validity, the view that mental distress claims are 'less deserving' reflects instrumentalist as

---

[41] *Restatement* (note 1) § 46.
[42] Ibid. cmt i. ('An unusually susceptible person may not recover if an ordinary person would not have suffered serious emotional disturbance.')
[43] *Restatement* (note 1) § 47.
[44] Dobbs (note 15) § 309.
[45] *Engler v Illinois Farmers Ins. Co.* 706 NW 764 (Minn 2005).
[46] *Dillon v Legg* 441 P2d 912 (Ca 1968).
[47] *Thing v La Chusa* 771 P2d 814 (Ca 1989).
[48] Dobbs (note 15) § 309.
[49] *Restatement* (note 1) § 47 cmt. c.
[50] Ibid. cmt b.   [51] Ibid. cmt. f.   [52] Ibid.

well as normative considerations. From an instrumentalist viewpoint, the value judgement responds to two major concerns—the concern of opening the floodgates of litigation as well as the related concern of potentially crushing liability. Thus, recognizing the need to prioritize claims in a pool of limited funds, courts erected barriers to recovery for the less trustworthy emotional harm claim.[53] Further, courts suggested that extending liability for emotional harms based purely on foreseeability would be disproportionate to the tort involved, since a single tort could cause a large number of people to suffer some degree of distress. At bottom, this reflects a concern of crushing liability from the injurer's perspective as well as the concern, from the victim's perspective, that limited funds may not reach all deserving (i.e. physically harmed) plaintiffs. Thus, time and again, courts recite the need for caution in the face of opening the floodgates of litigation to unchecked claims of responsibility for negligent conduct.[54] By rejecting a general foreseeability test and replacing it with limitations such as the 'zone of danger' test, courts indicated their commitment to curtail widespread litigation rights. Comments to the Restatement draft section on bystander claims pick up on these themes:

> The general rule that pure emotional disturbance is not recoverable is not because the harm is not genuine or foreseeable, but because as a matter of policy it is an injury whose cost the legal system should not normally shift, even to someone who is negligent.[55]

From a normative viewpoint, the judicially crafted limitations on duty for recovery for emotional harm reflect a different concern. The limitations function as a policing mechanism between acceptable and unacceptable social behaviour, with the boundaries functioning at the extremes. In that way, certain hurtful social behaviours, such as infidelity or being an unfortunate eyewitness to a crime, are not compensatable but instead viewed as part of the normal course of life. Individuals are expected to 'just get over' whatever misfortunes come their way, or turn to other social mechanisms, such as family and friends, to help overcome their distress. In these situations, the concern is less about the validity of the emotional distress than the need to promote free-flowing social intercourse. This judgement also reflects the concern, from a moral point of view, that these harms do not deserve punishment. Comments to the Restatement (Third) draft echo these views, by stating that 'some minor or modest emotional harm is endemic in living in society and individuals must learn to accept and cope with such harm'.[56]

At bottom, these instrumentalist and normative concerns support the fundamental judgement that the free-standing claims of 'emotional harm' are 'less deserving' than those for physical consequences. If we change the equation so that we no longer need to question the validity of these claims, would these other

---

[53] Mark A. Geistfeld, *Tort Law: Essentials* (New York, Aspen Pub., 2008).
[54] *Metro-North Commuter Railroad Co. v Buckley* 521 US 424, 433 (1997) (listing reasons why common law courts have restricted recovery for emotional harm including the difficulty in separating valid claims from invalid ones; the threat of 'unlimited and unpredictable liability' and the 'potential for a flood' of relatively unimportant claims).
[55] *Restatement* (note 1) § 47 cmt. f.
[56] Ibid.

instrumentalist and normative values still win the day? One assumption underlying American tort law for emotional distress is that these harms do not have a physiological basis and that we cannot test empirically the existence and amount of emotional distress. This premise, however, is being eroded by advances in neuroscience that study cognitive disorders. It is not the province of this paper to be a comprehensive analysis of neuroscience, but rather to highlight available research that suggest the beginnings of an increased ability to detect and quantify emotional harm. A summary of some of these advances are set forth below, in particular focusing on studies that indicate how individuals with Post Traumatic Stress Disorders experience abnormalities.

12.2 Use of Neuroscience and Neuroimaging in Assessing Cognitive Dysfunction

Through the use of neuroscience and neuroimaging, scientists have begun to link cognitive disorders from exposure to trauma to neurological conditions.[57] While individual differences influence whether an individual will develop a cognitive disorder following a traumatic event, we are developing a deeper understanding of how these disorders come into existence.[58] Although the field is in its early stages, advances in neuroscience and neuroimaging have begun to allow scientists to look closely at the brain regions involved during and after a traumatic event to learn how they function and interact. Neuroscientists have come to understand that dysfunction occurs in the neural systems that regulate emotion when an individual experiences acute stress.

Neuroscientist and Professor Joseph E. LeDoux advocates the view that emotions, like other physical sensations, result from physiological processes and therefore can be studied objectively.[59] He states:

If we want to understand feelings, it is likely going to be necessary to figure out how the more basic systems work. Failure to come to terms theoretically with the importance of processing systems that operate essentially unconsciously has been a major impediment to progress in understanding the neural basis of emotion. To overcome this, brain researchers need to be more savvy about the nature of emotions, rather than simply relying on common sense beliefs about emotions as subjective feeling states.[60]

Researchers have heeded LeDoux's call to investigate the 'neural basis of emotion'.[61] Focusing on psychiatric disorders, this research has begun to shed light on

---

[57] Amir Garakani, Sanjay J. Mathew, and Dennis S. Charney, 'Neurobiology of Anxiety Disorders and Implications For Treatment' (2006) 73(7) *The Mount Sinai Journal of Medicine* 941.
[58] Kevin N. Ochsner, David H. Ludlow, Kyle Knierim, et al., 'Neural Correlates of Individual Differences in Pain-Related Fear and Anxiety' (2006) 120 *Pain* 69.
[59] Joseph E. LeDoux, 'Emotional Circuits in the Brain' (2000) 23 *Ann. Rev. of Neuroscience* 155.
[60] Ibid. 156–7.
[61] Naomi L. Eisenberger, 'Identifying the Neural Correlates Underlying Social Pain: Implications For Developmental Processes' 2006 49(5) *Human Development* 273; Kevin N. Ochsner, David H.

the specific neural circuitry dedicated to emotional function. Evidence indicates that individuals with psychiatric disorders have abnormalities in these neural circuitry systems.[62]

Witnessing or experiencing a traumatic event involves a state described as acute stress, which activates a number of hormonal and neurotransmitter systems.[63] The systems that are activated trigger a chain of chemical processes that result in alterations in the neural networks that regulate memory and fear.[64] These physiological changes can materialize in the form of emotional distress symptoms, particularly anxiety symptoms.[65]

Extensive and replicated research has revealed brain regions that are associated with emotional trauma. In particular, structural and functional neuroimaging results implicate specific subregions of the medial prefrontal cortex (MPFC), orbitofrontal cortex (OFC), anterior cingulated (ACC), and insular cortices, the amygdala, and the hippocampus in the processing of emotional information.[66] Research suggests that dysfunction in this circuitry triggers and maintains emotional disorders.[67]

At the outset, this part outlines the processes, structures, and neural networks underlying memory and fear that can be considered the foundation of emotional harm. Next, this part highlights studies indicating the dysfunction that occurs when an individual experiences classic anxiety disorders, focusing in particular on the disorder of PTSD.[68] The focus on PTSD is appropriate for a number of

---

Ludlow, Kyle Knierim, et al., 'Neural Correlates of Individual Differences in Pain-Related Fear and Anxiety' (2006) 120 *Pain* 69.

[62] Adriana Feder, Eric J. Nestler, and Dennis S. Charney, 'Psychobiology and Molecular Genetics of Resilience' (2009) 10 *Nature Reviews Neuroscience* 446; Roger K. Pitman, Lisa M. Shin, and Scott L. Rauch, 'Investigating the Pathogenesis of Posttraumatic Stress Disorder with Neuroimaging' (2001) 62 [suppl. 17] *J. Clin. Psychiatry* 47.

[63] Roger K. Pitman, Kathy M. Sanders, Randall M. Zusman, Anna R. Healy, Farah Cheema, Natasha B. Lasko, et al., 'Pilot Study of Secondary Prevention of Posttraumatic Stress Disorder with Propronolol' (2002) 51 *Biological Psychiatry* 189; Benno Roozendaal, Bruce S. McEwen, and Sumantra Chattarji, 'Stress, Memory and the Amygdala' (2009) 10 *Nature Reviews Neuroscience* 423; Yvonne Ulrich-Lai and James P. Herman, 'Neural regulation of endocrine and autonomic stress responses' (2009) 10 *Nature Reviews Neuroscience* 397.

[64] Pitman (note 63) 189; Roozendaal (note 63) 424; Ulrich-Lai (note 63) 398.

[65] 427 *Roozendaal* (note 63).

[66] Martin P. Paulus, 'The Role of Neuroimaging for the Diagnosis and Treatment of Anxiety Disorders' (2008) 25 *Depression and Anxiety* 350; Roger K. Pitman, Kathy M. Sanders, Randall M. Zusman, Anna R. Healy, Farah Cheema, Natasha B. Lasko, et al., 'Pilot Study of Secondary Prevention of Posttraumatic Stress Disorder with Propronolol' (2002) 51 *Biological Psychiatry* 189; Hidenori Yamasue, et al., 'Gender-Common and -Specific Neuroanatomical Basis of Human Anxiety-Related Personality Traits' (2007) 18 *Cerebral Cortex* 46.

[67] Paulus (note 66) 351–2.

[68] Post-traumatic Stress Disorder (PTSD) is a type of Anxiety Disorder. The essential element of PTSD is the development of certain symptoms after exposure to an extremely traumatic event or experience that involves actual or threatened death or serious injury to oneself or others. The immediate reaction to the event or experience must include intense fear, helplessness, or horror. The categories of symptoms resulting from exposure to the triggering event include persistent re-experiencing of the trauma, avoidance of stimuli associated with the trauma, and a numbing of general responsiveness, as well as a number of symptoms associated with increased arousal. These symptoms must be present for more than one month to receive a diagnosis of PTSD. Otherwise, the individual is

reasons. First, this is probably the most heavily researched anxiety disorder in neuroscience at this stage. Moreover, PTSD offers a good parallel to what we label emotional harm in law, since the disorder is trauma-induced and has a delayed manifestation of outward symptoms.

12.2.1 The neural foundations of memory and fear

Learning, memory, and perception are all involved in experiencing the state of acute stress.[69] Learning and memory are complex processes that result in cellular-level changes as neurons modulate the strength and structure of their interconnections.[70] These processes are a type of brain plasticity, which is the relatively rapid and reversible change in brain structure and function.[71] Significantly, studies in neuroscience indicate that even a single exposure to a traumatic event can cause long-lasting cellular changes (or stress- induced plasticity) in the amygdala, the structure believed to be central to both anxiety and memory.[72]

The brain is very efficient in creating long-term memories of emotionally significant events, both positive and traumatic.[73] As we know, 'emotionally significant experiences tend to be well remembered'.[74] Two key structures are associated with this process. The amygdala, or the 'emotion centre', stimulates the 'arousal system' when trauma and stress are experienced,[75] and the pre-frontal cortex is considered the controlling mechanism to keep our emotions in check. The pre-frontal cortex regulates our experience of emotion and naturally compensates for aversive events. When functioning properly the pre-frontal cortex facilitates the formation of new connections that overrides the traumatic memory, a process called extinction. This system can be maladaptive, however, and the retention of traumatic material in the brain can result in increased anxiety, clinical mood disorders, and overall emotional distress.[76]

---

eligible for a diagnosis of Acute Stress Disorder, which will change to PTSD if the symptoms are still present one month after the trauma occurred: American Psychiatric Association, *DSM-IV-TR* (note 3) 467.

[69] McGaugh and Roozendaal, 'Memory Modulation' in John H. Byrne (ed.), *Learning and Memory: A Comprehensive Reference* 521, ch. 3.26 (Boston, Elsevier, 2009).

[70] Ibid.

[71] Rudi DeRaedt, 'Does Neuroscience Hold Promise for the Further Development of Behavior Therapy? The Case of Emotional Change After Exposure In Anxiety and Depression' (2006) 47 *Scandinavian J. of Psychology* 225; Feder (note 62) 453.

[72] Roozendaal (note 63) 429.

[73] J. L McGaugh, 'Memory Consolidation and the Amygdala: A Systems Perspective' (2002) 25(9) *Trends in Neuroscience* 465, McGaugh (note 69) 521.

[74] Roozendaal (note 63) 423; McGaugh (note 73) 465; Oliver T. Wolf, 'Stress and Memory in Humans: Twelve Years of Progress?' (2009) 1 *Brian Res.* doi:10.1016/j.brainres.2009.04.013.

[75] Roozendaal (note 63) 423; McGaugh (note 73) 465. (The BLA of the amygdala is activated by emotional arousal and helps made significant experiences memorable by enhancing the consolidation of long-lasting memory in other brain regions.)

[76] Roozendaal (note 63) 427.

When an individual experiences stress, the experience is encoded into the working memory, short-term memory, and finally consolidated into the long-term memory.[77] Memories are stored in the form of an increase in synaptic strength or in the pattern of the synapses themselves.[78] Short-term memory is the modification of already synthesized molecules, which strengthens existing connections.[79] Long-term memory involves the synthesis of new messenger RNA and new proteins, a process that can result in the induction and stabilization of long-lasting forms of entirely new synthesized neural connections.[80] The consolidation process involves the transfer of information from labile, short-term memory into long-term memory.[81]

The consolidation of a traumatic memory forms the basis for classic anxiety disorders, including PTSD.[82] The amygdala/prefrontal cortex circuitry is central to this process.[83] The consolidation process enables the interpretation of emotional information as well as controls the mechanisms that influence what individuals perceive in their environment and how they interpret that information (the attentional and interpretive processes).[84] Research suggests that when this circuitry is disrupted, anxiety results.[85] This is because the amygdala is hyperactive and the controlling mechanisms in the pre-frontal cortex are inadequately recruited. In other words, acute stress impairs the pre-frontal cortical function.[86] This malfunction or dysregulation leads to alterations in interpretive processes, or more precisely, a threat-oriented bias in anxious individuals. As a result of this bias, individuals with anxiety disorders react to stimuli that would objectively be interpreted as neutral or only mildly aversive with distress, hyperarousal, and attempts to avoid the anxiety-provoking object or situation.[87]

Scientists have begun to study these disproportionate fear responses, which provide valuable insight into their origins and neural bases.[88] Using Pavlovian

---

[77] Amy F. T. Arnsten, 'Stress Signaling Pathways that Impair Prefrontal Cortex Structure and Function' (2009) 10 *Nature Reviews Neuroscience* 410; Robert S. Blumenfeld and Charan Ranganath, 'Dorsolateral Prefrontal Cortex Promotes Long-term Memory Formation through its Role in Working Memory Organization' (2006) 26(3) *Journal of Neuroscience* 916.

[78] Gary Lynch, 'Memory Consolidation and Long-term Potentiation' in Michael S. Gazzaniga (ed.), *The New Cognitive Neurosciences* (Boston, MIT Press, 2000) 139.

[79] Kelsey C. Martin et al., 'Molecular Mechanisms Underlying Learning-Related Long-Lasting Synaptic Plasticity' in Michael S. Gazzaniga (ed.), *The New Cognitive Neurosciences* (Boston, MIT Press, 2000) 121.

[80] Lynch (note 78) 139.

[81] Garakani (note 57) 941; Wolf (note 74) 3.

[82] Pitman (note 63) 189.

[83] Amit Etkin and Tor D. Wager, 'Functional Neuroimaging of Anxiety: A Meta-Analysis of Emotional Processing in PTSD, Social Anxiety Disorder, and Specific Phobia' (2007) 164 *Am. J. Psychiatry* 1476, 1484; Ulrich (note 63) 401; Kevin S. LaBar, J. Christopher Gatenby, John C. Gore, Joseph E. LeDoux, and Elizabeth A. Phelps, 'Human Amygdala Activation during Conditioned Fear Acquisition and Extinction: A Mixed-trial fMRI Study' (1998) 20 *Neuron* 937, 939.

[84] LeDoux (note 59) 174; Wolf (note 74) 7.

[85] Sonja J. Bishop, 'Neurocognitive Mechanisms of Anxiety: An Integrative Account' (2007) 11(7) *Trends in Cognitive Science* 307.

[86] Arnsten (note 77) 410–11.

[87] Bishop (note 85) 307.

[88] Garakani (note 57) 941; LaBar (note 83) 937.

fear conditioning in healthy control subjects,[89] the fear system is treated as a set of processing circuits that detect and respond to danger.[90] This paradigm provides an objective basis to study fear in an experimental setting.[91] The lateral amygdala is believed to be the site where memory consolidation and plasticity in fear-conditioning occur, thus making the amygdala the primary structure of interest in the study of memory and fear.[92]

During and following a stressful event, the brain is flooded with stress hormones, which result in a number of physiological changes to the neural networks that regulate memory and fear.[93] The flooding of stress hormones solidifies the memory of the trauma by enhancing the consolidation process of the mental and emotional experience of the event.[94] In addition, the neurotransmitter norepinephrine has a central role in regulating stress effects on memory consolidation.[95] Evidence reveals that noradrenergic[96] activity in the basolateral complex of the amygdala (BLA) is critical to modulating other hormones and neurotransmitters involved in memory consolidation.[97]

Studies show that this series of events—adrenal stress hormones that trigger the regulation and consolidation of memory—holds true for emotionally arousing information, such as trauma and stress, but does not affect the consolidation of memory of emotionally neutral information.[98] When memory is 'retrieved', the fear response can be retriggered. The consolidation and retrieval of a traumatic

---

[89] The Pavlovian fear conditioning paradigm is the classic way to study fear conditioning in healthy control subjects: Garakani (note 57) 941; Diane B. Paul and Arthur L. Blumenthal, 'On the Trail of Little Albert' (1989) 39(4) *Psychological Record* 547; John B. Watson and Rosalie Rayner, 'Conditioned Emotional Reactions' (1920) 3 *J. of Experimental Psychology* 1. The classic example of conditioned fear is known as the Little Albert study. Albert, an infant, was given a rat to play with. Prior to the study, Albert demonstrated no aversion to furry animals or objects. Then presentation of the rat was repeatedly paired with a loud noise and the noise caused Albert to cry. When subsequently presented with the rat, instead of playing with it as he had done previously, Albert became distressed and anxious, presumably because he associated the rat with the loud noise he did not like. This reaction, when a previously neutral stimulus elicits a fear response, has become known as a conditioned fear. While today, due to issues of informed consent and minimum age requirements for participation in experimental studies, the Little Albert study would be highly unethical, and despite the fact that critics have pointed out the flawed methodology and questionable veracity of the study's findings, it retains its roots in psychology lore as the famous example of classically conditioned fear.

[90] LeDoux (note 59) 159.
[91] Ibid.
[92] Ibid. 161, 167; Roozendaal (note 63) 424.
[93] Roozendaal (note 63) 424–25.
[94] McCaugh and Roozendaal (note 69) 205.
[95] Roozendaal (note 63) 423–24.
[96] A neurotransmitter system comprised of noradrenergic neurons is how norepinephrine is transmitted through the brain. Noradrenergic neurons act on adrenergic receptors located in the amygdala, hippocampus, hypothalamus, thalamus, as well as numerous other brain structures and the spinal cord. Because of the large number of areas that contain adrenergic receptors, when the norepinephrine system is activated, a significant area of the brain is affected.

[97] The BLA influences the consolidation of memory through its many connections to other brain structures. The BLA projects directly to the caudate nucleus and both directly and indirectly to the hippocampus. The BLA also has connections to the insular cortex, which other studies have revealed is a common denominator in the manifestation and maintenance of anxiety disorders: *Roozendaal* (note 63) 424; McCaugh and Roozendaal (note 69) 205.

[98] Roozendaal (note 63) ; McCaugh and Roozendaal (note 69) 208.

memory forms the basis for classic anxiety disorders, including PTSD, the most severe disorder on the spectrum of anxiety disorders.[99]

12.2.2 Neuroscience studies of PTSD

Research studying fear conditioning in both animals and humans supports the hypothesis that a common element of PTSD specifically, and emotional distress in general, may be amygdalar dysfunction.[100]

Neuroimaging results in human subjects using functional magnetic resonance imaging ('fMRI') show that fear conditioning leads to increases in amygdalar activity.[101] Once the amygdala detects danger, it can activate various 'arousal' networks, which can then influence sensory processing.[102] Following fear conditioning, the information transmitted from the amygdala results in the individual experiencing and exhibiting a fear response.[103]

Recent neuroimaging studies have investigated the relationship between specific brain regions and both auditory and visual cues in PTSD patients.[104] Functional MRI studies have also revealed the relationship between the amygdala and medial prefrontal regions in PTSD.[105] A 2004 study compared PET scans of thirty-six Vietnam veterans with diagnosed PTSD to PET scans of Vietnam veterans without PTSD.[106] The researchers used script-driven imagery to conduct the study. All of the male participants had served in combat and all of the female participants had served as nurses in Vietnam. None of the veterans had a history of head injury, neurological disorders, or other major conditions.

---

[99] Using the DSM-IV-TR classifications of Anxiety Disorders as a reference, specific clusters of symptoms—represented by the particular Anxiety Disorder—can be placed as points along the spectrum. At the less severe (yet still clinically significant) end of the spectrum are generalized anxiety symptoms and on the opposite end of the spectrum are Acute Stress Disorder and PTSD: *DSM-IV-TR* (note 3).
[100] Elisabeth A. Murray, 'The Amygdala, Reward, and Emotion' (2007) 11 *Trends in Cognitive Science* 489; Garakani (note 57) 942; Bishop (note 85) 307; Gleb P. Shumyatsky, Evgeny Tsvetkov, and Gael Malleret, et al., 'Identifying a Signaling Network in the Lateral Nucleus of Amygdala Important for Inhibiting Memory Specifically Related to Learned Fear' (2002) 111 *Cell* 905; M. Davis and P. J. Whalen, 'The Amygdala: Vigilance and Emotion' (2001) 6 *Molecular Psychiatry* 13; LeDoux (note 59) 171; Wolf (note 74) 7.
[101] Christian Buchel et al., 'Brain Systems Mediating Aversive Conditioning: An Event-related fMRI Study' (1998) 20 *Neuron* 947; LaBar (note 83) 937.
[102] LeDoux (note 59) 177; Etkin and Wager (note 83) 1482. (Information is transmitted from the amygdala to the behavioural, autonomic, and endocrine response control systems located in the brainstems.)
[103] Shumyatsky (note 100) 905.
[104] Etkin and Wager (note 83) 1480. The recollection of traumatic events in PTSD patients has been associated with activation of the amygdala, orbitofrontal cortex, anterior temporopolar cortex, and insular cortex and decreased activation in anterior cingulated gyrus, medial frontal gyrus, and subcallosal gyrus.
[105] Lisa Shin et al., 'Regional Cerebral Blood Flow in the Amygdala and Medial Prefrontal Cortex during Traumatic Imagery in Male and Female Vietnam Veterans with PTSD' (2004) 61 *Arch. Gen. Psychiatry* 168.
[106] Ibid.

Results of the study indicated hyperresponsivity of the amygdala and hyporesponsivity of medial prefrontal regions and that these responses are reciprocally related.[107] The more hyper- and hypoactive these regions were, the more severe the symptoms. These results support the hypothesis that PTSD symptoms reflect extreme dysregulation in these regions and neural mechanisms. While such a relationship between the amygdala and medial prefrontal regions in clinically diagnosed PTSD patients had been suspected, no previous studies in the literature had documented data in support of such a relationship.

Studies have also begun to document and distinguish at the chemical and structural level between PTSD, the most extreme anxiety disorder, and other anxiety disorders. In 2007, Etkin and Wager conducted a meta-analysis of studies that had used brain scans to investigate emotional processing in patients with anxiety disorders.[108] The meta-analysis compared fMRI and PET scans of individuals with one of three anxiety disorders—PTSD, social anxiety disorder, and specific phobia—with the scans of healthy individuals who had undergone fear conditioning. The results indicated that patients with the anxiety disorders showed consistently greater activity in the amygdala and insula.[109] Even more significant, the dysregulation in the neural circuitry of PTSD patients was more exaggerated than that of patients suffering from the other anxiety disorders. Only patients with PTSD showed hypoactivation in the dorsal and rostral anterior cingulated cortices and ventromedial prefrontal cortex—additional structures linked to the experience and regulation of emotion.

The effects unique to PTSD suggest that emotional dysregulation in that situation extends beyond an exaggerated fear response or beyond the fear response demonstrated in other diagnosable (medically significant) anxiety disorders. The results of this study have revealed both that the amygdala and insula are critical structures in the common neurobiological pathway in anxiety disorders, and support the view that a core fear system exists and when it is activated, anxiogenic[110] symptoms result.[111]

---

[107] The reciprocal or inverse relationship between the two regions means the more active, or hyperresponsive, the amygdala is, the more the inhibited, or hyporesponsive, the prefrontal cortex is.

[108] Etkin and Wager (note 83) 1476.

[109] Ibid. 1480. Hyperactivation of amygdala and insula was more pronounced in the scans of subjects with social anxiety disorder and specific phobia than in subjects with PTSD. Hypoactivation was seen in four areas—ventromedial prefrontal cortex, rostral and dorsal anterior cingulated cortices and the thalamus—in PTSD patients but not in patients with the other anxiety disorders or healthy matched control subjects.

[110] Anxiogenic is synonymous with reflecting, causing, or producing anxiety; not every anxiety disorder reaches the level of PTSD. Some individuals may simply experience generalized disturbances in anxiety or mood. But neuroscience advances now indicate that that these neural changes occur for anxiety disorder in general, with individuals diagnosed with PTSD displaying the most dramatic alterations in neural circuitry and consequently the most severe symptoms.

[111] Anxiogenic is synonymous with reflecting, causing, or producing anxiety; not every anxiety disorder reaches the level of PTSD—some individuals may simply experience generalized disturbances in anxiety or mood. But neuroscience advances now suggest that that these neural changes occur for anxiety disorder in general, with individuals diagnosed with PTSD displaying the most dramatic alterations in neural circuitry and consequently the most severe symptoms. Etkin and Wager (note 83) 1485.

Other research reveals that a brief exposure to stress triggers a series of cellular changes that take time to come to an end, meaning there is a delay in the time it takes for the cellular changes in the BLA to be completed.[112] The result of this time delay is that once triggered, the plasticity mechanisms continue after the event, despite restoration of normal levels of neurotransmitters and hormones.[113] This means that a single brief exposure to stress results in some modest structural changes at the synaptic level that take time to build up and to slow down—in other words, they have delayed anxiogenic effects at the behavioural level.[114]

This finding—that even a single exposure to a traumatic event can cause long-lasting cellular changes, or stress-induced plasticity, in the amygdala—is highly significant. As discussed above, acute stress results in neuronal remodelling through the creation of new synaptic connections and dysregulation in neural circuitry in the BLA and medial amygdala. The process of storing the memory of the emotional event may be an important mechanism in the development of anxiety disorders such as PTSD. Retrieving the memory of emotionally arousing information induces greater activity in and connectivity between the amygdala and the hippocampus than retrieving the memory of emotionally neutral information.[115] This evidence suggests that emotionally relevant and emotionally neutral information are treated differently and stored through different mechanisms and processes in the brain.[116] The fact that we have begun to detect these changes through neuroscience gives us the opportunity to quantify an event that we long considered a subjective state.

In sum, the physiological changes that occur in the brain after an individual experiences or witnesses a traumatic event can result in a dysfunction of the neural networks that regulate memory and fear.[117] We rely on the prefrontal cortex inhibitory function to override adverse effects from emotional stress.[118] However, when the prefrontal cortex is prevented from carrying out this function, that dysfunction manifests itself in the symptoms of anxiety disorders. Even though the only symptoms the individual may demonstrate are emotional in nature (and misleadingly believed to be solely subjective symptoms), scientists may now begin to document and observe the physiological changes that occur in the brain after experiencing trauma as a result of advanced neuroimaging techniques.

Long before the neuro-scientific advances in understanding the physiological basis of emotional harm began, English courts took a different approach to resolving emotional distress claims. Because the English approach can easily accommodate our evolving understanding of emotional distress, the next part describes the English law approach.

---

[112] Roozendaal (note 63) 429.　[113] Ibid.
[114] Ibid.　[115] Ibid. 426.
[116] LaBar (note 83) 942.
[117] Roozendaal (note 63) 430; Bishop (note 85) 307; Arnsten (note 77) 410; Etkin and Wager (note 83) 1476; Shumyatsky (note 100) 905.
[118] Arnsten (note 77) 411.

12.3 English Law Approach: Recognizable Psychiatric Illness

The English courts approach the issue of emotional harm differently from American courts. Instead of drawing a sharp distinction between emotional and physical harm,[119] English courts impose a threshold requirement on the plaintiff to prove a diagnosable psychiatric illness as an aspect of bodily harm. Distinguishing between the American approach as exemplified by the Restatement Third and the English approach, Professor Martin Matthews explains:

> More significant than differences in specific outcomes, however, is the difference in tone that becomes possible with a de-emphasis on distinctions between 'physical' and 'emotional'; with a recognition of all illnesses as aspects of bodily harm; and with the substitution of medical diagnoses of psychiatric disorders to define free-standing compensable conditions, in place of the vagaries of jury understandings about the meaning of the word 'severe.'[120]

This difference in tone and approach may create the ground work for reconceptualizing the American tort claim for emotional harm.

The English courts distinguish, like the American courts, between primary and secondary (or bystander) victims. For primary victims, absent a claim of physical harm,[121] the plaintiff must satisfy an absolute threshold requirement that they are suffering from a recognizable psychiatric illness, or 'nervous shock' as it is referred to in common legal terminology.[122] Specific illnesses are not required, as long as it is a recognized psychiatric disorder.[123] Damages have been awarded for wide-ranging illnesses, including morbid depression,[124] hysterical personality disorder,[125] post-traumatic stress disorder,[126] pathological grief disorder,[127] and chronic fatigue syndrome.[128]

---

[119] Restatement, § 4, cmt. a (note 1). ('Emotional disturbance is distinct from bodily harm and means harm to a person's emotional tranquility.').

[120] Martin Matthews, 'Remarks Presented at a Symposium on the Third Restatement of Torts, Wake Forest University School of Law: Negligent Infliction of Emotional Distress: A View of the Proposed Restatement Provisions from England' (2–3 April 2009). See generally, M. H. Matthews, 'Negligent Infliction of Emotional Distress: A View of the Proposed Restatement (Third) Provisions from England' (2009) 44 Wake Forest L. Rev. 1177.

[121] A person pursuing a claim for physical injury will be entitled to compensation for mental distress that arises as a consequence of the injury. Michael A. Jones, 'Liability for Psychiatric Illness: More Principle, Less Subtlety?' (1995) 4 Web JCLI <http://webjcli.ncl.ac.uk/articles4/jones4.html> ('The courts... see no difficulty in compensating psychiatric damage produced as a direct consequence of physical injury to the plaintiff.').

[122] Ibid. 2 note 1. While courts use different terms such as recognized psychiatric injury, illness or disorder, the general term 'medically recognized mental disorder' has been adopted by the Scottish Law Commission on Damages for Psychiatric Injury: Report on Damages for Psychiatric Injury (Scot Law Com. No. 196, 2004) [1.7].

[123] For a full discussion of the various potential categories of psychiatric disorders giving rise to a claim, see Law Commission, 'Liability of Psychiatric Illness' (Law Com. No. 249, 1998) <http://www.lawcom.gov.uk/homepage.htm>.

[124] Hinz v Berry [1970] 2 QB 40.
[125] Brice v Brown [1984] 1 All ER 577, CA.
[126] Frost v Chief Constable of South Yorkshire Police [1997] 1 All ER 540.
[127] Vernon v Bosley (No. 1) [1997] 1 All ER 577, CA.
[128] Page v Smith [1996] AC 155.

Courts distinguish between 'mere feelings' and 'injury'. Anxiety, fear, or even short-term symptoms of shock are not compensable conditions, if they are not a diagnosed psychiatric illness.[129] Lord Bridge, in *McLoughlin v O'Brien*,[130] explained:

> The common law give no damages for the emotional distress which any normal person experiences when someone he loves is killed or injured...Anxiety and depression are normal human emotions. Yet an anxiety neuroses or a reactive depression may be recognizable psychiatric illnesses, with or without psychosomatic symptoms. So, the first hurdle which a plaintiff claiming damages of the kind in question must surmount is to establish that he is suffering, not merely grief, distress, or any other normal emotion, but a positive psychiatric illness.[131]

Lord Grieve added in a separate but contemporaneous case:

> While it is well established that damages can be recovered on the ground of nervous shock, as well as the ground of actual physical injury, there must be some evidence that the 'nervous shock' complained of is a condition which can reasonably be described as an illness.[132]

Thus, under English law, the plaintiff must affirmatively prove the appearance of a psychiatric illness, rather than simply arguing the absence of mental well-being.

The threshold requirement of a recognizable psychiatric illness can be traced to a 1970 decision, *Hinz v Berry*.[133] In that case, the plaintiff and her husband and children went for a picnic in the country. The pregnant Mrs Hinz took her three-year-old daughter for a short walk, leaving her husband and the other seven children next to the roadway.[134] The defendant's out-of-control car left the road, striking and killing the plaintiff's husband and injuring nearly all of her children.[135] The plaintiff heard the sound of the impact and rushed to her family.[136] Mrs Hinz was compensated for the pecuniary loss of her husband as well as the children's injuries.[137] The question remained, though, whether she could recover for the emotional shock and long-term psychiatric harm she experienced by witnessing the accident. In deciding the case, the court distinguished between general emotional harms and psychiatric injury.[138]

> In English law no damages are awarded for grief or sorrow caused by a person's death. No damages are to be given for the worry about the children, or for the financial strain or stress, or the difficulties of adjusting to a new life. Damages are, however, recoverable for nervous shock, or to put it in medical terms, for any recognizable psychiatric illness caused by the breach of duty by the defendant.[139]

The court found that the plaintiff's state was not merely a 'harassed state owing to the difficulty of looking after [the plaintiff's] family in this situation, but something

---

[129] *Hinz v Berry* [1970] 2 QB 40; *Page v Smith* [1996] AC 155 (HL).
[130] [1983] 1 AC at 431.
[131] Ibid.
[132] *Simpson v Imperial Chemicals Ltd* [1983] SLT 601 (HL Sc) 609.
[133] *Hinz v Berry* [1970] 2 QB 40.
[134] Ibid. [135] Ibid. [136] Ibid. [137] Ibid.
[138] Ibid. [139] Ibid. 42.

more than that: she has been and still is in a positively morbid state. There is a recognizable psychiatric illness.'[140]

The requirement of proving a recognizable psychiatric illness is imposed in all cases of free-standing mental harm claims, whether the victim is a primary victim or a secondary (bystander) victim. In the bystander case, however, the plaintiff must also show that the psychiatric injuries result from 'shock', or the 'sudden appreciation by sight or sound of a horrifying event, which violently agitates the mind'.[141] In *Alcock v Chief Constable of South Yorkshire Police*,[142] sixteen relatives of victims in the Hillsborough stadium disaster of 1989 brought claims for psychiatric injuries suffered due to negligent police actions.[143] Some of the claimants were at the stadium that day, while others witnessed the events live on television.[144] The police admitted negligence for the deaths and injuries, but denied the existence of duty to the plaintiffs for psychiatric harm. The court unanimously rejected a simple foreseeability test and instead imposed various limiting tests for recovery on plaintiffs suffering psychiatric illness as a result of seeing or hearing the incident, including the shock requirement.[145] Requiring a showing of shock may represent an outmoded view of causation for mental harm or it may simply be another version of preventing disproportionate liability in the bystander context, but it remains the current law in English courts.

Soon after *Alcock*, the court considered *Page v Smith*.[146] At issue was whether a driver who was not physically harmed in a traffic accident could recover for a recurrence of his chronic fatigue syndrome, considered an emotional harm claim by the court, allegedly caused by the accident.[147] On the facts of the case, it seemed clear that the accident could have been reasonably foreseen to cause physical injury, but not psychiatric injury.[148] The Court of Appeal found that the lack of foreseeability barred recovery for his emotional harm claim.[149] The House of Lords disagreed. It considered the plaintiff to be a primary victim, since he was involved in the accident, and therefore found that the *Alcock* limitations for bystander liability did not apply.[150] As a primary victim, the court reasoned that the plaintiff did not need to prove foreseeability of psychiatric harm when physical harm was foreseeable.[151] As Lord Lloyd wrote:

---

[140] *Hinz v Berry* [1970] 2 QB 44.
[141] *Alcock v Chief Constable of South Yorkshire Police* [1992] 1 AC 310 (HL) 401 (Lord Ackner).
[142] [1992] 1 AC 310 (HL).
[143] Ibid. *Alcock* was a test case for approximately 150 similar claims for the Hillsborough disaster working their way through the court system. Gerald Schaefer, 'The Development of the Law on Psychiatric Injury in the English Legal System' *New Zealand Postgraduate Law E-Journal* at <http://nzpostgraduatelawejournal.auckland.ac.nz/PDF%20Articles/Issue%204%20(2006)/GeraldPsychInjury.pdf>.
[144] All of the plaintiffs were relatives of people who were in the disaster area, but none were spouses or parents of one of the ninety-five people who died, or of the over 400 who sustained injuries.
[145] *Alcock* imposed three limiting tests: the incident must have created a 'sudden shock'; there must be a loving relationship between the claimant and the victim; and the claimant must have had either a presence at the scene or the immediate aftermath: ibid. (citing Lord Wilberforce's opinion in *McLoughlin* (note 131).
[146] [1996] AC 155 (HL).   [147] Ibid.   [148] Ibid.
[149] Ibid.   [150] Ibid.   [151] Ibid.

In the case of physical injury there is no such [normal fortitude] requirement. The negligent defendant, or more usually his insurer, takes his victim as he finds him. The same should apply in the case of psychiatric injury. There is no difference in principle... between an eggshell skull and an eggshell personality.[152]

Thus, a primary victim may recover for an unforeseeable psychiatric injury if some physical injury was foreseeable, even if the physical injury did not occur.[153]

The *Page* decision has come under some criticism. Critics argue that regardless of whether *Page* was a primary or secondary victim, he should still have been required to show that the mental illness was contracted in circumstances where a person of reasonable fortitude would have suffered similarly. Before *Page* was decided, a claimant would have to make a separate showing of foreseeability of a psychiatric disorder before the claim could proceed. Thus, according to these critics, when *Page* was decided, it threatened to open the proverbial floodgates.[154]

Under *Page*, physical and psychiatric harm would simply be two sides of the same injury coin for primary victims.[155] This represents an approach that aligns with advances in neuroscience, suggesting that drawing a sharp distinction between physical harm and mental harm may be outmoded. To argue for a requirement of ordinary fortitude to withstand mental harm—when we do not require the same ordinary fortitude for physical harm—implies an assumption about the value or validity of one claim over the other that may be eroding.

The *Page* approach is not without limits, however. Although *Page* made clear that the psychiatric illness did not have to be reasonably foreseeable, as long as some bodily harm was foreseeable, the court did suggest that the danger of bodily harm had to be immediate. A later ruling picked up this same theme and aligned it closely with the Restatement (Third) position. In *Rothwell v Chemical and Insulating Co. Ltd*,[156] the claimants had been exposed to asbestos and developed pleural plaques. The House of Lords found that this did not qualify as a physical injury. One of the

---

[152] Ibid. 189.
[153] Causation must be immediate between the physical risk and the psychiatric harm, however, or the claimant is not considered a primary victim. *Grieves v F.T. Everard & Sons Ltd* [2007] AC 281 (distinguishing *Page* since the claimant's injury was not caused immediately by the exposure of the asbestos but by the doctor telling him of his possible pleural plaques years later).
[154] *White v Chief Constable of South Yorkshire Police* [1999] 2 AC 455. Lord Goff explored the criticisms in his opinion: 'Furthermore, as the Law Commission record in para. 5.14 of their Report, the revolutionary thesis in Page v. Smith has provoked severe criticism by a number of scholars with a special interest in this branch of the law, notably by Nicholas Mullany in "Psychiatric Damage in the House of Lords—Fourth Time Unlucky: Page v. Smith" (1995) 3 Journal of Law and Medicine 112, and Dr. Peter Handford in "A New Chapter in the Foresight Saga: Psychiatric Damage in the House of Lords" (1996) 4 Tort L.Rev. 5; but also by Professor Tan Keng Feng in "Nervous Shock to Primary Victims" [1995] Singapore Journal of Legal Studies 649; F. A. Trindade in "Nervous Shock and Negligent Conduct" (1996) 112 L.Q.R. 22; and Alan Sprince in "Page v. Smith—being 'primary' colours House of Lords' judgment" (1995) 11 Professional Negligence 124. Most of them deplore the abandonment of the previously accepted general requirement of foreseeability of psychiatric injury.'
[155] In *Page*, both Lord Browne-Wilkinson and Lord Lloyd argued that a distinction should no longer be made between physical and psychiatric injury.
[156] [2007] UKHL 39; [2008] 1 AC 281.

claimants claimed to have suffered a psychiatric illness after having an x-ray many years after the exposure and being informed of the presence of pleural plaques and the risk of future illness. Attempting to invoke *Page v Smith*, the plaintiff sought damages for his psychiatric illness, but the court distinguished *Page* on two grounds: 1) any future illness would not be the immediate result of the exposure to asbestos; and 2) the psychiatric illness was due to information received after the x-ray, not from the exposure itself. Thus, mere distress suffered through fear of future injury absent some physical harm does not give rise to a cognizable claim for damages.[157]

If American courts borrowed from the *Alcock/Page* line of cases to impose a threshold showing of a 'recognizable psychiatric illness' most significantly it would require a medical diagnosis of illness, which is not currently required in the vast majority of American jurisdictions nor proposed by the Restatement (Third). But medicalizing the claim should lead more naturally to inclusion of neuroscience evidence and de-emphasize the distinction between physical and emotional. It also may satisfy the instrumentalist need for bright lines in a more satisfactory and less arbitrary way than the tests put forth by the Restatement.

It is arguable that the English threshold requirement both broadens and restricts liability. On one hand, it may broaden liability by recognizing psychiatric injury from witnessing damage to property, such as a pet or a house,[158] whereas it is unlikely that a plaintiff could recover under those circumstances under the Restatement view. On the other hand, it imposes a higher threshold requirement that could eliminate recovery for mental effects that do not amount to medical disorders, a more restrictive approach than that of the Restatement. But if there are legitimate reasons to include these other (lesser) mental effects in recovery, then courts should be forced to articulate the reasons for that inclusion. Imposing a threshold requirement of a medical disorder will also assist in addressing the floodgates concern by curtailing widespread litigation rights. Leaving the question of 'serious emotional disturbance' to the jury, especially in light of advances in neuroscience, seems to offer a greater chance of extending the scope of liability.

Most significant for our purposes, the English approach points in the direction of viewing emotional harm as part of the rubric of bodily injury. The threshold requirement of recognizable psychiatric disorder allows us to distinguish between 'mere feelings' and 'injury' in a more satisfying way, based on medical evidence. It may impose a higher standard of proof than 'serious emotional disturbance', but once the hurdle is overcome, it no longer requires the plaintiff to pursue a track different from that for physical injury.

---

[157] *Hicks v Chief Constable of Yorkshire Police* [1992] 1 AC 310.
[158] *Attia v British Gas Plc* [1998] QB 304 (house on fire).

12.4 Rethinking the American Approach to Free-Standing Emotional Distress Claims

Neuroscience has not yet reached the point where it enables us to quantify distress claims. Yet advances in neuroimaging are real, and it is clear that emotional harm is not the enigma it once was. These advances challenge our traditional doctrine and policy reasons for limiting compensation in this area, forcing us to clarify our reasons for allowing or disallowing compensation for emotional harm. We need to rethink whether other policy reasons—besides the physical/mental dichotomy—justify treating claims for emotional harm as less worthy than those for physical injury.

Regardless of whether we retain the current distinction in tort law between physical and emotional harm, at the very least, advances in neuroimaging may give us another, more probative avenue to verify claims for injury related to mental distress. For example, the advances may enable us to enhance the analysis of both the objective and subjective elements of the free-standing emotional distress tort. With regard to the objective element—would a reasonable person suffer mental distress under similar circumstances—neuroscience may help us quantify how an average person generally would react in a given situation. Further study of reactions of individuals in certain circumstances may give us the opportunity to quantify levels of distress actually experienced in response to certain stimuli and in certain circumstances.

Advances in neuroimaging should also improve proof on the subjective element of the tort—did this plaintiff actually suffer harm in this instance—one of the chief concerns of the free-standing emotional distress tort. As described above, we have already begun to correlate the subjective feelings of patients diagnosed with PTSD with specific imaging findings. This suggests we may be able to apply the correlation in the other direction and verify a claim for PTSD on the basis of neuroimaging. Indeed, at least one court suggests that we have already reached this point.[159]

Moreover, the availability of neuroimaging evidence should argue in favour of abandoning the more artificial and arbitrary tests for limiting emotional harm claims such as physical impact, physical manifestation and zone of danger.[160] Psychic injury should no longer raise the concerns it once did of fraud or lack of predictability. This should allow us to substitute foreseeability of harm as the main determinant of duty in this area, bringing it in line with the fundamental basis of duty for most other torts.

On a larger scale, advances in neuroscience may invite us to rethink generally our approach to the tort of emotional distress once we begin to document the

---

[159] *Allen* (note 5).
[160] Use of fMRI evidence should not be a return to the 'physical manifestation' test. Unlike physical manifestations of emotional harm such as stomach aches and headaches, the neuroimaging is not a condition itself. Instead, it corresponds more closely to a picture of a condition, similar to an x-ray of a broken bone.

physiological changes that occur in the brain from stress and fear. The English approach to these claims would facilitate this. Adopting the diagnosis-based system of the English courts would shift the focus of the claim from a normative decision to a medically based one. Although we may end up with the same result under either approach—the symptoms traditionally offered as evidence by plaintiffs advancing an NIED claim often correlate very strongly to the symptoms present in mood and anxiety disorders in the DSM-IV-TR[161]—the allowance or disallowance of the claim will be on a more supportable basis. It would allow us to move away from the relentless distinction between physical and mental harm maintained by the American courts to a broader based conception of bodily harm.

There are three major stumbling blocks with regard to practical applications of neuroimaging to the area of emotional harm specifically: 1) establishing the 'before' picture, or the plaintiff's baseline; 2) extrapolating information gleaned in generalized studies to a specific instance, or 'individuation'; and 3) dealing with the different paces at which science will document different disorders.[162]

Establishing a baseline goes to the issue of causation; that is, whether the plaintiff's emotional harm is really due to the events the plaintiff has experienced as opposed to their prior history or some other event. We must continue to emphasize the distinction between correlation and causation in using neuroimaging data in court. Finding a correlation often misleads us into believing that a causative effect exists, but discovering the neural correlates of cognitive phenomena does not tell us that the defendant's negligence caused the dysfunction.[163] Without some evidence of the plaintiff's condition prior to the accident, it is hard to evaluate whether it was the negligence that actually caused the plaintiff's psychological harm or aggravated it further, or whether it was a pre-existing condition.

It is unlikely that the plaintiff will have had previous brain scans to compare to the current scan. We have other measures, however, to help give us a 'before' picture. In other areas, such as assessing damages for traumatic brain injuries, we rely on circumstantial evidence, such as school, employment, and medical records,

---

[161] For example, the *DSM-IV-TR* lists the following symptoms for Major Depressive Episode: 1) Depressed mood most of the day; 2) Marked diminished interest or pleasure in all or most activities; 3) Significant weight loss; 4) Difficulty concentrating; 5) Hypervigilance; 6) exaggerated startle response *DSM-IV-TR* (note 3) 468.

[162] Of course, the same concerns exist here that surround admission of other scientific evidence and the evidence will be subjected to the same threshold requirements of reliability and testing by cross-examination. William R. Uttal, 'Neuroscience in the Courtroom: What Every Lawyer Should Know about the Mind and the Brain' 175, 240–2 (Tucson, Lawyers & Judges Pub. Co., 2009) (neuroscience of cognitive dysfunction still does not meet the *Daubert* criteria for admissibility). Furthermore, neuroimaging may give the appearance of scientific neutrality but it is actually 'the product of a complex set of techniques, subjective decisions, technical choices, and informed interpretations'. Joseph H. Baskin, Judith G. Edersheim, and Bruce H. Price, 'Is a Picture Worth a Thousand Words? Neuroimaging in the Courtroom' (2007) 33 *Am J. L. & Med* 239, 250. For example, the neuroimaging specialist makes many subjective decisions such as the level of detail to use, how many images to order, how thick or thin to make the 'slices', how much contrast to use between differing types of tissue, and how to filter out background noise from the signal: ibid. Different neuroimaging technologies vary as to spatial and temporal resolution and how the results are interpreted: ibid.

[163] Uttal (note 162) 240.

to establish the baseline to help evaluate the claim. It may be that we will have to follow a similar avenue here.

In this regard, however, it is significant to note how specific brain scan technology has become. Neuroscience is so advanced that we are not examining whole brains, but rather individual brain structures, chemical levels, individual receptors and synaptic connections in neural circuitry. Scientists have begun to determine whether the individual's brain structures are performing properly or whether there is an abnormality somewhere in the system. For example, in a patient with depression, scientists can examine the chemical structure of the brain, and in particular try to measure their serotonin level. This specificity may eventually help us in establishing baselines, as we learn how to measure, for example, how long a patient has been experiencing dysregulation in their serotonin system.

Even assuming that we can establish a baseline through relatively trustworthy evidence, however, another related problem remains—that of individuation. There exists great variability within the normal population of both brain structure, in terms of size and shape, and brain ability to compensate for and adjust to injury.[164] Thus, individualization of neuroimaging data is difficult to apply—it is hard to extrapolate from a normative template to the scan of an individual.[165] All of the brain studies are based on averages. At best, the brain scans show us the average brain states of individuals who suffer from, say, PTSD usually exhibit. It will be difficult to extrapolate from those neural correlations to prove injury in an individual case. This also requires us to determine the boundaries of the tort. Since the use of brain scan evidence is so new, we have yet to define the level of correlation we will require—i.e. how many different brain states must the plaintiff's scan correlate to before we qualify it as a recognized harm under the law?[166]

Further, it is likely that neuroscience evidence will develop at different paces for different disorders. And it is likely that there will be disorders that we will not be able to measure pending years of scientific development. If we have neuroscientific support for some disorders, like PTSD, but not yet (or ever) for others, should we disallow the claim for the scientific laggards? Presumably, we do not want to limit compensable injury to emotional harm that is measurable by neuroimaging if other forms of reliable proof exist. Similarly, it is unlikely that we would preclude evidence we currently use to verify emotional harm, such as insomnia, nightmares and nausea. How will this evidence relate to brain scan evidence—will it be considered prerequisite, superfluous, or corroborative evidence?

Along the same line, our legal system currently compensates for other kinds of emotional harms besides medically recognized anxiety disorders. Neuroscience has not even begun to document these types of emotional disturbances. We will need to determine how to integrate proof requirements for requirements for medically

---

[164] Ibid.
[165] Turhan Canli and Zenab Amin, 'Neuroimaging of Emotion and Personality: Scientific Evidence and Ethical Considerations' (2002) 50 *Brain & Cognition* 414, 424. Moreover, we do not yet have enough data to determine 'which brain measure should be used as a normative standard' (ibid).
[166] A. M. Viens, 'The Use of Functional Neuroimaging Technology in the Assessment of Loss and Damages in Tort Law' (2007) 7 *AJOB Neuroscience* 63, 64.

recognized cognitive disorders with those that do not reach that level of recognition. We also can anticipate the gap between evidence supporting claims of medically recognized cognitive disorders and non-medically recognized disorders will grow larger as science focuses more intensively on the former type of claim.

If some type of neuroimaging evidence does exist in a given instance, can the plaintiff be penalized for not proffering such evidence? For example, in one case, the US Court of Appeals for the Eighth Circuit suggested that the plaintiff's expert should have ordered a positron emission tomography or single photon emission computed tomography scan of the plaintiff's brain to support the claim that post-traumatic stress disorder caused injury to the plaintiff's brain.[167] Because the expert failed to do so, the court rejected the argument that the plaintiff suffered physical injury to her brain.[168]

On a larger scale, advances in neuroimaging may force us to rethink how we approach certain normative questions in tort law. The use of norms in measuring liability for emotional harm depends on the type of tort involved. On one end of the tort spectrum, defining outrageous and extreme behaviour ultimately is a social normative issue that is less focused on the individual's reaction to the behaviour than on the behaviour itself. Thus, in torts like assault, false imprisonment, libel and slander, and IIED, we define broad categories of behaviour as unacceptable, and assume that these types of torts are very likely to evoke emotional distress. On the other end of the spectrum, as in the tort of NIED, the plaintiff's mental state in relation to the behaviour is more significant, as we depend more on an individual's as well as a reasonable person's reaction to the behaviour as the definitional base.

With regard to the torts that define certain categories of behaviour as unacceptable, it is unlikely that advances in science will change those definitions, except to the extent that neuroimaging may be able to provide evidence of what brain states usually are exhibited in individuals who experience certain trauma in specific situations. This may then inform the normative question of defining acceptable social behaviour. With regard to the latter—determining what the hypothetical reasonable person would experience in a given situation—neuroimaging advances may invite us to change the question of how a reasonable person would respond to certain behaviour to the question of how brain scans of the individual subjected to the behaviour will deviate from the statistical norm of the brain scan of the average individual.[169] As our information becomes more exact, we need to rethink whether we seek to compare our plaintiff against the average individual or against a more idealized hypothetical reasonable person.

Ultimately, advances in neuroscience should bring us clearly to the question whether we should treat distress claims like other negligence claims or continue to treat them like second class citizens. By removing traditional doubts about the

---

[167] *In re Air Crash at Little Rock* 291 F3d 503, 511 (8th Cir 2002) ('[The plaintiff] was not given a magnetic resonance spectroscopy, a positron emission tomography (PET) scan or a single positron emission computed tomography (SPECT) scan, all tests which ... could have been utilized to show the functioning of [her] brain.'). The court made no mention of the cost of such tests.

[168] Ibid. (rejecting claim under the Warsaw Convention).

[169] Viens (note 166) 64.

validity of the claim, the classic arguments in favour of tort recovery will come to the foreground. Policies promoting the general principles of negligence law—such as deterring negligent behaviour, shifting the cost of the injury from the plaintiff to the culpable defendant, and compensating the victim of the injury—are strengthened as the validity of the claim becomes stronger. Yet if other instrumentalist concerns—such as the avoidance of disproportionate, crushing liability and distribution of limited funds in a just and fair manner—are the fundamental justifications for limiting liability in this area, then advances in neuroscience may not sway the equation.[170]

In the end, advances in technology and science promise more—not less—debate in the courts on the proper scope of emotional distress claims. Advances in neuroscience merely shift the debate away from the validity of the claim to the more important legal and policy questions of the permissible scope of liability. With this shift, courts will increasingly be called upon to articulate new boundaries of liability in the area of emotional harm. If there is to be no or lesser recovery for mental distress claims, then this choice should be better explained by policy concerns about ruinous liability and a desire to reserve funds for victims of other harms rather than based on an unexamined mental-physical boundary.

---

[170] I leave for another day the question why liability for stand-alone emotional distress claims may be considered excessive or disproportionate compared to traditional physical injury claims. It may be that 'as a categorical matter (the relevant frame for duty analysis), an individual's ability to act in the world is more restricted by physical harm than emotional distress.' Geistfeld (note 53) 164. But once we remove questions about the validity or foreseeability of the stand-alone emotional distress tort, we are confronted with other, less legitimate reasons, for devaluing the harm. For arguments suggesting that the emotional harm claim in tort law is devalued for gender-bias reasons, see Leslie Bender, 'Feminist (Re)Torts: Thoughts on the Liability Crisis, Mass Torts, Power, and Responsibilities' (1990) *Duke L. J.* 848, 851–3; Martha Chamallas, 'The Architecture of Bias, Deep Structures in Tort Law' (1998) 146 *U. Pa. L. Rev.* 463, 499.

13

Neuroscience and Ideology: Why Science Can Never Supply a Complete Answer for Adolescent Immaturity

*June Carbone**

For those who study adolescent development and the transition to adulthood, the lure of innovations in neuroscience is considerable. Magnetic resonance imaging (MRIs) and CT scans offer the illusion of peering into the teen brain and discovering what many have long suspected—while all the parts are there, they are not necessarily fully connected. Increasingly sophisticated imagery and analysis suggests that full adult maturity may not occur until well past the legal age of majority at eighteen, and we are acquiring greater insight into the pathologies that make some teens less likely to control inappropriate impulses than others.

These innovations tempt us—to declare that we know with certainty that which is merely suggested, to see what we want to see in the face of ambiguity. The combination of a misplaced certainty—that we can connect intriguing patterns to specific behaviour—and the possibility of oversimplification and bias lead many to declare that the neuroscience evidence is not ready for prime time; that it should be of questionable admissibility in individual cases and dubious as a basis for legal policy.

Dismissing, or simply failing to reference these innovations, however, carries a different set of risks. First, we may overlook explanations that add depth or nuance to the deliberation. Second, to the extent that the scientific discourse influences popular or legal perceptions, the failure to recognize that influence courts dishonesty. Finally, we need to ask what the alternatives are. If we are not to consider the imperfect and still developing scientific evidence about adolescent decision-making, on what do we base our decisions—our own, highly unscientific, experience as parents?

The question is certainly not unique to issues about adolescent decision-making. Instead, it raises questions about how we deal with complexity in any context. We have entire disciplines (economics comes immediately to mind) determined to use

---

* Edward A. Smith/Missouri Chair of Law, the Constitution and Society, at the University of Missouri, Kansas City.

simplifying assumptions that reduce complex issues to single variables. Moreover, we have begun to recognize that the appeal of certainty itself can be a matter of ideology and personality—economics and business students do answer questions differently from liberal arts students.

What we do not have is a fully developed alternative approach that allows balancing multiple factors, weighing odds that cannot be precisely calculated, and rendering judgments, whether judicial or legislative, that need to reassure multiple constituencies. In short, innovations in neuroscience and adolescent decision-making say more about our existing decision-making processes than they do about the adolescents who are the subject of the decisions.

This paper will argue that to resolve the issues about the role of neuroscience, we need to question the framework in which it arises. That is, the increasing complexity of scientific determinations raises issues of institutional capacity. Recognizing innovations in the science of adolescent development may change not so much our view of adolescence as the calculus underlying institutional functions.

Consider the issue of the juvenile death penalty or the permissibility of a life sentence for a crime committed by a fifteen-year-old. The fact that adolescent reasoning has not yet matured may or may not make the sentence cruel. But the ability to determine whether a particular fifteen-year old is capable of reasoned deliberation may be a more difficult task than judging whether fifteen-year-olds as a group have such capacity. Abolishing the juvenile death penalty may therefore be the right answer not so much because adolescent decision-making is necessarily flawed, but because deciding whether it is in individual cases is practically impossible. At the same time, such a decision should not be resolved, in any absolute sense, on the basis of neuroscience findings. Instead, they can be at best a strand in a complex decision that situates the idea of justice, rather than cognitive capacity, in an appropriate societal framework.

To consider the appropriate construction of such frameworks, this paper will begin by describing the 'lure of neuroscience', that is, the promise and limitations of the scientific advances, comparing legal decision-making capacity in individual cases versus broader matters of constitutional doctrine or public policy, analysing the recent US Supreme Court decisions on the juvenile death penalty in such terms, and assessing the role of neuroscience in the different possible outcomes of that case. The paper will conclude that Justice Kennedy's majority opinion gave appropriate weight to the neuroscience findings in exactly this sense; it is an element supporting, but not dictating, a conclusion the court reached on broader grounds.

13.1 The Lure of Neuroscience

The law, in a relatively arbitrary fashion, sets an age of majority; that is, an age when young men and women acquire the full rights of citizenship—the ability to vote, to enter into contracts, and to make certain decisions on their own. Today, in the US, it is set at eighteen; before the late sixties, most states set it at twenty-one. The immediate impetus for the legal change was the Vietnam War and the desire to give the full rights of citizenship to young men who were going off to combat. What did not enter into the equation in a formal way was cognitive development.[1] Is there a significant difference between the mental function of an eighteen-year-old and a twenty-one-year-old and should that difference matter legally?

We certainly have an intuition that decision-making at eighteen and at twenty-one (and perhaps beyond) differs and that it should matter for some things. The insurance industry collects extensive data, for example, on driving behaviour and sets the age for lower rates at twenty-five. Rental car companies follow suit. Can we say more than that automobile accidents peak in the late teens, decline significantly though the early twenties, and begin to level off by twenty-five?[2]

Increasingly, we can. That is, even aside from the more detailed research into the causes of teen traffic accidents, we are learning more about brain maturation more generally. A significant part of these insights (pun intended) comes from neuroscience; that is, from the ability to use brain imaging techniques to measure changes in brain composition and function. These findings can be summarized by the differences in teen and adult responses to an invitation to go 'swimming with sharks'.

Ask an adult, 'would you like to go swimming with sharks?', and you tend to get a very quick, 'hell, no'. Ask a teen, and they are more likely to be curious, think it over, and focus on whether they are eager to be included in the group going, and how you will respond to their answer. Psychologists Abigail Baird and Jonathan Fugelsang quantify the effect, finding that adolescents take significantly more time (measured in milliseconds) to questions such as 'Is it a good idea to set your hair on fire?', 'Is it a good idea to drink a bottle of Drano?', and 'Is it a good idea to swim with sharks?'[3]

The neuroscience images suggest why. The overall size of the human brain does not change during late adolescence.[4] Instead, brain composition evolves. Baird observes that 'Perhaps the most consistently reported finding associated with adolescent brain development is the decrease of grey matter and the increase of white matter throughout the cortex, but most significantly within frontal cortex.'[5] The cortex generally and the prefrontal cortex, in particular, play a major role in higher-level reasoning and the coordination of different brain regions. Grey matter, which is responsible for information processing, grows dramatically at

---

[1] Claudia Goldin and Lawrence F. Katz, 'The Power of the Pill: Oral Contraceptives and Women's Career and Marriage Decisions' (2002) 110 *J. Pol. Econ.* 730, 754–5.

[2] The most dangerous period is between the ages of sixteen and nineteen, with teen drivers experiencing double the accident rate of older drivers. See Center for Disease Control and Prevention (CDC), Teen Drivers: Fact Sheet, <http://www.cdc.gov/ncipc/factsheets/teenmvh.htm>; see also H.Y. Berg, 'Reducing Crashes and Injuries Among Young Drivers: What Kind of Prevention Should We Be Focusing On?' (2006) 12 *Injury Prevention* i15, i15 (Supp. I).

[3] Abigail A. Baird and Jonathan A. Fugelsang, 'The Emergence of Consequential Thought: Evidence from Neuroscience' (2004) 359 *Philosophical Transactions of the Royal Society of London, Series B: Biological Sciences* 1797–804.

[4] Ibid. at 1800.

[5] Abigail A. Baird, 'Moral Reasoning in Adolescence: The Integration of Emotion and Cognition' in W. Sinnott-Armstrong (ed.), *Moral Psychology* forthcoming, <http://faculty.vassar.edu/abbaird/about/publications/pdfs/2006-Baird-Morality.pdf> (accessed 5 October 2009).

the beginning of adolescence, and is pruned back thereafter.[6] White matter contributes to the pruning, reinforcing connections in accordance with new experiences, and improving overall brain efficiency.[7] Baird and Fugelsang suggest that these effects more closely integrate reason and emotion, improving response times and incorporating the lessons from past experiences.[8] In a summary of these developments, Baird explains that:

> Synaptic pruning is the elimination of synapses in the brain that are used less frequently facilitating growth of a more efficient brain. One striking difference regarding the development of the prefrontal cortex relative to other cortical areas is the continuation of synaptic pruning into young adulthood. This decrease in synaptic density during adolescence coincides with the emergence of newly entwined cognitive and emotional phenomena. The secondary process that is taking place during this time is the fortification of synaptic connections that will remain into adulthood. There has been further speculation that this 'use it or loose it' process may represent the behavioral, and ultimately, the physiological suppression of immature behaviors that have become obsolete due the novel demands of adulthood... The delayed maturation of this brain region allows the individual to adapt to the particular demands of their unique environment.[9]

This fascinating account links brain development—and the relationship between white matter and grey matter—to the process of maturation in decision-making. Moreover, it does so in a way that explains how that process responds to individual experience. As Baird acknowledges, however, the emerging account contains a measure of speculation, and it is not yet, at least, at the point where it can establish that a particular brain composition causes any given behaviour.

Instead, small-scale studies have provided intriguing clues about the correlations. Research demonstrates, for example, that adults and adolescents use different parts of their brains in performing certain tasks.[10] An early study showed adult and adolescent subjects fearful faces and asked them to identify the emotions. The adults displayed significant activation of the prefrontal cortex in performing the task, while the adolescents showed no activation in that region. Conversely, the teenagers responded with greater activation of the amygdala, an evolutionary older part of the brain that develops at earlier ages, suggesting that, for adults, the activity in the prefrontal cortex may have suppressed the response of the amygdala.[11] A later study demonstrated that the use of different parts of the brain also affects overall

---

[6] Craig M. Bennett and Abigail A. Baird, 'Anatomical Changes in the Emerging Adult Brain: A Voxel-based Morphometry Study' (2005) 27 *Human Brain Mapping* 766, <http://www3.interscience.wiley.com/cgi-bin/fulltext/112162278/main.html,ftx_abs>.

[7] Ibid. See also Staci A. Gruber and Deborah A. Yurgelun-Todd, 'Neurobiology and the Law: A Role in Juvenile Justice?' (2006) 3 *Ohio St. J. Crim. L.* 321, 324.

[8] (note 3 above) at 1801.

[9] Baird, *Moral Reasoning*, at 5.

[10] See, e.g., Nitin Gogtay et al., 'Dynamic Mapping of Human Cortical Development during Childhood through Early Adulthood' (2004) 101 *Proc. Nat'l Acad. Sci.* 8174, 8174 (studying thirteen healthy children for eight to ten years and showing that the frontal cortex and the areas of the brain responsible for higher-level thinking and behaviour management develop later than other parts of the brain).

[11] Gruber and Yurgelun-Todd (note 7 above) at 328.

accuracy. Shown a series of facial expressions, the adults correctly identified fear every time while the teens did so only about 50 per cent of the time. Accurate responses corresponded with activation of the prefrontal cortex, while the less accurate respondents relied more on use of the amygdala.[12] Researchers hypothesize that greater reliance on the amygdala corresponds to more emotional, erratic, and impulsive decision-making. It may also explain teens' greater difficulty in interpreting emotional signals.[13] Baird maintains that part of the explanation for these differences lies with the fact that reason simply does not do very well in the absence of emotion, and reason in turn makes the emotional parts of the brain more accurate.[14] Yet, the last part of the brain to develop is the white matter connections that more thoroughly integrate cognitive and emotional processing.

Baird provided the most comprehensive map of these developments by examining the change in the brain patterns of college freshman in their first year away from home. She found continuing changes in five different brain regions associated with conscious awareness of emotion, the integration of sensory information into higher-order processes, and formulation of strategies.[15] Both Baird and Dr Ruben Gur, who is the Director of the University of Pennsylvania Medical Center, agree that there is strong and growing evidence that the brain continues to mature into the early twenties 'in those relevant parts that govern impulsivity, judgment, planning for the future, foresight of consequences, and other characteristics'.[16]

The neuroscience findings complement the conclusions drawn from more traditional social science research. Law professor Elizabeth Scott and psychologist Laurence Steinberg, in examining juvenile crime and behaviour, refer to this older data in distinguishing between what they call 'cognitive' capacity and 'psychosocial' development.[17] They draw on studies that observe teen behaviour, often in laboratory settings. They observe that by mid-adolescence, teens resemble adults in their cognitive capacity; that is, their ability to learn through education and experience, their information-processing skills, such as attention, short- and long-term memory, and organization, and their overall ability to understand and reason; in short, teens have the same ability as adults to judge the approach of that speeding car as it enters the intersection.[18] What develops more slowly than cognitive ability

---

[12] Paul Arshagouni, 'Introduction' (2007) 9 J. *Health Care L. and Pol'y* 315, 350.

[13] Lucy C. Ferguson, 'Comment: The Implications of Developmental Cognitive Research on "Evolving Standards of Decency" and the Imposition of the Death Penalty on Juveniles' (2004) 54 *Am. U L Rev.* 441, 455.

[14] Abigail A. Baird, 'Adolescent Moral Reasoning: The Integration of Emotion and Cognition' *Moral Psychol.* (forthcoming), <http://theteenbrain.com/about/publications/pdfs/2006-Baird-Morality.pdf> (accessed 2 April 2008).

[15] Bennett and Baird (note 6 above) at 9.

[16] Arshagouni (note 12 above) at 351 (quoting Am. Bar Ass'n Juvenile Justice Ctr., *Adolescence, Brain Development, and Legal Culpability, Cruel and Unusual Punishment: The Juvenile Death Penalty* (2004) 3.

[17] Elizabeth Scott and Laurence Steinberg, 'Blaming Youth' (2003) 81 *Tex. L. Rev.* 799, 812–3.

[18] Ibid. at 812; see also ibid. at n. 54 (stating that 'the key advances during this period are gains in deductive reasoning, the ability to think about hypothetical situations, the ability to think simultaneously in multiple dimensions, the ability to think abstractly, and the ability to think about the process of thinking ("metacognition")'); John H. Flavell, Patricia H. Miller and Scott A. Miller,

is 'psycho-social' development; that is, the contextual judgement to determine whether it is worth the risk to attempt to enter the intersection before the approaching car. Scott and Steinberg conclude that four factors critical to mature judgement continue to develop into early adulthood: 1) peer orientation; 2) attitudes toward and perception of risk; 3) temporal perspective; and 4) capacity for self-management.[19] Teens react differently from adults in weighing the small chance of an accident against long-term catastrophic consequences, the reaction of peers present in the car versus the impact on loved ones who are not physically present, and in incorporating experience—those who drive regularly or have witnessed an accident may respond differently from those who have not.[20]

The neuroscience findings appear to provide a causal mechanism that explains the psycho-social research—and everyday parental observations. Yet, the very persuasiveness of the account masks the uncertainties that neuroscience cannot resolve. The neuroscience studies themselves, which are at a relatively early stage in the development of the science, are small-scale studies, with subjects who are not randomly chosen.[21] Bennett and Baird's innovative research on the brain development of college students, for example, deliberately selects college freshmen at selective schools who are away from home for the first time in an effort to map the effect of changing experience on brain development.[22] They would not necessarily get the same results from other subjects at the same age who had been living on the streets or were still at home with their parents.

Moreover, while the MRI results suggest intriguing correlations between brain development and observed behaviour, the studies do not attempt to prove causation. Neuroscientists still know too little to suggest that the activation of a particular region in an MRI necessarily means that the development of that part of the brain *per se* causes a particular behaviour.[23] Professors Tancredi and Brodie explain that:

Cognitive Development (Englewood Cliffs, Prentice Hall, 1993) (1963) (outlining Piaget's theory and updating it based on new data); Barbel Inhelder and Jean Piaget, *The Growth of Logical Thinking From Childhood to Adolescence* (Ann Parsons and Stanley Milgram trans., New York: Basic Books, 1958); Jean Piaget, *Genetic Epistemology* (Eleanor Duckworth trans., New York, Columbia University Press, 1970); Robert S. Siegler, *Children's Thinking* (Englewood Cliffs, Prentice-Hall, 1991) (1986).

[19] Scott and Steinberg (note 17 above) at 813.

[20] Cass R. Sunstein, 'Adolescent Risk Taking and Social Meaning: A Commentary' (John M. Olin Law & Econ. Working Paper No. 386, 2d Series, 2008), <http://www.law.uchicago.edu/Lawecon/index.html>, at 8 (explaining the role of experience in making judgements both faster and more accurate). Experience, of course, can make individuals overly cautious. Someone who has experienced an accident may have a strong association between the decision to enter an intersection and a visceral sense of fear that may interfere with efficient decision-making. Moreover, the social meaning of a given act may also influence its emotional valence. Sunstein gives the example of the different social images associated with smoking in African-American versus white communities that produce significantly lower smoking rates among African-American than white teens: ibid. at 10–11.

[21] Jay D. Aronson, 'Neuroscience and Juvenile Justice' (2009) 42 *Akron L. Rev.* 917, 917, 924 (observing that the studies are in their infancy, sample sizes are small, and the samples are not randomly chosen, with one study group, for example, having an average IQ of 125).

[22] Bennett and Baird (note 6 above) *Anatomical Changes*, at 5–6.

[23] Jay D. Aronson, 'Neuroscience And Juvenile Justice' (2009) 42 *Akron L. Rev.* 917, 917 ('we do not yet understand the actual link between brain structure and behavior').

An abnormal image does not tell us what is happening causally between the abnormality and the brain region, or the abnormality and the behavior in question. Hence, the image is not in a one-to-one relationship with the brain. To illustrate, a brain image does not show us what criminal intent, or a 'bad' thought, looks like. It does not provide a causal connection; the variables of image, abnormal brain region, and specific behavior, therefore, do not have a linear relationship, i.e. one characterized by cause and effect. Rather, the relationship can at best be characterized as a statistical correlation.[24]

The fact that researchers are describing a statistical correlation (those with less activation in the prefrontal cortex at age eighteen may exhibit less mature judgement than those with more activity in the same region) rather than a causal factor (those with less activation in the prefrontal cortex *will* fail consider the consequences of their acts) makes this research less useful in predicting, explaining, or judging individual actions.

These brain states, after all, even as they describe tendencies rarely describe incapacities.[25] Less mature teens have the ability to distinguish right from wrong; they even have the capacity to engage in reasoned deliberation. They are just less likely than adults to do so, particularly in the context of stressful, peer-dominated, or alcohol- or drug-influenced settings.[26] Perhaps as intriguingly, male and female brains develop somewhat differently, with adult woman having a lower white-to-grey matter ratio than men, and teen girls experiencing a more gradual increase in white matter, beginning at earlier ages, than teen boys.[27] Yet, no one suggests that higher white/grey ratios mean that teen boys are more mature than teen girls.

Neuroscience results accordingly provide evidence, sometimes very powerful evidence, that increases insight into brain development, and helps diagnose medical abnormalities.[28] The question of how extensively this research should influence legal determinations and policy is another matter.

---

[24] Laurence R. Tancredi and Jonathan D. Brodie, 'The Brain and Behavior: Limitations in the Legal Use of Functional Magnetic Resonance Imaging' (2007) 33 *Am. J. L. and Med.* 271, 288–9.
[25] See Stephen J. Morse, 'Determinism and the Death of Folk Psychology: Two Challenges To Responsibility from Neuroscience' (2008) 9 *Minn. J.L. Sci. and Tech.* 1, 7.
[26] Valerie F. Reyna and Frank Farley, 'Risk and Rationality in Adolescent Decision-Making: Implications for Theory, Practice, and Public Policy' (2006) 7 *Psychological Science in the Public Interest* 1. They observe that: 'In principle, barring temptations with high rewards and individual differences that reduce self-control (i.e., under ideal conditions), adolescents are capable of rational decision making to achieve their goals. In practice, much depends on the particular situation in which a decision is made. In the heat of passion, in the presence of peers, on the spur of the moment, in unfamiliar situations, when trading off risks and benefits favors bad long-term outcomes, and when behavioral inhibition is required for good outcomes, adolescents are likely to reason more poorly than adults do.'
[27] See generally R. J. Haier, R. E. Yung, R. A. Yeo, K. Head, and M. T. Alkire, 'The Neuroanatomy of General Intelligence: Sex Matters' (2005) 25 *Neuroimage* 320–7. See also Ruben C. Gur, Bruce I. Turetsky, Mie Matsui, Michelle Yan, Warren Bilker, Paul Hughett, and Raquel E. Gur, 'Sex Differences in Brain Gray and White Matter in Healthy Young Adults: Correlations with Cognitive Performance' (1999) 19 *The Journal of Neuroscience* 4065–72.
[28] See, e.g., Tancredi and Brodie (note 24 above) at 290.

13.2 Legal Frames for Decision-Making

13.2.1 Context, content, and decision-making

The role of neuroscience in assessing decision-making capacity is multidimensional, and its influence may be dependent on the framework for the decision. The competence of adolescent decision-making may arise in the context of: 1) individual determinations of culpability or capacity such as those involved in criminal adjudications; 2) group-based judicial determinations of appropriate process or standards such as a constitutional ruling on the validity of the juvenile death penalty or life imprisonment for crimes created by teens; 3) statutory approaches to group capacity, such as legislative determinations of the age of majority or age-based criminal sanctions; and 4) policy determinations, which may include legislative or executive approach to public issues, such as restrictions on the sale of firearms to minors or the appropriate age for access to contraception or marriage.

These contexts vary significantly from each other. First, the decision-makers differ. They range from lay juries in the context of criminal actions, to legislatures that may choose to draw upon (or ignore) any manner of expert advice, to executive officials who may be selected because of their expertise in the relevant fields. Second, the significance of neuroscience findings may not be the same. The question at hand may be structured in a way that neuroscience findings are decisive. If, for example, the question is whether a given party suffers from a disease or defect eligible for health insurance coverage, an MRI showing a brain tumour may be dispositive evidence.[29] If, on the other hand, the issue is whether seventeen-year-olds should be allowed to buy guns, MRI studies *per se* are unlikely to be conclusive. Instead, the legislative judgment may turn on the importance of a variety of factors such as the importance of hunting in a given locale. Finally, the risk of bias may vary not only with the decision-makers (expert or lay) and the nature of the issue, but the strength of the decision-makers' predispositions.[30] If the issue is the requisite age for gun purchases, for example, the influence of neuroscience findings suggesting teen immaturity may correspond to the decision-makers' commitments on gun ownership; National Rifle Association members are more

---

[29] Indeed, in one of the rare cases to show a direct relationship between the state of the brain and criminal behaviour, a man who engaged in uncontrollable sexual behaviour regained the ability to resist such impulses once doctors diagnosed and removed a brain tumour. See Jeffrey M. Burns and Russell H. Swerdlow, 'Right Orbitofrontal Tumor with Pedophilia Symptoms and Constructional Apraxia Signs' (2003) 60 *Archives Neurology* 437, 437–8. Even studies that suggest correlations between specific forms of brain damage and observed behaviour, however, may not necessarily establish a standard for distinguishing 'normal' from 'abnormal' brain scans, or that all subjects with the abnormal scans engage in criminal behaviour. Jennifer Kulynych, 'Comment, Brain, Mind, and Criminal Behavior: Neuroimages as Scientific Evidence' (1996) 36 *Jurimetrics J.* 235, 238–40.

[30] See Roger J. R. Levesque, *The Psychology and Law of Criminal Justice Processes* 299 (New York, Nova Science Publishers, 2007) (authoritarian personalities are more likely to take a punitive approach to low status defendants and to notice different facts about the individual, while those with other values preferences are more likely to notice situational information).

likely to discount the findings than gun-control advocates.[31] Conversely, decision-makers may have more open minds to novel or less ideological issues, such as individual eligibility for medical insurance coverage (at least if the decision-makers are not insurance company employees).

To date, most of the discussion of law and neuroscience has focused on the first (individual determinations of criminal responsibility) or second (group-based standards of legal capacity) issues. The focus is understandable. The death penalty routinely commands disproportionate attention,[32] and the juvenile justice system, which overreacted to the image of youthful predators in the nineties, is ripe for reform.[33] Nonetheless, the claims at the core of the juvenile justice debate frame the relationship between law and neuroscience in ways that may not be applicable to other contexts. It is time to separate the challenges involved in proving individual states of mind from those involved in crafting policy more generally.

13.2.2 Retribution v desert

The most controversial applications of neuroscience involve individual criminal capacity. These claims are controversial, in part, because they are intrinsically divisive and in part because of the far-reaching claims some proponents have made. Joshua Greene and Jonathan Cohen, for example, have famously written that 'free will, as conceptualized by the folk psychology system, is an illusion'.[34] Carter Snead describes Greene and Cohen's view (and that of similar minded neuroscientists) as one 'to work a radical conceptual revision of criminal punishment itself; more specifically, ... to use the premises and tools of neuroscience—and neuroimaging in particular—to embarrass, undermine, and ultimately overthrow retributive justice as a principle of punishment'.[35]

These claims have inspired two kinds of responses. The first is technical—neuroscience is not at a point where it delivers on such claims. While the neuroimaging studies provide evidence for those predisposed to view behaviour as the product of neural states, no one can yet prove that particular brain signals cause particular acts. At best, they suggest that particular behaviour is more likely. Stephen Morse explains:

---

[31] See, e.g., Donald Braman and Dan M. Kahan, 'Overcoming the Fear of Guns, the Fear of Gun Control, and the Fear of Cultural Politics: Constructing a Better Gun Debate' (2006) 55 *Emory L.J.* 569; Dan M. Kahan and Donald Braman, 'More Statistics, Less Persuasion: A Cultural Theory of Gun-Risk Perceptions' (2003) 151 *U. Pa. L. Rev.* 1291; Dan M. Kahan, 'The Gun Control Debate: A Culture-Theory Manifesto' (2003) 60 *Wash. and Lee L. Rev.* 3.

[32] Hence, Samuel Johnson's oft misquoted observation, 'Depend upon It, Sir, when a man knows he is to be hanged in a fortnight, it concentrates his mind wonderfully': James Boswell, *The Life of Samuel Johnson* (Oxford World's Classic ed., Oxford, Oxford Univ. Press, 1998) 849.

[33] See, e.g., Laurence Steinberg and Elizabeth S. Scott, *Rethinking Juvenile Justice* (Cambridge, Harvard University Press, 2008); Elizabeth S. Scott and Thomas Grisso, 'Developmental Incompetence, Due Process, and Juvenile Justice Policy' (2005) 83 *N.C. L. Rev.* 793, 811–16.

[34] Joshua Greene and Jonathan Cohen, 'For the Law, Neuroscience Changes Nothing and Everything' (2004) 359 *Phil. Transactions Royal Soc'y: Biological Sci.* 1775, 1776.

[35] O. Carter Snead, 'Neuroimaging and the "Complexity" of Capital Punishment' (2007) 82 *N.Y. U. L. Rev.* 1265, 1316.

Note that the law requires possession of a general capacity at the time in question rather than an exercise of that capacity. Failure to exercise a capacity does not necessarily mean that one lacks that capacity. Indeed, acting irrationally and foolishly is common even among people with the greatest capacity for rational conduct. Under the law, if a person is capable of exercising the capacity for rationality if there is good reason to do so—as there always is when important interests are at stake—then that person may be held responsible even if she failed to exercise that capacity.[36]

In Morse's view, for a person to escape criminal responsibility, neuroscience would have to be able to show not just that a person is unlikely to engage in mature reasoning, but that they are incapable of doing so—a claim that existing neuroscience findings, at least on the issue of teen development,[37] do not support.

Instead, this type of discussion inflames deep ideological divisions about the role of punishment. Most researchers define conservative versus liberal values orientations to include different attitudes toward punishment.[38] Empirical studies demonstrate that authoritarian preferences often predict more punitive jury approaches;[39] while more liberal political orientations correspond to more contextualist approaches to criminal responsibility and more emphasis on rehabilitation.[40] The broader claims for neuroscience, like those of Greene and Cohen, recreate these divisions without necessarily better informing individual determinations. It is unsurprising therefore that neuroscience to date has had relatively little impact on criminal punishment, with courts routinely upholding death sentences, for example, even after the introduction of compelling neuroscience evidence that a defendant's behaviour may have been influenced by cognitive abnormalities.[41]

Indeed, Carter Snead has argued that rather than excuse criminal defendants, neuroscience findings of cognitive abnormalities may result in harsher sentences

---

[36] Stephen J. Morse, 'Determinism and the Death of Folk Psychology: Two Challenges to Responsibility from Neuroscience' (2008) 9 *Minn. J.L. Sci. and Tech.* 1, 7.

[37] See, e.g., Tancredi and Brodie (note 24 above) at 288–9, explaining the neuroscience findings in terms of statistical correlations. The conclusion might be different in the much rarer case of the use of neuroscience to establish the existence of a specific abnormality, such as a brain tumour.

[38] See, e.g., John R. Alford, Carolyn L. Funk and John R. Hibbing, 'Are Political Orientations Genetically Transmitted?' (2005) 99 *Am. Pol. Sci. Rev.* 153, 164–5 (using twin studies to suggest a hereditary component to liberal v conservative political orientations, including attitudes toward the role of punishment); John T. Jost, 'The End of the End of Ideology' (2006) 61 *American Psychologist* 651 (distinguishing left v right political orientations in terms of attitudes toward hierarchy and order).

[39] Levesque, (note 30 above) at 299 (jurors with authoritarian values orientations tend to favour more punitive punishment).

[40] See, e.g., George Lakoff, *Moral Politics: How Liberals And Conservatives Think*, 2nd edn (Chicago, University of Chicago Press, 2002).

[41] See, e.g., Richard E. Redding, 'The Brain-Disordered Defendant: Neuroscience and Legal Insanity in the Twenty-first Century' (2006) 56 *Am. U.L. Rev.* 51, 56, who observes that: 'Indeed, neuropsychological studies show that the prevalence rate of brain dysfunction among criminal populations is extremely high, with prevalence rates of ninety-four percent among homicide offenders, sixty-one percent among habitually aggressive adults, forty-nine to seventy-eight percent among sex offenders, and seventy-six percent among juvenile offenders (by comparison, the prevalence rate in the general population is only three percent).' See also Jonathan H. Pincus, 'Neurologist's Role in Understanding Violence' (1993) 50 *Archives Neurology* 867, 868 (observing that every death row inmate has received APD (antisocial personality disorder) diagnosis). Cf. Terry A. Maroney, 'The False Promise of Adolescent Brain Science in Juvenile Justice' (2009) 85 *Notre Dame L Rev* 89.

because they may persuade juries that the defendant is dangerous and unlikely to be rehabilitated. Snead observes that:

Studies have shown that capital juries often regard evidence of future dangerousness as the most important aggravating factor in their sentencing calculus. Indeed, two commentators have noted that 'Future dangerousness takes precedence in jury deliberations over any mitigating evidence, such as remorse, mental illness, intelligence, or drug/alcohol addiction, and any concern about the defendant's behavior in prison.' In fact, it has been observed that even in those jurisdictions that do not explicitly direct the capital jury to consider future dangerousness as an aggravating factor, jurors do so anyway.[42]

Ironically, while neuroscience results rarely show that a defendant could not form the requisite mental intent for a crime, they more frequently provide evidence of future dangerousness in the sentencing phase, an inquiry that permits the introduction of more probabilistic or speculative testimony.[43] Even in the juvenile context, where evidence of immaturity suggests greater hope that the defendant's capacity for moral judgement will improve over time, youth may be seen an aggravating factor where the crime is particularly brutal.[44]

It is accordingly unsurprising that neuroscience testimony has had relatively little effect in mitigating criminal responsibility in criminal cases where it has been introduced into evidence. First, as Morse demonstrates, the existing state of the science can rarely show a complete lack of capacity to control behaviour or to form the requisite mental intent.[45] Second, the influence of the neuroscience testimony may depend on the predispositions of the decision-makers, with many judges and jurors inclined to give little weight to such testimony in the face of heinous crimes.[46] Finally, such evidence competes with other purposes of punishment, such as securing the safety of the public. It is simply too hard, in the face of a violent and senseless murder, to excuse the conduct, particularly when the same 'mitigating' evidence suggests that the person is likely to repeat the offence.[47]

13.3 The Juvenile Death Penalty: What Role Does Neuroscience Play?

Consider the role of neuroscience in resolving the issue of the juvenile death penalty (that is, execution in the US of defendants who committed capital offences before

---

[42] Snead (note 35 above) at 1328.
[43] Ibid. at 1325–6.
[44] See discussion of *Roper v Simmons*, below at 242.
[45] See also Jennifer Kulynych, 'Comment, Brain, Mind, and Criminal Behavior: Neuroimages as Scientific Evidence' (1996) 36 *Jurimetrics J.* 235, 241.
[46] See discussion of Justice Scalia's personal views on this issue, below.
[47] Even in the juvenile context, defendants who commit the extreme acts that qualify for a death sentence often exhibit symptoms of brain damage that may limit their prospects for rehabilitation. See generally Redding (note 41 above).

their eighteenth birthdays). In *Roper v Simmons*,[48] the US Supreme Court, in a controversial 5–4 decision, declared the juvenile death penalty unconstitutional.[49] The case involved a number of issues about constitutional approaches to the Eighth Amendment, which forbids cruel and unusual punishments, and one of them addressed the appropriate way to resolve the issue: through individual determinations of teen capacity, a blanket constitutional ruling precluding such executions, or state-by-state legislative judgments. In each setting, the decision-makers might consider neuroscience research suggesting teen immaturity likely to affect the circumstances in which minors commit crimes, but the role of that research would vary considerably in each setting.

The Supreme Court's deliberations in *Roper v Simmons* did not expressly take these factors addressing institutional competence into account, but it could have done so. The case arose fifteen years after a deeply divided Supreme Court had upheld the constitutionality of execution for those who committed crimes at the age of sixteen or over.[50] Numerous *amicus* briefs, including one by the American Bar Association, asked the court to consider the neuroscience literature on brain development that had been compiled during the intervening fifteen years. Jay Aronson concludes that:

at the heart of the *Simmons* legal argument was the claim that a direct link existed between the behavioral characteristics of adolescents and immature brain structure and function. Because the areas of the brain most responsible for impulse control, emotional response, decision making, and risk assessment generally occur only after the 18th birthday, 16- and 17-year-olds lack the accountability for their actions that is required to be charged with the death penalty, and they do not grasp future consequences, negating any deterrent effect of the death penalty.[51]

Roper could therefore be seen as a case squarely presenting the issue of adolescent decision-making capacity.

The Supreme Court split in the case 5–4,[52] and the majority and dissent differed as much on the standard to apply in resolving the issue as the substance of the

---

[48] 543 US 551 (2005).
[49] The case is controversial for many reasons that have nothing to do with the use of neuroscience. For a more complete discussion of the statistical evidence in the case, see Deborah W. Denno, 'The Scientific Shortcomings of *Roper v Simmons*' (2006) 3 *Ohio St. J. Crim. L.* 379.
[50] See *Stanford v Kentucky* 492 US 361, 106 L Ed 2d 306, 109 S Ct 2969 (1989) (Scalia J announced the judgment of the court and delivered the opinion of the court with respect to Parts I, II, III, and IV-A, in which Rehnquist CJ, and White, O'Connor, and Kennedy JJ, joined, and an opinion with respect to Parts IV-B and V, in which Rehnquist CJ, and White and Kennedy JJ, joined. O'Connor J, filed an opinion concurring in part and concurring in the judgment, at 380. Brennan J, filed a dissenting opinion, in which Marshall, Blackmun, and Stevens JJ, joined, at 382.) A year earlier the court had invalidated execution of those who committed crimes before the age of sixteen. See *Thompson v Oklahoma* 487 US 815 (1988) (Stevens J announced the judgment of the court and delivered an opinion in which Brennan, Marshall, and Blackmun JJ, joined. O'Connor J, filed an opinion concurring in the judgment, at 848. Scalia J filed a dissenting opinion, in which Rehnquist CJ and White J joined, at 859. Kennedy J took no part in the consideration or decision of the case.).
[51] Jay D. Aronson, 'Brain Imaging, Culpability and the Juvenile Death Penalty' (2007) 13 *Psych. Pub. Pol. and L.* 115, 132.
[52] Ibid. at 133–6.

Neuroscience and Ideology 243

dispute. Justice Kennedy rested the majority ruling on the juvenile death penalty on 'the evolving standards of decency that mark the progress of a maturing society', and concluded that the execution of juveniles was 'so disproportionate' as to be cruel and unusual.[53] His assessment of these evolving standards referred to three factors affecting the propriety of capital punishment, but in doing so he only tangentially acknowledged neuroscience studies of teen development.

First, he observed that a majority of states prohibit the juvenile death penalty, 'comprising 12 that have rejected the death penalty altogether and 18 that maintain it but, by express provision or judicial interpretation, exclude juveniles from its reach'.[54]

Second, he rejected Justice Scalia's statement for the *Stanford* majority (the case that had upheld the juvenile death penalty fifteen years earlier) that the court's independent judgment had no bearing on the acceptability of a particular punishment under the Eighth Amendment. Instead, he returned to the pre-*Stanford* rule that 'the Constitution contemplates that in the end our own judgment will be brought to bear on the question of the acceptability of the death penalty under the Eighth Amendment'.[55] Justice Kennedy then acknowledged:

As any parent knows, and as the scientific and sociological studies respondent and his amici cite tend to confirm,' youth lack maturity and have an underdeveloped sense of responsibility; juveniles are more susceptible to peer pressure and outside influences; and the character of juveniles is not as well formed as that of an adult.[56]

Third, he concluded that: 'Our determination that the death penalty is disproportionate punishment for offenders under 18 finds confirmation in the stark reality that the United States is the only country in the world that continues to give official sanction to the juvenile death penalty.'[57]

In short, he based his decision on a consensus emerging from multiple decision-makers, considering multiple factors.

Three of the four dissenting justices, the most conservative on the court, joined in an opinion by Justice Scalia.[58] Justice Scalia, expounding on his theory that constitutional interpretation should always remain faithful to the original intent of the drafters, caustically observed that:

What a mockery today's opinion makes of Hamilton's expectation [that the judiciary would be 'bound down by strict rules and precedents'], announcing the Court's conclusion that the

---

[53] 543 US at 561.
[54] Ibid. at 564.
[55] Ibid. at 463 (quoting *Coker v Georgia*, 433 US 584, 597 (1977) (plurality opinion)).
[56] Ibid. at 569–70. The Supreme Court, however, did not base its decision on the *science* of brain development alone, and the legal and psychological experts who follow this issue disagree as to the whether the scientific findings have reached the point where it would be appropriate to base legal or policy decisions on them. See Aronson (note 52 above) summarizing the limitation of the scientific evidence on brain development, noting in particular that the neuroscience studies involve relatively small sample sizes and that they do not purport to prove a causal link between the observed patterns of brain development and individual behaviour.
[57] 543 US, at 575.
[58] Rehnquist and Thomas joined in the dissent. *Roper* (note 48 above) at 607 et seq.

meaning of our Constitution has changed over the past 15 years—not, mind you, that this Court's decision 15 years ago was *wrong*, but that the Constitution *has changed.*[59]

He dissented with particular vehemence from the majority's conclusion that the court bring its own judgment to bear, observing that the 'Court thus proclaims itself sole arbiter of our Nation's moral standards—and in the course of discharging that awesome responsibility purports to take guidance from the views of foreign courts and legislatures.'[60] Instead, Justice Scalia tied his theory of original intent to an insistence that the court should ordinarily defer to democratically elected legislatures absent clear constitutional language consistent with the intent of the founders.[61]

Justice O'Connor, the other dissenting justice, in contrast, agreed with Kennedy that the court should assess 'evolving standards of decency', but disagreed with Kennedy's conclusion that these standards had changed over the last fifteen years, and favoured instead individual determinations about the capacity of each juvenile to be executed.[62]

13.3.1 O'Connor and the role of neuroscience in individual determinations

These three approaches have fundamentally different implications for the role of neuroscience in judicial decision-making, with implications both for the constitutional decision before the court and for subsequent decisions in individual cases. Though she did not acknowledge it, the debate over the role of neuroimaging would have become most prominent if the court had followed Justice O'Connor's suggestion that the constitutional issue of appropriate punishment for juveniles be resolved in the context of individual determinations of mental capacity. At the same time, Justice O'Connor herself showed little appreciation for its significance in the determinations she recommended.

Justice O'Connor's opinion rests on the premise that courts should be expected to draw the line between those seventeen-year-olds who deserve the death penalty, and those who do not. She writes:

Adolescents *as a class* are undoubtedly less mature, and therefore less culpable for their misconduct, than adults. But the Court has adduced no evidence impeaching the seemingly reasonable conclusion reached by many state legislatures: that at least *some* 17-year-old murderers are sufficiently mature to deserve the death penalty in an appropriate case. Nor has it been shown that capital sentencing juries are incapable of accurately assessing a youthful defendant's maturity or of giving due weight to the mitigating characteristics associated with youth.[63]

---

[59] Ibid. at 608 (italics in original).   [60] Ibid. at 608.
[61] Scalia J, however, along with Kennedy and Thomas JJ, is among the least likely of the justices on the court to defer to Congress. Paul Gewirtz and Chad Golder, 'So Who Are the Activists?' (2005) *New York Times,* 6 July, A23.
[62] Ibid. at 587, et seq.
[63] Ibid. at 588.

How exactly should courts be expected to conduct such an inquiry? Justice O'Connor simultaneously observes that adolescent maturity is of constitutional significance in determining the validity of executions while engaging in analysis that appears to deny (or at least fail to recognize) the elements that corresponds to adolescent immaturity. Her analysis in *Roper v Simmons*, which does not acknowledge either the neuroscience or psychosocial research on adolescent development, focuses instead on the facts that make the crime in that case particularly heinous; she writes, 'Whatever can be said about the comparative moral culpability of seventeen-year-olds as a general matter, Simmons' actions unquestionably reflect "a consciousness materially more 'depraved' than that of"... the average murderer.'[64] These facts, however, provide textbook examples of immaturity, and illustrate Carter Snead's argument that consideration of the facts of individual cases associated with cognitive impairment or lack of development are more likely to result in harsher punishment than mitigation.

In reciting the facts, Justice O'Connor emphasizes that Simmons wanted to kill someone, and justified his willingness to do so by the belief that he and his friends could '"get away with it" because they were minors'.[65] The fact that he ties his willingness to go ahead with the crime to the likelihood of punishment, rather than its intrinsic acceptability, is shocking, but it is also a sign of profoundly flawed judgement, and of a flaw that would place Simmons on a younger age of moral development than the average seventeen-year-old (much less the average adult). Lawrence Kohlberg, a Harvard psychologist building on Piaget's development theories, constructed six stages in the development of moral judgement.[66] The first, characteristic of young children, rests moral judgement on obedience and fear of punishment; the sixth, final stage involves principled conscience.[67]

Second, Justice O'Connor refers to Simmons's acts in planning the crime with two teenage friends. Again, the planning behaviour makes the crime more blameworthy than an impulsive act. Yet, psychosocial studies of teen behaviour suggest that teens are considerably more likely to commit crimes in groups, and one of the critical factors that distinguishes teen from adult behaviour is the role of peer influence.[68]

---

[64] Ibid. at 601.
[65] Ibid. at 601.
[66] Carol Gilligan, *In a Different Voice: Psychological Theory and Women's Development* (Cambridge, MA, Harvard University Press, 1982) 18. The six stages are:

Level	Stage	Social Orientation
Pre-conventional	1	Obedience and Punishment
	2	Individualism, Instrumentalism, and Exchange
Conventional	3	'Good boy/girl'
	4	Law and Order
Post-conventional	5	Social Contract
	6	Principled Conscience

[67] Ibid.
[68] Scott and Steinberg (note 17 above) at 813.

Finally, Justice O'Connor refers to Simmons' lack of reflection on the impact that killing would have on his own life or on others as evidence of the depravity of his conduct. It *is* evidence of depravity—*and* evidence of the lack of mature reasoning that distinguishes adult cognitive development from that of teens. Abigail Baird, in her studies of the teen brain, emphasizes that what has not yet developed are the emotional connections between the different parts of the brain that allow a person to contemplate an act and immediately feel, at a visceral level, a reaction that follows from consideration of the consequences of that conduct.[69] Mature adults associate murder with incalculable grief; the immature are more likely to be able to contemplate murder without emotion.[70] Thus, Simmons's failure to anticipate and react to the pain his actions would impose are simultaneously evidence of depravity *and* immaturity.

In short, Justice O'Connor's armchair suggestion that the facts of the conduct can provide a basis for a determination of individual blameworthiness replicate the flaws that have limited the influence of neuroscience studies in individual cases. The neuroscience available to date, at its best, produces probabilities—teens are less likely than adults to have developed the neural pathways that integrate emotion with reason.[71] They are accordingly less likely to appreciate the significance of their acts, to factor in not only the immediate consequences of their acts, but the long-term impact on themselves, their victims, and others, and to reach conclusions independently of peer influence. In short, teens are more likely than adults to act like Christopher Simmons.

Nonetheless, however accurate the probabilities, neither neuroscience, nor for that matter psycho-social empirical studies are likely to demonstrate that Christopher Simmons's murdered someone *because* he was seventeen rather than twenty-three. Simmons conduct is, of course, unusual for anyone, and unlikely to be simply a reflection of immaturity.[72] The result of Justice O'Connor's analysis is therefore doubly ironic. It simultaneously suggests that a court must consider Simmons's immaturity for his execution to pass constitutional muster and that the facts consistent with immature judgement make his crime so heinous as to warrant his execution. Neuroscience studies that bolster the former conclusion (poor judgement as a result of immature development) are unlikely, in the context of this type of individual determination, to be sufficient to counter the impact of the latter

---

[69] Abigail A. Baird, 'Moral Reasoning in Adolescence: The Integration of Emotion and Cognition' in W. Sinnott-Armstrong (ed.), *Moral Psychology* forthcoming, <http://faculty.vassar.edu/abbaird/about/publications/pdfs/2006-Baird-Morality.pdf>. Indeed, Baird's study focuses on a murder not so different from the one Simmons committed, where three teenage girls brutally murdered a twelve-year-old over a period of time.

[70] Ibid.

[71] Ibid.

[72] Indeed, the petition for clemency on Simmons' behalf maintained that no evidence had been presented at trial about the systematic abuse to which Simmons had been subjected, his alcohol, marijuana and LSD abuse, or his mental functioning. A psychological evaluation conducted after he was sentenced to death showed that Simmons suffered from Schizotypal Personality Disorder. See 'Petition for a Commutation of, or reprieve of, a Sentence of Death', <http://www.internationaljusticeproject.org/pdfs/CSimmons-clemency.pdf>.

(heinous conduct) both because they are unlikely to be able to isolate the cause of the conduct and because the suggestion that the defendant was not entirely responsible for the conduct may make him seem that much more dangerous.[73] Indeed, the prosecutor in the Simmons's cases point to the defendants' youth as an aggravating, not a mitigating factor.[74]

13.3.2 Justice Kennedy and the role of neuroscience in categorical decision-making

Justice Kennedy's opinion for the court, precisely because it does not depend on individual determinations, allows him to recognize the brutality of Simmons' conduct and still conclude that his execution is constitutionally impermissible. Within the opinion, Justice Kennedy states almost off-handedly, 'As any parent knows, and as the scientific and sociological studies respondent and his amici cite tend to confirm, youth lack maturity...'[75] Justice Kennedy's analysis rests more heavily on the psychosocial than the neuroscience studies, and he does not describe any of them in detail.[76] After summarizing the conclusions, he observes:

> These differences render suspect any conclusion that a juvenile falls among the worst offenders. The susceptibility of juveniles to immature and irresponsible behavior means 'their irresponsible conduct is not as morally reprehensible as that of an adult.' Their own vulnerability and comparative lack of control over their immediate surroundings mean juveniles have a greater claim than adults to be forgiven for failing to escape negative influences in their whole environment.[77]

Justice Kennedy links these observations about the character of adolescence to the emerging consensus against the juvenile death penalty and it is that consensus by scientists, state legislatures, and foreign governments that justifies his decision. The scientific factors become simply one of a number of elements contributing to the conclusion, and the importance of the studies, in the legislative and research arenas Justice Kennedy discusses, is their analysis of the *probabilities* that adolescents will be immature rather than proof of the mental condition of a given individual.

---

[73] Snead, (note 35 above).   [74] *Roper* (note 48 above) at 558.

[75] *Roper* (note 48 above) at 569–70. The Supreme Court, however, did not base its decision on the *science* of brain development alone, and the legal and psychological experts who follow this issue disagree as to the whether the scientific findings have reached the point where it would be appropriate to base legal or policy decisions on them. See Aronson, above, summarizing the limitation of the scientific evidence on brain development, noting in particular that the neuroscience studies involve relatively small sample sizes and that they do not purport to prove a causal link between the observed patterns of brain development and individual behaviour.

[76] He refers, in particular, to Steinberg and Scott, 'Less Guilty by Reason of Adolescence: Developmental Immaturity, Diminished Responsibility, and the Juvenile Death Penalty' (2003) 58 *Am. Psychologist* 1009, 1014 ('[A]s legal minors, [juveniles] lack the freedom that adults have to extricate themselves from a criminogenic setting') and older studies such as Arnett, 'Reckless Behavior in Adolescence: A Developmental Perspective' (1992) 12 *Developmental Review* 339 ('adolescents are overrepresented statistically in virtually every category of reckless behavior') and E. Erikson, *Identity: Youth and Crisis* (New York, W. W. Norton & Company, 1968); *Roper* (note 48 above) at 569–70.

[77] Ibid. at 570 (citations omitted).

Indeed, in writing for the court, Justice Kennedy emphasizes the difficulty of individual determinations as a factor in the decision. He writes that:

> It is difficult even for expert psychologists to differentiate between the juvenile offender whose crime reflects unfortunate yet transient immaturity, and the rare juvenile offender whose crime reflects irreparable corruption... As we understand it, this difficulty underlies the rule forbidding psychiatrists from diagnosing any patient under 18 as having antisocial personality disorder,... which is characterized by callousness, cynicism, and contempt for the feelings, rights, and suffering of others... If trained psychiatrists with the advantage of clinical testing and observation refrain, despite diagnostic expertise, from assessing any juvenile under 18 as having antisocial personality disorder, we conclude that States should refrain from asking jurors to issue a far graver condemnation—that a juvenile offender merits the death penalty.[78]

Justice Kennedy's opinion ultimately rests on a conclusion that the Court can consider evolving standards of decency and should do so to reach a result compatible with Greene and Cohen's desire to undermine the strength of retribution as an appropriate purpose of punishment. He states that: 'Retribution is not proportional if the law's most severe penalty is imposed on one whose culpability or blameworthiness is diminished, to a substantial degree, by reason of youth and immaturity.'[79] The decision, however, rests less on its direct assessment of any individual study or scientific conclusion than it does on the relationship between such notions and the direction of state—and international—legislation.

13.3.3 Scalia and the role of neuroscience in deference to the legislature

The most pointed dissent in this 5–4 case was Justice Scalia's, which Justices Rehnquist and Thomas joined. Neuroscience, of course, did not exist at the time of the Constitution and, consistent with Justice Scalia's theory of original intent, is irrelevant to the type of deliberation in which he would have the Court engage.[80] The dissenters would accordingly leave the matter to state legislatures, who could consider whatever evidence they liked in prescribing capital punishment.

Justice Scalia could not resist, however, objecting to Justice Kennedy's analysis of youthful immaturity. First, Justice Scalia observes that 'the Court looks to scientific and sociological studies, picking and choosing those that support its position. It

---

[78] *Roper* (note 48 above) at 573.
[79] Ibid. at 571. He also observes (ibid.) that 'because the same characteristics that render juveniles less culpable than adults suggest as well that juveniles will be less susceptible to deterrence'.
[80] Justice Scalia, of course, is particularly incensed because *Roper v Simmons* overturned an opinion he wrote fifteen years earlier. He writes (ibid. at 608): 'What a mockery today's opinion makes of Hamilton's expectation, announcing the Court's conclusion that the meaning of our Constitution has changed over the past fifteen years—not, mind you, that this Court's decision fifteen years ago was *wrong*, but that the Constitution *has changed*... It then finds, on the flimsiest of grounds, that a national consensus which could not be perceived in our people's laws barely fifteen years ago now solidly exists... The Court thus proclaims itself sole arbiter of our Nation's moral standards–and in the course of discharging that awesome responsibility purports to take guidance from the views of foreign courts and legislatures.'

never explains why those particular studies are methodologically sound; none was ever entered into evidence or tested in an adversarial proceeding.[81] Second, he cites contradictory evidence, noting the *amicus* briefs filed on behalf of adolescent abortion rights (sometimes by the same *amici* opposing the death penalty in *Roper*), who argued in support of adolescent decision-making capacity.[82] Finally, he emphasizes that:

> Even putting aside questions of methodology, the studies cited by the Court offer scant support for a categorical prohibition of the death penalty for murderers under 18. At most, these studies conclude that, *on average*, or *in most cases*, persons under 18 are unable to take moral responsibility for their actions. Not one of the cited studies opines that all individuals under 18 are unable to appreciate the nature of their crimes.[83]

Justice Kennedy, of course, crafted the majority opinion to rely on the lesser capacity of youths as a general matter and the difficulty of individual proof as a reason for a categorical ruling.[84] Nonetheless, it is important to recognize that he and Justice Scalia differ less in the way they read the studies of adolescent behaviour than the factors they see as important in determining appropriate punishment. When Justice Scalia states that not one of the studies opines that all youngsters 'are unable to appreciate the nature of their crimes', he is of course right. No study maintains that older adolescents all have the same characteristics; moreover, no study maintains that adolescents are incapable of 'appreciating' the nature of their crimes, either in terms of understanding the nature of their actions or their illegality or immorality.[85]

Justice Kennedy and Scalia are using the studies to support different conclusions in accordance with different standards that in turn reflect different objectives. Neither Justice Kennedy nor the neuroscience studies claim that teens are incapable of distinguishing right from wrong, nor that they cannot or should not be held responsible for their actions. Instead, Justice Kennedy, in writing for the majority in *Roper*, engages in a comparative analysis. He emphasizes teens' 'underdeveloped' sense of responsibility, their greater inclination to engage in 'impetuous or ill-considered actions',[86] and their less well-formed and more transitory characters.[87] His conclusion is that '[r]etribution is not proportional if the law's most severe penalty is imposed on one whose culpability or blameworthiness is diminished, to a substantial degree, by reason of youth and immaturity'.[88] An extraordinary number of words in the sentence—'proportional', 'most severe', 'diminished', 'to a substantial degree'—are words of comparison.

Justice Scalia's analysis, in contrast, is cast in absolutes. He categorically rejects Kennedy's analysis of the impropriety of retribution as 'transparently false'.[89] He refers to Justice Kennedy's 'generalization that youth *always* defeats culpability',

---

[81] Ibid. at 617.   [82] Ibid. at 617–8.   [83] Ibid. at 618.
[84] See *Roper* (note 48 above) at 573.
[85] Indeed, Scott and Steinberg conclude that a teen defendant 'rarely is so deficient in his decisionmaking capacity that he cannot comprehend the immediate harmful consequences of his choice or its wrongfulness': Steinberg and Scott (note 33 above) at 131.
[86] Ibid. at 569.   [87] Ibid. at 570.
[88] Ibid. at 571.   [89] Ibid. at 621.

as though conviction and life imprisonment were not culpability. More critically, he sees the case as one about passing judgment on heinous acts. He objects that 'the studies the Court cites in no way justify a constitutional imperative that prevents legislatures and juries from ... determining that some murders are not just the acts of happy-go-lucky teenagers, but heinous crimes deserving of death'[90]—as though there were no intermediate position. He further quotes his own opinion in *Stanford*[91] to reiterate that it is:

> absurd to think that one must be mature enough to drive carefully, to drink responsibly, or to vote intelligently, in order to be mature enough to understand that murdering another human being is profoundly wrong, and to conform one's conduct to that most minimal of all civilized standards.[92]

In short, Justice Scalia argues that for the court to rule out execution for crimes committed before the age of eighteen, it would necessary to conclude the teens are insufficiently mature 'to understand that murdering another human being is profoundly wrong' or to conform their conduct to such a standard.[93] No study makes such a claim of incapacity on the basis of adolescent age alone; yet, without such proof, Scalia would leave the determinations to state legislatures, and leave the legislatures free to prescribe the requisite punishments on any grounds they choose.

13.4 Neuroscience in an Age of Division

At the end of the day, the individual justices' philosophies may have more to do with how they interpret scientific studies—and apply them to the juvenile death penalty—than the scientific studies have to do with the nature of criminal responsibility. Justice Kennedy's opinion for the court emphasized changing context, the weight of *international* opinion, and juvenile offenders' lesser culpability as constitutional limitation on the role of retribution. Justice Scalia, with his emphasis on original intent, rejected the relevance of context, denounced Justice Kennedy's willingness to consider international opinion,[94] and viewed retribution in absolute rather than relative terms. Indeed, in a speech exploring his personal views on the death penalty, he explained that:

> the Christian is also more likely to regard punishment in general as deserved. The doctrine of free will, the ability of man to resist temptations to evil is central to the Christian doctrine of salvation and damnation, heaven and hell. The post-Freudian secularist, on the other hand, is more inclined to think that people are what their history and circumstances have made them, and there is little sense in assigning blame.[95]

---

[90] Ibid. at 619.   [91] 492 US at 374.
[92] Ibid.   [93] Ibid.
[94] Ibid. at 624 (rejecting 'out of hand' the idea that the US should consider the laws of other countries).
[95] Address of Antonin Scalia at the Pew Forum Conference at the University of Chicago Divinity School (25 January 2002), <http://pewforum.org/deathpenalty/resources/transcript3.php3>. Scalia further observed (ibid.) that many people oppose the death penalty because they conflate the authority

With attitudes toward punishment rooted in religious worldviews and corresponding to deep divisions between left and right, scientific studies, however, rigorous or persuasive are unlikely to bridge the chasm.

This suggests, further, that such studies, at least in the context of multifaceted issues such as teen maturity should not be dispositive in themselves, but rather should be part of the multifaceted decision-making Kennedy's opinion demonstrates at the Supreme Court level and Scalia's opinion would leave to the states.

First, the role of neuroscience evidence *per se* is likely to be most persuasive in individual cases where it is causal. The best example is the case of the brain tumour, which was not only present in an area of the brain associated with sexual conduct, but which lent itself to what otherwise be an ethically dubious experiment—the defendant's behaviour changed when the tumour was removed.[96] Teen capacity for mature judgement, on the other hand, does not lend itself to individual determinations based on neuroscience.[97] No lab experiment is likely to replicate the circumstances involved in a decision to commit murder; moreover, some of the most common factors—e.g. drug and alcohol abuse—are difficult to test ethically in controlled circumstances.[98] Yet, the most persuasive argument is that a teen like Simmons who seriously abuses drugs is more likely to suffer from impaired judgement than an adult engaged in the same activity.

Second, neuroscience studies in themselves, at least at this point in the development of the science, are unlikely to be persuasive without other studies that test the links between brain conditions and behaviour.[99] Justice Kennedy, after all, cited Scott and Steinberg's evidence of the psycho-social research on adolescent behaviour to a greater degree than the neuroimaging studies. The more multidisciplinary the studies, however, the more sophisticated decision-makers need to be to parse their significance. As Emily Buss observes, 'Common to the law's use of all social science is the risk of bad data or misused data, and the danger that lawmakers will not have the sophistication or the inclination to assess the data closely and limit its use accordingly.'[100]

---

of the people with the authority of a democratic government, but this obscures 'the divine authority behind government'.

[96] See, e.g., Burns and Swerdlow (note 29 above).

[97] See Barry C. Feld, 'A Slower Form of Death: Implications of *Roper v Simmons* for Juveniles Sentenced to Life without Parole' (2008) 22 *ND J. L. Ethics and Pub.Pol'y* 59 ('Despite developmental differences, clinicians lack the tools with which to assess youths' impulsivity, foresight, or preference for risk in ways that relate to maturity of judgment and criminal responsibility.')

[98] Indeed, Simmons is alleged to have ingested a sufficient quantity of illegal substances at the time of the murder to endanger the health of a lab volunteer. See note 72 above.

[99] See Emily Buss, 'Review: Rethinking the Connection between Developmental Science and Juvenile Justice' (2009) 76 *U. Chi. L. Rev.* 493, 509 ('Brain imaging studies cannot, however, explain why adolescents with presumably similar brain structures behave so differently in different cultures around the world, nor have they yet captured the influence adolescents' different life experiences might have on the maturation of their brains.')

[100] Ibid. at 509. Buss observes further (ibid. at 510) that 'even the highest quality interdisciplinary work... will only be as good as the currently available data, which we know will be improved upon with additional research'.

Third, consideration of teen immaturity as a statistical matter (as opposed to individualized determinations of cognitive capacity) is a matter of policy rather than proof. As Justice O'Connor's opinion indicates, the facts of a particularly heinous crime are likely to overwhelm more dispassionate consideration of the flaws in teen judgement; yet, those who commit the most heinous crimes are most likely to suffer from cognitive limitations that preclude fully mature judgement. The balance between the importance of retribution versus the limits of teen responsibility *is* a general issue that should be resolved in relatively uniform ways rather than turn on the predispositions of the panel that decides an individual case.[101]

Fourth, placing too much weight on individual determinations of cognitive capacity may displace the underlying dispute that Kennedy and Scalia's different views represent. Scalia accuses Kennedy of selectively emphasizing the studies that support his view of proportional punishment; Scalia tries to contrast *amicus* briefs from abortion cases concluding that teens have the capacity to make mature decisions.[102] The Cultural Cognition Project suggests that many decision-makers are likely to interpret evidence through the cultural lens that reflect their predispositions.[103] If what actually drives decision-making about capital punishment is the justice's view of the importance of retribution in criminal sentencing, let the policy debate focus directly on that issue rather than on arcane interpretations of neuroimaging studies or queries about statistical significance.

Fifth, a more honest policy debate might inspire prospective as well as retributive reactions. Recognizing the correlation between auto accidents and youth, for example, has led to a number of targetted interventions, such as restricted licences, that have cut the teen accident rates. Debating individual culpability for the accidents is much less productive.

Taken together, these observations suggest that the capacity of decision-makers to evaluate the empirical evidence is a reason for categorical decisions, and the categorical decisions, whether by the Supreme Court or state legislatures, are unlikely to turn on sophisticated reading of the neuroscience literature. Nonetheless, ignoring that literature, either in terms of its support for a changing view of adolescence or in terms of the limitations of such inquiry, impoverishes the discourse as well. At the very least, the discussion should include consideration of the limitations of brain science, including the limited capacity of legal decision-makers to apply the findings to individual cases.

The next round in the teen capacity debate, at least within the context of the American criminal justice system, addressed the constitutionality of a sentence of life imprisonment without parole for crimes short of homicide committed while the

---

[101] *Roper* recognized that 'the brutality or cold-blooded nature of any particular crime would overpower mitigating arguments based on youth as a matter of course, even where the juvenile offender's objective immaturity, vulnerability, and lack of true depravity should require a sentence less severe than death.' (note 48 above) at 573.
[102] *Roper* (note 48 above) at 617.
[103] See Braman and Kahan (note 31 above).

defendant was a minor.[104] Justice Kennedy emphasized in *Roper* that a significant factor in the unconstitutionality of the juvenile death penalty 'is that the character of a juvenile is not as well formed as that of an adult'[105] and he reiterated in *Graham* that 'developments in psychology and brain science continue to show fundamental differences between juvenile and adult minds'.[106] The court further emphasized that juveniles 'are more capable of change than are adults, and their actions are less likely to be evidence of 'irretrievably depraved character' than are the actions of adults'.[107] The court accordingly rejected the constitutionality of a categorical life sentence for a crime that did not involve a homicide for reasons very similar to those articulated in *Roper*.

Yet, the court's rejection of the possibility of life without parole leaves open what considerations the states may bring to bear on the later determination of parole. The majority observed that:

> a categorical rule gives all juvenile nonhomicide offenders a chance to demonstrate maturity and reform. The juvenile should not be deprived of the opportunity to achieve maturity of judgment and self-recognition of human worth and potential. In Roper, that deprivation resulted from an execution that brought life to its end. Here, though by a different dynamic, the same concerns apply. Life in prison without the possibility of parole gives no chance for fulfillment outside prison walls, no chance for reconciliation with society, no hope. Maturity can lead to that considered reflection which is the foundation for remorse, renewal, and rehabilitation.[108]

The states must now individually address what considerations to use in determining 'maturity' and 'reform'. Presumably, however, these standards will not turn on wholesale differences between adolescent and adult reasoning, but on the individual determinations currently made in the context of other parole decisions. These determinations, while intrinsically subjective, need not involve relitigating the issues of individual culpability the court avoided in *Roper*.[109]

13.5 Conclusion

The advent of neuroscience studies has increased the claims that we can peer into the brain and discover its workings. Some commentators, like Greene and Cohen, believe that better understandings of the biological bases of behaviour will

---

[104] *Graham v Florida*, 130 S. Ct. 2011 (2009). See also Barry C. Feld, 'A Slower Form of Death: Implications of *Roper v Simmons* for Juveniles Sentenced to Life without Parole' (2008) 22 *ND J. L. Ethics and Pub. Pol'y* 9.

[105] *Roper* (note 48 above) at 570.

[106] *Graham*, (note 104 above) at 2026.

[107] Ibid. Indeed, in a decision in 1968, the Kentucky Supreme Court invalidated a sentence of life without parole for a juvenile, observing that 'it is impossible to make a judgment that a fourteen-year-old youth, no matter how bad, will remain incorrigible for the rest of his life': *Workman v Commonwealth* 429 SW2d 374, 378 (Ky 1968).

[108] Ibid., at 2032.

[109] For discussion of the implications, see Robert Smith and G. Ben Cohen, 'Redemption Song: Graham v. Florida and the Evolving Eighth Amendment Jurisprudence' (2010) 108 *Mich L. Rev.* 86.

transform our understandings human responsibility, and dethrone criminal liability tied to notions of desert. Others despair that any demonstration of limited capacity will overcome juror fear of the violent or distaste for the depraved, whatever the source of the urges that produce such behaviour. Others still would reject the relevance of neuroimaging for criminal determinations unless it rises to a level that can show a defendant lacked the ability to discern the significance of his acts (mistaking the victim's head for a coconut) or the ability to control his actions.

I have tried to argue in this paper that the nature of the new scientific findings more readily belongs in the debate between Justices Kennedy and Scalia than in the individual determinations O'Connor championed. Justice Kennedy, in his embrace of relative notions of culpability, and Scalia, at least in his defence of the primacy of retribution, are relatively honest in acknowledging the worldviews that drive their approaches. In accordance with these world views, neuroscience studies become one element in a multifaceted approach.

The lessons for other potential applications of such findings require acknowledging the difficulty and limitations of multidisciplinary legal research. As we discover more about the workings of the brain, we may become more pressed to combine such findings with an empirical testing of their impact on behaviour, and consideration of their implications, not just for our scientific understandings of human society, but of their interaction with the societal structure that shapes behaviour and ultimately contributes to understandings of justice.

14

Adolescent Brain Science and Juvenile Justice

*Terry A. Maroney**

14.1 Introduction

Recent scientific findings about the developing teen brain have both captured public attention and begun to percolate through legal theory and practice. Many believe that developmental neuroscience contributed to the US Supreme Court's elimination of the juvenile death penalty in *Roper v Simmons*.[1] Post-*Roper*, scholars and advocates alike increasingly assert that the developmentally normal attributes of the teen brain counsel differential treatment of young offenders. Courts and legislatures have begun to take note—US Senator Edward Kennedy, for example, convened a 2007 hearing on the juvenile justice implications of brain development.[2] Popular culture, too, has caught on: magazines, books, and even insurance companies now offer brain-based explanations for why teens are 'primal', 'crazy', and irresponsible.[3]

When one tests the legal impact of developmental neuroscience by analysing cases in which juvenile defendants have attempted to put it into practice, though, a surprising trend emerges: most such efforts fail.[4] Doctrinal factors hamstring most claims—for example, that persons with immature brains are incapable of forming the requisite *mens rea* for serious crimes. Limitations intrinsic to the science itself—for example, individual variation—also hinder its relevance and impact. Moreover,

---

* Associate Professor, Vanderbilt University Law School. An expanded treatment of the study analyzed herein was published as 'The False Promise of Adolescent Brain Science in Juvenile Justice' (2009) 85 *Notre Dame L. Rev.* 89, available at <http://ssrn.com/author=447138>, from which this paper is adapted.

[1] 543 US 551 (2005).
[2] *Hearing on Adolescent Brain Development and Juvenile Justice Before the Subcomm. on Healthy Families and Communities of the S. Comm. on Education and Labor and the Subcomm. on Crime, Terrorism, and Homeland Security of the S. Comm. on the Judiciary*, 110th Cong. (2007).
[3] Michael J. Bradley, *Yes, Your Teen Is Crazy!* (Gig Harbor, Harbor Press, 2002); Barbara Strauch, *The Primal Teen* (New York, Anchor, 2003); David Walsh, *Why Do They Act That Way?* (New York, Free Press, 2004); C. Wallis, 'What Makes Teens Tick?' (2004) *Time*, 10 May, 56; S. Begley, 'Getting Inside a Teen Brain' (2000) *Newsweek*, 28 February, 58; Allstate Insurance Co., 'Why Do Most 16-Year-Olds Drive Like They're *Missing A Part of Their Brain*? Because They Are' (Advertisement advocating restrictions on teenage driving), <http://www.allstate.com/content/refresh-attachments/Brain-Ad.pdf> (accessed 12 January 2010).
[4] This paper focuses on cases within the US. However, its analysis should apply to any legal system adopting a similar doctrinal approach to juvenile justice.

direct reliance on neuroscience as the metric for juvenile justice policy may jeopardize equality and autonomy interests, and brain-based arguments too frequently risk inaccuracy and overstatement. Finally, neuroscience is unlikely materially to shape legal decision-makers' beliefs and values about youthful offenders; it will instead be read through the lens of those beliefs and values.

This paper first traces the ascendance of developmental neuroscience within juvenile justice. It then demonstrates that, despite optimistic projections, adolescent brain science has had, is likely to have, and should have, only moderate impact in the courts. Neuroscience can, however, play a limited role in juvenile justice policy. It reinforces the (once) noncontroversial idea that, as a group, young people differ from adults in systematic ways directly relevant to their relative culpability, deterrability, and potential for rehabilitation. Therefore, legal decision-makers exercising a policy-making role—usually legislatures but sometimes the courts—ought to consider developmental neuroscience one source among many upon which to draw when making legally relevant assumptions about adolescents as a group. To go further is unwarranted and unwise.

14.2 Adolescent Brain Science and Juvenile Justice: An Overview

Adolescent brain science came to occupy its current prominence within juvenile justice because of the confluence of three distinct phenomena. Developmental psychology became far more sophisticated; neuroscientific technology improved dramatically; and scholars began a dialogue over the legal implications of neuroscience.

14.2.1 Developmental psychology and neuroscience

Theories of adolescence as a developmental stage importantly distinct from both childhood and adulthood always have been central to juvenile justice, underlying not only the core idea—that of having a separate system at all—but also the attributes of that system. However, for most of the twentieth century developmental psychology was in a fairly primitive state and focused primarily on young children.

It was not until the 1980s that a sustained programme of relevant empiricism took hold. Scientists began to study teens' risk-taking behaviours; sensation-seeking; ability to adopt a future-time perspective; perceptions of personal vulnerability; attitudes toward authority; self-concept; peer orientation; and decision-making. Research generally showed that teenagers are indeed distinct from both children and adults. For example, normal teens show a marked increase in risk-taking behaviour, though they often display adult-level cognitive understanding of risk; they also display far higher levels of peer orientation and sensation-seeking.[5] Of

---

[5] Elizabeth S. Scott and Laurence Steinberg, *Rethinking Juvenile Justice* (Cambridge, Harvard University Press, 2008) 38–44; B. J. Casey and others, 'The Adolescent Brain' (2008) 1124 *Annals N.Y. Acad. Sci.* 111, 112, 122.

particular importance for juvenile justice, research demonstrated that some level of delinquent behaviour is normal, particularly for boys, and that the vast majority of teens 'age out' of such offending.[6] Psychologists and legal scholars began in the 1980s a collaborative effort to define and measure teens' law-relevant psychological attributes, such as competence to waive *Miranda* rights or choose abortion. Nevertheless, in the early 1990s juvenile justice policy was still largely being 'devised in a context of empirical uncertainty', and scholars undertook a concerted effort to close that gap.[7] By the late 1990s a respectable body of research was in place, more research was underway, and advocates increasingly cited such research.

At precisely this same time, a revolution was taking place in neuroscience. Technological breakthroughs allowed for increasingly sophisticated observation of human brains *in vivo*, including those of young people. Widely publicized structural imaging studies demonstrated that the adolescent brain is still developing.[8] Adolescent structural maturation appeared to revolve around myelination (insulation of neural axons with a fatty substance referred to as 'white matter') and changes in the volume and density of 'gray matter' (neuron cell bodies and synapses). Healthy brains showed linear increases in white matter from childhood until adulthood, indicating a progressive increase in potential for fast, efficient communication among brain systems. Scientists also identified a preadolescent overproduction of neurons similar to one previously observed in very early childhood. Following this second wave of 'exuberance,' neural connections are over the course of adolescence 'pruned back'—likely because of relative use, dependent on life experiences, and reflecting a fine tuning of ability. Further, pruning and myelination affect different regions of the brain at different times; the brain's evolutionarily new frontal cortices (responsible for higher-order reasoning and 'executive control') are the last fully to achieve structural maturity. A small number of functional imaging studies additionally suggested that adolescents might tend to employ different brain processes than adults when carrying out identical tasks.[9]

By the early 2000s, then, neuroscience supported the notion that teen brains are structurally and functionally different from those of both children and adults. A link between behaviour and brain development appeared logical. Normal development results in an adult brain that is better equipped quickly and efficiently to

---

[6] Franklin E. Zimring, *American Juvenile Justice* (New York, Oxford University Press, 2005) 63, 91–103.

[7] E. S. Scott, 'Judgment and Reasoning in Adolescent Decisionmaking' (1992) 37 *Vill. L. Rev.* 1607, 1663.

[8] J. N. Giedd and others, 'Brain Development During Childhood and Adolescence: A Longitudinal MRI Study' (1999) 2 *Nature Neuroscience* 861, 861–2; T. Paus and others, 'Structural Maturation of Neural Pathways in Children and Adolescents: In Vivo Study' (1999) 283 *Sci.* 1908, 1908; E. Sowell and others, 'In Vivo Evidence for Post-Adolescent Brain Maturation in Frontal and Striatal Regions' (1999) 2 *Nature Neuroscience* 859, 860–1. See also N. Gogtay and others, 'Dynamic Mapping of Human Cortical Development during Childhood through Early Adulthood' (2004) 101 *Proc. Nat. Acad. Sci.* 8174, 8174.

[9] A. Baird and others, 'Functional Magnetic Resonance Imaging of Facial Affect Recognition in Children and Adolescents' (1999) 38 *J. Am. Acad. Child Adolescent Psychiatry* 195, 198–9.

respond appropriately to life's challenges and perform the types of tasks for which a person has trained. While the average normal adolescent's physical capacity for such maturity far exceeds that of a child, it falls short of that of the average normal adult. Neuroscience and developmental psychology, taken together, support a hypothesis that to 'the extent that transformations occurring in the adolescent brain contribute to the characteristic behavioral predispositions of adolescence, adolescent behaviour is in part biologically determined'.[10]

This narrative emerged against the backdrop of a dialogue then taking shape over the legal implications of neuroscience. Scholars predicted that emerging brain science would be particularly relevant to criminal law, given the centrality of mental states to criminal responsibility.[11] The most aggressive claim was that neuroscience would upend entrenched concepts of free will and responsibility. A more modest prediction was that neuroscience might improve identification and understanding of the types of irrationality—for example, incompetence and insanity—already relevant to criminal law. Because juvenile justice explicitly invites insights from the mind sciences, this particular brain–law connection appeared especially promising to both scholars and advocates.

It also appeared to be much needed, for the law of juvenile justice began in the 1990s to undergo a convulsive change of its own. Prompted by a spike in gun-related youth homicides in the US, commentators and policymakers warned that a new breed of juvenile 'superpredators' would be responsible for a 'coming bloodbath' of youth crime.[12] States responded—indeed, during the 1990s nearly every state amended its juvenile code.[13] States made it far easier to transfer ever-younger children to adult court for an ever-growing list of offences; eroded confidentiality protections; and de-emphasized rehabilitation. The 'coming bloodbath' never materialized; the youth homicide spike fell off quickly, and juvenile crime has been at historic lows for some time. The deep systemic changes enacted in response to those fears, though, remain largely in place. In important respects, the juvenile system became indistinguishable from the adult one, and the benefits it retained became available to fewer young persons.

Scholars and advocates in the late 1990s therefore correctly perceived that science and law were moving in opposite directions: the former was solidifying around the view that adolescents are different from adults in ways directly relevant to their culpability and capacity for change, while the latter was solidifying around the view that adolescents, particularly older ones or those accused of very serious crimes, ought to be treated like adults.

---

[10] L. P. Spear, 'The Adolescent Brain and Age-Related Behavioral Manifestations' (2000) 24 *Neuroscience & Biobehavioral Revs.* 417, 447.

[11] Semir Zeki and Oliver R. Goodenough, 'Introduction' *to Law and the Brain* (New York, Oxford University Press, 2006) xi, xiii–xiv.

[12] J. J. Dilulio, Jr, 'The Coming of the Super-Predators' (1995) *Weekly Standard,* 27 November, 23.

[13] P. Torbet and L. Szymanski, 'State Legislative Responses to Violent Juvenile Crime: 1996–1997 Update' (November 1998) *Juv. Just. Bull.* 1, <http://www.ncjrs.gov/pdffiles/172835.pdf> (accessed 13 January 2010).

14.2.2 The brain-based challenge to the juvenile death penalty

Scholars and juvenile advocates soon saw an opportunity to use brain science to break the tension and move law in their preferred direction: a challenge to the juvenile death penalty. In 2004 the US Supreme Court granted certiorari in *Roper v Simmons*, in which the Missouri Supreme Court had defied a 1989 decision upholding the death penalty for sixteen- and seventeen-year-olds[14] and ruled the juvenile death penalty unconstitutional.

Christopher Simmons's lawyers chose to highlight adolescent brain science in their briefs, arguing that 'the parts of the brain that enable impulse control and reasoned judgment', as well as 'competent decision-making, control of emotions, and moral judgment', are 'not yet fully developed in 16- and 17-year-olds', deficits rendering them less culpable, less deterrable, and less than the '"fully rational, choosing agent[s]" presupposed by the death penalty'.[15] Simmons's counsel, at oral argument, devoted more time to neuroscience than to any other issue. This focus was complemented by a number of *amicus* parties, notably the American Medical Association, whose brief urged that '[a]dolescents' behavioral immaturity mirrors the anatomical immaturity of their brains'.[16]

The *Roper* court's decision closely tracked many of Simmons's arguments about adolescent maturity. It noted '[t]hree general differences between juveniles under 18 and adults': greater propensity to 'immaturity and irresponsibility', resulting in overrepresentation in 'virtually every category of reckless behavior'; increased vulnerability and susceptibility to negative influences, including 'peer pressure'; and 'more transitory, less fixed' personalities, reflective of less 'well formed' character. These attributes of youth, the court held, 'render suspect' both the notion that the death penalty effectively deters teens and 'any conclusion that a juvenile falls among the worst offenders'.[17] For these and other reasons it struck down the juvenile death penalty.

However, the influence of neuroscience was unclear. The court drew most of its language from prior decisions, none of which had relied on brain science, and remarked that 'any parent knows' that teenagers are immature. It buttressed this experiential observation by noting that 'the scientific and sociological studies respondent and his *amici* cite tend to confirm' it, but nowhere specified which *amicus* briefs it found relevant and persuasive. These ambiguous signals, though, were seen in light of the prominence of neuroscience in briefing and argument, and were further coloured by an emerging societal fascination with the teen brain.

---

[14] *Stanford v Kentucky* 492 US 361 (1989).
[15] Brief for Respondent at 10, 23, *Roper* 543 US 551 (2005) (No. 03-633), 2004 WL 1947812 (quoting *Thompson v Oklahoma* 487 US 815, 825 note 23 (1989) (plurality opinion)) (internal quotes and citations omitted).
[16] Brief of the American Medical Ass'n et al. as Amici Curiae Supporting Respondent at 10, *Roper* 543 US 551 (No. 03-633).
[17] *Roper* 543 US at 569–70.

Developmental neuroscience thus came to be regarded—accurately or not—as a major influence on the highest-profile juvenile case in decades.

14.2.3 Adolescent brain science beyond *Roper*

Since *Roper* many scholars and advocates have urged that such science holds enormous potential to transform juvenile justice well beyond the death penalty.[18] The vast majority of post-*Roper* claims are based on a combination of developmental psychology and neuroscience, with the findings of the latter being invoked generally to buttress the reliability of the former. Scholars and advocates regard that buttressing as critically important, on the theory that it lends a 'hard science' edge to behavioural findings that might otherwise be dismissed as inordinately 'soft'.[19]

The most generalized claim is that evidence of population-typical brain immaturity during the teenage years both reinforces the original impulse to create a separate system of adjudication and treatment for juveniles and counsels recommitment to that system. Perhaps the most prominent contemporary scholars of developmental science and juvenile justice, Elizabeth S. Scott and Laurence Steinberg, articulate this notion in *Rethinking Juvenile Justice*.[20] Their claims may be synopsized as follows. First, structural immaturity in normal teenagers' frontal lobes may explain their relative deficiency in imagining the future, including the long-term consequences of their actions. Second, puberty-linked changes in the brain's reward circuitry and hormone production predispose teens to seek novelty and to value the rewards of risky behaviour more than its risks. Third, the relative weakness of neural connections between frontal cortices and areas associated with primary social and emotional processing contributes to their poor impulse control and emotional regulation. Fourth, because brain regions associated with executive function fully mature only in late adolescence and early adulthood, while those associated with primary emotional arousal and social information mature shortly after puberty, that teenager will experience a 'maturity gap' during which they are attracted to risky or irresponsible behaviours that they lack full capacity to appreciate or control. Thus, the normal attributes of the teen brain add up to 'a prescription for bad choices', reflective more of normative developmental process than of bad character. These aspects of adolescent brain development, as manifested in behaviour, should mitigate the law's response to juvenile offending.

---

[18] See, e.g., MacArthur Foundation Research Network on Adolescent Development and Juvenile Justice, 'Development and Criminal Blameworthiness' (2006), <http://www.adjj.org/downloads/3030PPT%20Adolescent%20Development%20and%20Criminal%20Blameworthiness.pdf> (accessed 13 January 2010); Scott and Steinberg, note 5, above; D. M. Bishop and H. B. Farber, 'Joining the Legal Significance of Adolescent Developmental Capacities with the Legal Rights Provided by *In re Gault*' (2007) 60 *Rutgers L. Rev.* 125, 172–3; J. D. Aronson, 'Brain Imaging, Culpability and the Juvenile Death Penalty' (2007) 13 *Psychol. Pub. Pol'y & L.* 115, 117 ('[J]uvenile justice advocates are currently seeking to expand the scope of the *Roper* decision and to use neuroscientific evidence for a variety of non-death penalty related issues.'); N. Cahn, 'Poor Children: Child "Witches" and Child Soldiers in Sub-Saharan Africa' (2006) 3 *Ohio St. J. Crim. L.* 413, 430.

[19] Aronson, note 18, above, 133.

[20] Scott and Steinberg, note 5, above, 28, 44–50.

Scott and Steinberg's theory has been endorsed by virtually every scholar, advocate, and defender now seeking to expand the influence of neuroscience within juvenile justice. Specific claims, synopsized here, fall at every point along the life course of a juvenile proceeding.

14.2.3.1 Waiver of rights

Adolescents' impulsivity and relatively deficient decision-making processes, particularly when under stress, render them less able knowingly, intelligently, and voluntarily to consent to searches, participate in identification procedures, waive *Miranda* rights, confess, waive counsel, or enter a guilty plea.

14.2.3.2 Competence

Neuroscience buttresses research showing that younger juveniles are less likely than adults to demonstrate adjudicative competence—that is, the ability to understand proceedings, consult with and assist counsel, and make critical decisions in a minimally rational and self-protective manner. Normal developmental immaturity ought to provide a basis for finding a juvenile incompetent, particularly in adult court, even if he or she cannot demonstrate a psychiatric disorder, developmental disability, or neurological abnormality relative to other teens.

14.2.3.3 Transfer to adult court.

Transferring a minor to adult court is out of step with developmental reality. A juvenile's crime may cause as much harm as an adult's, but that equivalency does not obviate brain-development differences relevant to both culpability and amenability to reform. Transfer should be abolished or, if allowed, be triggered only by individualized findings by a juvenile court judge.

14.2.3.4 Mens rea and mental-state defences

Because of brain immaturity, juveniles are less able or likely to form specific intent to carry out a particular action or to cause a particular result. Even when a juvenile can and does form the requisite mental state, that *mens rea* is a relatively poor proxy for culpability and future dangerousness. Further, assessment of both criminal intent and defences based on a 'reasonable person' standard should adopt the perspective of someone with an age-typical brain.

14.2.3.5 Imposition of adult punishment

Adult-like sentences—those that appear on the juvenile's public, permanent record, include state control for longer periods of time than permitted in the juvenile system, and/or are at least partially served in adult institutions—never should be imposed, whether as a result of transfer or a 'blended sentencing' scheme. Adult sentences are disproportionate to juveniles' diminished culpability and ignore the developmental reality that most will desist naturally as their brains mature. Such sentences also are unlikely to deter other minors, who inadequately consider consequences. Finally, incarceration (particularly with adults) can distort juveniles' growth at a critical juncture in brain development.

These claims are not radically different in kind from those regularly made by scholars and advocates on the basis of developmental psychology and 'common sense'. They are different in so far as they purport to rest on a different empirical basis and to result in more unshakeable conclusions, as a biological basis for immaturity ostensibly shows immaturity to be more deeply rooted and involuntary than does a psychological basis. They are also different to the extent they suggest that adolescent maturation takes longer than once was thought. Those differences, though, have not proved as consequential in legal practice as some predicted.

14.3 The Limited Impact of Adolescent Brain Science in the US Courts

Before *Roper* scholars and advocates had begun to envision a powerful role for developmental neuroscience within juvenile justice. Buoyed by apparent success, such theories have since proliferated and are being tested in cases. To measure the extent to which reality is conforming to predictions, I conducted a study of such cases. The study demonstrates that the range of neuroscientific arguments before the courts—state and federal, juvenile and criminal—is both wide and deep.[21] Their impact, however, has been shallow.

This shallow impact cannot be explained fully on the grounds that the science is new or the effort early. Rather, the courts' response to adolescent brain science reflects a frequent disconnect between the questions asked by law and those answered by science. Though courts sometimes cite the science approvingly, they do so only to buttress conclusions otherwise fully explained. The shallow impact also reflects scientific limitations that are genuine and likely to persist. These factors explain how courts generally have responded to developmental neuroscience arguments, but also show why that response has some basis. Two additional factors demonstrate why courts should not unduly privilege such claims. First, juvenile justice cannot directly track neuroscience without implicating equality and autonomy concerns, and no adequate limiting principle has yet been articulated. Second, the pressures of legal advocacy incentivize overstatement and often result in inaccuracy; while this tendency can be controlled, it cannot be eliminated. Adolescent brain science therefore has not been (and is unlikely to be) a transformative force in juvenile justice, at least in the courts.

---

[21] In-depth case discussion, as well as a discussion of the methodology, may be found in the article referenced in note *. A chart of the cases, of which there were 57, is available at <http://ndlawreview.org/archive/issue.php?vol=85&num=1>.

14.3.1 Doctrinal obstacles

The most frequent shoal upon which post-*Roper* adolescent brain science claims founder is that of existing legal doctrine, which tends to render them either irrelevant or unpersuasive. In some instances, courts perceive that the issue has been foreclosed by legislatures; in others, doctrine directs a relatively narrow inquiry and scientific insights fall largely outside its boundaries.

14.3.1.1 Adult punishment

Contemporary Eighth Amendment doctrine, under which non-death sentences will be invalidated only if so 'extreme' as to be 'grossly disproportionate' to the crime, frequently is fatal to juveniles' neuroscientific claims that particular punishments are unconstitutional. Similarly, courts have tended to uphold adult-sentencing schemes against brain-science challenges, hewing to doctrine directing deference to facially reasonable legislative and judicial choices as to which youths, or categories of youths, may or must be tried and punished as adults. The only punishment context in which neuroscience has had discernable, if marginal, impact is in a small number of individual sentencing proceedings, a context in which—unless mandatory sentences apply—judges have considerable latitude.

14.3.1.1.1 Juvenile life without parole

Largely repeating arguments made in *Roper*, juvenile justice advocates are now challenging the most extreme remaining sentence—juvenile life without parole (JLWOP). Two cases making such claims are currently pending before the US Supreme Court; the court's treatment of developmental neuroscience in those cases may provide valuable insight, largely absent in *Roper*, as to how it regards such evidence.[22] Even if no such insight is forthcoming, its decisions will alter the doctrinal landscape within which JLWOP claims are decided.

---

[22] Shortly before this paper went to press, the US Supreme Court decided the pending JLWOP cases. In *Graham v Florida*, 130 S. Ct. 2011 (2010), the court extended the developmental reasoning of *Roper* and held that the Eighth Amendment prohibits a sentence of life without possibility of parole for a non-homicide crime committed when the offender was under the age of eighteen. See also *Sullivan v Florida*, 130 S. Ct. 2059 (2010) (per curiam) (dismissing certiorari). In so doing, the court stated its understanding that 'developments in psychology and brain science continue to show fundamental differences between juvenile and adult minds,' and cited evidence that 'parts of the brain involved in behavior control continue to mature through late adolescence.'

The explicit nod to neuroscience demonstrates that the *Graham* court (and almost certainly the *Roper* court) believed neuroscience relevant to general propositions as to the normal developmental course of adolescence. The decision provides welcome support for legal policy-makers—whether in courts or legislatures—who seek to draw modestly on such science in reinforcing commitments to the special legal status of youth. This is precisely the weight, this paper argues, that the science presently can bear. All the cautions and limitations articulated in this paper remain unaltered. For a detailed discussion, see T. A. Maroney, 'Adolescent Brain Science after *Graham v Florida*' 86 *Notre Dame L. Rev.* (forthcoming 2011).

Under the existing doctrinal framework, though, such arguments have been almost uniformly unsuccessful. The most commonly articulated justification for rejection is *Roper* itself, in which the court appeared (in *dicta*) to endorse the Missouri Supreme Court's resentencing of Simmons to JLWOP. Many courts have relied on this *dictum*. The second major justification is the oft-repeated mantra that 'death is different': many courts have stated that *Roper* applies only in the death penalty context, and have instead judged JLWOP under the grossly disproportionate standard that, long before *Roper*, underlay the failure of most Eighth Amendment challenges.[23] These long odds have not changed with invocation of brain science. Courts often ignore the brain-science arguments, and those that have directly addressed neuroscientific claims in the JLWOP context generally have treated the issue as either doctrinally irrelevant or as replicating more traditional immaturity arguments.

14.3.1.1.2 Lengthy sentences

Juveniles also have used brain science to challenge other lengthy or harsh sentences. Such challenges stand on even less secure doctrinal footing, as the possibility of parole (even if remote) weighs in favour of constitutionality. Accordingly, few have prevailed. As in the JLWOP context, courts have tended to take a narrow view of substantive sentencing oversight. They also have tended to dismiss arguments based in developmental neuroscience, often under the rationale that it fails to offer anything meaningfully new, but also because it fits poorly with record evidence as to *mens rea* or aggravating factors.

It is worth noting a small counter-trend. In two cases, state courts relied on developmental principles—possibly including neuroscience—to limit extraordinarily long sentences.[24] Two different state courts also appear to have relied in small part on brain science to invalidate juvenile sex-offender registration, once by removing a juvenile from the list and once by invalidating the entire registration scheme.[25]

This small group of cases indicates that developmental neuroscience sometimes may work to solidify a holding where a court regards developmental principles as both persuasive and relevant to punishment and sees brain science as informing, in

---

[23] *State v Allen* 958 A2d 1214, 1234–5 (Conn 2008) (stating that 'in the past twenty years, courts have consistently rejected Eighth Amendment claims made by juvenile murderers attacking their life sentences' and citing dozens of pre-*Roper* cases). After *Graham*, though, cases upholding JLWOP for non-homicide offenses are no longer good law. Those involving homicide cases are unaltered. However, *Graham* also rejected the 'death is different' separation between capital and non-capital cases. It therefore undermines those cases whose reasoning was heavily dependent on that separation.

[24] One court allowed the eventual possibility of discretionary parole. *Cotting v State* No. A-9909, 2008 WL 4059580, at *3 (Alaska Ct App 3 September 2008) (citing the 'scientific and sociological studies' language of *Roper* and allowing parole eligibility after thirty-three years). The other imposed a term of years well below the maximum. *State v Carrasquillo* 962 A2d 772, 775–76 (Conn 2009) (discussing Carrasquillo's expert evidence about 'significant differences between the adolescent brain and the adult brain' and imposing sentence of thirty-five years instead of life).

[25] *Fletcher v State* No. 0404010688, 2008 WL 2912048, at *1 (Del Fam Ct 16 June 2008) (approvingly quoting an unpublished opinion by a Nevada family court judge invalidating application of that state's registration scheme to juveniles, in part because of brain science).

some way, those principles. Most punishment cases, though, indicate that courts tend to view the findings of developmental neuroscience as either irrelevant to the specific determination before them or as insufficiently persuasive to invalidate schemes for imposition of non-death sentences.

14.3.1.2 Transfer to adult court

Brain-based challenges to the transfer of minors to adult court also have been relatively ineffective. This area of law is closely related to adult punishment, as such punishment—for example, incarceration beyond the twenty-first birthday—generally may be imposed only following transfer. Historically, transfer decisions were left to juvenile court judges. Legislatures provided broad parameters within which that discretion was exercised—for example, by setting an age below which transfer was unauthorized—bounded on the outside by due process principles. Increasingly, though, states allow prosecutors to determine the court in which to proceed, or provide for legislative transfer, in which adult jurisdiction follows automatically from the state's selection of a particular charge against a person of a prescribed age.

14.3.1.2.1 Nonindividualized transfer

Well before the advent of developmental neuroscience, young people had argued that these newer schemes unconstitutionally exposed them to adult punishment without the benefit of an individualized hearing on their maturity, culpability, and potential for rehabilitation. Virtually all such challenges failed. Courts overwhelmingly deferred to legislatures' choices as to what combinations of age and charged offence categorically warrant adult treatment; they also affirmed prosecutors' power to make that determination, either by choosing the charge or by choosing the court. Brain science has not altered these tendencies.

A number of youths have urged post-*Roper* that developmental neuroscience shows the irrationality of nonindividualized transfer and counsels reversal of this doctrinal trend. A teenager named David Garcia, for example, offered expert testimony on adolescent brain development to support his claim that New Mexico's transfer law was 'a rejection of biology';[26] youths in California and Illinois made similar claims.[27] These appeals, like their pre-*Roper* predecessors, appear to have failed because of courts' broad deference to the legislative scheme.

14.3.1.2.2 Judicial transfer

There is no evidence that juveniles have on the basis of neuroscience persuaded individual judges to retain juvenile-court jurisdiction; nor have they managed to overturn transfer decisions on appeal. The case of Christopher Pittman, a

---

[26] *State v Garcia* No. CR 2005-422 (NM Dist Ct 14 December 2007), Child Defendant's Closing Remarks at 2, 5.
[27] Petition for Review Following Denial of Petition for Writ of Habeas Corpus at 5–6, *People v Gonzales* Nos. E036344, E037793, 2005 WL 1799520 (Cal Ct App 1 September 2005); Motion to Declare Defendant's Transfer to Adult Court Unconstitutional at 7–8, 10, *People v Jones* (Ill. Cir Ct, Cook County, 7 April 2006), available at <http://www.njdc.info/2006resourceguide/start.swf>.

twelve-year-old convicted of killing his grandparents, is exemplary. Pittman argued that a juvenile court judge lacked authority to transfer him because 'recent scientific data' shows that twelve-year-olds lack capacity to be tried as adults. The South Carolina Supreme Court instead held that the 'rules of statutory construction do not allow the Court to determine legislative intent based on scientific data' and relied on the fact that the statute contained no minimum age for judicial transfer.[28]

Thus, developmental neuroscience has to date proved no match for the strong doctrinal pull toward courts' deference to transfer schemes, and has failed materially to influence individual transfer determinations.

14.3.1.3 Mental states

Defenders' efforts to use developmental neuroscience in the context of mental-state assessment—whether going to *mens rea*, mental-state defences, or to the ability competently, knowingly, and intelligently to assert or waive constitutional rights— also have largely fallen short, primarily because of the generally adult-like tests by which juveniles are judged. Substantive criminal law generally is adopted wholesale by the juvenile justice system; the special attributes of that system cluster around adjudication procedures and dispositional consequences, not standards for determining guilt. Thus, the same mental-state concepts are used in juvenile and adult court. More, while age clearly matters to assertion of Fourth Amendment rights and to competence determinations, courts have yet to reach any consensus over how this is so, and tend to use adult-like tests despite brief nods to the impact of youth. Juveniles' mental-state challenges therefore often are perceived as going to the legitimacy of the rules themselves rather than their application.

14.3.1.3.1 Intent

In a number of homicide cases defenders have claimed that the young person, because of brain immaturity, did not consciously desire, or realize to a substantial certainty, that someone would die as a result of his actions. They sometimes also argue that the young person did not consciously deliberate over whether to act, defeating any element of premeditation.

Pittman, for example, argued that 'the portion of the brain that gives one the cognitive capacity to satisfactorily perform acts such as forming malice... is underdeveloped in a twelve-year-old'.[29] The court found the argument 'unconvincing given the nature of the criminal acts', pointing to evidence that the child acquired a gun, waited until his grandparents were asleep, 'executed an escape plan, and

---

[28] *State v Pittman* 647 SE2d 144, 161 (SC 2007).
[29] Ibid. 163. In a similar vein, a commentator suggested that fifteen-year-old Andy Williams—who shot and killed several persons at his high school—probably 'did not think about the specifics of shooting at all,' and that 'if a gun is put in the control of the prefrontal cortex of a hurt and vengeful 15-year-old, and it is pointed at a human target, it will very likely go off'. D. Weinberger, 'A Brain Too Young for Good Judgment' (2001) *New York Times,* 10 March, A13. Williams had an MRI taken of his brain and now is seeking to vacate his guilty plea. *Williams v Ryan* No. 05-cv-0737, 2007 WL 925834, at *1 (SD Cal 2 March 2007).

concocted a false story' to mislead police. As such actions by an adult would be sufficient to infer either a conscious plan to cause death or an awareness that death would (and did) result, it was considered *a fortiori* to allow the same inference for a child.[30]

In like fashion, Garcia invoked brain science to assert that older teens are so generally incapable of forming a 'willful, deliberate and premeditated' *mens rea* as to invalidate their wholesale transfer to adult court when charged with first degree murder. In a relatively unusual step, he was granted a hearing at which to present expert testimony on brain development. However, Garcia was not so much arguing that teens cannot satisfy the legal test for specific intent as he was arguing for a different conception of the mental state morally justifying conviction of a teen for intentional murder. His own experts agreed that adolescents are capable of forming specific intent. Their main point was that though a typical teenager literally is capable of intending his actions and their consequences, because of brain immaturity his technically sufficient mental state is substantively irrational. For example, an adolescent might intend the victim to die, but he lacks a meaningful conception of what it means for a person to be dead. The experts' secondary point was about odds: that planning and forethought are far less common in adolescents than is acting impulsively, particularly in emotionally intense situations. Even if true (and they may be true), these points are irrelevant unless a court were willing to adopt a substantively deep concept of the applicable *mens rea* and to find that specific intent is so rare in teenagers who kill as to upset the transfer scheme on its face. The *Garcia* court instead hewed closely to traditional *mens rea* definitions and deferred to legislative choices.

14.3.1.3.2 Reasonableness

If brain-based challenges to specific intent have been unsuccessful because of the relatively undemanding prevailing conception of that *mens rea*, challenges going to reasonableness might fare better. Failure to foresee consequences is culpable only where such failure constitutes a gross deviation from what a reasonable person in the actor's situation would have foreseen, and the developmental attributes of one's age are part of one's situation. Notwithstanding this relatively open space, however, adolescent brain science generally has failed to persuade.

Courts' first rationale is that the legislature has allowed them less interpretive room than advocates urge. *State v Heinemann* makes this point.[31] Gabriel Heinemann, sixteen, asked that the adult-court jury considering his duress defence be instructed on attributes of the 'reasonable adolescent', informed in part by the lessons of brain science. Dismissing as irrelevant 'literature about the developing adolescent mind', the trial court determined that whether a person of 'reasonable firmness' in Heinemann's position would have been unable to resist a threat was 'a community

---

[30] Similarly, a Tennessee court rejected expert testimony about adolescent brain development in determining that a fifteen-year-old premeditated the killing of her grandparents. *State v Daniel* No. M2005-01211-CCA-R3, 2006 WL 3071329, at *10–11 (Tenn Crim App, 30 October 2006).
[31] 920 A2d 278 (Conn 2007).

objective standard'. On appeal Heinemann and his *amici* again presented developmental neuroscience evidence. The Connecticut Supreme Court 'acknowledge[d] that juveniles often have more immature decision-making capability and recognize[d] the literature supporting the notion that juveniles are more vulnerable to all sorts of pressure, including, but not limited to, duress'. However, developmental science was not sufficient to persuade the court to characterize as 'clearly irrational and unreasonable' the legislature's decision to confine its 'appreciation of the different mental abilities and susceptibilities of younger persons' to those under sixteen, and to express that appreciation not through differential definition of reasonableness but through maintenance of a separate juvenile justice system.

A further rationale, previously noted in the JLWOP cases, is that brain science adds little or nothing to fact-finders' existing knowledge. A Minnesota appellate court, for example, stated that 'every parent and person who has gone through adolescence is familiar with and can understand the immaturity and impulsive responses of adolescents'.[32]

14.3.1.3.3 Felony murder

Courts also have turned aside efforts to invalidate application to juveniles of the felony murder doctrine. Under that doctrine, the state generally need not prove intent to kill if it is able to prove intent to commit the predicate felony and a causal link to the death. Like reasonableness, the doctrine necessarily relies on group-level assumptions about what people do and should foresee; the doctrinal space is similarly somewhat open. Therefore, JB, an Ohio thirteen-year-old convicted of the felony murder of his infant brother, argued (unsuccessfully) before the US Supreme Court that it is unreasonable to assume that minors, particularly very young ones, would or should foresee a risk of death when committing predicate felonies.[33] Though this assertion, if true, would indeed undermine the felony murder doctrine, courts have stuck by the doctrine. In so doing they have relied on the legislatures' choices to apply the same responsibility standards to adults and juveniles.

14.3.1.3.4 Ability to assert or waive rights

Few courts have been directly presented with neuroscientific claims going to minors' competence to waive rights or to face prosecution. In one such case (involving interrogation of out-of-state runaways) a teenager argued unsuccessfully that juveniles' undeveloped brains render them a suspect class.[34] In a small handful

---

[32] *State v Alford* No. A07-1025, 2008 WL 4006657, at *5–6 (Minn Ct App 2 September 2008).

[33] Petition for Writ of Certiorari at 25–29, *J.B. v Ohio* 549 US 1246 (2006) (No. 06-7611); *J.B. v Ohio* 549 US 1246 (2007) (No. 06-7611). JB, left home alone in charge of four younger siblings, was convicted of 'reckless abuse' of his brother, JR. JB testified that he had accidentally hurt JR; lost his temper when JR would not stop crying; and then injured him further. He and another sibling attempted CPR and tried to call for help, but their mother had removed the phone. They lay JR in a blanket and prayed next to him until their mother came home. Garcia, note 26, above, made a similar (and also unsuccessful) claim that, because of brain immaturity, teens generally lack the level of forethought justifying a felony-murder conviction.

[34] *People v Blankenship* 119 P3d 552, 555 (Colo App 2005).

of other cases, defendants and *amici* have raised brain science as one reason why evidence—statements to police or the fruits of a consent search—should have been suppressed, and courts have simply ignored or rejected the assertion as insufficiently developed.[35]

As in the sentencing context, though, there is a small countertrend. In one case, the Wisconsin Supreme Court (nowhere relying on neuroscience) used the well-established 'totality-of-the-circumstances' test to conclude that a fourteen-year-old's written confession was involuntary. Abrahamson CJ wrote a lone concurrence in which she asserted eight reasons why she would go further and 'adopt a per se rule, excluding in-custody admissions from any child under the age of 16 who has not been given the opportunity to consult with a parent or interested adult'; reason number three was that '[e]merging studies demonstrate that the area of the brain governing decision making and the weighing of risks and rewards continues to develop into the late teens and the early twenties'.[36]

In addition, at least one competence challenge succeeded in part because of neuroscience. A California appellate court ordered competency hearings for two boys, aged eleven and twelve, holding that simple 'developmental immaturity' (rather than a mental or cognitive abnormality) might provide a basis for an incompetence finding.[37] While the court relied primarily on psychological findings, it—unlike the trial court—also credited expert testimony about the brain immaturity of very young adolescents. In each of these cases the role of brain science appears to have been small, but it is noteworthy that it was mentioned as one of many reasons to grant a juvenile defendant relief.

* * *

The impact of adolescent brain science on juvenile justice thus has been strongly cabined by the extrinsic reality of legal doctrine. Doctrine is not, of course, a full independent measure of a claim's merit. For example, if a procedural default bars pursuit of an actual innocence claim, that holding says far more about the doctrinal valuation of procedural bars than it does about innocence as an exculpatory factor. The point at this juncture is not to endorse the status quo but, rather, to demonstrate how it currently is operating to diffuse neuroscientific claims. Courts generally perceive developmental neuroscience either as proving nothing new or as raising a challenge to the rules themselves, rather than informing an inquiry properly falling within the confines of the rules. While they sometimes are 'troubled by' the rules and follow them 'reluctantly',[38] courts generally do believe themselves to be bound to them. Though the science has been positively received by a small number of courts and judges, usually in the context of sentencing, in no instance has it been outcome-determinative.

---

[35] Pittman, note 28, above, 166 ('Appellant has presented no evidence, other than his age, supporting his claim that his confession was involuntary. Appellant instead relies exclusively on abstract scientific data and rhetorical questions for his argument. This evidence is not probative of coercion.').
[36] *In re Jerrell C.J.* 699 NW2d 110, 139–40 (Wis 2005) (Abrahamson CJ concurring).
[37] *Timothy J. v Superior Court* 58 Cal Rptr 3d 746, 754 (Ct App 2007).
[38] *People v Pratcher* No. A117122, 2009 WL 2332183, at *44 (Cal Ct App 30 July 2009).

Doctrine can change and therefore represents a relatively soft target. But in this area of law it is not very soft. Because the above-described doctrinal forces are so entrenched and of such broad applicability within criminal law, adolescent brain science is inadequate to provoke deep change, at least within the courts.

14.3.2 Scientific limitations

The challenge for brain science in juvenile justice, though, goes deeper than doctrine. Certain limitations that inhere in the science itself show some of the courts' general reticence to be well placed.

14.3.2.1 Individual differences

The most significant current limitation of developmental neuroscience is inability to inform individual assessment. Imaging studies that show group trends in structural maturity—such as relative levels of myelination in prefrontal cortex—do not show that all individuals in the group perfectly reflect the trend.[39] Normal brains follow a unique developmental path bounded roughly by the general trajectory; while all humans will pass through the same basic stages of structural maturation at more or less the same stages of life, the precise timing and manner in which they do so will vary. Moreover, such variation cannot be detected or interpreted in any legally meaningful way. Neither structural nor functional imaging can determine whether any given individual has a 'mature brain' in any respect, though imaging might reveal gross pathology.[40] Researchers consistently agree that developmental neuroscience cannot at present generate reliable predictions or findings about an individual's behavioural maturity. Courts thus have a strong basis for deeming brain science irrelevant to many highly individualized claims, such as whether a defendant was able to form specific intent.

Indeed, the cases reflect the difficulties posed by individual variation. Legal decision-makers display incredulity, even annoyance, when general lessons about the adolescent brain appear to conflict with evidence about the individual juvenile. One particularly vivid account was offered by a Delaware judge who presided over a juvenile capital case while *Roper* was pending.[41] The teen defendant, Michael Jones, presented expert testimony 'that juveniles are less criminally culpable than adults because the area of their brains controlling foresight, goal setting, and ability to plan are not yet fully developed'. However, he also offered the testimony of a psychologist who stated that Jones was 'an exceptionally gifted planner', whose ability to plan and foresee consequences was 'off the charts'. As recounted by the court, this admission:

---

[39] Casey and others, note 5, above, 119–21; S. J. Morse, 'Brain Overclaim Syndrome and Criminal Responsibility: A Diagnostic Note' (2006) 3 *Ohio St. J. Crim. L.* 397, 403–4, 404 note 4.
[40] *See* Casey and others, note 5, above, 104.
[41] See *State v Jones* No. 9911016309, 2005 WL 950122 (Del Super Ct 10 April 2005).

annihilated Jones' only viable defense: that, as a juvenile, he was too young to reasonably calculate the possible outcomes of his murderous rampage, and to plan accordingly... The State used [the psychologist's] testimony to suggest that Jones would use his exceptional gift for planning to formulate an escape, endangering corrections officers and the public at large... When [Jones' brain-science expert] took the stand as the next defense witness, explaining the complicated science of brain development and its nexus to planning ability, the jury appeared disinterested. Their courtroom demeanor, as well as their sentencing recommendation, made it clear that the jury viewed the medical evidence as mere 'psychobabble' meant to mislead them into excusing an inexcusable crime.[42]

Similarly, in *Garcia* the state was able to rebut the notion that anatomical immaturity necessarily manifests itself in a lack of meaningful appreciation of death by showing that Garcia himself had such appreciation; he was deeply affected by the recent death of his grandmother and frequently worried that his gravely ill mother would die.[43]

Neuroscience may provide marginal support for categorically limiting the sanctions that may be imposed on juveniles, but it has little to offer in assessing the mental state, capacity for rehabilitation, or other law-relevant attributes of any given juvenile.

14.3.2.2 Structure v Behaviour

A related difficulty stems from the reality that structural and functional differences between individual brains may not correspond with predictable or discernable differences in behaviour. Just as scientists cannot look at an individual teen's brain and conclude that she has a particular level of behavioural maturity, observers cannot look at a teen's behaviour and deduce the structural or functional maturity of her brain. This is not an issue only for individual determinations, for even at the group level there are few data demonstrating a clear link between structural immaturity and immature behaviour.[44] The structure–behaviour hypothesis is a strong one, as brain attributes often correlate with specific behaviours, and a significant developmental stage is highly likely to manifest in behaviour. Developmental psychology provides a picture of the attitudes and behaviours that typify adolescents; neuroscience provides a picture of the brain maturation processes that typify adolescence; and the latter can be interpreted in such a way as to provide a plausible, partial explanation for the former. But though it is highly plausible that '[a]dolescents' behavioral immaturity mirrors the anatomical immaturity of their brains',[45] science has not determined the nature or extent of that mirroring.

---

[42] Ibid. *1, 4–6.
[43] *Garcia*, note 26, above, Transcript 225. The *Gonzales* court, too, remarked that '[r]egardless of whether the nature of the adolescent brain produces behavior that is more impulsive than an adult's... [Gonzales's] conduct in this case reveals a high degree of individual culpability.' 2005 WL 1799520, at *7.
[44] See, e.g., J. N. Giedd, 'Structural Magnetic Resonance Imaging of the Adolescent Brain' (2004) 1021 *Annals N.Y. Acad. Sci.* 77, 83.
[45] AMA Brief, note 16, above, 10.

Advocates, commentators, and defenders unnecessarily overstate the case when they claim that imaging studies explain adolescent behaviour, let alone any given adolescent's behaviour. Courts also have a basis for believing neural explanations to be less probative than behavioural ones.[46]

14.3.2.3 Relative deficiency

Even if one credits the strongest hypotheses about the behavioural impact of brain immaturity, that impact cannot automatically claim legal significance. Psychological studies show that adolescents are consistently less able than adults to implement fast, appropriate, and mature responses to environmental challenges; neuroscience suggests that these relative deficiencies are partly attributable not to bad character but to biological constraints attending developmental processes. But relative deficiency—for example, in impulse control—does not establish that the deficiency is legally meaningful or that any individual failure of control is excusable. It instead implies that, compared to a similar failure in an adult, it is less blameworthy to the extent that its avoidance would have required more effort, through no fault of the child's own. Relative deficiencies do not necessarily take juveniles below a legal threshold but may instead show that they exceed it by a lower margin. Where to set that threshold relative to juvenile deficits is, at its core, a moral and legal determination, not a scientific one.

Unfortunately, defenders and experts often treat the legal significance of the science as a given; indeed, they occasionally bypass the relative-deficiency point altogether and devolve into hard biological determinism. They sometimes argue, for example, that because of their immature brains adolescents *cannot* make good decisions under stress, control their emotions, suppress violent impulses, foresee consequences, or defy antisocial peers.[47] Legal decision-makers are, by and large, unprepared to accept flat assertions of inability. Such assertions conflict with everyday observations (and, often, record evidence) that most teenagers make good choices most of the time and that offenders, too, make socially beneficial, self-protective, or strategic choices, sometimes within the context of the offence behaviour itself.

---

[46] E. A. Phelps and L. A. Thomas, 'Race, Behavior, and the Brain: The Role of Neuroimaging in Understanding Complex Social Behaviors' (2003) 24 *Pol. Psychol.* 747, 748 ('Although brain science can inform our understanding of complex human behaviors, it cannot help us predict human behavior with any more certainty than can be derived from examining behavior itself.').

[47] Casey Jones Motion, note 27, above, 9, 19–20 (stating that 'science tells us that Casey did not have the logical reasoning and decision-making skills' to comprehend the import of carrying a gun near school, and asserting same claim for juveniles in general); *Garcia,* note 26, above, Motion to Dismiss 9, 12 ('juveniles under 18 are incapable of possessing the mens rea required for capital offenses' and the 'inability of juveniles to modulate their emotional responses and make rational decisions is a biological fact'); Wisconsin Council on Children & Families, *Rethinking the Juvenile in Juvenile Justice* (Madison, WCCF, 2006) 10 ('[D]eterrence does not work with juveniles.').

14.3.2.4 Age limits

Neuroscience also tends to run headlong into a perennial difficulty in juvenile justice: the search for a stable justification for pegging law's relative solicitude to the eighteenth birthday. Because it is implausible to posit that any given date constitutes a maturational tipping point, courts and theorists historically have relied on practical concerns justifying line-drawing. States' choices are not consistent: while most terminate juvenile court jurisdiction at age eighteen, others choose seventeen or sixteen; all allow adult treatment of younger children in some circumstances; and all recognize different age milestones for benefits and responsibilities such as driving, voting, and drinking. Adolescent brain science has not offered a theory by which this erratic line-drawing might be harmonized and may have further muddied the waters.

Developmental neuroscience consistently indicates that structural brain maturation is incomplete at age eighteen. Though estimates vary, many scientists have opined that structural maturation is not complete until the mid-twenties.[48] Some also have opined—including in court testimony—that just as brain maturation is completed by the mid-twenties, it starts to decline in middle age, perhaps as early as age forty-five.[49] Taking neuroscience as the proper benchmark therefore would suggest that the criminal justice system systematically should recognize the brain deficiencies of both young adults and the elderly. Not only would such a position be politically untenable, particularly because young men between eighteen and twenty-four have a high criminal offence rate, it would dilute any argument that there is something so developmentally special about age eighteen as to justify juvenile treatment for all below that age.

Scholars and advocates understandably have conceded the date's artificiality but point to a societal consensus as to its significance. Some articulate a deeper rationale: that eighteen is a reasonable guess as to when most people will have crossed an important developmental threshold even though they will continue to mature significantly. However, as with relative deficiency, science cannot define that threshold, nor can it tell us precisely when it is likely to have been crossed. Further, other evidence suggests that most adolescents achieve intellectual and cognitive maturity, though not psychosocial maturity, by the mid-teenage years.[50] There is, therefore, *some* law-relevant decisional maturation before eighteen, and it is not yet clear how to harmonize those findings with brain maturation.

---

[48] L. Sabbagh, 'The Teen Brain, Hard at Work, No, Really' (2006) *Sci. Am. Mind* 24 (researchers were 'surprised' at 'how long [the brain] changes into young adulthood').

[49] Declaration of Dr Ruben C. Gur, PhD, *Patterson v Texas* 536 US 984 (2002) (No. 02-6010) at 12–13 (stating also that men experience 'age-associated decline' earlier than women).

[50] L. Steinberg, 'Risk Taking in Adolescence: What Changes, and Why?' (2004) 1021 *Annals N.Y. Acad. Sci.* 51, 54. Competence studies reliably show, for example, that sixteen-year-olds have greater capacity than younger teens for understanding *Miranda* warnings. Thomas Grisso, 'What We Know about Youths' Capacities as Trial Defendants' in Thomas Grisso and Robert G. Schwartz (eds), *Youth on Trial* (Chicago, University of Chicago Press, 2000) 139, 149–50.

Just as neuroscience is not responsible for the difficulties of line-drawing, it does not resolve them.

Courts therefore rightly tend not to see in brain science significant support for a sharp dividing line at age eighteen. Generally this inures to youths' disadvantage, as when courts refuse to second-guess the legislatures' choice of the age at which children face transfer. Sometimes, though, this inures to a defendant's advantage. An unexpected finding of the case analysis is that a good number of the cases reflecting successful brain-based defence arguments involve young adults. An Illinois appellate court, for example, reduced an eighteen-year-old's forty-four-year sentence to thirty-six years, pointing to his great 'rehabilitative potential', and in so doing appeared to endorse expert testimony on brain development.[51] Similarly, in *United States v Gall* a federal district court noted that 'human brain development may not become complete until the age of twenty-five' before granting a downward departure to a man whose offence behaviour occurred before he turned twenty-one and who had demonstrated rehabilitative potential.[52] Other courts have cited developmental neuroscience when granting sentencing concessions to young adults, including one case in which the judge noted—in *dicta*, as the defendant was twenty-two—that he had 'conducted a review of the scientific literature in this area and believes there is compelling evidence that the judicial system's longstanding principle of treating youth offenders differently than adult offenders is justified in part based on the unformed nature of the adolescent brain'.[53]

The fact that such evidence is having at least as much, if not more, influence in young-adult than juvenile cases is striking. Several explanations suggest themselves. First, many of the juvenile challenges were broader, taking on (for example) entire sentencing schemes, while these adult cases were narrow appeals to an allowable exercise of mercy at sentencing. This cannot be the entire story, as some unsuccessful juvenile claims share that characteristic; however, it is buttressed by the fact that most of the marginal juvenile successes also fit that model. Second, perhaps these judges would have taken the same position had the defendants been juveniles, but also believe that evidence of continuing neural development counsels that the relative solicitude historically limited to those under eighteen ought also extend to young adults. Third, and on a deeper level, perhaps juveniles asserting such claims appear to courts to be *unusual juveniles*, that is, more calculating, callous, and dangerous, while these young adults appear to be *unusual adult offenders*, that is, *less* calculating, callous, and dangerous. The developmental attributes thought to stem from brain maturation may seem to conflict with perceptions of the former and to cohere with perceptions of the latter; that is, the perceived relevance of brain science may stem not from its inherent persuasive power but from the degree to which it challenges or confirms perceptions based on other factors.

---

[51] *People v Clark* 869 NE2d 1019, 1042 (Ill App Ct 2007), app. denied, 875 NE2d 1116 (Ill 2007).

[52] 374 F Supp. 2d 758 (SD Iowa 2005), rev'd and remanded, 446 F3d 884 (8th Cir 2006), rev'd, 552 US 38 (2007). That language was approvingly cited by the Supreme Court in upholding the departure.

[53] *United States v Stern* 590 F Supp 2d 945, 953 (ND Ohio 2008).

Thus, the lack of clear age-limit implications for developmental neuroscience poses a challenge to those who seek thus to justify sharp dividing lines benefiting minors.

14.3.2.5 Equality and autonomy commitments

Finally, direct reliance on developmental neuroscience implicates commitments to equality and teen autonomy.

Just as developmental neuroscience might, if taken literally, counsel special treatment of the elderly, it might counsel differential treatment of girls and boys. Brain maturation is importantly linked to puberty, and girls tend to reach puberty significantly earlier than boys.[54] Though physical and sexual maturity are poor proxies for either brain maturity or cognitive development, there is a clear gender differential, likely linked to pubertal onset. Girls, on average, experience early-adolescence neural exuberance—particularly in the frontal lobes—at least a year before boys, and possibly more.[55] If structural brain maturity were the correct legal metric, it would counsel that boys and girls become subject to juvenile-court jurisdiction, and age out of it, at different times; indeed, one testifying expert has conceded as much.[56]

The behavioural implications of brain-level gender differences are largely unknown. Whatever they may be, law should not track them. Indeed, behavioural research already shows that boys and girls have markedly different propensities for violence and lawbreaking, and law rightly does not officially impose more severe punishment for girls' violent acts because they are less normative. While the equality concern is most evident for gender, it is not confined to it. It would apply to any group for whom a statistically significant developmental trend could be identified, including racial or socioeconomic groups. As race is strongly linked to age of pubertal onset—it is well documented, for example, that African-American girls tend to begin puberty much earlier than white American girls—boys and girls of different races might be subject to different rules.[57] Any argument that law's treatment of children should track developmental neuroscience must demonstrate why such inequality is not its logical outcome, and the only way to do so is to concede that neuroscience (and, for that matter, developmental science generally) must sometimes give way to other values.

Undue emphasis on the immature brain also might alter our societal commitment to allow teens incrementally greater control over important aspects of their

---

[54] R. E. Dahl, 'Adolescent Brain Development: A Period of Vulnerabilities and Opportunities' (2004) 1021 *Annals N.Y. Acad. Sci.* 1, 12–16.

[55] Louann Brizendine, *The Female Brain* (New York, Broadway Books, 2006) 44 (female brain 'matures two or three years earlier the male brain'); Strauch, note 3, above, 54 (citing study showing girls' faster myelination); Giedd and others, note 8, above, 862–3 (earlier gray-matter peak in girls corresponds with 'earlier age of onset of puberty').

[56] *Garcia*, note 26, above, Transcript 65 (testimony of Ruben Gur) (because girls' brains mature faster, 'biology would say' that they should be held to a different standard for accountability than boys).

[57] Dahl, note 54, above, 12–13 and fig. 3.

lives—whether to access health services, leave school, marry, exercise their right to free speech, and the like. This issue has been transparent since *Roper*, in which Scalia J, in dissent, excoriated the American Psychological Association for taking what he saw as inconsistent stances on teen maturity in death penalty and abortion cases.[58] Justice Scalia's point was not quite accurate, and the decisional settings are importantly different. As others persuasively have argued, the state can, does, and should distinguish between the competence necessary to make certain critical choices about one's fate—such as whether to have an abortion—and the relative culpability that may justify differential treatment when accused of a crime.[59] But Justice Scalia's rebuke signals an important danger. A strong and simple message about brain immaturity poses a challenge to making complicated and contingent claims about autonomy, and the former easily is interpreted to be in irreconcilable tension with the latter.[60] Adolescent brain science appears (wrongly) to offer far too simple an answer, one that points in most instances away from autonomy.

14.3.3 Advocacy pressures

The previous discussion delineated the intrinsic limitations of developmental neuroscience for juvenile justice. It also raised reasons to be concerned were neuroscience to be given the influence some have urged. One additional concern is intrinsic to all efforts to link law to science. The realities of advocacy, in which nuance and complexity are difficult to convey without compromising effectiveness, incentivize advocates to oversimplify. All scientific data must be simplified for legal or policy arguments, if for no other reason than to render them comprehensible. But simplification easily can creep into oversimplification, creating a risk that legal decisions will be based on incorrect premises. This danger is not unique to juvenile justice, but it has manifested in this context, and its presence counsels great caution.

Consider, for example, how advocates, experts, and commentators tend to characterize teenagers' recruitment of the amygdala, an evolutionarily old brain structure often described as the seat of primitive, aggressive impulses. They consistently assert that teenagers act more 'emotionally' than adults, who are more 'rational', and that

---

[58] *Roper v Simmons* 543 US 551, 617–8 (2005) (Scalia J dissenting).
[59] D. L. Beschle, 'Cognitive Dissonance Revisited: *Roper v Simmons* and the Issue of Adolescent Decision-Making Competence' (2006) 52 *Wayne L. Rev.* 1, 28–9; N. W. Chernoff and M. L. Levick, 'Beyond the Death Penalty: Implications of Adolescent Development Research for the Prosecution, Defense, and Sanctioning of Youthful Offenders' (2005) 39 *Clearinghouse Rev.* 217–18; K. M. Mutcherson, 'Minor Discrepancies: Forging a Common Understanding of Adolescent Competence in Healthcare Decision-Making and Criminal Responsibility' (2006) 6 *Nev. L. J.* 927, 948–53; L. Steinberg and others, 'Are Adolescents Less Mature than Adults? Minors' Access to Abortion, the Juvenile Death Penalty, and the Alleged APA "Flip-Flop"' (2009) 64 *Am. Psychologist* 583, 592–3.
[60] *In re D.L.* No. B205263, 2009 WL 43513, at *3–4 (Cal Ct App 8 January 2009), provides a cautionary tale. A child-dependency judge partially justified his decision to remove an infant from the custody of the twenty-two-year-old father by referencing judicial education programmes, in which the judge learned that brain development is not complete in early adulthood. The judge opined that the father would not have adequate 'judgment' to know what is 'age appropriate' for his child until he was twenty-six years old, at which time he would have a fully mature brain.

such emotionalism explains teens' criminal behaviours. They explain this tendency toward unchecked emotionalism as the consequence of an overactive amygdala that has not yet been tamed by mature, rational frontal lobes.[61] In addition to oversimplifying the complex role of the amygdala, this narrative overstates the behavioural implications of the small number of relevant studies. In these studies, subjects' brains are typically scanned while they view photographs of unfamiliar persons displaying stylized 'fearful' facial expressions; they then are asked to identify the emotion being displayed.[62] This task bears little relation to juvenile offending. Further, though the studies report that teens display more amygdala (and less frontal-lobe) activation than adults when completing the task, the only reported behavioural outcome is teens' higher rate of misidentification of the emotion. That differential may be erased by using colour photographs and including images of people the teens know.[63] Indeed, other studies show that when presented with different tasks teenagers tend to display *greater* frontal-lobe activity than adults;[64] that aggression and violence sometimes correlate with *low* levels of amygdala activation;[65] and that teens have great variation in amygdala response.[66]

In short, the brain's emotional circuitry is highly complex. Teens unquestionably have distinctive emotional experiences. They may well have distinctive neural patterns of emotional activation and of emotion-cognition interaction, and those patterns may well be linked to maturation processes, but to date we know little about these phenomena or their behavioural implications. Teens' emotional lives, and their patterns of criminal behaviour, cannot be reduced to the relative strength of amygdala response; they are shaped by a rich set of factors including social goals and expectations, as well as relative lack of life experience.

It may be tempting to regard the frequently flattened or even distorted portrayal of neuroscience as harmless if it appears to come 'close enough' to the truth for

---

[61] *State v Ninham* 767 NW2d 326, 330 (Wis Ct App 2009) (citing defendant's contention that recent research 'shows adolescents' amygdalas are more active than adults'. The amygdala is closely related to emotionally-laden responses.').

[62] D. A. Yurgelun-Todd and W. D. S. Killgore, 'Fear-Related Activity in the Prefrontal Cortex Increases with Age during Adolescence: A Preliminary fMRI Study' (2006) 406 *Neuroscience Letters* 196 (providing similar data); Frontline, 'Interview: Deborah Yurgelun-Todd' (Public Broadcasting Service Interview, January 2002), <http://www.pbs.org/wgbh/pages/frontline/shows/teenbrain/interviews/todd.html> (accessed 13 January 2010).

[63] Aronson, note 18, above, 123 (stating that researcher questioned extent to which studies revealed 'anything relevant about impulse control'); M. Beckman, 'Crime, Culpability and the Adolescent Brain' (2004) 305 *Science* 596, 599; B. Bower, 'Teen Brains on Trial' (28 May 2004) *Sci. News* 299, 300.

[64] B. Luna and J. A. Sweeney, 'The Emergence of Collaborative Brain Function: fMRI Studies of the Development of Response Inhibition' (2004) 1021 *Annals N.Y. Acad. Sci.* 296, 302; Beckman, note 63, above, 597–9.

[65] Abigail Baird, 'The Developmental Neuroscience of Criminal Behavior' in Nita A. Farahany (ed.), *The Impact of Behavioral Sciences on Criminal Law* (New York, Oxford University Press, 2009) 115–6 (citing, *inter alia*, A. Raine, 'Biosocial Studies of Antisocial and Violent Behavior in Children and Adults: A Review' (2002) 30 *J. Abnormal Child Psychol.* 311).

[66] T. Parker-Pope, 'The Brain of a Bully' (2008) *New York Times Well Blog*, 12 November, <http://well.blogs.nytimes.com/2008/11/12/the-brain-of-a-bully> (accessed 13 January 2010).

legal, not laboratory, purposes. This temptation must be resisted. Inaccuracy has costs. Some are immediate: it may, for example, prompt one's opponent either to offer an equally inaccurate counterclaim (which a court might accept), or successfully to impeach evidence that might have been persuasive were it not being oversold. Some costs, though, cannot presently be anticipated and have wider reach. If, for example, courts were routinely to endorse the 'unchecked-amygdala' explanation for teen behaviour, that endorsement would lend undeserved support to an underlying theory about the low value of 'emotion' relative to 'reason'. That pernicious distinction already infects legal analysis, and it should receive no further encouragement.[67]

* * *

Contrary to many predictions, adolescent brain science has had no deep impact on juvenile justice in the US courts. It has proved generally insufficient to uproot doctrine that tends to disfavour juveniles' claims, particularly when they are accused of serious crimes. While most courts have ignored neuroscientific arguments, some have soundly rejected them, particularly where the individual juvenile appears to be an exception to the argued rule. Those courts that have reacted favourably to arguments about the adolescent brain have done so to buttress conclusions reached on other grounds.

More, these trends are far from irrational. Developmental neuroscience does not shed direct light on the highly individualized determinations that are so commonly at issue in specific criminal cases. Its implications cannot be fully grasped until its link to behaviour is better understood. To tether law to that science creates dangers—inequality, diminished autonomy, and inaccuracy—with no intrinsic hedge. Therefore, adolescent brain science should not on its own meaningfully shift doctrine, even if that shift is normatively desirable. These conclusions closely parallel those other scholars have reached in theorizing the role of neuroscience in adult criminal law.[68] Though insights from that literature have not before now significantly penetrated debates within juvenile justice, they should.

This is a sobering picture. But adolescent brain science nonetheless can play a real—albeit limited—role in moving juvenile justice away from the destructive trends of the last two decades.

---

[67] T. A. Maroney, 'Law and Emotion: A Proposed Taxonomy of an Emerging Field' (2006) 30 *Law & Hum. Behav.* 119, 121–3, 135; T. A. Maroney, 'Emotional Competence, "Rational Understanding," and the Criminal Defendant' (2006) 43 *Am. Crim. L. Rev.* 1375, 1434.

[68] T. R. Brown and E. R. Murphy, 'Through a Scanner Darkly—Functional Neuroimaging as Evidence of a Criminal Defendant's Past Mental States' (2010) 62 *Stan. L. Rev.* 1119; R. M. Sapolsky, 'The Frontal Cortex and the Criminal Justice System' in Semir Zeki and Oliver Goodenough (eds.), *Law and the Brain* (New York, Oxford University Press, 2006) 227–8, 238–40; O. C. Snead, 'Neuroimaging and the "Complexity" of Capital Punishment' (2007) 82 *N.Y.U. L. Rev.* 1265, 1280–99, 1338–9.

14.4 A Limited Role for Adolescent Brain Science within Juvenile Justice

This study's findings confirm the scepticism that many developmental neuroscientists have themselves expressed about the legal relevance of their research.[69] That research is not, however, utterly irrelevant. It contributes marginally to our understanding of general principles about the distinctiveness of adolescence as a developmental stage. General principles matter. The general principle that, as a group, normal young people differ from normal adults in systematic ways directly relevant to their relative culpability, ability to be deterred, and potential for rehabilitation, was for most of the last century invoked to justify differential treatment of juvenile offenders in virtually all instances, with only narrow exceptions. Always supported (if erratically) by everyday observation, these beliefs are now well supported by behavioural and criminological research. In the last two decades the juvenile justice landscape has shifted dramatically as our collective commitment to those principles has eroded, though (as *Roper* showed) that commitment is far from extinguished. The erosion can and should be reversed, and to the limited degree to which brain science helps remind us of these first principles, it is useful.

Adolescent brain science therefore is appropriately considered by legal decision-makers performing a policy-making function. '[A]ggregate data' about youth should be considered when formulating 'policy that will optimize the costs and benefits of treating a large similar group in a particular way.'[70] Because neuroscience generally corroborates the beliefs traditionally undergirding a strong and separate juvenile justice system, it somewhat strengthens the confidence policy-makers can have in those beliefs. If this minor buttressing role is less spectacular than some would hope, it is a real one. More, this role could expand if the science eventually were to show stronger connections between neural structure, neural functioning, and externalized behaviours.

Because legislatures unquestionably are in the best position to reverse the sweeping policy changes of the last two decades, they should be acknowledged as the primary audience. To be sure, legislatures are a tough audience. It is an unfortunate political reality that modern crime policy in the US tends to be a one-way ratchet consistently trending in the direction of more punishment, less judicial discretion, and fewer chances for serious offenders, including young ones. But though such political forces remain strong, very recent developments at the state level show that directing juveniles to the legislatures is far from a fool's errand. Even before *Roper* some states apparently had relied in part on developmental neuroscience to eliminate the juvenile death penalty. Since *Roper*, states have taken

---

[69] Several prominent developmental neuroscientists have taken 'a dim view of the movement to apply neuroscience to the law', and even those who believe that '[b]rain data are eventually going to support reduced legal culpability for adolescents' believed as recently as 2004 that 'we're not quite there yet'. Bower, note 63, above, 301.

[70] S. Pinker, 'My Genome, My Self' (2009) *New York Times Magazine*, 11 January, 24, 50.

additional steps to roll back certain other punitive policies, and some have looked to brain science. Washington State, for instance, in 2005 abolished mandatory sentencing of juveniles convicted as adults, relying in part on a legislative finding 'that emerging research on brain development indicates that adolescent brains, and thus adolescent intellectual and emotional capabilities, differ significantly from those of mature adults'.[71] Most recently, Texas abolished juvenile life without parole after legislative hearings that included testimony about juvenile brain development.[72]

Legislatures, though, are not the only relevant audience. Though it is unfashionable to say so, the courts also are entrusted with a policy-making role.[73] Not all of the extreme deference to legislatures reflected in the study's findings is warranted. Courts must make judgments about youth as a class when they determine, for example, what mental states are 'reasonable' for adolescents; whether the factual assumptions about foresight undergirding the felony-murder doctrine and accomplice liability are irrational when applied to youth; and whether youth are so different from adults as to warrant categorical protection under the Eighth Amendment from extreme, life-long, irrevocable punishments. As such legal determinations inevitably include policy judgments, courts should feel free to take from developmental neuroscience the same modest messages as would a legislature.

Whether directed at courts or legislatures, though, adolescent brain science never should be the primary argument for juvenile justice reform. The real struggle lies elsewhere, and always will.

First, neuroscience is not inherently persuasive. Legal decision-makers (like all people) filter factual assertions, including scientific ones, through their prior beliefs, values, and commitments.[74] Though a number of recent studies show that people unduly credit neuroscientific explanations, even bad ones,[75] this study shows little evidence of such an effect. Instead, it shows that—as this author and others have demonstrated elsewhere—fact-finders tend to accept evidence as relevant and plausible where it aligns with implicit views and judgements and to reject it when it does not.

Sometimes the fact-finder's views are grounded in record evidence. In the Delaware capital case, for example, evidence that the defendant was a good planner made the brain science seem implausible; in contrast, where sentencing courts saw

---

[71] *In re Hegney* 158 P3d 1193, 1208 (Wash Ct App 2007) (quoting Wash. Rev. Code Ann. § 9.94A.540 (West 2003 & Supp. 2008) (historical and statutory notes).
[72] See *Hearing on S.B. 839 Before the Comm. On Criminal Justice*, 2009 Leg., 81st Sess. (Tex 2009) (testimony of Isela Gutiérrez, Juvenile Justice Initiative Director, Texas Criminal Justice Coalition), available at <http://www.criminaljusticecoalition.org/files/userfiles/publicpolicy/SB_839_Testimony.pdf>.
[73] Richard A. Posner, *How Judges Think* (Cambridge, Harvard University Press, 2008) 81–8 (judges are '*occasional legislators*' whose policymaking powers are greatest in 'legalistically indeterminate' cases).
[74] D. M. Kahan and others, 'Whose Eyes Are You Going to Believe? *Scott v Harris* and the Perils of Cognitive Illiberalism' (2009) 122 *Harv. L. Rev.* 837, 842; T.A. Maroney, 'Emotional Common Sense as Constitutional Law' (2009) 62 *Vand. L. Rev.* 851, 885–6.
[75] D. P. McCabe and A. D. Castel, 'Seeing is Believing: The Effect of Brain Images on Judgments of Scientific Reasoning' (2008) 107 *Cognition* 343, 344; D. Skolnick Weisberg and others, 'The Seductive Allure of Neuroscience Explanations' (2008) 20 *J. Cognitive Neuroscience* 470, 470.

credible evidence that particular young-adult defendants had matured, they credited the brain science and found in it a plausible explanation. But more abstract background beliefs, too, play a filtering role. Legal actors evaluate brain science through implicit political, cultural, or role-based perspectives that predispose them to favour or disfavour juveniles' claims. That phenomenon may explain why juvenile advocates and defenders have tended wholeheartedly to embrace neuroscience and to take a broad view of its implications, while prosecutors have tended to take just the opposite tack, acknowledging the basic empirical points about structural maturation but displaying extreme scepticism as to its relevant behavioural implications. Judges and juries, too, necessarily come to juvenile cases with implicit views. It is noteworthy that in every instance in which a court positively cited developmental neuroscience, it did so as part of a roster of reasons why it would reach a particular result. Not only were the other items on the roster sufficient to justify the result, the fact that the court credited them is one reason why it also found the science relevant.[76] Developmental neuroscience is not materially shifting beliefs and values; it is instead being read through the lens of those beliefs and values.

The real task, then, for those seeking juvenile justice reform is to influence such beliefs, values, and inclinations directly, rather than expect such influence to flow naturally from explanation of neuroscience. While there is no simple formula for that task, it has long been the bread and butter of juvenile justice scholarship and advocacy. It includes demonstrating the ways in which teens are developmentally distinct, but also educating the public and legal decision-makers about the real-world effects of juvenile policy and what 'works' from a utilitarian perspective. Such messages suffer from few of the vulnerabilities attending brain science. The strongest challenge to transfer schemes, for example, has nothing to do with the juvenile brain and everything to do with robust data consistently showing that transfer to adult court increases recidivism and that many youth transferred to adult court are accused not of serious interpersonal violence but of property crimes. The public might be even more moved if they fully understood how frequently incarceration with adults leads to physical and sexual abuse. Strict 'zero tolerance' policies in US schools are becoming increasingly unpopular because they lead to patently absurd results. Attitudes about JLWOP might be swayed by stories of youth who have grown into different people, and yet necessarily will die in prison; juvenile sex offender registration may seem less palatable if the public were to learn about the range of youth on such lists (including, for example, pre-adolescents who engaged in inappropriate 'play' and have responded well to treatment) and what registration does to their futures.

Developmental principles, in short, tend to draw our attention inward. We need, too, to maintain a clear view of the world within which youth develop. Societal factors such as stable families, safe housing, medical and mental health care, good

---

[76] Steinberg has argued that the same phenomenon is true for legislatures, who 'often look to science for evidence that supports a position they have take for other reasons'. L. Steinberg, 'Should the Science of Adolescent Brain Development Inform Public Policy?' (2009) 64 *Am. Psychologist* 739, 745.

schools, and economic opportunities—all subject to relatively direct societal control—will continue to be the most important contributors to juvenile offending, and they should continue to receive the lion's share of attention. This is particularly so because a disproportionate focus on the teen brain tends to support a false notion that teens' propensity to offend is 'hard-wired,' a view that not only makes societal reform seem pointless but, by implying the impossibility of deterrence, could support needless incapacitation of many youth until their brains 'grow up'.[77]

Educating the public and policy-makers about teen brain development need not devolve into such counterproductive reductionism; instead, understanding the brain's 'biological processes can actually enhance the importance of behavioural or social policy interventions' by highlighting the extraordinary impact of environment during a critical period of development.[78] Conceptualizing neuroscience as background rather than foreground keeps us collectively focused on creating the conditions necessary for youth to become healthy, productive adults—including those youth who have committed serious offences.

14.5 Conclusion

This paper tells a cautionary tale. Relying aggressively on developmental neuroscience in legal theory and practice might wear out its welcome early, even though it now offers some law-relevant insights and in the future might offer more. The courts' early cold shoulder shows this to be a real danger. Nor is such reliance necessary, as we already have all the information we need to construct a rational juvenile justice policy. Adolescent brain science does not provide an independent basis to recommit to traditional juvenile justice values; it merely reinforces the wisdom of doing so. The bulk of that wisdom comes not from understanding what is going on inside the teen brain but from understanding the impact of the legal and social environments we create for young people.

---

[77] B. Smaller, 'Cartoon' (2006) *The New Yorker*, 24 April, 129 (showing parent disciplining teenage son by saying, 'Young man, go to your room and stay there until your cerebral cortex matures'); see also E. Buss, 'Rethinking the Connection Between Developmental Science and Juvenile Justice' (2009) 76 *U. Chi. L. Rev.* 493, 509–10 (noting the danger that by deferring too heavily to developmental principles, including brain science, law 'will lock in a developmental status quo', and asserting that 'law can shape development instead of the other way around').

[78] Dahl, note 54, above, 4.

15

The Neuroscience of Cruelty as Brain Damage: Legal Framings of Capacity and Ethical Issues in the Neurorehabilitation of Motor Neurone Disease and Behavioural Variant Frontotemporal Dementia

*Robin Mackenzie and Mohamed Sakel**

15.1 Introduction

This paper explores some ethical and legal issues arising from the impact of the neuroscience of decision-making and emotion on end-of-life decisions of patients with certain neurodegenerative disorders.

The neurorehabilitation of patients with neurodegenerative disorders raises specific ethical and legal issues, including perceptions of past, present and future identity together with the impact of neurological damage upon family/carers' burden and end-of-life decision-making processes. Clinical factors impact on these in unexpected fashions. To illustrate these issues, we focus on what in Britain is designated motor neurone disease (MND), known in the US as amyotrophic lateral sclerosis (ALS). We prefer British terminology throughout this article. Despite advances in neuroimaging, aetiology and mechanisms of MND remain uncertain. Nonetheless, diagnosis is relatively straightforward, involving spreading upper and lower motor neuronal degeneration with eventual complete loss of neuromuscular control. MND is associated with frontotemporal dementia (FTD), in that both conditions are associated with a familial component, in that where one family member has MND or FTD, others are more likely to develop one or both.[1] Frontal lobar dysfunction is present in 28-48

---

* Robin Mackenzie, Director, Medical Law & Ethics Centre, University of Kent; Dr Mohamed Sakel, Director, Neurorehabilitation Services and Director of Research and Development, East Kent University Hospital Trust.

[1] John Hardy et al., 'Fronto Temporal Dementia: Dissecting the Aetiology and Pathogenesis' (2006) 129 *Brain* 830.

per cent of MND patients,[2] and up to half will have mild or more serious cognitive impairment.[3] A subtype of FTD, behavioural variant FTD (bvFTD), is an under-diagnosed or misdiagnosed condition where symptoms include the likelihood that patients' behaviour is experienced by others as cruel, generally unpleasant and lacking in empathy. The symptoms of the other subtypes of FTD, progressive non-fluent asphasia (PNFA) and semantic dementia (SD), present as linguistic difficulties, easily distinguishable from MND or bvFTD.

Although the common perception of a person with MND is that of someone with an unaffected, lucid mind in a degenerating body, cognitive difficulties are prevalent in around half of cases.[4] Cognitive deficits and language impairments may be present before, after or at the onset of neuromuscular degeneration symptoms in MND.[5] These frequently remain undiagnosed, overshadowed by the more dramatic symptoms of neuromuscular degeneration or masked by confounding factors such as depression, side-effects of medications, sleep difficulties, fewer task demands, communication difficulties, lack of standardized measures and rationalisations, denial or ignorance of the part of family/carers and healthcare providers. The linguistic difficulties associated with PNFA and SD may key inquiries into cognitive impairment. When bvFTD is present, either in combination with MND or alone, however, cognitive impairment is unlikely to be either immediately evident or investigated. Those with bvFTD alone do not experience neuromuscular degeneration. Under-recognition and under-diagnosis of bvFTD means that those afflicted with it may appear to be people who may have become more disagreeable but whose decision-making capacity is unchanged. Nonetheless, they typically suffer from characteristic and measurable brain damage.

This is likely to have a significant impact on neurorehabilitation and end-of-life decision-making. Family/carers involved in rehabilitation treatment at home and in interpreting or reporting patient wishes are placed in a position of ongoing disproportionate stress. Healthcare professionals involved in neurorehabilitation and end-of-life decision-making typically regard family/carers' participation in both as critical for success.[6] Legal and policy frameworks covering capacity and treatment envisage and promote family/carer involvement.[7] We explore some legal difficulties, ethical issues and conflicts of interest that arise. We suggest that the current means of assessing capacity to make decisions over medical treatment

---

[2] C. Lomen-Hoerth and M. J. Strong, 'Frontotemporal Dysfunction in Amyotrophic Lateral Sclerosis' in H. Mitsumoto et al. (eds.), *Amyotrophic Lateral Sclerosis* (New York, Oxford University Press, 2006) 117.

[3] G. M. Ringholz et al., 'Prevalence and Patterns of Cognitive Impairment in Sporadic ALS' (2005) 65 *Neurology* 1546.

[4] Hiroshi Mitsumoto and Judith C. Rabkin, 'Palliative Care for Patients with Amyotrophic Lateral Sclerosis: "Prepare for the Worst and Hope for the Best"' (2007) 298 *Journal of the American Medical Association* 207.

[5] D. Irwin et al., 'Cognition and Amyotrophic Lateral Sclerosis (ALS)' (2007) 22 *American Journal of Alzheimer's Disease & Other Dementias* 300.

[6] Note 4, above.

[7] See, *inter alia*, Department of Health, *Carers at the Heart of 21st Century Families and Communities* (London, Department of Health, 2008).

should be read as measuring emotional as well as cognitive factors, that the ethics of clinician/carer relations should be monitored carefully to minimize conflicts of interest and exploitation and that bvFTD exemplifies certain difficulties arising with advance decisions to refuse treatment (ADRTs) that must be resolved. We conclude with a case study demonstrating these issues.

15.2 Clinical Aspects of MND, bvFTD and the Dementias

Neuroimaging has transformed diagnostic and nosological possibilities for many clinical conditions. MND and the dementias exemplify this in that distinctions may now increasingly be made between conditions previously grouped together on the basis of overlapping clinical features and post-mortem findings.[8] This is particularly welcome as previously diagnoses were problematic in so far as they depended upon clinicians distinguishing various forms of dementia from age-typical cognitive decline and/or memory loss according to symptoms presented. Behavioural and cognitive deficits characterize all dementias; i.e. Alzheimer's disease (AD), vascular dementia (VAD), vascular cognitive impairment (VCI), dementia with Lewy bodies (DLB) and Parkinson's disease with dementia (PDD) as well as frontotemporal lobe dementias (FTLDs).[9] AD accounts for between 50–75 per cent of all late-life dementias, VCI/VAD causes or contributes to up to 80 per cent, DLB is found in up to 25 per cent of autopsies of older adults and FTLDs are the second most common cause of early onset dementia and the third most common neurodegenerative syndrome after AD and DLB.

Each form of dementia has a characteristic cognitive-behavioural profile,[10] which diagnostic criteria compare with AD as the gold standard. AD patients exhibit memory loss and an inability to encode and consolidate new information. Despite this, appropriate social behaviour is maintained and affectionate bonds sustained until the last stages, when a third may suffer from mood disturbances or delusions such as depression, anxiety and paranoia. VCI/VAD post-stroke dementia is associated with psychomotor slowing and executive deficits such as difficulties with understanding or appreciating task demands until the task is completed, reframing situations, initiating, planning and self-regulation. This may be accompanied by flattened emotions, depression, withdrawal, lack of motivation and anxiety. DLB, which overlaps with AD and PDD, is characterized by fluctuating cognition, varying attention and alertness, visual, auditory and olfactory hallucinations, delusions and depression. PDD shares the difficulties associated with cognition and attention but these are accompanied by fewer hallucinations. FTLD may

---

[8] Kurt A. Jellinger, 'Criteria for the Neuropathological Diagnosis of Dementing Disorders: Routes Out of the Swamp' (2009) 117 *Acta Neuropathologica* 101.
[9] James A. Levy and Gordon J. Chelune, 'Cognitive-Behavioural Profiles of Neurodegenerative Dementias: Beyond Alzheimer's Disease' (2007) 20 *Journal of Geriatric Psychiatry and Neurology* 227.
[10] Alissa H. Wicklund et al., 'Rate of Cognitive Change Measured by Neuropsychologic Test Performance in Three Distinct Dementia Syndromes' (2007) 21 *Alzheimer's Disease and Associated Disorders* S70.

be subdivided into PNFA, SD and bvFTD. There is some overlap with AD, in that up to 20 per cent of AD patients have neuropathological markers for FTLD,[11] but MRI scans show distinct patterns of brain atrophy.[12] PNFA and SD are characterized by linguistic difficulties while bvFTD, compared to AD, has higher levels of apathy, disinhibition, emotional flatness and lack of empathy. Measures of social function enable distinctions between patients with bvFTD and psychiatric or AD patients.[13] Nonetheless, a degree of overlap exists not only between MND and FTD, but also between FTD and AD, in that up to 20 per cent of patients with AD possess neuropathological markers for FTD, providing more evidence for the under-diagnosis or misdiagnosis of FTD.[14]

MND and associated conditions are currently in a process of taxonomic reform.[15] Traditionally MND has been conceptualized as a neurodegenerative disorder targeting specific aspects of the motor system leading to decline and death, with a clinical overlap where FTD manifests as well or instead in some patients on a familial basis.[16] This framing of MND as solely a motor system disorder is now seen as inadequate and misleading.[17] The very recent current understanding, based on consensus criteria established in 2007, is of MND as a multi-system disorder that may manifest in motor system or non-motor system deficits. This clinicopathological spectrum of disease is associated with TDP-43 proteinopathies.[18] Consequent impairments in frontotemporal functions may manifest in a spectrum of cognitive and behavioural deficits collectively known as syndromes of frontotemporal dysfunction, or FTD. FTD incorporates behavioural variant FTD (bvFTD), progressive non-fluent aphasia (PNFA) and semantic dementia (SD).[19] Each has characteristic features which may be captured by neuroimaging, neuropathology and molecular genetics.[20] The new consensus criteria recommend that 'frontotemporal lobar degeneration (FTLD) be retained as general terminology for pathological conditions that are commonly associated

---

[11] Jill S. Goldman et al., 'New Genes, New Dilemmas: FTLD Genetics and its Implications for Families' (2007/8) 22 *American Journal of Alzheimer's Disease & Other Dementias* 507.

[12] G. D. Rabinovici et al., 'Distinct MRI Atrophy Patterns in Autopsy Proven Alzheimer's Disease and Frontotemporal Lobar Degeneration' (2008) 22 *Journal of Alzheimer's Disease & Other Dementias* 474.

[13] Dana Wittenberg et al., 'The Early Neuropsychological and Behavioural Characteristics of Frontotemporal Dementia' (2008) 18 *Neuropsychological Review* 91.

[14] Note 11, above.

[15] Felix Geser et al., 'Amyotrophic Lateral Sclerosis, Frontotemporal Dementia and Beyond: the TDP-43 Diseases' (2009) *Journal of Neurology* 1205.

[16] Ian R. Mackenzie, 'The Neuropathology of FTD Associated with ALS' (2007) 21 *Alzheimer's Disease and Associated Disorders* S44.

[17] Katya Rascovshy et al., 'Diagnostic Criteria for the Behavioural Variant of Frontotemporal Dementia (bvFTD): Current Limitations and Future Directions' (2007) 21 *Alzheimer's Disease and Associated Disorders* S14.

[18] M. Strong, 'Consensus Criteria for the Diagnosis of Frontotemporal Cognitive and Behavioural Syndromes in Amyotrophic Lateral Sclerosis' (2008) 9 *Amyotrophic Lateral Sclerosis* 252.

[19] Keith A. Josephs et al., 'Evaluation of Subcortical Pathology and Clinical Correlations in FTLD-U Subtypes' (2009) 117 *Acta Neuropathologica* 349.

[20] Keith A. Josphs, 'Frontotemporal Dementia and Related Disorders: Deciphering the Enigma' (2008) 64 *Annals of Neurology* 4.

with the clinical entities of FTD, PNFA and/or SD, and in which degeneration of the frontal and temporal lobes is a characteristic feature'.[21]

In order to find a common framework to investigate FTDs in MND, the latest consensus criteria agreed on four axes: 1) the motor neurone diseases variant, characterized by progressive deterioration of the motor neurones; 2) cognitive and behavioural dysfunction; 3) additional non-motor disease manifestations; and 4) presence of disease modifiers. Cognitive and/or behavioural dysfunction may not afflict all MND patients, but is commonly present as impairment or as dementia signalled by FTLD. Both are under-diagnosed in MND, as variation in daily decision-making, impulsivity and emotional lability are likely to be overshadowed by more dramatic neurodegenerative symptoms. In addition, they may not be recognized as symptoms but viewed as responses to the diagnosis or as personality characteristics. Given that two sets of criteria for the diagnosis of bvFTD (now considered outmoded)[22] include dramatic and florid symptoms, which are seldom found together,[23] more subtle manifestations that are within the range of normal are less readily discerned, so may not be identified as symptomatic of frontotemporal dysfunction. Diagnostic difficulties are increased by the fact that the profound behavioural and personality changes associated with bvFTD tend to be unique to each individual.[24]

Established criteria for the diagnosis of bvFTD are constraining and ambiguous in so far as they are open to clinical interpretation and rely upon family/carer reports. A multi-pronged approach incorporating behavioural features, measures of brain atrophy, neuropsychological abnormalities and impaired activities in daily living would promote accurate diagnosis.[25] The most frequent frontal lobe degenerations are associated with both cognitive and behavioural dysfunction, i.e. executive dysfunction impacting on abilities to organize information, shift attention and inhibit behaviour. FTD typically manifests as bvFTD, 'a progressive behavioural syndrome marked by insidious onset, altered social conduct, impaired regulation of interpersonal conduct, emotional blunting and loss of insight' (p. 6).[26] There is some evidence that bvFTD may manifest in two ways. Poor planning and/or judgement are associated with tau-positive pathology and impaired personal

---

[21] Ian R. Mackenzie et al., 'Nomenclature for Neuropathologic Subtypes of Frontotemporal Lobar Degeneration: Consensus Recommendations' (2009) 117 *Acta Neuropathologica* 15.

[22] Note 17, above.

[23] The Neary criteria include decline in personal hygiene and grooming, mental rigidity and inflexibility, distractibility and impersistence, hyperorality and dietary changes, perseverative and stereotyped behaviour and utilization behaviour, while Hodge's criteria include loss of insight, disinhibition, restlessness, distractibility, reduced empathy or unconcern for others, lack of foresight or planning, impulsiveness, social withdrawal, apathy or loss of spontaneity, reduced verbal output, verbal stereotypes or echolalia, verbal or motor perseveration, poor self care, gluttony and sexual hyperactivity; note 18, above, 7. It is anticipated that the consensus criteria will supercede previous criteria; note 13, above.

[24] Jennifer Merrilees, 'A Model for the Management of Behavioural Symptoms in Frontotemporal Lobar Degeneration' (2007) 21 *Alzheimer's Disease and Associated Disorders* S64.

[25] O. Piguer et al., 'Sensitivity of Current Criteria for the Diagnosis of Behavioural Variant Frontotemporal Dementia' (2009) 72 *Neurology* 732.

[26] Note 18, above.

conduct and a decline in personal hygiene with tau-negative.[27] Less common varieties of FTD, PNFA and SD, are associated with speech and language disorders and left-sided temporal atrophy. Anxiety, irritability and apathy may develop over time in these patients.[28] Atrophy of the right-sided temporal lobe, the right hippocampus and the right amygdale is associated with bvFTD.[29]

Caring for bvFTD patients is stressful in that they typically show little insight into their personality changes, exaggerating their positive qualities and minimizing any negative qualities.[30] Behavioural symptoms in bvFTD may include 'disinhibition marked by impulsivity, distractibility, and/or poor quality of social interactions... notable apathy or an alteration in emotional expressiveness. Caregivers may describe personality changes such as irritability, selfishness or disinterest that are inconsistent with pre-illness disease states (7)'.[31] For diagnosis of bvFTD, according to the consensus criteria, reports from family/carers are essential in order to establish a baseline in terms of previous ways of behaving for comparison.

Nonetheless, diagnostic difficulties remain. Frontotemporal atrophy is considered a normal part of ageing and there is considerable overlap between right-sided atrophy in the medial middle frontal region of diagnosed bvFTD patients and normal age-matched controls.[32] Diagnosis of bvFTD on the basis of behaviour can take up to five years.[33] All the required behavioural features for a bvFTD diagnosis may be lacking in up to two-thirds of patients subsequently diagnosed with bvFTD.[34] Patients with bvFTD, particularly younger people with early-onset bvFTD,[35] may present with psychiatric symptoms distinct from behaviours triggering diagnosis as bvFTD.[36] These manifest as dysthymia and anxiety, associated with right temporal hypoperfusion and moria or frivolous behaviour associated with other right temporal lobe changes. Neuropsychological measures and consensus criteria may thus lack diagnostic sensitivity for FTD.[37]

---

[27] William T. Hu et al., 'Clinical Features of Pathologic Subtypes of Behavioural-Variant Frontotemporal Dementia' (2007) 64 *Archives of Neurology* 1611.

[28] Sarah J. Banks and Sandra Weintraub, 'Neuropsychiatric Symptoms in Behavioural Variant Frontotemproal Dementia and Primary progressive Aphasia' (2008) 21 *Journal of Geriatric Psychiatry* 133.

[29] Dennis Chan et al., 'The Clinical Profile of Right Temporal Lobe Atrophy' (2009) 132 *Brain* 1287.

[30] E. Salmon et al., 'A Comparison of Unawareness in Frontotemporal Dementia and Alzheimer's Disease' (2007) 79 *Journal of Neurology, Neurosurgery and Psychiatry* 176.

[31] Note 18, above.

[32] Tiffany W. Chow et al., 'Overlap in Frontotemporal Atrophy Between Normal Aging and Patients with Frontotemporal Dementias' (2008) 22 *Alzheimer's Desease & Associated Disorders* 327.

[33] Tor Atlee Rosness et al., 'Frontotemporal Dementia: a Clinically Complex Diagnosis' (2008) 23 *International Journal of Geriatric Psychiatry* 837.

[34] M. F. Mendez and K. M. Perryman, 'Neuropsychiatric Features of Frontotemporal Dementia: Evaluation of Consensus Criteria and Review' (2002) 14 *Journal of Neuropsychiatry and Clinical Neuroscience* 424.

[35] D. Velakoulis et al., 'Frontotemporal Dementia Presenting as Schizophrenia-Like Psychosis in Young People: Clinicopathological Series and Review of Cases' (2009) 194 *British Journal of Psychiatry* 298.

[36] M. F. Mendez et al., 'Functional Neuroimaging and Presenting Psychiatric Features in Frontotemporal Dementia' (2006) 77 *Journal of Neurology, Neurosurgery and Psychiatry* 4.

[37] Mario F. Mendez et al., 'Accuracy of the Clinical Evaluation for Frontotemporal Dementia' (2007) 64 *Archives of Neurology* 830.

While neuroimaging may promote specificity,[38] it also reveals a conundrum. Diagnosis of bvFTD is based upon the reports of family/carers reporting changes in behaviour, yet this is not always borne out by neuroimaging results. MRI scans show two groups of those diagnosed with bvFTD on the basis of identical behavioural characteristics, those with frontotemporal abnormalities and progressive degeneration and those without either whose condition remains stable.[39] Those with abnormalities are impaired in their recognition of sarcasm and emotion, whereas those without are not.[40] This applies only to those diagnosed with bvFTD: while MRI scans revealed abnormality in a majority of FTD patients (75 per cent), focal atrophy was present in 100 per cent of SD patients but 47 per cent of bvFTD patients had scans in the normal range.[41] Patients diagnosed with bvFTD with normal MRI results did not show the shortened life expectancies of the group with frontotemporal lobe atrophy. This does not conform to the typical pattern of selective vulnerability progressing over time.[42] Moreover, patients with identical pathologies may present clinically in very different ways throughout the progression of the disease.[43]

This suggests that the overlap between the behaviour characterizing bvFTD and impolite or impulsive behaviour regarded as being within the norm, particularly as part of the ageing process, may contribute to the under-diagnosis, misdiagnosis or over-diagnosis of bvFTD. The difficulties of diagnosing bvFTD are thus distinct from those of diagnosing AD, as the memory loss that inevitably accompanies AD is less readily able to be seen as normal or subsumed under cognitive decline accompanying ageing. Without diagnostic biomarkers, core or necessary behaviours must be recognized to ground a clinical diagnosis of bvFTD, yet clinicians often misdiagnose symptoms as indicating psychiatric psychopathologies.[44] Reliance on family/carers as informants for diagnosis of bvFTD is thus problematic in several ways. Some who have bvFTD will not be diagnosed, as their unpleasant behaviour will be seen as part of the ageing process and not drawn to clinicians' attention. Even where clinical opinion is sought, some will be misdiagnosed as mentally ill. Others, even where bvFTD is considered, will remain undiagnosed as if around half of those diagnosed with bvFTD show no MRI abnormalities, the

---

[38] Norman L. Foster et al., 'FDG-PET Improves Accuracy oin Distinguishing Frontotemporal Dementia and Alzheimer's Disease' (2007) 130 *Brain* 2626.

[39] Rhys R. Davies et al., 'Progression in Frontotemporal Dementia: Identifying a Benign Behavioural Variant by Magnetic Resonance Imaging' (2006) 63 *Archives of Neurology* 1627.

[40] C. M. Kipps et al., 'Understanding Social Dysfunction in the Behavioural Variant of Frontotemporal Dementia: the Role of Emotion and Sarcasm Processing' (2009) 132 *Brain* 592.

[41] Christopher M. Kipps, 'Clinical Significance of Lobar Atrophy in Frontotemporal Dementia: Application of an MRI Visual Rating Scale' (2007) 23 *Dementia & Geriatric Cognitive Disorders* 334.

[42] Jillian J. Kril et al., 'Distribution of Brain Atrophy in Behavioural Variant Frontotemporal Dementia' (2005) 232 *Journal of the Neurological Sciences* 83.

[43] William T. Hu et al., 'Clinical Features and Survival of 3R and 4R Taupathies: Presenting as Behavioural Variant Frontotemporal Dementia' (2007) 21 *Alzheimer's Disease and Associated Disorders* S39.

[44] Mario F. Mendez et al., 'An Evidence Based Review of the Psychopathology of Frontotemporal Dementia: a Report of the ANPA Committee on Research' (2008) 20 *Journal of Neuropsychiatry and Clinical Neurosciences* 130.

presence of the underlying neuropathology is cast into doubt.[45] In addition, ageing people who wish to conduct their lives in ways that seem frivolous, irresponsible or otherwise undesirable may be wrongly diagnosed with bvFTD on the basis of reports from family/carers who are unsympathetic to these ways of being.

Ethical and legal issues arise as a result of the diagnostic difficulties associated with the dementias and MND as outlined above. These centre on changes in personality and identity of patients, family/carer interests and decision-making competence and capacity, especially at end-of-life. As the memory loss and linguistic difficulties associated with AD, VCI/VAD, DLB/PDD, PNFA and SD do not exemplify these difficulties as clearly as bvFTD, we have chosen to focus on bvFTD and MND. These will now be considered.

15.3 Changes in Personality and Identity of Patients and Family/Carer Interests

Changes of personality and identity resulting from cognitive and affective changes taking place through illness or injury are common within neurorehabilitation. Both patients and family/carers may struggle with accepting that the patient has lost skills or personal characteristics that may not return. This is distinct from the research within medical sociology scholarship that documents patients' adopting illness as an identifying characteristic.[46] Rather, neurorehabilitation treatment aims to achieve the highest possible level of functioning post-injury or during each stage of a neurodegenerative illness. How the patient and family/carers conceive of this ongoing transformation of identity and personality within neurorehabilitation affects the treatment process. More specifically, the finding of meaning in the adoption of new identities, or a narrative that links rehabilitation with an heroic struggle overcoming odds, is associated with more successful treatment outcomes.[47]

Where the changes in personality and identity result from incidents such as traumatic brain injury, spinal cord injury or a stroke, the crafting of a narrative of transformation is easier in that it is tied to recovery from a specific event. Cognitive and/or affective changes are expected as a result of that event. They may be measured and monitored as part of neurorehabilitation treatment. For instance, psychiatric symptoms like depression or anxiety occurring after the event may be acknowledged as a potential outcome, investigated and treated if necessary. Neurodegenerative illnesses associated with alterations in personality and identity provoked by cognitive and/or affective changes are in a different category. Where motor symptoms, memory loss or linguistic disorders are part of the process, as in MND, AD or VCI/VAD, DLB/PDD, PNFA and SD, these key their recognition

---

[45] Note 40, above.
[46] Gayle A. Sulik, 'Managing Biomedical Uncertainty: the Technoscientific Illness Identity' (2009) 31 *Sociology of Health & Illness* 1059.
[47] J. Landau and J. Hissett, 'Mild Traumatic Brain Injury: Impact on Identity and Ambiguous Loss in the Family' (2008) 26 *Family Systems & Health* 69.

as symptoms of illness, distinct from manifestations typical of the ageing process. Hence diagnosis is likely to be sought and obtained. When cognitive and/or affective alterations accompany more dramatic changes, like motor neurodegeneration in MND, or underpin behavioural changes within the range of normal, if unpleasant, social conduct, as in bvFTD, they are significantly less likely to be recognized as pathological so are likely to remain under-diagnosed.

Cognitive and/or affective changes occurring in MND may be interpreted, if they are acknowledged at all, as understandable depressive reactions to the diagnosis of a fatal neurodegenerative disease and treated, if at all, with anti-depressants. In bvFTD, as the behavioural symptoms keyed by affective changes initially fall within the range of unpleasant social and personal conduct, those with the condition are likely to be viewed by family/carers as behaving badly or in an unfeeling fashion deliberately. Unlike symptoms indicating psychiatric distress or bodily dysfunction, these behavioural symptoms are unlikely to key a search for diagnosis until or unless they become clearly abnormal. As those with bvFTD lack insight and do not recognize or acknowledge their affective changes,[48] they are unlikely of their own accord to initiate clinical consultations and to resist family/carers suggesting that they may be ill.

One consequence of this is that the narratives constructed by family/carers as a result of the cognitive and/or affective changes associated with MND and bvFTD are likely to feature those suffering from them not as heroes overcoming the odds but as the opposite, i.e. as having deliberately chosen to behave in selfish, cruel, cold and uncooperative fashion. When behaviour is framed as intentional unpleasantness rather than as an indication of illness, neurorehabilitative input based on diagnosis is foreclosed. Moreover, family/carers are unlikely even to raise the question of whether those afflicted are unable to feel empathy, altruism and presumably affection, as opposed to choosing not to do so. Thus the changes in identity and personality which occur are not perceived by those afflicted with them, while family/carers view them in pejorative terms. The salient changes and how they affect family/carer interests will now be considered in more depth.

How far the cognitive and the affective may legitimately be regarded as separate will be explored further below in relation to capacity and competence. In the following discussion, the distinction accepted within clinical assignment of symptoms amongst the MND spectrum and the dementias will be followed. Clinical ascription of symptoms accompanying MND and bvFTD reveal an increasing acknowledgement of a mixture of cognitive and affective dysfunctions accompanying at least half of MND patients. For those with bvFTD, however, cognition typically remains intact while behavioural changes manifest as a result of affective alterations. These centre in atrophies in the right frontotemporal network driving emotional processing.

Higher emotional abilities associated with self-processing, such as the ability to monitor the self, self-awareness and the capacity to place the self in a social context

---

[48] F. M. O'Keeffe et al., 'Loss of Insight in Frontotemporal Dementia, Corticobasal Degeneration and Progressive Supranuclear Palsy' (2007) 130 *Brain* 753.

and behave appropriately are lost.[49] These constitute metacognitive, as opposed to cognitive defects; i.e. the capacity to reflect upon one's own thinking and actions is impaired while the ability to exercise cognitive skills unrelated to the self, like arithmetic calculations remains unaltered. Interpreting the behaviour of others also becomes problematic: bvFTD patients are unable to interpret the facial or vocal expressions of others, particularly those associated with distress, such as sadness, fear and anger.[50] Thus they lack the feedback from others' reactions to them which is considered to form the basis of emotional socialization and empathy. Without access to their own and others' emotional states and pain, they become characterised by emotional shallowness, cold-heartedness and indifference to others.[51] However, their ability to experience their own emotions, or emotional reactivity, remains unimpaired.[52]

They are often said to lack empathy. The right anterior temporal and medial frontal regions are essential for empathetic behaviour. Empathy involves three steps, firstly, where the other's emotion is experienced in an activation of brain areas involved in subjective emotional experience; secondly, where a theory of mind (the ability to represent the thoughts and feelings of others) means that the other can be recognized as the source of the emotion; and thirdly, where one's own viewpoint is suppressed intentionally and the other's considered.[53] All three elements are compromised in bvFTD. Moreover, hypometabolism in specific areas is associated with unsuitable social behaviours: bvFTD patients with right anterior temporal hypometabolism have impaired ability to comprehend and manifest appropriate interpersonal behaviour.[54] In some patients, these characteristics of bvFTD may impact on executive dysfunction in the sense of impairments in planning, attention, organization, abstraction, problem-solving, judgement and mental flexibility, although some suggest these difficulties may manifest as a result of deficits in the theory of mind, or the ability to conceive of others as separate entities, independent of the level of executive functioning.[55] Certainly those with bvFTD lose cognitive empathy, or the theory of mind that enables the recognition of the perspectives of others, as opposed to healthy people of the same age, or those with AD who retain theory of mind capabilities.[56] The ability of those with bvFTD

---

[49] Virginia E. Sturm et al., 'Self-conscious Emotion Deficits in Frontotemporal Lobar Degeneration' (2006) 129 *Brain* 2508.
[50] Howard J. Rosen et al., 'Emotional Comprehension in the Temporal Variant of Frontotemporal Dementia' (2002) 125 *Brain* 2286.
[51] Matthias L. Schroeter et al., 'Neural Networks in Frontotemporal Dementia: a Meta-analysis' (2006) 29 *Neurobiology of Aging* 418.
[52] K. H. Werner et al., 'Emotional Reactivity and Emotion Recognition in Frontotemporal Lobar Degeneration' (2007) 69 *Neurology* 148.
[53] Katherine P. Rankin et al., 'Structural Anatomy of Empathy in Neurodegenerative Disease' (2006) 129 *Brain* 2945.
[54] Roland Zahn et al., 'Social Conceptual Impairments in Frontotemporal Lobar Degeneration with Right Anterior Temporal Hypometabolism' (2009) 132 *Brain* 604.
[55] Sinclair Lough et al., 'Dissociation of Social Cognition and Executive Function in Frontal Variant Frontotemporal Dementia' (2001) 7 *Neurocase* 123.
[56] Katherine P. Rankin et al., 'Patterns of Cognitive and Emotional Empathy in Frontotemporal Lobe Degeneration' (2005) 18 *Cognitive & Behavioral Neurology* 28.

to engage in moral reasoning is compromised through an inability to judge the seriousness of moral and conventional transgressions.[57] Moreover, as temporal lobe damage typifying bvFTD impairs understanding of social concepts such as 'stingy' or 'polite',[58] those with bvFTD may find incomprehensible remonstrations aimed at amending the behavioural consequences of this loss of cognitive empathy.

As a consequence of all these dysfunctions, those afflicted with bvFTD are significantly impaired in their ability to make moral judgements, particularly immediate interpersonal emotional decisions. Moreover, they fail to recognize and acknowledge when their actions or judgements incorporate transgressions of conventional or moral behaviour. They are likely to experience, but remain unaware of, changes in personality traits, their sense of self and political, social and spiritual values.[59] Executive function and social reasoning are thus likely to be significantly impaired.[60] Nonetheless, their self-esteem remains high as they over-rate themselves in multiple social, emotional and cognitive domains, since they are unable to process feedback from others that might lead them to revise these judgements more realistically.[61] Given that the right frontal-temporal network crucial for mediating social cognition is disrupted by bvFTD, the condition may be regarded as a social and executive disorder characterized by impairments in judgements of social dilemmas, theory of mind, self-awareness, empathy and cognitive flexibility.[62]

These characteristic behavioural disturbances include personality and social conduct disorder.[63] Knowledge of the difference between right and wrong is typically retained, but not observed, in so far as bvFTD patients are likely to approve of moral violations in full knowledge of moral values and conventional rules.[64] This indicates impaired emotional moral judgement manifesting in a lack of concern for the consequences for others of their transgressions. Mendez and colleagues have labelled these consequences of right frontotemporal dysfunction in bvFTD patients as 'acquired sociopathy'.[65] Acquired sociopathy has been linked with a reduced ability to generate expectations of other's negative emotional reactions generated by trauma to right frontal regions of the brain and distinguished from developmental psychopathy.[66] Current models of moral emotional decision-making in cognitive and affective neuroscience postulate a central role for the ventromedial prefrontal cortex

---

[57] Sinclair Lough et al., 'Social Reasoning, Emotion and Empathy in Frontotemporal Dementia' (2005) 44 *Neuropsychologia* 950.
[58] Note 34, above.
[59] Note 31, above.
[60] Note 57, above.
[61] Paul J. Enslinger et al., 'Metacognitive Deficits in Frontotemporal Dementia' (2005) 76 *Journal of Neurology, Neurosurgery & Psychiatry* 1630.
[62] Paul J. Enslinger, 'Oops! Resolving Social Dilemmas in Frontotemporal Dementia' (2007) 78 *Journal of Neurology, Neurosurgery & Psychiatry* 457.
[63] Note 44, above.
[64] Mario F. Mendez and Jill S. Shapira. 'Altered Emotional Morality in Frontotemporal Dementia' (2009) 14 *Cognitive Neuropsychology* 165.
[65] Note 44, above.
[66] R. J. R. Blair and L. Cipolotti, 'Impaired Social Response Reversal: a Case of Acquired Sociopathy' (2000) 123 *Brain* 1122.

(VMPFC).[67] Patients with focal VMPFC lesions, especially on the right, are characterized by loss of concern for others, reduced responsivity and lack of empathy for others' suffering. They retain the ability to engage in logical reasoning, knowledge of social and moral standards of conduct and the ability to assess and predict future outcomes. When faced with moral dilemmas, they are more likely to find immoral conduct acceptable. Mendez and Shapira summarize this research to link the altered emotional morality characterizing bvFTD with the disproportionate affects of early bvFTD on the VMPFC.[68] They draw upon earlier work to hypothesize that this lack of strong moral emotions may lead to an inability to override compulsions, disinhibitions, drives and tendencies leading to behaviours offending against normative standards.[69] Instead, those with bvFTD may act in an opportunistic and impulsive manner to commit sociopathic acts that they recognize as wrong. Though they may make remorseful statements, they are highly unlikely to act in remorseful fashion by apologizing, making amends, expressing guilt and shame or amending their behaviour. As bvFTD progresses, these infringements of social standards are likely to become more serious.[70] This raises the issue of capacity and competence, which will be explored further below after a consideration of family/carers' place in the neurorehabilitation of MND and bvFTD.

15.4 Family/Carer Involvement in MND and bvFTD Neurorehabilitation

MND is a progressive condition where neuromuscular weakness increases, with median survival around three years after diagnosis. Loss of respiratory capacity usually signifies that life expectancy is less than six months. Neurorehabilitation encompasses palliative care delivered by a multidisciplinary team that provides symptom management by maximizing motor function and minimizing drool, spasticity and secretions. Possible interventions also include non-invasive ventilation through face-masks, percutaneous enteral gastronomy (PEG) and long-term mechanical ventilation with tracheostomy (LTMV). Clinicians are advised to discuss each option with the patient and family/carers, along with end-of-life preferences, the appointment of a trusted person to make healthcare decisions and advance directives. In so far as the progression of MND is predictable and inexorable, each stage may be recognized, explained and subjected to choice in a relatively straightforward fashion.[71] MND patients are likely to have documented advance care planning as a result.[72]

---

[67] D. Kliemann et al., 'The Influence of Prior Record on Moral Judgement' (2008) 46 *Neuropsychologia* 2949.
[68] Note 64, above.
[69] Mario F. Mendez et al., 'Pedophilia and Temporal Lobe Disturbances' (2000) 12 *Journal of Clinical Psychiatry and Clinical Neuroscience* 71.
[70] Note 65, above.
[71] Note 4, above.
[72] A. B. Astrow et al., 'Decision-making in Patients with Advanced Cancer Compared with Amyotrophic Lateral Sclerosis' (2008) 34 *Journal of Medical Ethics* 664.

Family/carers' role in MND neurorehabilitation carried out at home involves feeding, which can take several hours a day once swallowing capacity has been lost without PEG, daily maintenance activities like dressing, and specific mechanical tasks such as removing secretions associated with LTMV, which may need to be done hourly.[73] In one study, despite home care assistance, primary family/carers spent a median of eleven hours daily caring for MND patients, with half stating that they felt 'physically and psychologically unwell'.[74] While most MND patients report fatigue, especially as the condition progresses, few are depressed. Indeed, the quality of life of family/carers may be affected more by MND than is the patients'.[75] In one study, 10 per cent were depressed, around double that of the rate in the general population.[76] In another, neither patients nor family/carers were depressed, although distress in family/carers was reflected in consequent levels of patient distress. Rabkin and colleagues' research suggests that depression and distress need not be a common or inevitable consequence of life-threatening illness or disability for patients or family/carers. Family/carer resilience was associated with finding positive meaning in caregiving.[77] This suggests that a positive transformation in the identity of family/carers as well as patients within neurorehabilitation is possible and desirable where MND is concerned. Mitsumoto and Rabkin suggest that the increasing family/carer burdens as MND progresses may be substantially offset by the patients' retaining their capacity to appreciate the caregiving, express affection and continue to participate emotionally in family life and relationships, even when they have lost mobility and the ability to speak.[78]

Both the inevitability of the disease progression through specific stages and the ability to sustain affectionate and appreciative bonds distinguish the situation of family/carers of MND patients and family/carers of those with bvFTD. These factors render the position of family/carers for those with bvFTD immeasurably more difficult until the last stages, where the lack of awareness of self and others or daily surroundings and inability to meet the requirements of daily living characterize those suffering from all the dementias so that the difficulties experienced by family/carers are relatively equivalent.[79] Yet, in the early and mid-stages, family/carers of those with bvFTD experience unique and significant stressors. They are involved in the diagnosis, rehabilitation and day-to-day care of those with bvFTD.

---

[73] Note 4, above.
[74] L. S. Krivikas et al., 'Home Care for Patients with ALS'. Abstract of Proceedings of the Seventh International Symposium on Amyotrophic Lateral Sclerosis and Motor Neurone Disease, 30 October–1 November 1995, Dublin.
[75] J. M. Wolley and S. Ringel, 'Caregiver Quality of Life in ALS: Relationship Between Patient Health Status and Caregiver Health Related Quality of Life'. Abstract in Proceedings of Eighth International Symposium on Amyotrophic Lateral Sclerosis and Motor Neurone Disease, 16–18 November 1997, Glasgow.
[76] M. C. McElhiney et al., 'Prevalence of Fatigue and Depression in ALS Patients and Change Over Time' (2009) 80 *Journal of neurology, Neurosurgery and Psychiatry* 1146.
[77] Judith C. Rabkin et al., 'Resilience and Distress Among Amyotrophic Lateral Sclerosis Patients and Caregivers' (2000) 62 *Psychosomatic Medicine* 271.
[78] Note 4, above.
[79] Samantha Riedijk et al., 'Sense of Competence in a Dutch Sample of Informal Caregivers of Frontotemporal Dementia Patients' (2009) 27 *Dementia and Geriatric Cognitive Disorders* 337.

All are problematic. Given that cortical atrophy is expected as part of normal ageing, that diagnosis of bvFTD is based on family/carer reports of behaviour and that MRI scans of around half of those diagnosed with bvFTD do not exhibit abnormalities, diagnosis to a large extent rests upon family/carer reports of monitoring behaviour. Consequent difficulties of uncertainty, under-diagnosis and misdiagnosis have been mooted above. Reliance on family/carers continues during neurorehabilitation. Treatment is problematic in so far as there are currently no medications approved by the Food and Drug Administration specifically for FTD. While selective serotonin reuptake inhibitors (SSRIs) and atypical antipsychotic agents may be used to manage behavioural symptoms, using psychiatric medications to treat bvFTD is not necessarily appropriate.[80] Neuroleptics may promote negative behaviours.[81] Indeed, the use of psychiatric medications to treat symptoms of dementia has been frowned upon in the UK by the National Institute for Health and Clinical Excellence (NICE). This means that family/carers may be expected to provide daily support for those whose difficult behaviour is unable to be modified pharmacologically. Nor do support services aimed at other types of dementia meet the specific needs of bvFTD family/carers.[82] Typically, those with bvFTD are callous, unemotional, impulsive, opportunistic, unable to acknowledge or care about the needs of others, likely to offend against moral norms while knowing that this is wrong and unable to recognize that their own behaviour has altered or to experience affection or remorse. Moreover, more florid symptoms such as gluttony and cravings for sweet and stodgy food may emerge.[83] These are associated with profound changes in cortical serotonin levels, altering satiety and food preferences.[84]

Recent understandings of the neural and genetic mechanisms associated with bvFTD are likely to stimulate clinical trials targeted to develop pharmacological treatments specifically for bvFTD. Non-pharmacological therapies currently focus upon behaviour management techniques to target socially inappropriate actions, or to draw upon the fact that procedural memory remains intact to engage in retraining.[85] Unlike neurorehabilitation for neuromuscular difficulties, as in MND, clinical expertise where bvFTD is concerned focuses on the enforcement of structured daily life activities that manipulate motivation, inappropriate behaviours and decision-making.[86] While informing family/carers that behavioural

---

[80] Adam Boxer and Brafley F. Boeve, 'Frontotemporal Dementia Treatment: Current Symptomatic Therapies and Implications of Recent Genetic, Biochemical and Neuroimaging Studies' (2007) 21 *Alzheimer's Disease & Associated Disorders* S79.

[81] Valeria Santoro Bahia et al., 'Behavioural and Activities of Daily Living Inventories in the Diagnosis of Frontotemporal Lobar Degeneration and Alzheimer's Disease' (2008) 2 *Dementia & Neuropsychologia* 108.

[82] Note 24, above.

[83] Mario F. Mendez et al., 'Changes in Dietary or Eating Behaviour in Frontotemporal Dementia Versus Alzheimer's Disease' (2008) 23 *American Journal of Alzheimer's Disease & Other Dementias* 280.

[84] M. Ikeda et al., 'Changes in Appetite, Food Preference and Eating Habits in Frontotemporal Dementia and Alzheimer's Disease' (2002) 73 *Journal of Neurology, Neurosurgery and Psychiatry* 371.

[85] Mario F. Mendez, 'Frontotemporal Dementia: Therapeutic Interventions' in P. Giannakopoulos and P. R. Hof (eds), *Dementia in Clinical Practice* (Basel, Karger, 2009) 168.

[86] Lauren Masimo, 'Patient Care and Management of Frontotemporal Lobar Degeneration' (2008) 23 *American Journal of Alzheimer's Disease & Other Dementias* 125.

symptoms have a neurological basis is viewed as a means to help them accept and adjust to changes, diagnosis is likely to be delayed or uncertain as traits like callousness and lack of empathy are liable to be interpreted as deliberate unfeeling behaviour rather than as symptomatic of illness. Even after diagnosis, given that bvFTD patients lack insight and do not perceive changes in themselves and are unable to perceive themselves through others' eyes,[87] family/carer involvement in restricting their movements and decision-making involves inherently highly stressful monitoring and negotiations. Family/carers are likely to find caring for bvFTD patients so demanding that their personal lives become sacrificed, with consequent psychological complaints and reduced mental and physical quality of life.[88] Together, these factors render bvFTD family/carers subject to the highest levels of stress amongst carers.

What is generally termed family/carer burden is assessed in terms of the impact of the strain of caregiving on psychological, physical, spiritual, social and financial well-being. Although the word 'burden' has been criticized as oversimplifying the complexities of the relationship,[89] it will be used here as accepted terminology. The fact that research into the burdens of family/carers of those diagnosed with bvFTD demonstrates the highest stress and depression levels amongst those involved in caring for the demented[90] is hardly surprising. Their needs are not met by supportive services directed towards the family/carers of other types of dementia, such as memory clinics, and institutional placement can prove more difficult.[91] The lack of clinical certainty over the progression and prognosis of bvFTD may lead not only to lack of or delayed diagnosis, as explored above, but also to increased stress for family/carers who do not know what to expect once diagnosis has taken place.[92] Feelings of guilt once diagnosis has taken place may increase burden and stress once unpleasant behaviour is recognized as symptomatic rather than blameworthy.[93] Family members who regard the behaviour of the family member who has bvFTD as deliberately unpleasant rather than as symptomatic of illness may become alienated rather than offering the primary carer the support that fosters experiencing caring as a positive experience.[94] Cognitive tests do not capture the devastating impact of bvFTD on the activities of daily living,

---

[87] Note 44, above.
[88] Note 79, above.
[89] Kevin McKee et al., 'Components of the Difficulties, Satisfactions and Management Strategies of carers of Older People: a Principal Component Analysis of CADI-CASI-CAMI' (2009) 13 *Aging & Mental Health* 255.
[90] Eneida Mioshi et al., 'Factors Underlying Caregiver Stress in Frontotemporal Dementia and Alzheimer's Disease' (2009) 27 *Dementia and Geriatric Cognitive Disorders* 76.
[91] Jennifer Merrilees, 'A Model for Management of Behavioural Symptoms in Frontotemporal Lobar Dementia' (2007) 21 *Alzheimer's Disease and Associated Disorders* S64.
[92] Catherine Quinn et al., 'The Experience of Providing care in the Early Stages of Dementia: an Interpretive Phenomenological Analysis' (2009) 12 *Aging & Mental Health* 769.
[93] Judith C. Gonyea et al., 'Adult Daughters and Aging Mothers: the Role of Guilt in the Experience of Caregiver Burden' (2008) 12 *Aging & Mental Health* 559.
[94] Yumi Shirai et al., 'Reaping Caregiver Feeelings of Gain: the Roles of Socio-emotional Support and Mastery' (2009) 13 *Aging and Mental Health* 106.

which is of considerable importance for family/carers.[95] Poor financial decisions made by those with bvFTD have depleted family assets.[96]

Riedijk and colleagues comment that 'in our clinical experience, caregivers sometimes literally cry for help, because the care-giving situation is overwhelmingly demanding. It seems that the concept of caregiver burden does not capture sufficiently the intrusive and never-ending nature of the (FTD) caregiver task'.[97] Measures of caregiver burden fail to provide an adequate assessment of the sufferings of bvFTD family/carers, which are 'often poorly evaluated and underestimated in clinical practice'.[98]

Moreover, the support offered bvFTD family/carers is frequently inadequate and inappropriate.[99] Most existing interventions show moderate benefits for family/carers at best.[100] This reflects the lack of an evidence-based consensus on outcome measures of psychosocial interventions in dementia care.[101] Services have been geared towards family/carers of patients with Alzheimer's disease, whose needs are very different. In addition, information over diagnosis and prognosis, together with that on available pharmacological and non-pharmacological treatments, is shrouded with uncertainties that have yet to be resolved through clinical trials.[102] Changes in personality and inappropriate and insufficient support for bvFTD family/carers are associated with increased and earlier institutionalisation of patients.[103] Yet family/carers who place those with bvFTD in institutions after relatively short periods of dementia feel more burdened and experience a poorer quality of life than those who continue to care for them at home.[104] Nonetheless, those who care for them at home as the condition progresses rather than institutionalizing them feel more emotionally burdened than those who place them in institutions, with worse physical and mental health when they are doing so on the basis of love.[105]

The difficulties experienced by family/carers of those with MND or bvFTD correspond with the differing neurodegenerative symptoms associated with the

[95] E. Mioshi et al., 'Activities of Daily Living in Frontotemporal Dementia and Alzheimer's Disease' (2007) 68 *Neurology* 2077.
[96] Note 24, above.
[97] Note 79, above, 338.
[98] Claire Boutoleau-Bretonniere et al., 'Zarit Burden Inventory and Activities of Daily Living in the Behavioural Variant of Frontotemporal Dementia' (2008) 25 *Dementia and Geriatric Cognitive Disorders* 272, at 276.
[99] Tor Atle Rosness et al., 'Support of Family Carers of Patients with Frontotemporal Dementia' (2008) 12 *Aging & Mental Health* 462.
[100] S. H. Zarit and E. E. Femia, 'A Future for Family Care and Dementia Intervention Research? Challenges and Strategies' (2008) 12 *Aging & Mental Health* 5.
[101] E. Moniz-Cook et al., 'A European Consensus on Outcome Measures for Psychosocial Intervention Research in Dementia Care' (2008) 12 *Aging & Mental Health* 14.
[102] Note 85, above.
[103] Tor Atle Rosness et al., 'Support of Family Carers of Patients with Frontotemporal Dementia' (2008) 12 *Aging & Mental Health* 462.
[104] Samantha R. Riedijk et al., 'Caregiver Burden, Health Related Quality of Life and Coping in Dementia Caregivers: a Comparison of Frontotemporal Dementia and Alzheimer's Disease' (2006) 22 *Dementia and Geriatric Cognitive Disorders* 405.
[105] Samantha R. Riedijk et al., 'Frontotemporal Dementia (FTD) Patients Living at Home and their Spousal Caregivers Compared with Institutionalised FTD Patients and their Spousal Caregivers' (2009) 8 *Dementia* 61.

conditions. Those for family/carers for MND patients will centre about catering for increasing physical frailty and dealing with the demands of treatment interventions, such as sucking out secretions from long-term mechanical intervention with tracheostomy (LMTV). Where there is cognitive dysfunction or decline as well as MND, or where bvFTD is also present, some of the difficulties will overlap.[106] Where bvFTD is concerned, many of the difficulties centre on unpleasant behaviour and impaired decision-making. Some ethical and legal implications of the symptoms of MND and bvFTD, particularly in terms of end-of-life decision-making, will now be considered. Although, as previously mentioned, there may be some overlap between those with the two conditions, they will be discussed separately to aid conceptual clarity.

15.5 Ethical and Legal Implications of MND and bvFTD: Decision-Making at End of Life and Capacity

Specific ethical and legal issues emerge as a result of the familial nature of MND and FTLD. Members of families where inherited diseases are common are likely to anticipate the possibility that they may develop them, although few choose to take steps to ascertain this, even where genetic tests are available, as in Huntington's disease. While technologies such as IVF and PGD provide means of ensuring reproductive choices eliminating this possibility for some inherited conditions, MND and FTLD are not among them. Ethical choices centre rather on end-of-life decisions, or once either condition has developed or been diagnosed. Legal issues hinge upon advance refusals of treatment and the determination of competence or capacity. These will now be discussed in turn for MND and bvFTD.

MND progresses inexorably, with ethical decisions to be made at each step of the neurorehabilitative process. Symptomatic relief may be gained through technologies permitting non-invasive assistance with breathing, provision of artificial nutrition and hydration *via* PEG once the swallowing reflex has deteriorated and finally invasive assistance to breathe through LTMV. Decisions to accept or refuse these procedures may be arrived at through consultations between clinicians, patients and family/carers. Given the predictable progression of MND, each stage is typically discussed amongst the multidisciplinary team, patients and family/carers over time so that a valid and considered consent or refusal of each intervention takes place as part of advance planning for end-of-life scenarios. The burden on both the patient and family/carers are likely to be explicitly acknowledged and evaluated as part of this process, e.g. the twenty-four hour hourly suction to remove secretions involved in LTMV may be considered to be too high a price to pay for extended life. Considerations of quality versus quantity of life should be subject to thoughtful resolution over time before the time for such interventions arises, in order to avoid

---

[106] Z. C. Gibbons et al., 'Inferring Thought and Action in Motor Neurone Disease' (2007) 45 *Neuropsychologia* 1196.

rushed decisions and missing out on windows after which the interventions are less likely to succeed. Hence, while advance refusals of treatment are subject to the well-known difficulties associated with lack of precision and changes of mind, the fact that there are established clinical guidelines and care pathways for refusal of treatment in MND puts protocols in place which provide safeguards. Moreover, patients' decision-making capacity will be assessed in relation to specific decisions over treatment. Thus, where cognitive decline is present,[107] it is liable to be detected and assessed appropriately.

End-of-life decisions for those with dementia are qualitatively different. While there is a clinical focus on the postponement or prevention of dementia, as in the popular memory clinics, current treatments for Alzheimer's disease and others with linguistic impairments such as PNFA and SD are aimed at symptomatic relief or curtailment once dementia has developed. End-of-life advance planning is not punctuated by the clear physical manifestations of decline that characterize MND. Rather, as cognitive powers fade, the sense of self and ability to engage independently in the activities of daily living are gradually lost. In debates over assisted dying, this is associated with a loss of autonomy that may key advance refusals of treatment or requests for assistance to hasten the dying process. Some consider that without autonomy quality of life is lost, while others disagree. There have been suggestions, most famously from Baroness Warnock,[108] that those who become demented should refuse treatment in advance, since as they constitute a burden on society they have a duty to die.

Different jurisdictions provide different legal frameworks for these ethical and clinical decisions. Assisted dying is lawful in some countries but unlawful in others. Space precludes reiterating the wealth of scholarship on this topic. One common factor is the ability to refuse medical treatment in keeping with the principle of autonomy, or the protection of bodily integrity. Competent adults, or those with decision-making capacity, have the right to exercise their autonomy by consenting to or refusing medical treatment, even when to do so would prove fatal. Many jurisdictions provide means whereby unwanted clinical interventions may be refused in advance. In England and Wales, for example, the Mental Capacity Act 2005 provides a mechanism whereby advance decisions refusing treatment (ADRTs) may be authenticated, placing an obligation upon clinicians to respect these wishes. The authority of such ADRTs rests on the presumption that in order to make autonomous decisions one must have the capacity to do so. This means that the decision-maker must be seen as competent in the eyes of the law, a status usually determined by clinicians or the court. Although capacity and competence may be considered not to be synonymous, in the literature they are often used interchangeably, and will be treated in this way here. The remainder of this paper will consider ethical and legal issues relating to competence and capacity in bvFTD in terms of decision-making at end of life.

---

[107] Note 5, above.
[108] J. Macadam, 'A Duty to Die?' [2008] *Life and Work* 23.

The central difficulty associated with judgements of whether those with bvFTD have the legal capacity to make decisions is the fact that tests of capacity measure cognitive functioning. Under the Mental Capacity Act 2005, there is a presumption of capacity that is otherwise to be assessed in relation to a specific decision at a specific time. Decision-making ability is to be maximized and decisions that may appear to be unwise are not to be interpreted as a sign of incapacity. In order for a person to be accepted as incapable, they must have an impairment of or a disturbance in the functioning of the brain or mind, and this defect must result in the inability to understand, retain, use or weigh information relevant to a decision. They must be able to communicate the decision in order to have capacity. Persons with capacity may grant another a lasting power of attorney to make healthcare decisions on their behalf should they lose capacity. They may also make ADRTs in the event of their losing capacity, which in general need not be in writing. Advance refusals of life-sustaining treatment must be written, signed and witnessed to be valid. All ADRTs may be altered while capacity is maintained or if it is regained. Where they apply to the circumstances in question, they must be followed by clinicians, who are obliged to investigate whether an advance refusal exists and to assess whether it is valid. In the absence of capacity and an advance refusal, those without capacity are to be treated in their best interests. Those with a reasonable belief in another's incapacity are protected from legal liability should they act in their best interests.

As surveyed above, frontotemporal dysfunction impacts on the decision-making of those with bvFTD in characteristic fashions. The lack of the ability to recognize others as having a point of view of their own (theory of mind or TOM), impulsivity or lack of appreciation for the consequences of their actions, emotional flatness, lack of insight, tendency to inflate their capabilities, recognition of right and wrong allied with tendency to make selfish decisions and lack of social inhibition all foster decisions that family/carers and others are likely to find problematic. These issues map onto debates within ethics, neuroscience and the law of how far brain damage or dysfunction should affect judgements of culpability and decision-making capacity.

Both criminal culpability and decision-making capacity are conceived of in the law as based on the ability to make rational decisions, where rational is conceptualized as cognitive but not affective. In the US, under the Comprehensive Control Act of 1984, the federal insanity defence is not available unless it can be proven that at the time of the act the accused was unable to appreciate the nature and quality or the wrongfulness of the act.[109] Psychopaths and sociopaths, who recognize the difference between right and wrong and offend against the criminal law, nonetheless have not been able to plead insanity successfully. Yet neuroscientists provide increasing evidence that envisaging rational decision-making as involving cognitive but not affective elements is unsustainable, as emotion guides rationality by providing a means of evaluating decisions.[110] Emotional and neural dysfunction

---

[109] R. Borum and S. M. Fulero, 'Empirical Research on the Insanity Defence and Attempted Reforms: Evidence Towards Informed Policy' (1999) 23 *Law & Human Behavior* 375.

[110] Terry A. Maroney, 'Law and Emotion: a Proposed Taxonomy of an Emerging Field' (2006) 30 *Law & Human Behavior* 119.

of the medial and the lateral prefrontal cortices have been put forward as diagnostic markers for deficits in decision-making.[111] Moreover, impaired decision-making processes following damage to the brain, particularly the prefrontal cortex, have been mapped repeatedly.[112] Hence whether or how far brain damage or dysfunction that affects the emotions should affect notions of culpability or capacity is problematic. While some have raised the issue of the role of emotion in judgements of capacity to consent to treatment,[113] most of the discussion has centred upon the impact of neuroscience on criminal culpability, in terms of the defence of insanity[114] or fitness to plead, adjudicative competence in decision-making,[115] or a comparison of medical assessments of competence and criminal responsibility.[116] Some suggest that those who lack the capacity for empathy should be treated differently by the law; that psychopaths should be excused from criminal responsibility[117] and children with conduct disorders seen as precursors to psychopathy should be classified as learning disabled.[118] This is in keeping with arguments that morality and hence responsibility is relational.[119] Space constraints preclude our exploring the intersection between criminal culpability, mental health assessment and competence to stand trial. Instead, we conclude with a consideration of the ethical and legal implications of bvFTD and end-of-life decision-making.

Impairments in emotional processing and moral judgement are associated with the ventromedial prefrontal cortex damage which characterizes bvFTD.[120] How far this should determine whether those with such damage are to be viewed as lacking in capacity is both an ethical and legal issue. Difficulties with the assessment of uncertain or impaired capacity are common, and were the most frequently reported ethical dilemma of 94.8 per cent of the 400 European doctors from Norway, Switzerland, the UK and Italy surveyed in a recent study.[121] While legal criteria for assessing capacity to consent to or refuse treatment vary across jurisdictions, most clinical tools measuring decision-making capacity test the elements identified by

[111] Georg Northoff, 'Neuroscience of Decision Making and Informed Consent: an Investigation in Neuroethics' (2006) 32 *Journal of Medical Ethics* 70.
[112] Facundo Manes et al., 'Decision Making Processes Following Damage to the Prefrontal Cortex' (2002) 125 *Brain* 624.
[113] Louis C. Charland, 'Is Mr Spock Mentally Competent? Competence to Consent and Emotion' (1998) 83 *Philosophy, Psychiatry & Psychology* 67.
[114] Laura Reider, 'Toward a New Test for the Insanity Defense: Incorporating the Discoveries of Neuroscience into Moral and Legal Theories' (1998) 46 *UCLA Law Review* 289.
[115] Terry A. Maroney, 'Emotional Competence, 'Rational Understanding' and the Criminal Defendant' (2006) 43 *American Criminal Law Review* 1376.
[116] Gerben Meynen, 'Exploring the Similarities and Differences Between Medical Assessments of Competence and Criminal Responsibility' (2009) *Medicine Health Care & Philosophy* DOI 10.1007/s11019-009-9211-1.
[117] Stephen J. Morse, 'Rationality and Responsibility' (2005) 74 *South Carolina Law Review* 251.
[118] Robin Mackenzie and John Watts, 'Callous/unemotional Conduct Disorder as a Learning Disability' (2009) 14 *Tizard Learning Disability Review* 38.
[119] R. A. Duff, 'Strict Responsibility, Moral and Criminal' (2009) 43 *Journal of Value Inquiry* 295.
[120] Liane Young and Michael Koenigs, 'Investigating Emotion in Moral Cognition: a Review of Evidence from Functional Neuroimaging and Neuropsychology' (2007) 84 *British Medical Bulletin* 69.
[121] S. A. Hurst, 'Ethical Difficulties in Clinical Practice: Experiences of European Doctors' (2007) 33 *Journal of Medical Ethics* 51.

Appelbaum and Grisso.[122] In relation to a specific decision at a specific time, these are the abilities to: 1) understand information as it relates to the choice, such as the nature, purpose and potential risks and benefits of the proposed treatment, 2) appreciate the relevance of the information for their condition or situation, such as recognizing they have a condition for which effective treatment could potentially be helpful, 3) consider, compare and reason about potential consequences of the options, including that of no treatment, and 4) communicate a choice. This is comparable with that in the Mental Capacity Act 2005, where the capacity to understand, retain, use or weigh information and to communicate decisions is assessed. While it would be possible to read emotional comprehension into these tests, e.g. to suggest that appreciation and understanding imply an affective apprehension,[123] this is not what standard tests measure. Rather, the tools assessing decision-making capacities test cognition alone.[124]

Appelbaum cites research demonstrating that impaired decision-making in hospitalized patients is likely to go undetected, with up to 48 per cent lacking in competence but only a quarter of these being identified by clinicians, even when treatment was refused.[125] Given that he cites figures of more than half of patients with mild to moderate dementia as typically lacking in decision-making capacity, with all of the severely demented doing so, these figures are alarming. Raymont and colleagues speculate that the low levels of detection of incapacity they found are associated with the fact that many patients passively acquiesce with recommended treatments so the need to assess capacity is overlooked unless treatment is refused, something that did not take place in their survey of 300 patients.[126] Moreover, in England and Wales, in so far as the Mental Capacity Act is aimed at the assessment of the capacity to make specific decisions in specific circumstances, it presumes not only capacity until proven otherwise, but also a level of expertise in assessing capacity that is not necessarily widespread. Patients' capacity tends to be taken for granted, often tested only when patients disagree with clinicians rather than truly consenting or passively acquiescing.[127] This also impacts upon the role of family/carers, who may be expected to assess capacity under the Act but are likely to lack the requisite expertise.

These difficulties in assessing capacity map onto the situation of those with bvFTD, their family/carers and clinicians in various ways. Given that capacity is unlikely to be perceived as needing to be tested even in hospital settings, it renders fairly marginal the likelihood of those with bvFTD's coming to the attention of

---

[122] Paul S. Appelbaum and Thomas Grisso, 'Assessing Patients' Capacities to Consent to Treatment' (1988) 319 *New England Journal of Medicine* 1635.
[123] Louis C. Charland, 'Appreciation and Emotion: Theoretical Reflections on the MacArthur Treatment Competence Study' (1998) 8 *Kennedy Institute of Ethics Journal* 359.
[124] R. J. Gurrera et al., 'Cognitive Performance Predicts Treatment Decisional Abilities in Mild to Moderate Dementia' (2006) 66 *Neurology* 1367.
[125] Paul S. Appelbaum, 'Assessment of Patients' Competence to Consent to Treatment' (2007) 357 *New England Journal of Medicine* 1834.
[126] Vanessa Raymont et al., 'Prevalence of Mental Incapacity in Medical Inpatients and Associated Risk Factors: Cross Sectional Study' (2004) 364 *Lancet* 1421.
[127] Ibid.

clinicians before symptoms become florid. Even once capacity is assessed, the bias towards cognitive rather than affective measurement leaves those with bvFTD likely to be seen as competent, although their social and emotional bases for interpersonal relations are significantly compromised. Given their lack of insight and tendency to overestimate their abilities, they are unlikely to perceive changes in themselves, or to agree that they should be subject to clinical intervention. The overlap between unpleasant behaviour that is regarded as a consequence of character defects like selfishness and that caused by frontotemporal dysfunction lessens the likelihood of diagnosis initiated by family/carers, who are more likely to blame them for their behaviour.

15.6 Conclusion

We wish to conclude with a case study to illustrate the complexities of end-of-life decision-making in the contexts explored above:

Arthur and Cuthbert belong to a family where some relatives are known to have become weak and bedridden before dying, while others become mean, nasty and cruel. Arthur is married to Betty and Cuthbert to Dora. The marriages are happy and loving. After seeing a television programme on dementia, MND, FTD and advance decision to refuse treatment (ADRT), all four decide to approach their lawyer and doctor sometime soon to construct ADRTs should either condition develop as they are adamant that their spouses should not be burdened should they become ill. Each signs a witnessed ADRT stating that should they develop a neurodegenerative disorder or dementia, they wish to refuse life-sustaining treatment. Arthur develops bvFTD and Cuthbert MND three months later.

Cuthbert's muscle weakness is readily diagnosed as MND. The MND Association provides information, support and sample ADRTs allowing Cuthbert and Dora to anticipate how the disease is likely to progress, assess the impact of interventions like PEG and discuss when, if at all, Cuthbert might wish to refuse treatment. Ongoing discussions of advance care planning including potential end-of-life scenarios take place with the multidisciplinary neurorehabilitation team. Cuthbert revises his ADRT after deciding provisionally that he wishes to live until he can no longer communicate, so accepts PEG and NIV on the understanding, recorded in his signed and witnessed ADRT, that NIV should be withdrawn under sedation once this stage is reached. He refuses LTVT. Cuthbert dies peacefully three years later. Ongoing discussions over the three years focus on whether his wishes have changed and whether he retains capacity. He retains capacity and his wishes do not change.

Arthur develops bvFTD at the same time but this goes unrecognized. Betty notices that he is becoming more difficult to be around but attributes this to his becoming a grumpy old man and to worry about Cuthbert. Arthur becomes increasingly selfish and cold, withdrawing emotionally in a way Betty experiences as cruel. He begins to behave inappropriately in social situations and alienates their friends and relations. He takes their savings out of safe government bonds, puts them into high-risk schemes and loses most of their retirement funds. When she tells him she is unhappy and says that she wants their marriage to go back to the way it was before, he says that he is happy enough and does not know what she is talking about. He disregards her wishes for more prudent investments. Betty thinks Arthur has bvFTD but he insists this is not so. She wants him to take oxytocin to promote social and emotional closeness but he refuses to do so, saying he has not changed and she is making a fuss about nothing. Arthur develops tingling down his arms as a result of repetitive strain injury from spending hours gambling on the internet. An MRI scan reveals some frontotemporal lobe shrinkage but no signs of MND. Arthur then develops pneumonia. He is conscious but will die without treatment. He insists that he is perfectly fine apart from the pneumonia and wants to be treated.

When Arthur signed the ADRT, he wished to avoid Betty's being burdened should he develop MND or any kind of dementia. However, like most of those with the condition, he does not realize that he has developed bvFTD. Nor does he notice that she is unhappy and burdened or consider that this is problematic. He has no wish to refuse treatment as he is convinced that he is well and is unable to appreciate her perspective or interests.

His clinicians have to decide whether to treat him, whether he has capacity or whether his ADRT is applicable and should be followed. Betty feels in a moral quandary. She does not wish to devote her life to caring for someone whose behaviour has changed so much he appears to be an unpleasant, cruel and ungrateful stranger, especially as if Arthur has bvFTD his behaviour will deteriorate. Given the uncertainties surrounding diagnosis of bvFTD, Betty is tempted to exaggerate her reports of Arthur's behaviour to make it certain that he will be diagnosed with bvFTD so his ADRT will be seen as valid and applicable. The clinicians are faced with ambiguous information to support a diagnosis of bvFTD, measures of capacity which test cognitive as opposed to affective dysfunction and are undecided over the applicability of the ADRT.

What is the best way forward?

We have concluded with this case study as we feel that it exemplifies the difficulties explored above. We see those with bvFTD as suffering from a condition Damasio has described as a 'moral misbehaviour syndrome', which he describes in the following terms:

Following neurologic damage to specific sectors of the pre-frontal cortex, previously healthy adults who are well acquainted with social conventions and ethical rules and who are known to have behaved accordingly, now fail to observe them. They deviate, more or less significantly, from broadly accepted norms. The contrast between the period before damage and the period after damage is a defining trait of the condition...

First, the misbehaviour caused by adult-onset prefrontal damage invariably is accompanied by some disturbance of emotional behaviour that includes both a diminished resonance of emotional reactions in general and a specific compromise of social emotions such as compassion and embarrassment. Second, the misbehaviour that can be qualified as moral is accompanied by other failures of decision making, for example, by poor planning in a variety of everyday activities and poor management of human relationships... Third, the misbehaviour I *not* accompanied by disorders of perception, movement, conventional memory, language and general reasoning ability. Impulsivity and perseveration, which can be caused by prefrontal damage, are not a necessary feature...

This syndrome includes assorted failures of decision making and an obligate disorder of emotion, and it spares the key instrumental resources of cognition (perception, movement, memory, language and reasoning). The term I proposed for this syndrome two decades ago—*acquired sociopathy*—still captures its flavour.[128]

While Damasio is discussing those with adult-onset frontotemporal lobe dysfunction caused by damage rather than atrophy, Mendez considers bvFTD to be an example of acquired sociopathy.[129] In that assessments of capacity hinge upon function rather than status, i.e. they relate to particular decisions at a particular time rather than a characteristic of the decision-maker like mental illness,[130] whether the presence of bvFTD would entail a loss of capacity would fall properly to be assessed in relation to specific decisions. Thus, it is quite plausible that Arthur would be competent to refuse to take oxytocin. Where the capacity of those with bvFTD becomes particularly problematic is in the obligation of clinicians to judge the validity and applicability of ADRTs.

Clinicians' decisions on such ADRTs hinge upon how emotion and emotional dysfunction is seen in relation to assessing capacity. Considering this issue in general, rather than proposing that a conception of emotional competence should replace or compete with traditional understandings of competence, Charland suggests that interpretations of 'appreciation' should take the influence of emotion into account.[131] Appelbaum agrees that emotions aid human decision-making, but has reservations over the feasibility and utility of a reorientation broadening the focus of competence assessment to include emotional capacities.[132] Drawing upon Damasio's research on patients with frontotemporal lobe brain damage, he expresses scepticism over whether such a rare condition could justify the requisite changes in capacity assessment practice. He calls for a precise specification of the degree of impairment that might render a person emotionally incompetent, wishing to eliminate false positives, where those with diminished emotional responses, such as schizoids or psychopaths, might become regarded as incompetent to make decisions for themselves.

As bvFTD, as evidenced above, may be now regarded as under-diagnosed and increasingly common, we believe that Appelbaum's view in 1998 that the problem of emotions' impact on capacity is relevant only in rare cases has been overtaken by events. Given the ever more common policies encouraging citizens to make advance decisions over end-of-life treatments, the issues raised by the case study appear likely to emerge as increasingly prevalent and problematic. This is likely

---

[128] Antonio Damasio, 'Neuroscience and Ethics: Intersections' (2007) 7 *American Journal of Bioethics* 3, 3.
[129] Note 64, above.
[130] *Re C (Adult, Refusal of Medical Treatment)* [1994] 1 WLR 190.
[131] Note 123, above.
[132] Paul S. Appelbaum, 'Ought We to Require Emotional Capacity as Part of Decisional Competence?' (1998) 8 *Kennedy Institute of Ethics Journal* 377.

to raise further issues of ethical relations between clinicians and family/carers. Although as canvassed above clinicians include family/carers within treatment planning and administration, their duty of care is to the patient rather than the family/carers. Family/carers provide significant services and input into neurorehabilitative treatment without the training, support and legal protections of healthcare professionals. Where those being cared for in the homes have conditions where diagnosis hinges upon family/carers' reports of previous and current behaviour, coupled with inadequate support systems, family/carers are place in an invidious position, open to moral hazard and exploitation. Potential conflicts of interest must be monitored carefully if relations between clinicians and family/carers are to remain ethical. As shown in the case study, ethical dilemmas are particularly likely to arise in relation to ADRTs.

We suggest that the current means of assessing capacity to make decisions over medical treatment should be read as measuring emotional as well as cognitive factors, that the ethics of clinician/carer relations should be monitored carefully to minimize conflicts of interest and exploitation and that bvFTD exemplifies certain difficulties arising with advance refusals of treatment that must be resolved.

16

The Carmentis Machine: Legal and Ethical Issues in the Use of Neuroimaging to Guide Treatment Withdrawal in Newborn Infants

Dominic Wilkinson* and Charles Foster**

16.1 Introduction

Neuroscience and neuroimaging raise a number of philosophical and ethical issues in adults, including questions with potential legal implications. Other papers in this collection, for example, discuss the use of neuroscience to ascertain the fact or degree of consciousness, to determine if a witness is lying, and to determine whether someone has capacity or responsibility. Such questions, informed by neuroscience, are the substrate of the new discipline of neuroethics.[1] But neuroscience has not, until now, been thought to raise particularly novel problems for newborn infants. This is largely because of the difficulty of detecting or interpreting current mental states or processes. Instead, neuroimaging in the newborn is concerned with the diagnosis of abnormality or injury and the prediction of future capacities. But, as we discuss in this paper, neuroimaging does raise a number of troubling and difficult questions.

In newborn intensive care neuroimaging is used for prognostication and to aid decisions about life-sustaining treatment. Where infants are predicted to have a very poor prognosis, parents and doctors may question whether continuing life support

---

* The Ethox Centre, Department of Public Health and Primary Health Care, the University of Oxford, Oxford Uehiro Centre for Practical Ethics, the University of Oxford. Dominic Wilkinson is supported by an Oxford Nuffield Medical Fellowship, Eric Burnard Fellowship, and Royal Australasian College of Physicians Astra-Zeneca Medical Fellowship. The funders had no involvement in this work.

** The Ethox Centre, Department of Public Health and Primary Health Care, the University of Oxford, Barrister. Outer Temple Chambers, London WC2.

Address for correspondence: The Ethox Centre, The University of Oxford, Badenoch Building, Headington, OX3 7LF; email: dominic.wilkinson@ethox.ox.ac.uk; tele: +44 (0) 1865 287 887; fax: +44 (0) 1865 287 884.

[1] Neil Levy, *Neuroethics* (Cambridge, Cambridge University Press, 2007); Sinnott-Armstrong, Roskies, Brown and Murphy, 'Brain images as legal evidence' (2008) 5 *Episteme* 359–73.

is in the best interests of the child. Assessments of the future quality of life influence decisions in a significant proportion of deaths in newborn intensive care.[2]

Currently the main forms of imaging used in newborns are ultrasound, computed tomography (CT), and magnetic resonance imaging (MRI).[3] These modalities provide imperfect guides to the severity and extent of brain damage, and there is often significant uncertainty about prognosis.[4] The courts have placed some emphasis on imaging results in a couple of recent cases.[5] But as neuroimaging techniques improve, predictions may become significantly more accurate.

In this paper we consider how such developments would influence legal judgments about the permissibility of withdrawing or withholding life support from newborn infants. In part 1 we consider a hypothetical form of neuroimaging able to predict accurately future impairments in newborn infants. We call this the 'Carmentis Machine'. How will and how should clinicians and the courts respond to such predictions? What conditions, predicted by the machine, will lead to treatment withdrawal decisions? We first describe the machine and the nature of its predictions. We then indicate what UK guidelines and law are currently applied to the relevant decision-making in neonates, and suggest how law, guidelines, and practice might respond to the more accurate prognostic data generated by the machine.[6]

There is currently little case law relating to such decisions; the vast majority of treatment withdrawal decisions do not come before the courts. Guidelines for clinicians, as we discuss further below, vary in the specificity of the criteria they provide, but none explicitly discusses the use of neuroimaging.

In part 2 of the paper we provide an analysis of two different approaches used in guidelines and by the courts to determine the best interests of infants. Both approaches are epistemically and conceptually problematic. In part 3 of the paper we will consider a further question raised by developments in neuroimaging and

---

[2] Roy, Aladangady, Costeloe, and Larcher, 'Decision making and modes of death in a tertiary neonatal unit' (2004) 89 *Archives of Disease in Childhood Fetal and Neonatal Edition* F527–30; Singh, Lantos, and Meadow, 'End-of-life after birth: death and dying in a neonatal intensive care unit' (2004) 114 *Pediatrics* 1620–6; Barton and Hodgman, 'The contribution of withholding or withdrawing care to newborn mortality' (2005) 116 *Pediatrics* 1487–91; Provoost, Cools, Mortier, Bilsen, Ramet, Vandenplas, Deliens, and Consortium, 'Medical end-of-life decisions in neonates and infants in Flanders' (2005) 365 *Lancet* 1315–20; Wilkinson, Fitzsimons, Dargaville, Campbell, Loughnan, McDougall, and Mills, 'Death in the neonatal intensive care unit: changing patterns of end of life care over two decades' (2006) 91 *Archives of Disease in Childhood Fetal and Neonatal Edition* F268–71.

[3] Robertson and Wyatt, 'The magnetic resonance revolution in brain imaging: impact on neonatal intensive care' (2004) 89 *Archives of Disease in Childhood Fetal and Neonatal Edition* F193–7.

[4] Shevell, Majnemer, and Miller, 'Neonatal neurologic prognostication: the asphyxiated term newborn' (1999) 21 *Pediatr. Neurol.* 776–84.

[5] For example the case of *NHS Trust A v MB* ([2006] 2 FLR 319, para. 31: '*Re MB*' (see below for further discussion)), or the case of *Re OT* [2009] EWHC 633 (Fam), para. 119, 127: '*Re OT*'), where brain imaging was taken as confirmatory evidence that the infant was unlikely to consciously control his movements: 'Dr. Pohl's view was that it was highly unlikely that a child of this age could kick one leg only, which is what is described, and his view, particularly taken in conjunction with his analysis of the MRI scan of the damage to OT's brain, and the lack of other similar observations, is that this was entirely consistent with a twitching movement and that it is highly unlikely to have been purposeful.'

[6] We will focus in this paper specifically on UK guidelines and UK case law, but the principles used in the UK are similar to those used elsewhere. The implications of the Carmentis machine are likely to be similar in other jurisdictions.

prognosis. What implications would the development of this machine have for the law and for practice?

16.2 Part 1: Thought Experiment: The Carmentis Machine

In ancient Rome, Carmentis was revered as a goddess of pregnancy and childbirth. It is said that newborn infants were brought to the temple of Carmentis, where the priestess would drink from a sacred spring, and then sing a song that disclosed the child's future.[7]

Now imagine: it is 2025. Newborn intensive care has come a long way since its earliest attempts to keep critically ill premature and term infants alive in the late 1960s and early 1970s. No longer are neonatal intensive care units (NICUs) places of intense noise, artificial light and feverish activity. Acoustic and luminescent shielding and thermal cocoons provide an environment that allows newborns undisturbed rest for large portions of the day, maximizing growth and neurodevelopmental potential. Sophisticated liquid ventilators provide gas exchange for infants unable to breathe independently without tearing and destroying fragile developing lung sacs. Built into the infants' clothing are sophisticated physiological monitors that feed back to and automatically adjust the systems regulating the environment and supporting the organs.

But despite improvements in the capacity to mitigate and prevent the complications of serious illness, some newborn infants still have irreversible brain injury. In the centre of modern units is a new machine—dubbed the Carmentis machine by the clinicians—that provides highly detailed images of the structure and functional connections of the newborn brain. Within a couple of days of life, or following major changes in condition, all infants in the NICU are taken to the machine. Like the goddess Carmentis in ancient Rome the machine is consulted for a vision of the future for a newborn. It provides detailed and highly accurate predictions of the future cognitive capacity of infants (including sub-scores for capacities such as language, abstract reasoning, creative thinking, and emotiveness), sensory deficits, and motor abilities. These predictions form the basis for decisions about the continuation or withdrawal of intensive care.[8]

In order to make the thought experiment plausible (and useful) it may be worthwhile setting out the limits of our hypothetical machine. It provides accurate predictions of future capacities, but it will not be able to determine whether or not other co-incident illnesses will occur that could affect the future of an infant (for example the infant might develop meningitis, or might suffer head injury after a fall), or whether or not future treatments (currently unavailable) will alter capacities or abilities. Furthermore it will not be able to determine the future mental states of an individual: it will not

---

[7] Christian Roy, *Traditional Festivals: A multicultural encyclopaedia* (Santa Barbara, California, ABC-Clio Ltd, 2005) 264; Pierre Grimal, *Dictionary of classical mythology* (Oxford, Wiley-Blackwell, 1990) 89.
[8] The killing of deformed infants and the abandonment of unwanted infants are thought to have been commonplace in ancient Rome: W. V. Harris, 'Child-exposure in the Roman-Empire' (1994) 84 *Journal of Roman Studies* 1–22. But such practices were not usually related to prophecies. There is, though, an intriguing story that an evil prophecy led the father of Augustus to consider destroying him shortly after his birth (ibid. 14, note 121).

be able to predict whether that individual will or (if allowed to live) would have pleasurable or painful experiences, or will or would (if it knew its own future) want to continue to live. Although we are attributing to it semi-magical powers of prediction, it will not be able to quantify interests, 'benefits', 'burdens', 'tolerability', or 'awfulness'.

It will be helpful to discuss some specific (hypothetical) cases:

Case 1: Baby Amelia was delivered in very poor condition following a planned home birth. She was resuscitated by the ambulance paramedics when they arrived at fifteen minutes of age, but had early evidence of hypoxic brain damage with severe encephalopathy and refractory seizures. Neuroprotective treatments were commenced on arrival in the NICU but the machine predicts that she will have profound motor impairment with severe spastic quadriplegic cerebral palsy and mild intellectual impairment (IQ 68). Expressive language will be limited by the degree of physical disability, but communication will be possible.

Case 2: Baby Angelos was noted before birth to have polyhydramnios (excess amniotic fluid) and to be growth restricted. After delivery he was noted to have multiple dysmorphic features, contractures, and to move less than normal. A severe congenital myopathy (a muscle disorder) is diagnosed. The Carmentis machine predicts that he will remain dependent on the mechanical ventilator and will require home ventilation if he survives. He will be moderately intellectually impaired with an IQ of 45. He will have severe muscle weakness and limited voluntary movements. Unless he has major surgery in mid childhood, he will develop progressive scoliosis and restrictive lung disease.

Case 3: Baby Philip was born prematurely. He was resuscitated and initially stabilized using partial liquid ventilation. On day three of life the Carmentis machine reveals evidence of widespread white-matter injury, and evolving changes in the periventricular area. The machine predicts that he will be severely cognitively impaired with a predicted IQ of 30, and mild motor impairment with spastic diplegic cerebral palsy. He will have epilepsy which is moderately difficult to control. He will have up to ten words of expressive language.

Case 4: Baby Chloe developed abnormal movements in the first days after birth. She was found to have hydranencephaly. The machine predicts profound intellectual impairment (IQ unmeasurable but <20), cerebral blindness, and deafness. She is not predicted to have significant motor impairment.

Box 1 Case summaries with details of physical and cognitive deficits[9]	
Case 1: Baby Amelia	Mild cognitive impairment (IQ 68), severe physical impairment
Case 2: Baby Angelos	Moderate cognitive impairment (IQ 45), profound physical impairment, ventilator dependent
Case 3: Baby Phillip	Severe cognitive impairment (IQ 30), mild-moderate physical impairment
Case 4: Baby Chloe	Profound cognitive impairment, minimal physical impairment

[9] These cases obviously do not encompass all possible combinations of impairments. The aim is that they represent a spectrum of realistic cases involving fairly substantial cognitive or motor deficits. The cases and names are entirely fictitious.

16.3 Guidelines

Specific guidelines may of course have changed by the time that a machine such as the Carmentis machine is available. But in this section we will look at how current professional guidelines in the UK for treatment withdrawal would apply to certain predictions of significant impairment such as those arising from the machine.

16.3.1 Royal College

The Royal College of Paediatrics and Child Health (RCPCH) published a revised framework for decisions about the withdrawal or withholding of life-sustaining treatment in children in 2004.[10] This framework has been cited by the courts[11] and is used by clinicians in practice.[12] It is based fundamentally upon a 'best interests' determination, but the guidelines delineate five settings where treatment limitation decisions could be appropriate. Of these, the fourth, (the 'no-purpose situation') is most relevant to the machine.[13] The guidelines suggest that this situation is present when survival is possible but the degree of physical or mental impairment will be so great that it is 'intolerable'.

The framework justifies this threshold for withdrawing or withholding treatment on the basis of the legal judgement in *Re J*.[14] The guidelines suggest two possible interpretations of intolerable: 'that which cannot be borne' or 'that which an individual should not be asked to bear', though seems to favour the latter. The guideline notes that: '. . . a quality of life which could be considered intolerable to an able-bodied person, would not necessarily be unacceptable to a child who has been born disabled'.[15]

Usefully, the guideline also provides some examples of what the EAP-RCPCH believes this to refer to in practice. The examples include 'when there is little or no

---

[10] Royal College of Paediatrics and Child Health, *Witholding and Withdrawing Life-saving Treatment in Children: A framework for practice* (London, Royal College of Paediatrics and Child Health, 2004).

[11] *Re OT* n 5 above, paras. 29–30; *An NHS Trust v D* [2000] 2 FLR 677 para. 77, *Re K* [2006] 2 FLR 883 paras. 37–9).

[12] Street, Ashcroft, Henderson, and Campbell, 'The decision making process regarding the withdrawal or withholding of potential life-saving treatments in a children's hospital' (2000) 26 *Journal of Medical Ethics* 346–52.

[13] Other categories of possible relevance include category one (if the machine predicted or diagnosed brain death), or category two (if the machine predicted a permanent vegetative state). Both of these diagnoses can be made currently on clinical grounds, though the latter is usually only diagnosed after a prolonged period of lack of clinical recovery (usually >12 months). The machine might accelerate this diagnosis. The categories defined by the RCPCH are confusingly named. The 'no purpose' situation refers to *impairment* so great that it is 'unreasonable to ask them to bear it'. (It is not explained why such treatment would lack purpose). The 'unbearable' situation refers to progressive and irreversible illness and *treatment* that is more than can be borne—for example children with chronic renal failure or malignancy: Doyal, Goldman, Larcher, and Chantler, 'Palliative medicine and children: ethical and legal issues' in Doyle, Hanks, Cherney, and Calman (eds.), *Oxford textbook of palliative medicine* (Oxford, Oxford University Press, 2005).

[14] *Re J (a minor) (wardship: medical treatment)* [1991] Fam 33: 'Re J'.

[15] Note 10 above, Royal College of Paediatrics and Child Health, 26.

prospect of meaningful interaction with others or the environment'.[16] The guideline suggests that spastic quadriplegia with severe associated cognitive and sensory deficits may be one qualifying condition.

Since the RCPCH guidelines are reasonably specific about the type of disability that would justify withdrawal of life-sustaining treatment, one might think that it would be reasonably straightforward to apply the guidelines to predictions of impairment such as those generated by the Carmentis machine. However, some difficulty in interpretation will remain. It is not clear when interaction with the environment becomes 'meaningful'. For example, does the capacity to communicate connote 'meaningful interaction'? Is the recognition of faces or voices sufficient? The examples given imply that that severe cognitive impairment is a necessary component of intolerable disability as envisaged by the RCPCH committee, but there is no attempt to specify what this means in practical terms. Conventionally severe intellectual disability refers to an IQ of 20–34.[17] Would it be permissible to withdraw life support if the Carmentis machine predicted an IQ of just less than 35, but there were no other major impairments (Case 3, Baby Phillip, is just such an example)? The example of severe cerebral palsy with concomitant cognitive and sensory deficit may be helpful, but the guidelines do not imply that this is the only situation in which withdrawal would be permitted. It will be harder to apply the RCPCH guidelines to other situations of severe impairment predicted by the machine (for example Cases 2 or 4).

16.3.2 General Medical Council[18]

More generic guidelines about decisions relating to treatment withdrawal have been provided by the General Medical Council (GMC) and the British Medical Association (BMA). GMC guidelines draw on case law, legislation, and the European Convention on Human Rights (ECHR). They set out general principles for decision-making emphasizing, in particular, the importance of best interests and a weighing up of benefits and burdens. There is little specific guidance about whether or when impairment might be sufficient to justify withholding treatment, though the guidelines acknowledge that it may be appropriate to assess quality of life.[19]

16.3.3 British Medical Association

The British Medical Association last revised its guidance document in 2007.[20] This guideline covers decision-making for both adults and children, and stipulates that

---

[16] Note 10 above, Royal College of Paediatrics and Child Health, 24.
[17] James Harris, *Intellectual Disability: Understanding its development, causes, classification, evaluation, and treatment* (Oxford, Oxford University Press, 2006) 54.
[18] General Medical Council, *Withholding and withdrawing life-prolonging treatments: good practice in decision-making* (London, GMC, 2006).
[19] Ibid. Appendix A. para. 1.
[20] British Medical Association, *Withholding and Withdrawing Life-prolonging Medical Treatment: Guidance for decision making* (Malden, Oxford, Blackwell, 2007).

the ethical principles covering infants should be the same.[21] According to this document, quality of life considerations are unavoidable, and may legitimately be taken into account. Specifically, criteria for assessing best interests in both infants and older individuals may include the capacity to develop awareness, to interact, and for self-directed action.[22] The guidelines discuss various legal cases, but do not provide examples of conditions that would justify withdrawal.

16.3.4 Nuffield Council on Bioethics

One of the most recent and relevant documents reviewing decision-making in newborn intensive care is the report by the Nuffield Council on Bioethics.[23] This report is not a professional guideline, but it has been extensively cited since publication,[24] and may influence decision-making. It covers a range of pre- and post-birth decisions for extremely premature and term newborn infants. The report, as with some guidelines, emphasizes the primacy of the best interests of the child in determining whether or not to provide treatment. But the authors also draw on the concept of intolerability:'It would not be in the baby's best interests to insist on the imposition or continuance of treatment to prolong the life of the baby when doing so imposes an intolerable burden upon him or her.'[25]

What sort of predictions from the Carmentis machine would impose an 'intolerable burden'? The report endorses the categories outlined in the RCPCH guideline and described above. In particular it suggests that providing burdensome treatment to a child predicted to have a life 'bereft of those features that give meaning and purpose to human life'[26] may impose an intolerable existence. These features are not elaborated, but in a separate part of the report the authors discuss the potential benefits of treatment to be included in a best interests determination: these include the capacity to establish relationships with others, the ability to experience pleasure, and independence from life support.[27]

In other parts of the report there is discussion of disabilities including cerebral palsy, sensory impairment, and learning disabilities. The authors did not believe that the future was 'intolerable' for a premature infant with predicted significant motor disability (spastic diplegia), but uncertain cognitive impairment.[28] On the other hand the report judged that it would have been permissible to withdraw treatment

---

[21] Ibid. 102.   [22] Ibid. 106.
[23] Nuffield Council on Bioethics, *Critical Care Decisions in Fetal and Neonatal Medicine: Ethical issues* (London, Nuffield Council on Bioethics, 2006).
[24] For example Ahluwalia, Lees, and Paris, 'Decisions for life made in the perinatal period: who decides and on which standards?' (2008) 93 *Archives of Disease in Childhood: Fetal and Neonatal Edition* F332–F335; Rennie and Leigh, 'The legal framework for end-of-life decisions in the UK' (2008) 13 *Seminars in Fetal and Neonatal Medicine* 296–300.
[25] Note 23 above, Nuffield Council on Bioethics 12. The report admits to some difficulty in defining this concept, admitting that people may disagree both about what constitutes intolerability, and whether or not a particular infant's condition is intolerable.
[26] Ibid. 12 para. 2.13.   [27] Ibid. 161 para. 9.33.
[28] Ibid. 99–100. It is not clear in the specific case whether the authors and doctors would have reached a different judgement if the infant had still been dependent on mechanical ventilation. It is also not clear how much of this judgement was related to uncertainty.

from a newborn predicted to be severely impaired, functionally dependent and unable to sit or eat,[29] perhaps implying that withdrawal in the case of Case 4 Baby Chloe would be permitted.

16.4 The Law

How would the courts respond to predictions from the Carmentis machine?[30] Where a patient cannot himself or herself weigh the benefits and burdens of treatment and come to their own conclusion about it, a decision is made on their behalf. The decision-maker is required to assess where the patient's best interests lie. In the case of adult incompetent patients the Mental Capacity Act 2005 will apply. The decision-maker is required to go through the criteria listed in s. 4 of the Act. In the case of children the Mental Capacity Act does not apply, but the Children Act 1989 does. The Children Act stipulates that the 'child's welfare shall be the court's paramount consideration'.[31] Further limits are provided by the Human Rights Act 1998 which for most practical purposes can be regarded as grafting the ECHR into the domestic law of the UK. Article 2 of the ECHR protects the right to life of infants, while Art. 3 protects against inhuman or degrading treatment. There is little case law relating to best interests determinations in very young infants; Table 1 summarizes the most relevant cases. We discuss those cases briefly below, and then discuss in more detail how the courts have more generally related severe disability to best interests.

16.4.1 General features of case law relating to treatment withdrawal in the UK

Of the cases that have come before the courts only two (*Re B*,[32] *Arthur*[33]) relate to decisions made in the newborn period (i.e. in the first month of life). This is perhaps understandable given the time-lag that often occurs before cases reach the court. Cases where parents and doctors agree about continuing or withdrawing treatment do not come before the courts. Where there is disagreement, there is usually significant reluctance to resort to legal proceedings: that reluctance often results in some time passing before doctors or parents seek legal intervention. But there is no legal precedent for treating decisions in newborn infants differently from older infants and children.

---

[29] Ibid. 101.
[30] There are potential differences between the three different jurisdictions within the UK (England and Wales, Scotland, and Northern Ireland). For the most part, however, these differences are not relevant to the decisions discussed here.
[31] CA 1989, s. 1(1).
[32] *Re B (A minor) (wardship: medical treatment)* [1981] 1 WLR 1421.
[33] *Re Arthur* (1981) 12 BMLR 1.

Table 1 Previous legal cases in the UK relating to treatment withdrawal in newborn infants.

Case	Medical condition	Age	Setting	Parents' views (supported or opposed doctors)	Type of decision (what was being sought?)	Ultimate decision
Re B (a minor) (wardship: medical treatment) [1981] 1 WLR 1421	Down syndrome, duodenal atresia.	newborn	Parents refusing consent for surgery, doctors seeking permission to operate.	Opposed	Continuation (surgery)	Permission given for surgery
R v Arthur (1981) 12 BMLR 1	Down syndrome, uncomplicated.	newborn	Review after death. Given analgesia and water only.	Supported	Legality of decision to withhold treatment (and administer painkillers)	Prosecuted but acquitted
Re C (a minor: wardship: medical treatment)(No.1) [1990] Fam 26	Congenital Hydrocephalus, predicted severe mental handicap, blindness, probable deafness, and spastic quadriplegic cerebral palsy. Unclear duration of survival.	4 months	In care, doctors sought leave to manage palliatively.	In care	Withholding	Antibiotics, nasogastric and intravenous feeding withheld
Re J (a minor) (wardship: medical treatment) [1991] Fam 33	Premature (27/40), blind, ?deaf, unlikely to be able to communicate or develop even limited intellectual abilities.	4 months	In care; doctors sought judgment on whether lawful to withhold ventilation if needed.	In care	Withholding	Treatment withheld

(continued)

Table 1 (Continued)

Case	Medical condition	Age	Setting	Parents' views (supported or opposed doctors)	Type of decision (what was being sought?)	Ultimate decision
Re J (A Minor) (Child in Care: Medical Treatment) [1993] Fam 15	Profoundly mentally and physically handicapped, microcephaly, cerebral palsy, cortical blindness, epilepsy (following head injury at 1/12).	17 months	Appeal—initially court granted injunction requiring doctors to resuscitate if C needed it.	Opposed (in care)	Withholding	Treatment withheld
Re C (A Baby) [1996] 2 FLR 43	Meningitis, severe brain damage, cortical blindness.	3 months	Doctors wanted to withdraw.	Supported	Withdrawal	Ventilation withdrawn
Re C (medical treatment) [1998] 1 FLR 384	SMA1.	16 months	Doctors sought authority to withdraw treatment, parents did not want to.	Opposed	Withdrawal/ withholding	Ventilation withdrawn
An NHS Trust v D [2000] 2 FLR 677	Ex 31-week premature infant, severe CLD, lissencephaly and Dandy Walker (severe developmental delay), heart failure, hepatic dysfunction, renal dysfunction.	19 months	Doctors wanted to withhold intensive care on the grounds that it was not in the child's best interests.	Opposed	Withholding	Treatment withheld

Case	Condition	Age	Facts	Parents	Withholding/Withdrawal	Outcome
Wyatt v Portsmouth NHS Trust v Wyatt [2005] 1 FLR 21, *Portsmouth Hospitals NHS Trust v Wyatt* [2005] 1 WLR 3995	Ex premature infant, blind, deaf, incapable of movement or response, chronic respiratory and kidney problems.	1–3yrs	Doctors sought to withhold ventilation if needed, parents disagreed.	Opposed	Withholding	Court initially sided with doctors—treatment could be withheld. Despite poor prognosis condition improved. At 2 yrs earlier order rescinded. At 2.5 yrs—deteriorated—order reinstated. Treatment withheld, but condition improved.
Re L (A Child) (Medical Treatment: Benefit) [2005] 1 FLR 491	Trisomy 18, multiple heart defects, chronic respiratory failure, gastroesophageal reflux, severe developmental delay, epilepsy, and hypertonia.	9 months	Doctors sought permission not to ventilate if required. Mother wanted it.	Opposed	Withholding	Treatment withheld
An NHS Trust v MB [2006] 2 FLR 319	SMA 1, normal intellect	18 months	Doctors sought authority to withdraw treatment, parents did not want to.	Opposed	Withdrawal	Treatment continued

(continued)

Table 1 (Continued)

Case	Medical condition	Age	Setting	Parents' views (supported or opposed doctors)	Type of decision (what was being sought?)	Ultimate decision
K (A minor) [2006] 2 FLR 883	Severe congenital myotonic dystrophy, very poor long term neurological prognosis, predicted short survival, total dependency if survives, dependent on intravenous nutrition.	5.5 months	Doctors and parents wanted to withdraw intravenous nutrition and manage palliatively, child in care of local authority—therefore came to court.	In care	Withdrawal	Treatment withdrawn
Re OT [2009] EWHC 633 (Fam)	Mitochondrial disorder, stroke, epilepsy, ventilator dependent.	9 months	Doctors wanted to withdraw ventilation; parents wanted it to continue.	Opposed	Withdrawal	Ventilation withdrawn

Most of the cases that have come before the courts have related to *withholding* potentially life-saving treatment.[34] But the courts have also authorized withdrawal of life-saving treatment in a number of recent cases.[35] Most cases arose when doctors wanted to limit or discontinue treatment but parents wanted to continue treatment. In several cases the court was asked to adjudicate because the infant was in state care. There is only one reported case (*Re B*) in which parents have sought limitation or discontinuation of treatment against the advice of doctors. There is also only one case in the UK in which there was an attempt to prosecute doctors or parents after decisions leading to the death of an infant (*Arthur*).

In almost all of the reported cases the courts ultimately endorsed the doctors' judgement. There is one recent exception. That is the case of *Re MB*[36] where the court declined to authorize withdrawal of mechanical ventilation. We discuss that case further below. Why have the doctors' judgements generally been adopted? We suggest that the main reason is that by and large the only cases to come to court are those where doctors have no doubt at all about what to do. Where there is doubt, doctors will comply with the wishes of the family, and accordingly continue treatment. All except two of the cases (*Re MB*[37] and *Re C* (1998)[38]) involved infants with predicted cognitive impairment, and in most cases the infants had additional illnesses that were likely to lead to death within a fairly short period irrespective of the treatment decision. Consequently in many cases decisions were based upon a judgement that further treatment was futile.[39] In a few cases, however, the courts explicitly made judgments about the infant's actual and/or predicted quality of life. These were linked to the underlying best interests determination in one of two ways: either in terms of 'intolerability', or in a weighted balance of benefits and burdens.

16.4.2 Intolerability

The notion that impairment could be so severe that it would be permissible to withhold life-saving treatment is first seen in a case reported in the law reports in 1981 (*Re B*).[40] There, the Court of Appeal authorized surgery for an infant with Down syndrome and duodenal atresia despite his parents' refusal of consent, on the grounds that the infant's life was not bound to be 'demonstrably awful' or 'intolerable'.[41] But one of the judges noted that: 'There may be cases, I know not, of severe proved damage where... the court might be driven to a different conclusion.'[42]

---

[34] Loane Skene, *Law and Medical Practice* (Chatswood, LexisNexis Butterworths, 2008). ch. 11.
[35] *Re C (A Baby)* [1996] 2 FLR 43, *Re C (medical treatment)* [1998] 1 FLR 384, *Re K, Re OT* note 12, above.
[36] Note 5, above.
[37] Ibid.
[38] Note 34, above.
[39] For example *Re OT* [2009], *Re D* 2000, *Re K* 2006, note 11, above.
[40] Note 32, above.
[41] *Re B*, note 32, above, 1424, para. B, H.
[42] Ibid. 1424, para. C.

Several subsequent cases drew on this judgment as providing a guide to decision-making. These cases acknowledged the presumption (embodied in the sanctity of life principle) that it is in the patient's best interests to live. But this presumption may be rebutted where there is evidence that continued existence would be intolerable.

16.4.2.1 Re C (1990)[43]

C was a four-month old infant who had been made a ward of court shortly after birth. She had severe congenital hydrocephalus, and, despite a shunt procedure in the newborn period, was predicted to develop 'severe mental handicap, blindness, probable deafness and spastic cerebral palsy of all four limbs'.[44] Because she was a ward of court doctors had applied for permission to manage her palliatively, and the court provided an order directing that this occur. The Official Solicitor appealed the decision and the case went to the Court of Appeal.

The Court of Appeal endorsed the earlier judgment that palliative care was appropriate, though there was some equivocation about the grounds for the decision. Lord Donaldson in several places said that C was dying, while Balcombe LJ noted that the child was terminally ill. Both observations were based on the following statement of the medical expert: 'I do not believe there is any treatment which will alter the ultimate prognosis, which is hopeless.'[45] It is not clear however, that this was actually the sense of hopelessness intended by the testifying doctor. The words immediately following 'hopeless' were 'She has a massive handicap as a result of a permanent brain lesion.' In other places the same doctor stated 'The high standard of care makes it difficult to forecast how long she will live.'[46]

The first instance judgment in *Re C* was more explicit about the decision being based upon impairment—specifically cognitive impairment—and linked this to the language used in the Court of Appeal judgment in *Re B*.[47]

> I adjudge that any quality to life has already been denied to this child because it cannot flow from a brain incapable of even limited intellectual function. In as much as one judges, as I do, intellectual function to be a hallmark of our humanity, her functioning on that level is negligible if it exists at all. Coupled with her total physical handicap, the quality of her life will be demonstrably awful and intolerable.[48]

Another decision in the Court of Appeal the following year is often cited in support of 'intolerability' and provides the clearest statement of the intolerability test.

---

[43] *Re C (a minor: wardship: medical treatment) (No. 1)* [1990] Fam 26: '*Re C*'.
[44] Ibid. 32.   [45] Ibid. 37.   [46] Ibid. 32.
[47] Note 32, above.
[48] *Re C* 1990 note 43, above, 35C.

16.4.2.2 Re J (1991)[49]

J was an ex-premature infant with severe brain injury, and, like C, was predicted to develop spastic quadriplegia, blindness, and deafness. J too was a ward of court (for reasons unrelated to the question at hand), and doctors sought the court's approval of withholding ventilatory support in the event of a cardiorespiratory collapse. In this case there was clear acknowledgement that J was not terminally ill, and that with treatment he may survive into late childhood or adolescence. But the court rejected the argument of the Official Solicitor that it was never justifiable to withhold life-saving treatment on the basis of a child's quality of life, saying that it could be in the best interests of a child to withhold such treatment. This conclusion was justified using the language of intolerability: 'where, viewed from... [his position], his future life might be regarded as intolerable to him the court acting solely on his behalf might properly choose a course of action which did not prevent his death'.[50]

In *Re B*, Down syndrome (which is usually associated with at most moderate cognitive impairment and mild physical impairment) was judged *not* to be intolerable. In contrast, in both *Re C* and *Re J* spastic quadriplegia and profound cognitive impairment combined with blindness and deafness were held to be intolerable. But in another more recent case with very similar impairments there was disagreement about whether or not this level had been reached. This was the case of Charlotte Wyatt.[51] She was an ex-premature infant with microcephaly, spastic quadriplegic cerebral palsy, and profound cognitive and sensory impairments. There was some equivocation about intolerability, since her day-to-day life was judged *not* to be intolerable,[52] though it would *become* intolerable were she to become ventilator-dependent. It is unclear whether courts would judge as intolerable other degrees of impairment (for example in the cases listed at the start of this paper).

More recently the intolerability test has been sidelined in a number of decisions relating to treatment withdrawal (notably *Wyatt*[53] and, in the case of an adult, *Burke*[54]). In fact in *Re J* only one of three judges (Taylor LJ) justified the decision on the grounds of intolerability. Lord Donaldson and Balcombe LJ explicitly rejected the idea that the phrases used in the initial *Re B* case should be applied as a test for determining treatment decision: 'I do not think that we are bound to, or should, treat Templeman L.J.'s use of the words "demonstrably so awful" or Dunn L.J.'s use of the word "intolerable" as providing a quasi-statutory yardstick.'[55]

---

[49] Note 14, above.
[50] *Re J*, note 14, above, 34 para. C.
[51] *Wyatt v Portsmouth Hospitals NHS Trust* [2005] 1 WLR 3995.
[52] Ibid. 4005 E.
[53] Ibid.
[54] *R (on the application of Burke) v General Medical Council* [2006] QB 273.
[55] *Re J*, note 14, above, 46F.

16.4.3 Balance sheet

It is now commonplace in cases involving the determination of best interests in both children and incompetent adults for the court to draw up a 'balance sheet'. This does not necessarily connote any substantive change in the law relating to the assessment of best interests: it is simply a convenient way of identifying the relevant issues. The approach has its origins in *Re A (Mental Patient: Sterilisation)* (which concerned the proposed sterilization of a twenty-nine-year-old man with Down syndrome). Thorpe LJ suggested that that in order to make an evaluation of best interests the first step was:

> [to] draw up a balance sheet. The first entry should be of any factor or factors of actual benefit... Then on the other sheet the judge should write any counterbalancing dis-benefits to the applicant.... At the end of that exercise the judge should be better placed to strike a balance between the sum of the certain and possible gains against the sum of the certain and possible losses. Obviously only if the account is in relatively significant credit will the judge conclude that the application is likely to advance the best interests of the claimant.[56]

This approach was a major part of the decision-making in the case of MB.

16.4.3.1 Re MB (2006)[57]

MB was an eighteen-month old infant with a severe congenital neuromuscular disorder (type 1.1 spinal muscular atrophy). He had been in hospital since seven weeks of age, and ventilator-dependent for six months. His condition led to progressive loss of muscle strength and tone, so that although he could initially cry audibly, smile, and move his limbs, by the time of the court hearing he could only move his eyes.[58] Although MB had profound motor impairment, he was not believed to have cognitive impairment. MB's doctors believed that continuing mechanical ventilation was 'cruel', and that it would be in his best interests to withdraw life-saving treatment and allow him to die. His parents on the other hand opposed the withdrawal of treatment.

Holman J asked the advocates on either side to draw up a list of the benefits and burdens of continuing or discontinuing mechanical ventilation, and included the list provided by the guardian in his judgment. He placed significant emphasis on the process of weighing up benefits and burdens, but he noted huge difficulties in reliably appraising the benefits of treatment, deciding what weight to give to future burdens, and in arriving at an overall balance.[59] The judge ruled that continued

---

[56] *Re A (Mental Patient: Sterilisation)* [2000] 1 FLR 549. The judge's reference to 'significant credit' may mean that this balancing exercise yields a similar answer to the intolerability test. In any event, as we have pointed out, the balance sheet is not intended to change the answer to the question 'What is in the patient's best interests?' It is simply intended to make that answer more easily ascertainable.

[57] Note 5, above.

[58] There was disagreement about whether or not he was able to initiate a flicker of movement in his eyebrows, corners of his mouth and toes.

[59] Ibid. para. 62.

mechanical ventilation (though not surgical tracheostomy) was in the best interests of MB. In reaching this decision the judge placed importance on the absence of cognitive and sensory impairment.

> So far as I am aware, no court has yet been asked to approve that, against the will of the child's parents, life support may be withdrawn or discontinued, ... [from] a conscious child with sensory awareness and assumed normal cognition and no reliable evidence of any significant brain damage.[60]
>
> As he can hear and see, I accept the evidence of his parents that he is attentive to TV, DVDs, CDs, stories and speech; and as all these things may give pleasure to other children of 18 months, I must and do assume they give pleasure to him.[61]
>
> But [MB's]... life does in my view include within it the benefits that I have tried to describe... Within those benefits, and central to them, is my view that on the available evidence I must proceed on the basis that M has age-appropriate cognition, and does continue to have a relationship of value to him with his family, and does continue to gain other pleasures from touch, sight and sound.[62]

The case is interesting. Its significance is debatable. The judge endorsed the idea of using a balance sheet of benefits and burdens to determine best interests, supported MB's parents against the unanimous opinion of medical experts, and placed significant emphasis on the presence of normal cognition and sensory abilities. If a similar approach were applied to predictions from the Carmentis machine, it might be anticipated that life-saving treatment would be judged in the best interests of Case 1 (Baby Amelia) predicted to have severe motor disability but relatively normal cognition. But given the substantial difficulties in the balancing process alluded to by Holman J it is simply not clear how courts would respond to other situations. Would mild or moderate cognitive impairment sufficiently reduce the benefits of treatment such that the balance became tipped in favour of withdrawal (for example case 2)? Would a single sensory impairment (for example blindness) significantly affect the balance?

To summarize this part: decisions in cases that have come before the courts have placed significant emphasis on the presence or absence of cognitive impairment. The language of that emphasis has varied. Some judges have found the notion of intolerability helpful. Others have not—preferring instead to refer solely to the overriding best interests test in auditing the benefits and burdens of life and treatment. The courts have not clearly indicated the severity of impairment that would justify the withdrawal or withholding of life-sustaining treatment.

16.5 Part 2: Application of the Best Interests Test to the Carmentis Machine

Reported cases do not provide enough information to know how courts would respond to predictions from the Carmentis machine. Nor do published guidelines.

---

[60] Ibid. para. 11. [61] Ibid. para. 65. [62] Ibid. para. 101.

In this section we apply a philosophical analysis to these two approaches, and assess how predictions from the Carmentis machine could or should be interpreted in terms of these two tests.

16.5.1 Intolerability

What might the Carmentis machine do that is of relevance to the courts? It might tell the decision-maker:

1) about the duration of survival for the child;
2) about the nature of impairment for the infant and the presence of painful symptoms;
3) about the timing of progression (or improvement for the infant).

But the output 'intolerable' would not appear on the machine, and accordingly it will be useful to clarify how intolerability would be determined from a specific prognosis.

The concept of 'intolerability' has two elements. First, there is the *subjective* nature of the judgement. In this it has some similarities with the use of substituted judgement in decision-making for incompetent patients.[63] Thus, Taylor LJ in *Re J*:

> I consider the correct approach is for the court to judge the quality of life the child would have to endure if given the treatment and decide whether in all the circumstances such a life would be so afflicted as to be intolerable to that child. I say 'to that child' because the test should not be whether the life would be intolerable to the decider. The test must be whether the child in question, *if capable of exercising sound judgment*, would consider the life tolerable. [emphasis added][64]

For adults, who were previously competent, one way of approaching medical decision-making is to determine (if possible) the decision that the patient would make were they competent now. This is the notion of 'substituted judgement', and it is an attempt to respect the autonomy of the patient, as assessed by reference to the patient's previously expressed views. This is most straightforward if the patient has previously made a written advance directive, but in other cases family and friends provide evidence of the preferences of the patient.

Decisions about patients who have never been competent (for example infants) are more complicated, since those patients have not been in a position to express preferences about treatment. An attempt to apply substituted judgement to these sorts of decisions is accordingly, and obviously, particularly difficult.

There are three possible ways of arriving at such an imagined judgement.[65] The first is to imagine that the infant now were able to choose. But since we have

---

[63] Buchanan and Brock, *Deciding for Others: The ethics of surrogate decision making* (Cambridge, Cambridge University Press, 1989) 10.
[64] *Re J*, note 14, above, 55F.
[65] David W. Archard, 'Children's rights' in Edward N. Zalta (ed.), *The Stanford Encyclopedia of Philosophy* (Winter 2008 edn), <http://plato.stanford.edu/archives/win2008/entries/rights-children/>. Archard highlights different versions of these potential perspectives but they have similar indeterminacies.

no knowledge of the opinions and preferences of infants, let alone infants with significant actual and future impairments, it is very difficult to know what judgement the infant in question would make.

The second possibility is to imagine the infant grown to maturity (with impairment), and to deploy, for the purposes of present decision-making, what it is supposed their future retrospective judgement would be (i.e. what they would have preferred to have happened). This possibility is somewhat more appealing, since we know something about what adults with some impairments think about the quality of their lives. But the views of such individuals may not always be helpful. For example, we have no way of knowing what the views of severely cognitively impaired adults would be about their treatment in the newborn period. By definition, only the views of those who have survived will be represented: the views of infants and children who died prior to reaching adulthood are unrepresented.

The third alternative is to imagine the infant as a competent adult reflecting on treatments that may or may not be provided for them in infancy. This alternative is somewhat akin to asking what a reasonable person would decide on behalf of the child, but it is also problematic since it is potentially subject to bias,[66] and requires the person making the judgement to imagine lacking (and never having had) capacities that are essential to their own identity (and hence, inevitably, to their judgement).[67] The notion of substituted judgement for neonatal treatment decisions therefore faces substantial epistemic problems.

The other conceptual component of 'intolerability' is the sense that it involves a *particularly bad* state of existence. Taylor LJ referred to 'extreme' circumstances and the 'cruelty' of life.[68] The judge in *Re B* referred to a life 'full of pain and suffering'.[69] Similarly the Nuffield report refers to 'extreme suffering or impairment'.[70]

When is future impairment intolerable? Which predictions of the Carmentis machine would warrant withdrawal of treatment on this basis? Many of those who have invoked the concept of intolerability have cited conditions involving severe physical suffering. But they have also referred particularly to conditions involving severe or profound cognitive impairment. For example Taylor LJ described 'a child ... so damaged as to have negligible use of its faculties and the only way of preserving its life was by the continuous administration of extremely painful treatment ... or ... sedated continuously as to have no conscious life at all'.[71] As noted earlier, Ward J in the case of baby C referred to 'intellectual function as the hallmark of our humanity'.[72] The RCPCH guideline argued that the lack of

---

[66] Boddington and Podpadec, 'Measuring quality of life in theory and in practice' in Kuhse and Singer (eds.), *Bioethics: An anthology* (Oxford, Blackwell, 1999).
[67] Wilkinson, 'Is it in the best interests of an intellectually disabled infant to die?' (2006) 32 *J Med Ethics* 454–9.
[68] *Re J*, note 14, above, 55.
[69] *Re B*, note 32, above, 1424C.
[70] Note 23, above, Nuffield Council on Bioethics, 13.
[71] *Re J*, note 14, above, 55c. Taylor LJ (at 56B) also listed as first amongst the factors justifying withholding treatment in the specific case of J that '... the severe lack of capacity of the child in all his faculties which even without any further complication would make his existence barely sentient ...'.
[72] *Re C* 1990, n 43 above, 35C.

the capacity for meaningful communication would make life intolerable,[73] and the Nuffield report refers to lives lacking those features that give life meaning and purpose.[74]

But one potential problem for these accounts is what we call the Tolerability Paradox: the paradoxical relationship between subjective tolerability and cognitive impairment.[75] While it is thought to be potentially intolerable for an individual to have a life lacking the capacities listed above, there is a sense in which it may be *better* for the individual to be severely cognitively impaired than to be mildly impaired or cognitively normal. Why should this be? The first reason is that the experience of cognitive limitation may be inversely proportional to its severity. Individuals with mild intellectual impairment may be aware of their limitations, and frustrated by their disability. They may be sensitive to the looks and attitudes of others, and be conscious of being treated differently. There is some evidence that clinical depression is more common in those with cognitive impairment.[76] But more severe cognitive disability may be *less* likely to cause this sort of distress.[77] For example in a large study of quality of life in children with cerebral palsy, those with an IQ <50 were less likely to have low ratings for mood, emotions, and self-perception than children with mild cognitive impairment or normal IQ.[78] Recall the case of MB mentioned earlier, the eighteen-month-old infant completely paralysed and dependent on a mechanical ventilator. The judgement in that case was not explicitly based on intolerability, but placed great emphasis on the lack of brain damage in his case. However his experience of life could be significantly worse if he were cognitively normal and aware of his surroundings than if he were impaired and (to some degree) unaware.[79] Although Taylor LJ thought that the life of a continuously sedated and permanently unconscious child would be intolerable, in fact such a life appears to be a classic example of a life that can be tolerated since it involves no negative experiences whatsoever. What is more, in severe cases of cognitive impairment, the concept of tolerability may lose traction. To tolerate implies a sense of trade-off—of enduring some experiences for the sake of others. It requires a minimum level of personal identity and psychological continuity. But at very severe levels of cognitive impairment the individual may not only be unable to judge whether or not their life is tolerable, but the concept itself may not apply.

---

[73] Note 10, above.   [74] Note 23, above.

[75] Note 67, above, Wilkinson, 'Is it in the best interests of an intellectually disabled infant to die?'

[76] J. A. McBrien, 'Assessment and diagnosis of depression in people with intellectual disability' (2003) 47 *Journal of Intellectual Disability Research* 1–13.

[77] At the other end of life, with progressive cognitive decline it is often observed that early stages are worse for the individual, but that as the dementia worsens, and the individual loses awareness of their illness it becomes easier to bear.

[78] Arnaud, White-Koning, Michelsen, Parkes, Parkinson, Thyen, Beckung, Dickinson, Fauconnier, Marcelli, McManus, and Colver, 'Parent-reported quality of life of children with cerebral palsy in Europe' (2008) 121 *Pediatrics* 54–64.

[79] Note 23, above, Nuffield Council on Bioethics, 139 A similar problem arises with the question of consciousness and individuals in a persistent vegetative state. It may be worse for them to have some degree of awareness than to be completely unconscious. Wilkinson, Kahane, Horne, and Savulescu, 'Functional neuroimaging and withdrawal of life-sustaining treatment from vegetative patients' (2009) 35 *J. Med. Ethics* 508–11.

In summary, the problem with using 'intolerability' as a criterion in relation to certain predictions of impairment such as those provided by the Carmentis machine, is that there are substantial epistemic problems in determining the tolerability of conditions in infants. Moreover it is difficult to take into account the apparent importance of severe cognitive impairment within the concept of intolerability.

16.5.2 Balancing benefits and burdens

In the case of MB the decision about continued treatment was based upon an objective weighing up of the benefits and burdens of different treatment alternatives. Is this approach more helpful than intolerability in deciding, on the basis of the Carmentis predications, whether treatment should be withdrawn?[80]

16.5.2.1 Physical impairment

Physical impairment or illness may lead to burdens for the child, particularly where those impairments are associated with significant pain or suffering. For example a congenital abnormality that will predictably require multiple surgical procedures would impose a definite burden. But one prognostic problem that the Carmentis machine would not solve is this: even if future impairment is known with certainty, the degree of pain or suffering associated with that impairment may not be predictable. For example, one child might have a relatively uncomplicated course and short hospital stays, while another child might develop a post-operative infection, require repeat surgery, and have long hospital admissions. Children also vary in their tolerance of pain, making it difficult to know how adversely affected they would be. Other impairments may cause limitations in activity without leading to physical suffering—for example muscle weakness or paralysis. They would not necessarily lead to burdens on the child,[81] but could be included instead on the benefits side of the equation (as a relative reduction in benefits). For example life may provide less benefit to the child if she is unable to walk or run and is confined to a wheelchair. But another problem with weighing future physical impairments is that the degree to which the individual is disabled by their impairment is contingent upon the society in which they live and the support that is provided to them.[82] If there is little provision for wheelchairs in society[83] an individual may be very limited in their ability to take part in social activities. On the other hand, if society provides a high level of support and provision for wheelchair users, the limitations inherent in such an impairment may be substantially attenuated. So an

---

[80] Is it, indeed, any different when examined properly? We can only pose this second question here. The question is interesting, complex, and possibly repercussive.

[81] Though they could if the child were frustrated, depressed or anxious as a result of the impairments.

[82] This is one component of the social model of disability: A. Silvers, 'On the possibility and desirability of constructing a neutral conception of disability' (2003) 24 *Theoretical Medicine and Bioethics* 471–87.

[83] Including provision of appropriate wheelchairs themselves.

attempt to include physical impairment in a weighted balance of benefits and burdens will need to take into account both the current level of support provided by society, and the anticipated future level of support. The final point to note about weighing up the benefits and burdens of physical impairment is that some individuals appear able to realize high levels of personal achievement and well-being despite overwhelming physical impairments. Writers Christy Brown and Christopher Nolan with severe cerebral palsy, or the physicist Stephen Hawking with amyotrophic lateral sclerosis are often cited. The evidence of these individuals might lead us to doubt whether physical impairment alone would ever tip the balance in favour of withdrawal of life support.[84]

16.5.2.2 Other impairments

How should other impairments be taken into account in the balancing exercise? Sensory impairment might be thought to affect the balance in a similar way to some physical impairments. Blindness or deafness could reduce the benefits of life but would not necessarily lead to burdens for the individual. Again, this would be contingent on the amount of support provided to the child/adult, and would not necessarily preclude a life that was of net benefit to its possessor. But what about cognitive impairment? The examples cited of individuals who had overcome severe physical impairment all had normal cognitive capacity. In the previous section, the analysis of intolerability suggests that cognitive impairment would not necessarily increase the burdens experienced by the future child or adult. But the other possibility is that cognitive impairment may reduce the benefits of life for the child.[85] By diminishing or preventing the individual from accessing a number of goods in life (for example deep personal relationships, the development and attainment of personal goals), cognitive impairment, especially if severe, may make it easier for burdens to outweigh benefits. This is reflected in the judgments in *Re MB*[86] and *Re K*.[87] While in the former case the presence of normal cognition was felt to outweigh the burdens of extreme physical impairment, in the latter case the absence of the benefits afforded by normal intelligence tipped the balance in the other direction.

But while it is clear that future physical and cognitive impairment may affect the balance of benefits and burdens for treatment it is far more difficult to know to what degree. Even if we knew with certainty the degree and nature of future impairments for a child, there is no straightforward way of deciding how much weight to give to different benefits or burdens, how they should be aggregated or combined, and how they should be weighed against each other. Although this type

---

[84] In the *Wyatt* case (in an earlier proceeding than the one noted previously) Hedley J noted that intolerability might provide an evidentiary role of this sort: 'the concept of 'intolerable to that child' should not be seen as a gloss on, much less a supplementary test to, best interests. It is a valuable guide in the search for best interests in this kind of case' [2005] 1 FLR 21, para. 24.

[85] Note 67, above, Wilkinson, 'Is it in the best interests of an intellectually disabled infant to die?'

[86] Note 5, above.   [87] Note 11, above.

of appraisal is sometimes referred to as an 'objective' best interests test[88] (because it involves a weighing up of different interests and does not rely on the preference or choice of the individual) there is no truly objective way of determining the balance.

Our answer to the question posed at the start of this paper may seem unsatisfactory. It is not possible to predict which predictions form the Carmentis machine would lead to treatment withdrawal decisions by clinicians or the courts. But this answer in itself is useful. It highlights some of the difficulties that existing guidelines present for decision-makers. Even if prognostic uncertainty is removed, uncertainty about the appropriate test to apply, about the concepts that are used, and about the evaluation of impairment remains. But would the development of the Carmentis machine have other effects on the way that decisions are made?

16.6 Implications

16.6.1 May force explicit criteria for withdrawal of life support

The small number of cases previously considered by the courts makes it hard to assess how consistently the courts deploy concepts such as 'intolerability'. In addition, uncertainty about the prognosis in some cases makes it hard to extrapolate from one case to another and perhaps makes it possible for the courts and for guidelines to dodge difficult questions. But if the Carmentis Machine is able to provide detailed and accurate predictions of impairment, the relationship between specific conditions and normative judgements will be necessarily more explicit. The courts would be forced to say, for instance, whether, and if so, why, a life with a certain level of cognitive impairment was on balance a deal worth taking. Professional guidelines would need to deal explicitly with how the outputs of the machine should relate to clinical decisions.

16.6.2 May decrease or increase withdrawal of life support

The development of highly accurate and specific predictions of future impairment may reduce the number of cases where withdrawal of life support takes place. This may be because, first, infants who would previously have been allowed to die because of a risk of a profound impairment are now saved (because the machine predicts that in fact impairment will not be so bad). This would relate particularly to those decisions that are currently made without court oversight, where uncertainty about future impairment is sometimes taken to justify giving parents the final say in decisions about continuation or withdrawal of treatment.[89] Second, it is possible that some parents at least will be less willing to withdraw if they can be given specific information about the nature of the impairments from which an

---

[88] For example *Re MB*, note 5, para. 16.
[89] For example, see American Academy of Pediatrics Committee on Fetus and Newborn, 'Non-initiation or withdrawal of intensive care for high-risk newborns' (2007) 119 *Pediatrics* 401–3; Kipnis, 'Harm and uncertainty in newborn intensive care' (2007) *28 Theor. Med. Bioeth.* 393–412.

infant will suffer. Knowing exactly how impaired a child will be may help them to come to terms with disability, as well as to seek advice from parents of similarly impaired children. And third, there may be a type of expressivist[90] concern about withdrawal where there is sure knowledge of the degree of impairment. If there are living children and adults with the same impairments, allowing parents/doctors to withdraw life-sustaining treatment may seem to express a judgement about the value of those other lives. Although this same concern applies to decisions made currently, greater prognostic certainty may make the expressivist objection more forceful. This type of expressivist concern may make courts reluctant to judge such lives as being intolerable, or may give doctors or parents qualms about choosing to withdraw treatment.

But specific predictions may also increase withdrawal of life support. It may result in more judicial determinations that life-sustaining treatment should not be started or should be stopped. This is because the certainty that the machine would bring would stop judges from erring on the side of caution. There is an inevitable tendency, embodied in the sanctity of life principle, to give life the benefit of the doubt. But where there is no doubt, death is more likely to be sanctioned. Finally it is possible that predictions from the machine may lead to a shift in attitude amongst parents about disability and the continuation of treatment. The availability of accurate neuroprognosis may make parents reluctant to continue treatment in the presence of predicted impairment. It might conceivably lead to parents being criticized if they decide to continue treatment.

16.6.3 May raise the question of two normative thresholds for withdrawal

If it is permissible to withdraw life support from an infant, is it obligatory? Some professional guidelines have discussed the idea of a 'grey zone' of prognosis, where prognosis is uncertain and the views of parents may determine whether or not treatment is withdrawn.[91] On the other hand, legal decisions have specifically discounted the idea that parental wishes could affect the interests of the child and so make a difference to whether or not treatment should be withdrawn.[92] The Nuffield report raised specifically the question of whether the decision in the case of MB would have been the same if the parents had been in agreement with the doctors.[93] The Carmentis machine may influence the framework within which treatment decisions are made by delineating two different types of quality of life assessment—each with a different legal corollary: first, a level of future impairment so severe that continued life support would represent an assault on the child, and, second, a level of impairment such that parents may request life support be continued, even if doctors disagree.

---

[90] Note 82, above, Silvers.
[91] Note 89, above, American Academy of Pediatrics Committee on Fetus and Newborn.
[92] For example the religious views of MB's father were not considered relevant in *Re MB*, note 5, above, para. 50.
[93] Note 23, above, Nuffield Council on Bioethics, 139.

16.6.4 May highlight questions of resource allocation

Finally, and perhaps more speculatively, if infants are predicted with a high degree of certainty to have severe impairment (but not necessarily severe enough for treatment withdrawal to be obligatory) questions of relative priority for scarce resources may be raised. There may, for instance, be competition for intensive care beds, cardiac surgery or transplantable organs. Currently uncertainty about the degree of impairment provides one reason not to prioritize on this basis, but it may be that the Carmentis Machine will put pressure on doctors and medical administrators to include future impairment in decisions about who should receive life-saving treatment.

16.7 Conclusions

The Carmentis machine does not exist—but machines like it are already in use. Ultrasound examination of the newborn brain is available at the bedside of virtually all neonatal intensive care units in developed countries. Magnetic resonance scanners have been incorporated and integrated into the design of contemporary neonatal units.[94] Professional guidelines recommend that all infants with perinatal asphyxia have magnetic resonance imaging performed in the first week of life,[95] and in recent years there has been debate about whether all premature infants should have similar scans.[96] Brain scans have played a small role in recent court cases involving withdrawal of life support, but it seems likely that they will play an increasing role in future decisions. They play a major role in clinical practice in decisions about life-sustaining treatment for newborn infants.[97]

Although it is unlikely that a perfect prognostic machine will ever be developed, technology continues to improve the ability of doctors to predict future impairment. Even if it is not possible to predict outcome for all infants, for at least some infants it will be possible with a high degree of certainty to predict that they will be impaired, the degree of their likely impairment, and the range of conceivable outcomes. It is evidentially, legally, metaphysically, and epistemically difficult to decide whether or not specific impairments lead to lives that are intolerable, or cause the burdens of existence to outweigh the benefits. These difficulties limit the

---

[94] Maalouf and Counsell, 'Imaging the preterm infant: practical issues' in Rutherford (ed.), *MRI of the Neonatal Brain* (London, Gulf Professional Publishing, 2002).

[95] Ment, Bada, Barnes, Grant, Hirtz, Papile, Pinto-Martin, Rivkin, and Slovis, 'Practice parameter: neuroimaging of the neonate: report of the Quality Standards Subcommittee of the American Academy of Neurology and the Practice Committee of the Child Neurology Society' (2002) 58 *Neurology* 1726–38.

[96] Woodward, Anderson, Austin, Howard, and Inder, 'Neonatal MRI to predict neurodevelopmental outcomes in preterm infants' (2006) 355 *N. Engl. J. Med.* 685–94; de Vries and Cowan, 'Should cranial MRI screening of preterm infants become routine?' (2007) 3 *Nat. Clin. Pract. Neurol.* 532–3.

[97] D. Wilkinson, 'We don't have a crystal ball': neonatologists' views on prognosis and decision-making in newborn infants with birth asphyxia' (2010) 29 *Monash Bioeth Rev* 5.1–5.19.

practical usefulness of current legal and ethical guidelines. Improvements in prognostic accuracy will raise difficult questions about the level of future impairment that are deemed to render a life not worth living. Developments in this technology may affect decision-making in a number of ways highlighted in this paper, and they may lead to the development of more specific guidelines for clinicians and parents. But in the absence of such guidelines the Carmentis machine will not necessarily make it easier for doctors, parents, or the courts to decide. In some ways it may make decision-making substantially harder.

17

The Right to Silence Protects Mental Control*

Dov Fox**

17.1 Introduction

This chapter examines the idea that individuals have a right of control, vis-à-vis the state, over the use of their thoughts. Advances in cognitive science and forensic neurotechnology could make it possible for government actors to acquire reliable information from a suspect's brain without requiring the suspect's participation. Such interrogation techniques would elude traditional right-to-silence concerns about physical or psychological harm. I analyse Fifth Amendment jurisprudence to argue that the right to silence protects a right of mental control.

Though neither judges nor scholars have defended this account in explicit terms, the notion of mental control underlies much that is assumed about the relation between the right to silence and the values of freedom and privacy. Brain imaging brings the moral and legal significance of mental control into sharp relief. I argue that a right of mental control prohibits the state either from extracting a suspect's thoughts without her meaningful consent or from making use of a suspect's compelled recall or recognition to lay criminal blame upon her.

This chapter proceeds in four parts. Part I lays out the normative stakes and shows why existing accounts of the constitutional privilege against self-incrimination are ill-equipped to address the doctrinal implications of safe and reliable forensic neuroscience. Part II argues that brain imaging is importantly different, for Fifth Amendment purposes, from all other forms of evidence, such as speaking, writing,

---

* The author is grateful to Michael Freeman for the invitation to contribute and to Lisa Penfold for organizing such a stimulating colloquium. For provocative discussion about these ideas, he owes a debt to participants at the Neuroscience and Law colloquium at the University College London, at the Law, Neuroscience, and Governance symposium at the Akron School of Law, and in the Neuroscience and the Law course at Yale Law School. Hank Greely, Dan Habib, Dan Kahan, and Michael Pardo provided helpful conversations and criticisms. For comments on multiple drafts, he owes a special debt of gratitude to Alex Stein. Thanks to the *Akron Law Review* for permission to reprint. An earlier version of this piece appeared in (2009) 42 *Akron Law Review* 761.

** Law Clerk to Hon. Stephen Reinhardt, US Court of Appeals for the Ninth Circuit. JD, Yale Law School, 2010; DPhil, Oxford University, 2007; AB, Harvard College, 2004.

and nodding;[1] photographic and video evidence;[2] DNA, fingerprint and breathalyser tests;[3] hair, saliva, and blood samples;[4] voice and handwriting exemplars for identification;[5] appearance in a line-up[6] or wearing certain clothing;[7] and even polygraph tests.[8] What makes brain imaging unique is that it enables state officials to obtain information directly from a suspect's brain, in a way that affords her no opportunity to control the transmission of that information.

Part III examines the doctrinal distinction between 'testimonial' evidence, which is privileged by the self incrimination clause, and 'physical' evidence, which is not. There are three elements—compulsion, incrimination, and testimony—that must be present to trigger Fifth Amendment protection.[9] Since brain imaging could straightforwardly qualify as both compelled and incriminating, as in the India murder case described in the next section, the critical question is whether it counts as 'testimonial'.[10] This part argues that the Supreme Court counts evidence as 'testimonial' only when it conveys a suspect's intention to communicate her thoughts.[11] The Court's traditional understanding of this distinction likely leaves a suspect without constitutional protection against the use of compelled brain imaging to extract incriminating information.

Part IV argues that the physical/testimonial distinction presupposes a flawed conception of mind/body dualism.[12] Exposing this dualism reveals that the normative significance we confer to a suspect's control over his thoughts against unwanted use by the government. It is for this reason that the constitutional

---

[1] See, for example, *Schmerber* 384 US 757 at 763 (1966).
[2] See, for example, *People v Hines* 938 P2d 388 (Cal 1997) (videos); *Edwards v Butler* 882 F2d 160 (5th Cir 1989) (photographs).
[3] See, for example, *Wisconsin v Santana-Lopez* 613 NW 2d 918 (Wis Ct App 2000) (DNA); *People v Shaw* 531 NE2d 650, 651 (1988) (breathalyzer); *Palmer v State* 604 P2d 1106 (Alaska 1979) (breathalyzer).
[4] See *Schmerber* 384 US at 757; *State v Athan* 158 P3d 27 (Wash 2007) (saliva).
[5] See, for example, *United States v Dionisio* 410 US 1 (1973) (voice); *United States v Wade* 388 US 218 (1967) (voice); *Doe v United States* 487 US 201 (1988) (handwriting); *United States v Mara* 410 US 19 (1973) (handwriting); *United States v Euge* 444 US 707 (1980) (handwriting); *Gilbert v California* 388 US 263 (1967) (handwriting).
[6] See, for example, *Wade* 388 US at 218; see also *Schmerber* 384 US at 764, 764 note 8 (affirming that Fifth Amendment does not protect against incriminating compulsion 'to stand, to assume a stance, to walk, or to make a particular gesture').
[7] See, for example, *Holt v United States* 218 US 245 (1910).
[8] See, for example, *United States v Scheffer* 523 US 303 (1998) (holding that a *per se* exclusion of polygraph evidence in a military court does not violate the Sixth Amendment right to present a defence).
[9] See *Fisher v United States* 425 US 391, 408 (1976). In addition to the compulsion, incrimination, and testimony elements, a Fifth Amendment case must also arise under a matter of criminal law. See *Chavez v Martinez* 538 US 760, 766–7 (2003). Therefore, the privilege does not apply when the compelled information would lead to exclusively non-criminal sanctions, such as the loss of a job or a licence.
[10] See *Schmerber v California* 384 US 757, 763–4 (1966).
[11] See *Pennsylvania v Muniz* 496 US 582, 589 (1990).
[12] See C.B. Nemeroff et al., 'Editorial, Functional Brain Imaging: Twenty-First Century Phrenology or Psychobiological Advance for the Millennium?' (1999) 156 *American Journal of Psychiatry* 671, 672 ('Neuroimaging offers a powerful probe of brain state, but we are now faced with metaphysical questions; i.e., what is a brain state, and how is it related to the outward manifestations of behavior? This has the potential for degenerating into the old mind-body duality of Descartes...').

right to silence protects against involuntary disclosures of knowledge or understandings from an unwilling suspect for use in criminal prosecution against him. The use of compelled neuroscientific evidence is illegitimate when it deprives the accused of control over her mental life. This suggests that defendants may opt for brain imaging to exculpate themselves, but that prosecutors may not comment on a suspect's decision to decline the testing, and that judges should instruct jurors not to draw adverse inferences from a choice to decline testing. Instructions against drawing adverse inferences are likely to be effective, however, only if jurors come to recognize legitimate reasons (whether moral, religious, or strategic) to decline testing.

17.2 The Privilege Against Compelled Self-Incrimination

Imagine a safe, painless, and non-invasive interrogation device that could ascertain thoughts and memories from an uncooperative suspect with virtually 100 per cent precision. Would the use of this incriminating information to prosecute the suspect violate his right to silence?[13] The prospect of reliable truth detection is, to date, an unqualified fiction, since even the most sophisticated technologies remain largely untested and prone to error. But advances in cognitive neuroscience, which make it possible to measure the properties and processes of the brain in high resolution imagery, invite us to consider this question. Brain imaging test results were admitted into evidence in a recent murder case in India.

In June 2008, police in Maharashtra, India took twenty-four-year-old student Aditi Sharma and her husband Pravin Khandelwal into custody as suspects in the murder of Sharma's former fiancé, Udit Bharati.[14] Sharma and Bharati had been living together in Pune when Sharma met another man and eloped with him to Delhi.[15] The following year, Sharma returned to Pune, where, according to prosecutors, she asked Bharati to meet her at a McDonald's Hotel and laced his food with arsenic.[16] Bharati died from the poisoning.[17] Sharma insisted she was innocent. Police asked Sharma to sit for a Brain Electrical Oscillations Signature (BEOS) test.[18] Officers strapped her onto a high-tech gurney and fastened a cap with thirty-two electrodes to her head.[19] The sensors measured electrical brainwaves in response to targeted stimuli.[20]

The trial court opinion indicates that investigators presented Sharma with factual statements along with their theory of the crime.[21] The questioning was meant to help the brain imaging software to distinguish between Sharma's actual memories and her normal cognition.[22] Though Sharma said nothing, the BEOS sensors were

---

[13] See Henry T. Greely, 'Prediction, Litigation, Privacy, and Property: Some Possible Legal and Social Implications of Advances in Neuroscience' in Brian Garland (ed.), *Neuroscience and the Law: Brain, Mind, and the Scales of Justice* (New York, Dana Press, 2004), 114, 146.
[14] *Maharastra v Sharma and Khandelwal* Sessions Case No. 508/07 (12 June 2008).
[15] Ibid. at ¶ 1–3.   [16] Ibid. at ¶ 4.   [17] Ibid. at ¶ 9.
[18] Ibid. at ¶ 114.   [19] Ibid. at ¶ 101.   [20] Ibid.
[21] Ibid. at ¶ 107.   [22] Ibid.

able to measure and display her brainwave patterns to suggest that she knew some of the statements were true and that some were false.[23] At the murder trial, the brain scans were admitted into evidence.[24] These neuroscientific test results persuaded Judge S. S. Phansalkar-Joshi that Sharma had 'experiential knowledge' of having committed the murder.[25] Sharma was convicted and sentenced to life in prison.[26]

There is strong reason to think that the BEOS test suffers from unacceptable error rates as a lie detection technique.[27] But assume, for the sake of argument, that the brain imaging technique used in this case was highly reliable. Many people may nevertheless think that police committed a grave wrong by coercing the disclosure of Sharma's thoughts.[28] The Supreme Court has speculated that the involuntary transmission of incriminating lie-detection evidence would violate a suspect's right to silence.[29] 'To compel a person to submit to testing in which an effort will be made to determine his guilt or innocence on the basis of physiological responses, whether willed or not', Justice Brennan wrote in dicta for a 5–4 majority in the 1966 case of *Schmerber v California*,[30] 'is to evoke the spirit and history of the Fifth Amendment.'[31] Although the compelled-response standard that Justice Brennan gestured toward in *Schmerber* was not binding then, and is not applied by courts today, its widespread appeal and influence remains.[32] Presuming that neuroscience test results were sufficiently reliable, existing accounts of the privilege against compelled self-incrimination provide no support for Justice Brennan's reflections. This mismatch, between right-to-silence principles and widely held intuitions that the Fifth Amendment should protect against compelled brain imaging, is the anomaly that propels this inquiry into the implications of forensic neuroscience for the self-incrimination clause.

Justification for the right to silence is fraught with murkiness. Many have observed an 'ambiguous and sometimes contradictory picture of the reasons underlying the Fifth Amendment's enactment'.[33] The privilege against compelled

---

[23] Ibid. at 101. [24] Ibid. [25] Ibid. at 11.
[26] For discussion, see A. Giridharadas, 'Brain Scan a New Wave in Criminal Evidence' (2008) *International Herald Tribune*, 8 September, 1; A. Giridharadas, 'India's Novel Use of Brain Scans in Courts is Debated' (2008) *New York Times*, 14 September, A8.
[27] See M. Raghava, 'Stop Using Brain Mapping for Investigation and as Evidence' (2008) *The Hindu*, 6 September, available at <http://www.hindu.com/2008/09/06/stories/9780199599844100.htm>.
[28] See S. E. Stoller and P. R. Wolpe, 'Emerging Neurotechnologies for Lie Detection and the Fifth Amendment' (2007) 33 *American Journal of Law and Medicine* 359, 371 ('[D]irect and unwanted government access to our mental life is a chilling concept.').
[29] The self-incrimination clause of the Fifth Amendment provides that '[n]o person ... shall be compelled in any criminal case to be a witness against himself ...' US Constitution, Amendment V.
[30] 384 US 757, 764 (1966).
[31] Ibid.
[32] See Part II, below.
[33] P. Arenella, '*Schmerber* and the Privilege Against Self-Incrimination: A Reappraisal' (1982) 20 *American Criminal Law Review* 31, 36 (citing Leonard W. Levy, *Origins of the Fifth Amendment: The Right Against Self-Incrimination* (New York, Clio Enterprises, Inc., 1968)); E. M. Morgan, 'The Privilege Against Self-Incrimination' (1949) 34 *Minnesota Law Review* 1; R. C. Pittman, 'The Colonial and Constitutional History of the Privilege Against Self-Incrimination in America' (1935) 21 *Virginia*

self-incrimination has origins as a safeguard against government abuses such as the Roman Inquisition, Star Chamber, and *ex officio* oaths of the ecclesiastical courts, long forsaken in the modern judiciary.[34] Commentators have referred to the privilege as 'schizophrenic',[35] an 'unsolved riddle of vast proportions, a Gordian knot in the middle of our Bill of Rights'[36] that is incapable of 'be[ing] squared with any rational theory'.[37] Even the Supreme Court has conceded that while the right to silence is 'an expression of the moral striving of the community[,] . . . a reflection of our common conscience',[38] it remains unclear 'just what it is supposed to do or just whom it is intended to protect'.[39]

Scholars and judges have articulated a number of ways to justify the right to silence. Five are most prominent and compelling: 1) non-reliance on presumptively unreliable evidence[40] which 'the accused reasonably believes . . . is within his power to affect the probative value of the evidence sought';[41] 2) preservation of the credibility of innocent defendants;[42] 3) protection of innocents who fear poor performance on cross-examination;[43] 4) pardon of suspects who make a reasonable decision to lie in their own defence,[44] and 5) avoidance of the 'cruel trilemma' that forces a suspect to choose among self-accusation, contempt of court, and perjury.[45]

Law Review 763)); see also R. H. Helmholz, 'Introduction' in R. H. Helmholz et al. (eds.), *The Privilege Against Self-Incrimination: Its Origins and Development* (Chicago, University of Chicago Press, 1997) 6 ('Despite its reputation as a foundation stone of common law jurisprudence, . . . the privilege as we know it is actually the product of relatively recent choice.') (citations omitted).

[34] See *Ullmann v United States* 350 US 422, 428 (1956); *Muniz* 496 US at 595–8; *Doe v United States* 487 US 201, 212 (1988); see generally F. Riebli, 'The Spectre of Star Chamber: The Role of an Ancient English Tribunal in the Supreme Court's Self-Incrimination Jurisprudence' (2002) 29 *Hastings Constitutional Law Quarterly* 807.

[35] S. J. Schulhofer, 'Some Kind Words for the Privilege Against Self-Incrimination' (1992) 26 *Valparasio University Law Review* 311, 311.

[36] A. R. Amar and R. B. Lettow, 'Fifth Amendment First Principles: The Self-Incrimination Clause' (1995) 93 *Michigan Law Review* 857, 857.

[37] W. J. Stuntz, 'Self-Incrimination and Excuse' (1988) 88 *Columbia Law Review* 1227, 1228.

[38] *Malloy v Hogan* 378 US 1, 9 note 7 (1964), quoting Erwin Nathaniel Griswold, *The Fifth Amendment Today* (Cambridge, MA, Harvard University Press, 1955) 73.

[39] *Murphy v Waterfront Comm'n of N.Y. Harbor* 378 US 52, 56 note 5 (1964) (quoting H. Kalven, Jr, 'Invoking the Fifth Amendment—Some Legal and Impractical Considerations' (1953) 9 *Bulletin of Atomic Scientists* 181, 182).

[40] See Amar and Lettow, note 36, above.

[41] B. M. Dann, 'The Fifth Amendment Privilege against Self-Incrimination: Extorting Physical Evidence from a Suspect' (1970) 43 *Southern California Law Review* 597, 598 (arguing that the self-incrimination clause protects against psychologically intrusive compulsion of evidence).

[42] See D. J. Seidman and A. Stein, 'The Right to Silence Helps the Innocent: A Game-Theoretic Analysis of the Fifth Amendment Privilege' (2000) 114 *Harvard Law Review* 430.

[43] See C. M. Bradley, '*Griffin v California*: Still Viable After All These Years' (1981) 79 *Michigan Law Review* 1290, 1294 (arguing that innocent defendants may choose not to testify at trial to avoid adverse inferences on cross-examination as a result of anxiety, confusion, or prior convictions).

[44] Stuntz, note 37, above, 1228–9.

[45] *Murphy v Waterfront Comm'n of N. Y. Harbor* 378 US 52, 55 (1964). See also *Chavez v Martinez* 538 US 760, 767 (2003); *Pennsylvania v Muniz* 496 US 582, 596 (1990). Scholars have referred to this worry about the 'cruel dilemma' as the 'original American justification'. See W. T. Pizzi and M. B. Hoffman, 'Taking Miranda's Pulse' (2005) 58 *Vanderbilt Law Review* 813, 843; see also P. Westen and S. Mandell, 'To Talk, To Balk, or To Lie: The Emerging Fifth Amendment Doctrine of the "Preferred Response"' (1982) 19 *American Criminal Law Review* 521, 535–40.

Assuming that neuroforensic technology was sufficiently reliable, however, none of these ways of thinking about the privilege would bar the compelled use of certain brain imaging techniques to extort information from a suspect's brain.[46]

Consider the implications of reliable brain imaging evidence for each of the justifications above. 1) Reliability concerns fall away with an accurate brain imaging technique because 'an effective and reliable lie detector test deprives the individual of any opportunity to deceive the questioner.'[47] 2) Anxiety about the integrity of statements made by innocent defendants is put to rest by an assurance that brain imaging could accurately detect memories and capture truthfulness.[48] 3) There would be no reason to worry about innocent defendants getting flustered by skilful prosecutors when neither the subject nor the examiner exercised any control over the results of the test. 4) Nor would the excuse theory protect suspects such as Sharma because 'there [was] no falsehood to excuse and therefore no need to immunize noncooperation'.[49] 5) And since neuroscience techniques could reliably access Sharma's responses in the absence of her participation, police were able to obtain the incriminating information without her being put to any choice among indictment, contempt, and perjury. So there was no threat that Sharma would incur the psychological cruelty thought to accompany a decision about whether to consent to her own punishment.

If we nonetheless recoil from the use of neurotechnology to extract thoughts from an unwilling suspect, it will not do simply to affirm that involuntary brain imaging 'shocks the conscience', like the stomach pumping in *Rochin v California*.[50] Alternatively, we could shake off our disquiet as so much confusion and try to fit forensic neuroscience into the existing jurisprudence. Some scholars apply common law distinctions to argue that brain imaging techniques are unlikely to qualify for Fifth Amendment protection,[51] while others make the case that compelled neuroforensics could be readily protected by reference to the line of right-to-silence opinions in the wake of *Schmerber*.[52] No matter how descriptive or predictive these approaches prove, as a matter of doctrine they remain morally unsatisfying to the extent that they are 'not normative or justificatory'.[53] Explanatory theories

---

[46] See M. S. Pardo, 'Neuroscience Evidence, Legal Culture, and Criminal Procedure' (2006) 33 *American Journal of Criminal Law* 301, 333.

[47] Arenella, note 33, above, 44–5 note 70.

[48] See Pardo, note 46, above, 334–5.

[49] Stuntz, note 37, above, 1276.

[50] 342 US 165, 172 (1952). For an example of this argument, see S. K. Thompson, 'A Brave New World of Interrogation Jurisprudence' (2007) 33 *American Journal of Law and Medicine* 341, 353.

[51] See Stoller and Wolpe, note 28, above, 369 ('[I]f control over results is a defining element of a testimonial act, brain fingerprinting would not qualify as such.'); R. J. Allen and M. K. Mace, 'The Self-Incrimination Clause Explained and its Future Predicted' (2004) 94 *Journal of Criminal Law and Criminology* 243, 246 (providing a theory of the self-incrimination clause that protects the 'substantive content of cognition').

[52] G. M. Dery, 'Lying Eyes: Constitutional Implications of New Thermal Imaging Lie Detection Technology' (2004) 31 *American Journal Criminal Law* 217, 248 (drawing on dicta in *Schmerber* to argue that the use of involuntarily transmitted information to incriminate a suspect in a criminal trial is 'squarely within the scope of the Fifth Amendment', while noting, however, that 'Justice Brennan's ruminations on lie detectors' were not within the Court's holding) (citation omitted).

[53] Allen and Mace, note 51, above, 248.

might account for Fifth Amendment doctrine in a convincing way; but to the extent that they accept extant case law as legitimate and fixed, these understandings of the right to silence must be tailored to capture anomalous or wrongly decided holdings. The possibility that reliable brain imaging would be admitted in the courtroom invites us to rethink prevailing understandings of the privilege against compelled self-incrimination.

17.3 Cognitive Neuroscience and Forensic Evidence

Attempts to detect lies in criminal suspects date back thousands of years.[54] In 900 BC, Hindu interrogators sought to identify deceitful behaviour in criminal suspects by looking for physiological markers—cagey movements, facial discoloration, or 'speak[ing] nonsense, rub[bing] the great toe along the ground, and shiver[ing]'— that were thought to correspond to a deceitful state of mind.[55] Centuries later, the ordeals of Christian Europe sought to expose the guilty through supernatural intervention that rarely saved the accused from hot irons, boiling water, or drowning.[56] Nineteenth-century Europe and the US saw the rise of phrenology to detect honest or deceptive character by measuring the shape and contours of a subject's head.[57] Brain imaging heralds a step in this project.

Neurotechnology cannot read minds.[58] Even the most advanced devices are not capable of ascertaining a subject's moral beliefs, religious convictions, or aspirations for the future.[59] Nor can they establish any objective fact about some state of affairs in the world.[60] What brain imaging can do, instead, is determine a subject's sensory recall or perceived recognition—what a subject believes he saw, heard, or knows—about a particular set of facts or about the state of past events.[61] In a landmark 2002 study, Dr Daniel Langleben and colleagues asked participants

---

[54] J. A. Blumenthal, 'A Wipe of the Hands, A Lick of the Lips: The Validity of Demeanor Evidence in Assessing Witness Credibility' (1993) 72 *Nebraska Law Review* 1157, 1166.
[55] P. V. Trovillo, 'History of Lie Detection' (1939) 29 *Journal of Law and Criminology* 848, 849.
[56] See G. Fisher, 'The Jury's Rise as Lie Detector' (1997) 107 *Yale Law Journal* 575, 585–6.
[57] See P. Schlag, 'Law and Phrenology' (1997) 110 *Harvard Law Review* 877, 880.
[58] See J. Adler, 'Mind Reading' (2004) *Newsweek*, 9 August, 42, available at <http://www.newsweek.com/id/54762> (regarding fMRI lie detection); D. O'Brien, 'Mind Readers: Scanning Technology Promises to Map the Brain's Pathways, but Some Fear its Ability to Expose a Patient's Secrets and Lies' (2004) *Baltimore Sun*, 10 December, 1E.
[59] See D. Kennedy, 'Neuroimaging: Revolutionary Research Tool or a Post-Modern Phrenology?' (2005) 5 *American Journal of Bioethics* 19.
[60] As Martha J. Farah and Paul Root Wolpe put it: 'Although brainwaves do not lie, neither do they tell the truth; they are simply measures of brain activity': M. J. Farah and P. R. Wolpe, 'Monitoring and Manipulating Brain Function: New Neuroscience Technologies and their Ethical Implications' (May-June 2004) *Hastings Center Report* 35, 40.
[61] See C. E. L. Stark and L. R. Squire, 'Functional Magnetic Resonance Imaging (fMRI) Activity in the Hippocampal Region during Recognition Memory' (2000) 20 *Journal of Neuroscience* 7776, 7776; J. D. E. Gabrieli, 'Functional Neuroimaging of Episodic Memory' in Roberto Cabeza and Alan Kingstone (eds.), *Handbook of Functional Neuroimaging of Cognition* (Cambridge, MA, MIT Press, 2001) 262.

in a magnetic resonance imaging (MRI) study to conceal the identity of a five of clubs, while images of different playing cards appeared on the screen along with the question whether the featured card was the one in the subjects' hand.[62] Langleben and colleagues found conspicuous differences between patterns of brain activity when subjects truthfully denied having other cards and when they falsely denied having the five of clubs.[63] Subsequent studies successfully replicated Langleben's results.[64] Further investigation has determined that truthful behaviour, because it is spontaneously facilitated, requires less cognitive control and the use of fewer brain regions (primarily the hippocampus and left prefrontal cortex, which store and regulate memory) than does deceptive behaviour, which utilizes greater cognitive control in discrete premotor brain regions to produce the inhibitory feedback and conscious calculation necessary to generate new (false) 'memories'.[65]

There are two classes of brain imaging devices that can be used for purposes of memory detection.[66] The first class seeks to measure blood flow patterns in the brain;[67] the second measures electrical activity patterns in the brain.[68] This first class of techniques relies on the fact that blood flow in the brain differs when people lie as compared with when they tell the truth.[69] Blood flow patterns reflect metabolic and circulatory adjustment resulting from increased brain activity,

---

[62] See D. D. Lengleben et al., 'Brain Activity during Simulated Deception: An Event-Related Functional Magnetic Resonance Study' (2002) 15 *NeuroImage* 727, 729.

[63] See ibid. at 731.

[64] See K. L. Phan et al., 'Neural Correlates of Telling Lies: A Functional Magnetic Resonance Imaging Study at 4 Tesla' (2005) 12 *Academic Radiology* 164, 171.

[65] See, for example, D. D. Langleben et al., 'Telling Truth from Lie in Individual Subjects with Fast Event-Related fMRI' (2005) 26 *Human Brain Mapping* 262, 262; J. M. Nunez et al., 'Intentional False Responding Shares Neural Substrates with Response Conflict and Cognitive Control' (2005) 25 *NeuroImage* 267, 267; F. A. Kozel et al., 'A Pilot Study of Functional Magnetic Resonance Imaging Brain Correlates of Deception in Healthy Young Men' (2004) 16 *Journal of Neuropsychiatry and Clinical Neurosciences* 295; F. A. Kozel et al. 'A Replication Study of the Neural Correlates of Deception' (2004) 118 *Behavioral Neuroscience* 852, 855; G. Ganis et al., 'Neural Correlates of Different Types of Deception: An fMRI Investigation' (2003) 13 *Cerebral Cortex* 830, 836; S. A. Spence et al., 'Behavioural and Functional Anatomical Correlates of Deception in Humans' (2001) 12 *NeuroReport* 2849, 2850.

[66] See R. M. Henig, 'Looking for the Lie' (2006) *New York Times*, 5 February, § 6, 47; J. O'C. Hamilton, 'Journey to the Center of the Mind: 'Functional' MRI Is Yielding a Clearer Picture of What Thoughts Look Like' (2004) *Business Week*, 19 April, 78; F. Flam, 'Your Brain May Soon Be Used Against You' (2002) *Philadelphia Inquirer*, 29 October, A01.

[67] See J.-D. Haynes et al., 'Reading Hidden Intentions in the Human Brain' (2007) 17 *Current Biology* 323; P. R. Wolpe et al., 'Emerging Neurotechnologies for Lie-Detection: Promises and Perils' (2005) 5 *American Journal of Bioethics* 39, 39; 'Investigative Techniques—Federal Agency Views on the Potential Application of "Brain Fingerprinting"' (2001) *GAO Report*, 31 October (Rep. No. GAO-02-22), available at 2001 WL 1560280.

[68] Carter Snead has distinguished 'anatomical' neuroimaging techniques (those that are 'limited to the observation of the brain's architecture') from 'functional' techniques (those that 'permit[ ] the construction of computerized images that measure the brain's activity with varying degrees of temporal and anatomical resolution, depending on the technology employed'.): O. C. Snead, 'Neuroimaging and the "Complexity" of Capital Punishment' (2007) 82 *New York University Law Review* 1265, 1281 (citation omitted).

[69] See L. Hernandez et al., 'Temporal Sensitivity of Event-Related fMRI' (2002) 17 *NeuroImage* 1018, 1025.

independent of whether the subject makes any effort to conceal a memory.[70] Brain imaging techniques take advantage of neural differences between truthful and deceptive behaviour by measuring how much blood is flowing to different parts of the brain in response to targeted stimuli.[71] The most commonly used method for measuring blood flow in the brain, called functional Magnetic Resonance Imaging (fMRI), uses a powerful magnet to identify changes in blood oxygenation that occur when a person performs a mental activity, such as viewing an image, answering a question, listening to a voice, or telling a story.[72] FMRI then projects a graphic representation of brain activity with resolution that is an order of magnitude greater than what was not long ago the most precise brain imaging technology, the Positron Emission Tomography (PET) scan.[73]

The reliability of fMRI is controversial.[74] While a private company called 'No Lie MRI' purports to achieve 90 per cent accuracy in detecting deception,[75] critics note the low number of research subjects studied in widely cited fMRI research experiments.[76] Some experts suggest that the changes in blood flow that fMRI measures are too small to yield any significant information about the content of activity in the brain.[77] Others argue that increased blood flow could result from neurological processes other than efforts to conceal, including a subject's anxiety with the claustrophobic machine or an emotional state that is heightened for some other reason unrelated to the cognitive processes for which the fMRI tests.[78] Still others claim that blood flow patterns do not actually signify output activity in the

---

[70] See S. D. Forman et al., 'Improved Assessment of Significant Activation in Functional Magnetic Resonance Imaging (fMRI): Use of a Cluster-Size Threshold' (1995) 33 *Magnetic Resonance Medicine* 636.

[71] F. A. Kozel et al., 'Detecting Deception Using Functional Magnetic Resonance Imaging' (2005) 58 *Biological Psychiatry* 605, 611.

[72] See D. G. Norris, 'Principles of Magnetic Resonance Assessment of Brain Function' (2006) 23 *Journal Magnetic Resonance Imaging* 794, 794; Richard B. Buxton, *An Introduction to Functional Magnetic Resonance Imaging: Principles and Techniques* (Cambridge, Cambridge University Press, 2002) 23–45.

[73] See F. B. Mohamed et al., 'Brain Mapping of Deception and Truth Telling about an Ecologically Valid Situation: Functional MR Imaging and Polygraph Investigation—Initial Experience' (2006) 238 *Radiology* 679.

[74] See M. Talbot, 'Duped' (2007) *New Yorker*, 2 June, 52–61.

[75] See H. T. Greely and J. Illes, 'Neuroscience-Based Lie Detection: The Urgent Need for Regulation' (2007) 33 *American Journal of Law and Medicine* 377, 392 (citing 'No Lie MRI, Market Opportunities', <http://www.noliemri.com/investors/MarketOpportunity.htm> (accessed 10 February 2009)).

[76] See D. I. Donaldson, 'Parsing Brain Activity with fMRI and Mixed Designs: What Kind of a State Is Neuroimaging In?' (2004) 27 *Trends in Neurosciences* 442.

[77] See S. Blakeslee, 'Just What's Going On Inside that Head of Yours?' (2000) *New York Times*, 14 March, F6.

[78] Orrin Devinsky and Mark D'Esposito explain: 'When a subject performs a task during imaging, it is difficult to demonstrate conclusively that he or she is differentially engaging a single, identified cognitive process. The subject may engage in unwanted cognitive processes that either have no overt, measurable effects or are perfectly confounded with the process of interest. Consequently, the neural activity measured by the functional neuroimaging technique may result from some confounding neural computation that is itself not necessary for executing the cognitive process under study.': Orrin Devinsky and Mark D'Esposito, *Neurology of Cognitive and Behavioral Disorders* (Oxford, Oxford University Press, 2003) 53–4.

brain at all, but rather input activity.[79] Furthermore, since the fMRI machine requires that a subject's head remain still for several hours, even a small physical movement can impede the scanner's ability to obtain data on blood flow patterns.[80] Finally, fMRI may call for voluntary participation from a subject who must be willing to answer a question or otherwise register a response to specific stimuli.[81] For the second class of neurotechnologies, however, there is no need for a subject's cooperation in order to acquire information from her brain.[82]

This second class of technologies relies on the fact that every piece of information in a person's brain is stored by specific neurons, which fire when the brain recognizes that information, producing electrical activity.[83] This electrical activity is a direct measure of information-processing in the brain.[84] Neurotechnologies can measure these electrical brainwaves through the use of electroencephalographic (EEG) sensors attached to the scalp.[85] By measuring electrical brainwaves in response to targeted stimuli, the EEG technique, also known as 'brain fingerprinting', can ascertain the presence or absence of information in a subject's brain, thereby confirming or denying the subject's familiarity with a particular event or image, such as a photograph of a crime scene, victim's face, or murder weapon under investigation.[86] Other neuroimaging techniques within this second class, including functional near-infrared light technology (fNIR), which reflects infrared light off the frontal cortex,[87] and thermographic technology, which detects heat emanating from the skin of the face,[88] are less reliable than EEG.

Brain fingerprinting suffers from limited testing outside of the laboratory[89] as well as empirical shortcomings, including the risk of false negative and false positive results.[90] Because the EEG sensors measure electrical activity associated with recognition, the subject needs to have retained the targeted information in her brain for brain fingerprinting to work.[91] If the subject has experience with

---

[79] See S. Cleland, 'What Does fMRI Actually Measure?' (2004) 17 *Psychologist* 388.

[80] See Randy L. Buckner and Jessica M. Logan, 'Functional Neuroimaging Methods: PET and fMRI' in Roberto Cabeza and Alan Kingstone (eds.), *Handbook of Functional Neuroimaging of Cognition* (Cambridge, MA, MIT Press, 2001) 28, 30.

[81] See S. M. Hayes et al., 'An fMRI Study of Episodic Memory: Retrieval of Object, Spatial, and Temporal Information' (2004) 118 *Behavioral Neuroscience* 885, 886.

[82] See J. Knight, 'The Truth About Lying' (2004) 428 *Nature* 692, 692–4.

[83] See K. R. Foster et al., 'Bioethics and the Brain' (2003) 40 *IEEE Spectrum* 34, 36.

[84] See H. Pearson, 'Lure of Lie Detectors Spooks Ethicists' (2006) 441 *Nature* 918, 918–9.

[85] Brain Fingerprinting Laboratories, Scientific Procedure, Research, and Applications, <http://www.brainwavescience.com/TechnologyOverview.php> (accessed 29 February 2009).

[86] Ibid.

[87] M. Izzetoglu et al., 'Functional Near-Infrared Neuroimaging' (2005) *IEEE Transactions on Neural Systems Rehabilitation Engineering* 153, 156–8.

[88] See D. A. Pollina et al., 'Facial Skin Surface Temperature Changes During a "Concealed Information" Test' (2006) 34 *Annals Biomedical Engineering* 1182, 1183.

[89] See *Slaughter v State* 105 P3d 832, 835 (Okla Crim App 2005) ('[W]e have no real evidence that Brain Fingerprinting has been extensively tested...').

[90] See Y. Kamitani and F. Tong, 'Decoding the Visual and Subjective Contents of the Human Brain' (2005) 8 *Nature Neuroscience* 679, 679.

[91] M. S. Beauchamp, 'Functional MRI for Beginners' (2002) 5 *Nature Neuroscience* 397, 398 (reviewing Richard B. Buxton, *An Introduction to Functional Magnetic Resonance Imaging: Principles and Techniques* (Cambridge, Cambridge University Press, 2002) and Peter Jezzard et al. (eds.),

the targeted information, but fails to remember it due to a head injury, for example, the EEG sensors will detect only the subject's incomplete familiarity, producing a false negative.[92] If, on the other hand, the targeted information is familiar to the subject for some reason other than its association with the crime, then a false positive result will follow.[93] So if the subject happened to collect the same kind of gun as the murder weapon, or if she saw a representation of the crime scene on television, then these stimuli would indicate the subject's familiarity with the information in question, even though this conclusion would be misleading.[94]

Despite these limitations, brain fingerprinting was admitted into evidence in a 2003 criminal case, *Iowa v Harrington*,[95] which involved voluntary EEG testing in a post-conviction relief action.[96] In 1977, seventeen-year-old Terry Harrington was found guilty of first degree murder in the shooting death of a retired police officer.[97] Harrington claimed he had spent the night in question at a concert with friends, several of whom confirmed his alibi at trial.[98] But one witness, Kevin Hughes, testified that he and Harrington had driven to a dealership that night to steal a car, and that Harrington had shot and killed the retired officer who was working as a night watchman at the lot.[99] A jury convicted Harrington and sentenced him to life in prison.[100]

Twenty-three years later, forensic neuroscience gave Harrington another chance.[101] In 2000, a trial judge admitted exculpatory EEG test results as grounds to open a post-conviction relief claim.[102] After a day-long hearing featuring three

---

Functional Magnetic Resonance Imaging: An Introduction to Methods (Oxford, Oxford University Press, 2001).

[92] See T. Buller, 'Can We Scan for Truth in a Society of Liars?' (2005) 5 *American Journal of Bioethics* 58, 59.

[93] See F. Fang et al., 'Lie Detection with Contingent Negative Variation' (2003) 50 *International Journal of Psychophysiology* 247, 252–3; Kelly Joyce, 'Appealing Images: Magnetic Resonance Imaging and the Production of Authoritative Knowledge' (2005) 35 *Social Studies of Science* 437, 450.

[94] J. P. Rosenfeld et al., 'Simple, Effective Countermeasures to P300-based Tests of Detection of Concealed Information' (2004) 41 *Psychophysiology* 205.

[95] 659 NW2d 509 (Iowa 2003).

[96] See ibid. at 515.

[97] See *State v Harrington* 284 NW2d 244, 245 (Iowa 1979).

[98] See *State v Harrington* 659 NW2d 509, 515 (Iowa 2003); see also L. A. Farwell, 'Supplement to Forensic Science Report: Brain Fingerprinting Test on Terry Harrington' (accessed 29 February 2009) *Re State of Iowa v Terry Harrington* in the Iowa District Court for Pottawattamie County at Council Bluff <http://www.brainwavescience.com/HarringtonSupplement.php>.

[99] See *Harrington*, 284 NW2d at 248.

[100] See ibid. at 245; see also *Harrington v State* 458 NW2d 874 (Iowa Ct App 1990) (initial post-conviction relief action); *Harrington v Nix* 983 F2d 872, 874 (8th Cir 1993) (federal habeas relief action).

[101] See A. A. Moenssens, 'Brain Fingerprinting—Can it be Used to Detect the Innocence of Persons Charged with a Crime?' (2002) 70 *University of Missouri-Kansas City Law Review* 891, 916.

[102] See *Harrington v Iowa* No. PCCV 073247 (Pottawattamie County DC Iowa, 14 November 2000). See also *Harrington* 659 NW2d at 515–16 (citing Iowa Code § 822.2(4) (1999)). Cf. *US v Taveras* 570 F Supp 2d 481, 484 (EDNY 2008) ('Confrontation rights and hearsay exclusions in practice are designed to protect defendants against inculpation, not to limit their use for exculpation.' (citing D. A. Nance, 'Allocating the Risk of Error: Its Role in the Theory of Evidence Law' (2007) 13 *Legal Theory* 129, 157 (reviewing Alex Stein *Foundations of Evidence Law* (2005)) ('[A] defendant is

expert witnesses, Judge Ternus cited several patents[103] and publications[104] to conclude that the brain fingerprinting technique satisfied reliability standards for legal admissibility under *Daubert v Merrell Dow Pharmaceuticals*.[105] The neuroscientific expert in Harrington was Dr Lawrence Farwell, the scientist who developed the EEG technique, and founder and chief scientist of Brain Fingerprinting Laboratories.[106] First, Farwell conducted a test based on details about the crime scene that would be known only to the perpetrator and investigators. The test concluded, with 99.9 per cent confidence, that the information was absent from Harrington's brain.[107] Then, Farwell tested unique details about the concert that Harrington claimed to have been at on the night of the murder for which he was convicted.[108] The test found with similar confidence that the information was present.[109]

While the EEG results were not dispositive in Harrington—the Iowa Supreme Court instead granted Harrington a new trial on the basis of a *Brady* violation[110]—when confronted with the test results, the single eyewitness, Kevin Hughes, recanted his testimony and confessed to perjuring himself at the original trial.[111] Upon reversal and remand, the prosecution announced that it was dismissing the case on the grounds that the 'admissible evidence which is left after 26 years is not

---

entitled to adduce in his or her defen[s]e any evidence—hearsay or non-hearsay—if that evidence is the best evidence available.')) (internal alterations omitted).

[103] Method and Apparatus for Multifaceted Electroencephalographic Response Analysis (MERA), US Patent No. 5,363,858 (filed 5 May 1993) (issued 15 November 1994); Method and Apparatus for Truth Detection, US Patent No. 5,406,956 (filed 11 February 1993) (issued 18 April 1995); Method for Electroencephalographic Information Detection, US Patent No. 5,467,777 (filed 15 September 1994) (issued 21 November 1995).

[104] J. J. Allen et al., 'The Identification of Concealed Memories Using the Event-Related Potential & Implicit Behavioral Methods: A Methodology for Prediction in the Face of Individual Differences' (1992) 29 *Psychophysiology* 504; L. A. Farwell and E. Donchin, 'The Brain Detector: P300 in the Detection of Deception' (1986) 24 *Psychophysiology* 434; L. A. Farwell et al., 'Optimal Digital Filters for Long-Latency Components of the Event-Related Brain Potential' (1993) 30 *Psychophysiology* 306; L. A. Farwell and E. Donchin, 'Talking Off the Top of Your Head: Toward a Mental Prosthesis Utilizing Event-Related Brain Potentials' (1988) 70 *Electroencephalography and Clinical Neurophysiology* 510; L. A. Farwell and E. Donchin, 'The Truth Will Out: Interrogative Polygraphy ('Lie Detection') with Event-Related Brain Potentials' (1991) 28 *Psychophysiology* 531; L. A. Farwell, 'Two New Twists on the Truth Detector: Brain-Wave Detection of Occupational Information' (1992) 29 *Psychophysiology* 20; L. A. Farwell, 'The Brain-Wave Information Detection (BID) System: A New Paradigm for Psychophysiological Detection of Information' (1992) (unpublished PhD dissertation, University of Illinois at Urbana-Champaign).

[105] 509 US 579 (1993). In addition to 'general acceptability' by the relevant scientific community, *Daubert* standards require rigorous testing, published peer-review, and demonstration of an acceptably low rate of error. See *Daubert* 509 US at 583, 591–4; see also, *Frye v United States* 293 F 1013, 1014 (CADC 1923) (discussing 'general acceptability'); *General Electric Co. v Joiner* 522 US 136 (1997); *Kumho Tire Ltd v Carmichael* 526 US 137 (1999).

[106] See Brain Fingerprinting Laboratories, <http://www.brainwavescience.com> (accessed 29 February 2009).

[107] See Farwell, note 98, above.

[108] Ibid.

[109] Ibid.

[110] See *Harrington v State* 659 NW2d 509, 522 (Iowa 2003).

[111] Brain Fingerprinting Laboratories, *Brain Fingerprinting Testing Helps to Exonerate Man Falsely Convicted of Murder*, <http://www.brainwavescience.com/HarringtonSummary.php> (accessed 29 February 2009).

sufficient to sustain a conviction against Mr. Harrington'.[112] After twenty-five years in jail, Terry Harrington was freed. It remains to be seen whether other state or federal courts will admit brain fingerprinting as evidence.

Having reviewed some of the capacities and limitations of the two major classes of neurotechnology and focused attention on the class that measures brainwaves, we are in a position to see what makes neuroforensic techniques such as electroencephalography different from traditional lie-detection methods like the polygraph machine. Polygraphy measures the physiological reactions that follow from emotional responses.[113] These reactions are thought to express themselves as detectable changes in the peripheral nervous system.[114] While the subject is being interviewed, therefore, the polygraph machine monitors changes in a subject's perspiration (by attaching metal sensors to the fingers or palms), breathing rate (using rubber tubes wrapped around the chest), and blood pressure and heart rate (with an arm cuff).[115] A problem with the polygraph is that a subject may be able to control and learn to manipulate their emotional responses, and hence the physiological reactions that the machine is able to account for.[116] Subjects have been able to 'fool' the polygraph machine by using simple countermeasures like stressors, such as flexing muscles or placing tacks in a shoe, to induce or inflate physiological responses, or sedatives, such as barbiturates or minor tranquilizers, to depress the central nervous system and dampen stress reactions.[117] That the polygraph is not immune to bluffing renders the test unreliable for purposes of lie detection and truth verification.[118]

Neuroscientific forensic technology is not just high-tech polygraphy. Unlike the polygraph test, which gauges a subject's emotional response to the act of lying, brain imaging techniques ascertain the neurological processes required for a subject's brain even to produce a lie.[119] The application of cognitive neuroscience enables interrogators to sidestep the peripheral nervous system in order to gain

---

[112] M. Siebert, 'Free Man' (2003) *Des Moines Register*, 25 October, B8.
[113] American Polygraph Association, Frequently Asked Questions, <http://www.polygraph.org/section/resources/frequently-asked-questions> (accessed 20 August 2009).
[114] Ibid.
[115] Ibid.
[116] See American Psychological Association, 'The Truth About Lie Detectors (aka Polygraph Tests)' (2004) *Psychological Matters*, 5 August. <http://www.psychologymatters.org/polygraphs.html>; Susan McCarthy, 'Passing the Polygraph' (2000) *Salon*, 2 March. <http://www.apa.org/research/action/polygraph.aspx>.
[117] See D. Eggen and S. Vedantam, 'Polygraph Results often in Question: CIA, FBI Defend Test's Use in Probes' (2006) *Washington Post*, 1 May, A1.
[118] See American Polygraph Association, 'Polygraph Validity Research', <http://www.polygraph.org/section/resources/polygraph-validity-research> (accessed 20 February 2009). The American Polygraphy Association (APA), which oversees scientific research and development, establishes methodologies and procedures, and maintains qualifying standards and licensing tests for all polygraph examiners in the US, suggests that reliability rates for modern polygraph techniques approach 92 per cent. See ibid. See also 'Concerns over Use of Polygraphs: Hearing before the Senate Committee on the Judiciary' 107th Cong. 5–8 (2001) (statement of Michael H. Capps, Deputy Director for Developmental Programs, Defense Security Service), available at <http://www.access.gpo.gov> ('There is common agreement in the scientific community that modern polygraph techniques do produce [results] usually in excess of 90% ... this compares favorably with many other common techniques in the behavioral sciences.').
[119] T. M. C. Lee et al., 'Lie Detection by Functional Magnetic Resonance Imaging' (2002) 15 *Human Brain Mapping* 157, 163.

direct access to the workings of the brain.[120] Whereas the polygraph measures controllable physical manifestations of emotional tension, brain fingerprinting techniques measure involuntary brain activation.[121] The extra 'work' that it requires for a subject's brain to create a new memory occurs independent of stress levels, and cannot be effectively controlled.[122] The primary cognitive processes that electroencephalography measures are thus more difficult to dissimulate than the physical manifestation of secondary emotional responses that are measured by the polygraph.[123] As one commentator puts it: '[E]ven a trained counter-spy must use creativity and calculation to formulate a new lie, and the most nervous Nellie will use memory to recount an event in her past.'[124]

Not only is intentionally bluffing the EEG machine not an option; neither is overriding memory detection unintentionally. Brain fingerprinting can detect even good faith mistaken responses, since the machine identifies the brain activity required to create the new memory, whether the lie was deliberate or inadvertent.[125] Moreover, the use of non-human examiners renders neurotechnologies less vulnerable to test bias than polygraphy.[126] Polygraph tests are performed by human interrogation and interpreted by human examiners, whose behaviour and judgement can affect test results.[127] With EEG, by contrast, the subject is presented test questions on a computer screen and the analysis of brain activity is displayed in high-resolution imaging and performed using computer software.[128] All this renders brain fingerprinting far more reliable, or at least potentially so, than traditional lie-detection techniques like polygraph testing.

17.4 The Distinction between Testimonial and Physical Evidence

Whether brain fingerprinting is privileged by right-to-silence jurisprudence turns on whether it counts as 'testimonial' evidence, which is protected by the Fifth

---

[120] G. Ganis et al., 'Neural Correlates of Different Types of Deception: An fMRI Investigation' (2003) 13 *Cerebral Cortex* 830, 830 (noting that fMRI techniques make it possible to 'examine directly the organ that produces lies, the brain').

[121] To the extent that fMRI, like polygraphy, requires the subject to speak his responses to external stimuli provided by the government, this type of neuroforensic technique is somewhat less interesting for purposes of Fifth Amendment inquiry, since spoken communications are generally protected under the privilege. See, for example, *Doe v United States* 487 US 201, 213 (1988) ('The vast majority of verbal statements thus will be testimonial' because '[t]here are very few instances in which a verbal statement, either oral or written, will not convey information or assert facts'.).

[122] See Lee et al., note 119, above, 163 ('controlling one's cerebral activity to avoid detection is unfeasible').

[123] See ibid.

[124] L. Kittay, Note, 'Admissibility of FMRI Lie Detection: The Cultural Bias Against "Mind Reading" Devices' (2007) 72 *Brooklyn Law Review* 1351, 1355.

[125] J. Illes and E. Racine, 'Imaging or Imagining? A Neuroethics Challenge Informed by Genetics' (2005) 5 *American Journal of Bioethics* 5, 9–10.

[126] See P. C. Giannelli, 'Polygraph Evidence: Post-*Daubert*' 49 *Hastings Law Journal* (1998) 895, 905.

[127] See ibid. ('The examiner's role cannot be overstated, because it is the examiner who decides whether there is sufficient indication of deception.').

[128] S. Silberman, 'Don't Even Think About Lying: How Brain Scans are Reinventing the Science of Lie Detection' (January 2006) *Wired Magazine* 142.

Amendment, or 'physical' evidence, which is not. Of the three factors—compulsion, incrimination, and testimony—that must be present to warrant constitutional scrutiny under the Fifth Amendment,[129] the third factor—testimony—is where the action is.[130] Courts have struggled to make sense of what counts as 'testimonial' evidence in this 'admittedly abstract and under-determined area of law'.[131] This challenge is exacerbated by forensic neuroscience.

It is not difficult to imagine that brain fingerprinting could readily qualify as compelled—in the sense that, 'considering the totality of the circumstances, the free will of the [subject] was overborne'[132]—or as incriminating—in the sense that it 'could be used in a criminal prosecution or could lead to other evidence that might be so used'.[133] What is less clear is whether EEG evidence is 'testimonial'.[134] With respect to this third prong, the Supreme Court has held that a person is a 'witness' against himself for purposes of the Fifth Amendment when he is 'compelled to testify... or otherwise provide the State with evidence of a testimonial or communicative nature'.[135] To count as testimonial, as opposed to physical, the communication must 'explicitly or implicitly, relate a factual assertion or disclose information'.[136]

The seminal case laying out the physical/testimonial distinction came more than forty years ago, in *Schmerber v California*.[137] *Schmerber* presented the Supreme Court with the issue of whether the use of compelled blood tests to incriminate a defendant violated his constitutional right to silence.[138] Armando Schmerber, after drinking at a local bowling alley, drove his car into a tree.[139] While at the hospital receiving treatment for injuries sustained in the crash, Schmerber was arrested for driving under the influence of alcohol and instructed to submit to a blood test to determine his level of intoxication.[140] When Schmerber refused, the arresting

---

[129] See *Fisher v United States* 425 US 391, 408 (1976) ('[T]he Fifth Amendment does not independently proscribe the compelled production of every sort of incriminating evidence but applies only when the accused is compelled to make a Testimonial Communication that is incriminating.'). In addition to the compulsion, incrimination, and testimony elements, a case must also arise under matter of criminal law for the privilege to apply. See *Chavez v Martinez* 538 US 760, 766–7 (2003). So there is no Fifth Amendment protection when the compelled information would lead to exclusively non-criminal sanctions, such as the loss of a job or a licence.

[130] *Schmerber v California* 384 US 757, 762–3 (1966) (noting that '[h]istory and a long line of authorities in lower courts have consistently limited' the right to silence to situations involving compelled testimonial evidence).

[131] *United States v Hubbell* 167 F3d 552, 570 (DC Cir 1999), aff'd, 530 US 27 (2000); see also *Schmerber* 384 US at 774 (Black J dissenting) ('These words [testimonial and physical] are not models of clarity and precision as the Court's rather labored explication shows.').

[132] *United States v Washington* 431 US 181, 188 (1977) (citing *Rogers v Richmond* 365 US 534, 544 (1961)).

[133] *Kastigar v United States* 406 US 441, 455 (1972).

[134] See *Schmerber* 384 US at 763–4.

[135] *Doe v United States* 487 US 201, 210 (1988).

[136] Ibid.

[137] 384 US 757 (1966).

[138] Ibid. at 758. Schmerber also raised Due Process, Fourth Amendment, and Sixth Amendment right to counsel claims. The Court rejected each: ibid. at 759–72.

[139] Ibid. at 758.

[140] Ibid.

officer directed the attending doctor to withdraw a blood sample from his body even so.[141] Laboratory analysis of Schmerber's blood 'indicated intoxication'.[142] At trial, California prosecutors presented the results of Schmerber's blood test, over his objections, as evidence that he had been driving drunk.[143] Schmerber was convicted.[144]

Schmerber appealed the case to the US Supreme Court, arguing that the government's use of the involuntary blood test to establish his guilt violated his Fifth Amendment privilege against compelled self-incrimination.[145] The Supreme Court dismissed his claim. In a 5–4 decision, the Court held that the use of the blood test to convict Schmerber did not violate the self-incrimination clause because that evidence, although it was obtained from Schmerber against his will, was properly characterized as 'physical', as opposed to 'testimonial'.[146] Writing for the majority, Justice Brennan explained that the Fifth Amendment privilege excluded 'physical' evidence, such as fingerprints, handwriting exemplars, appearance in a lineup, or, as in the case before the Court, a blood sample.[147] While Schmerber's blood may 'testify' to something, Professor Louis Seidman explains, ' it does so in a purely physical fashion without implicating his interior, mental life'.[148] Justice Brennan affirmed that the right to silence is limited in scope to evidence like a written confession or verbal communication, which is actively 'testimonial' in nature.[149]

Writing in dissent, Justice Black rejected the distinction between physical and testimonial evidence on conceptual and normative grounds.[150] On the coherency objection, Justice Black argued that so-called physical evidence like a blood test is the functional equivalent of oral testimony, in the sense that both are extracted from the suspect himself and that both may be comparably persuasive to the jury that will determine the suspect's verdict.[151] The distinction between physical and testimonial evidence was therefore a distinction without difference in Justice Black's view. On the desirability objection, Justice Black remarked that '[i]t is a strange hierarchy of values that allows the State to extract a human being's blood to convict him of a crime because of the blood's content but proscribes compelled production of his lifeless papers'.[152]

---

[141] Ibid. at 759.   [142] Ibid. at 757.
[143] Ibid. at 758–9.   [144] Ibid. at 757.
[145] Ibid. at 758.   [146] Ibid. at 765.
[147] Ibid. at 765 (arguing that the blood test and its results did not involve 'even a shadow of testimonial compulsion upon or enforced communication by the accused').
[148] Louis M. Seidman, 'Points of Intersection: Discontinuities at the Junction of Criminal Law and the Regulatory State' (1996) 7 *Journal of Contemporary Legal Issues* 69, 131–2.
[149] *Schmerber* 384 US at 764.
[150] Ibid. at 773 (Black J dissenting).
[151] Ibid. at 774.
[152] Ibid. at 775. 'Lifeless papers' referred to the Court's opinion in *Boyd v United States* 116 US 616 (1886), in which the Court held that the Fourth and Fifth Amendments create a zone of privacy into which the government cannot intrude by using his books and papers as evidence against him in a criminal proceeding: *Boyd* 116 US at 621-2. The holding in *Boyd* has since been rejected in *Fisher v United States* 425 US 391 (1976) and *United States v Doe* 465 US 605 (1984).

In defending the physical/testimonial distinction, Justice Brennan drew for support on John Henry Wigmore's classical *Evidence* treatise[153] and on Justice Holmes's opinion more than half a century earlier in *Holt v United States*.[154] Wigmore cited the case of *Block v People*[155] to argue that since 'the public interest in obtaining [incriminating] evidence is usually sufficient to outweigh by a clear margin the private interests sacrificed in the process', only testimonial compulsion was protected by the right to silence.[156] Wigmore limited the bounds of 'testimony' to those beliefs of the accused that have been extracted for government use:

> Unless some attempt is made to secure a communication—written, oral or otherwise—upon which reliance is to be placed as involving his consciousness of the facts and the operations of his mind in expressing it, the demand made upon him is not a testimonial one.[157]

Fingerprinting, blood tests, and appearance in a line-up did not qualify as testimonial, Wigmore argued, because these forms of evidence did not involve revelation of the suspect's knowledge.[158]

Holt involved the appeal of a murder conviction on the ground, among others, that the defendant's Fifth Amendment rights were violated when he was forced to try on an incriminating shirt to show that it fit him.[159] Writing for the Court, Justice Holmes rejected the defendant's claim as 'an extravagant extension of the 5th Amendment'.[160] He distinguished government compulsion as used, on the one hand, 'to extort communications from [the defendant]', and, on the other hand, to use the defendant's 'body as evidence when it may be material'.[161] Justice Holmes applied a *reductio ad absurdum* to argue that the right to silence does not protect against compelled evidence from the defendant's body.[162] To extend Fifth Amendment protection to such physical evidence would, 'in principle[,] ... forbid a jury to look at a prisoner and compare his features with a photograph in proof'.[163] To avoid this implausible conclusion, Justice Holmes restricted the privilege against self-incrimination to 'communications'.[164]

In *Schmerber*, the Court applied this same distinction between ' "communications" or "testimony" ',[165] and 'real or physical evidence'.[166] Justice Brennan explained that whether evidence counts as physical or testimonial depends on whether the process by which the evidence was acquired or evaluated 'implicated'

---

[153] 8 Wigmore, *Evidence* § 2263 (McNaughton rev. 1961).
[154] 218 US 245 (1910).
[155] 240 P2d 512 (Colo 1951). 'The purpose of the amendment against self-incrimination is to prevent a man from being compelled to utter words that will incriminate him, and not to obliterate all evidence of physical facts showing who and in what condition he is.': ibid. at 516.
[156] 8 Wigmore, note 153, above, § 2265.
[157] Ibid.   [158] Ibid.   [159] *Holt* 218 US at 246, 252–3.
[160] Ibid. at 252.   [161] Ibid. at 252–3.   [162] See ibid.
[163] Ibid. at 253.   [164] Ibid. at 252–3.
[165] *Schmerber v California* 384 US 757, 764 (1966).   [166] Ibid.

the suspect's 'testimonial capacities'.[167] The privilege applied to testimonial communications broadly, in 'whatever form they might take', whether word of mouth, written notes, or communicative body language such as pointing or nodding.[168] The drawing of Schmerber's blood, although both compelled and incriminating, did not involve his testimonial capacities and did not relate to other written or verbal communications by the defendant.[169] The blood test therefore qualified as 'physical' evidence, so it was not protected by the self-incrimination clause.[170]

Justice Brennan reaffirmed the basic distinction between physical and testimonial evidence in the Court's next term. *United States v Wade*[171] involved a suspect accused of robbing a bank at gunpoint, disguised with 'a small strip of tape on each side of his face'.[172] Police later arrested him and placed in a line-up alongside him several other prisoners, each of whom 'wore strips of tape such as allegedly worn by the robber', and repeated the robber's words: 'put the money in the bag'.[173] Witnesses to the robbery relied on the line-up to identify the suspect as the robber. Rejecting the suspect's Fifth Amendment claim, Justice Brennan wrote for the majority:

> We have no doubt that compelling the accused merely to exhibit his person for observation by a prosecution witness prior to trial involves no compulsion of the accused to give evidence having testimonial significance. It is compulsion of the accused to exhibit his physical characteristics, not compulsion to disclose any knowledge he might have.[174]

Nor did forcing the suspect to 'utter words purportedly uttered by the robber' constitute testimonial evidence, since his voice was used merely as 'an identifying physical characteristic, not to speak his guilt'.[175] Having found that neither forced appearance in the robber's disguise nor forced vocalization of the phrase used during the robbery counted as testimonial, Justice Brennan concluded that evidence provided in the line-up was 'not within the cover of the privilege'.[176]

Justice Brennan reached the same result in applying the physical/testimonial distinction to the handwriting exemplar the suspect in *Gilbert v California*[177] provided to the state to compare against a handwritten robbery note demanding money from the Mutual Savings and Loan Association of Alhambra.[178] Although a person's 'voice and handwriting are, of course, means of communication', Justice Brennan conceded, '[i]t by no means follows... that every compulsion of an accused to use his voice or write compels a communication within the cover of the privilege'.[179] Given that the 'voice recordings were... used solely to measure the physical properties of the witnesses' voices, [and] not for the testimonial or communicative content of what was to be said',[180] this mere exemplar, 'like the voice or body itself, is an identifying physical characteristic outside its protection'.[181] These exemplar cases do not qualify as testimonial because although the

---

[167] Ibid. at 765. [168] Ibid. at 763–4.
[169] Ibid. [170] Ibid. [171] 388 US 218 (1967).
[172] Ibid. at 218. [173] Ibid. [174] Ibid. at 222. [175] Ibid. at 219.
[176] Ibid. at 224. [177] 388 US 263 (1967). [178] Ibid. at 264.
[179] Ibid. at 265. [180] Ibid. at 269. [181] Ibid. at 266.

act of speaking specified words or signing one's name or trying on certain clothes, without more, conveys comprehension of and response to a command, no substantive information is thereby communicated to the government. By 1967, the self-incrimination clause excluded from its coverage appearance in a line-up or in certain clothes, and voice or handwriting exemplars. These were not testimonial kinds of evidence, the Court said, and so were, like Schmerber's blood test, beyond the purview of the Fifth Amendment.

Justice Brennan had noted in *Schmerber* that the 'privilege [against self-incrimination] has never been given the full scope which the values it helps to protect suggest',[182] but ruminated on further extensions of the right to silence. Speaking to circumstances that lay beyond the facts of the case, Justice Brennan addressed in *dicta* the question of whether the logic that makes compelled blood tests immune to the Fifth Amendment also applies to compelled polygraphs.[183] Justice Brennan reasoned that even though a polygraph, like a blood test, measures physical processes, lie-detection may nevertheless count as testimonial because '[s]ome tests seemingly directed to obtain "physical evidence," for example, lie detector tests measuring changes in body function during interrogation, may actually be directed to eliciting responses which are essentially testimonial'.[184] Justice Brennan continued: 'To compel a person to submit to testing in which an effort will be made to determine his guilt or innocence on the basis of physiological responses, *whether willed or not*, is to evoke the spirit and history of the Fifth Amendment.'[185] Were Justice Brennan's musings the privilege central to the holding in *Schmerber*, those words would presumably apply to protect a criminal suspect from the use of brain imaging to extract incriminating information from her brain, even though she exercised no control over the transmission of the information.

In subsequent decades, however, the Court interpreted the physical/testimonial distinction as turning on whether the *process* of disclosing incriminating information bears testimonial qualities, as opposed to whether the *product* of that information is testimonial.[186] For evidence to count as 'testimonial' on this account, the process by which the government acquires information from the suspect must convey his active and intended transmission of internal knowledge about the outside world. There are good reasons for emphasizing the process of

---

[182] *Schmerber v California* 384 US 757, 762 (1966).
[183] Ibid. at 764.
[184] Ibid.
[185] Ibid. (emphasis added). Justice Brennan's conclusion that involuntary lie detector evidence might be barred by the self-incrimination clause is again consistent with Wigmore's discussion of the physical/testimonial distinction. Wigmore noted the difficulty in classifying the use of truth serums or polygraphs in cases where 'not only is the person's affirmative participation essential (at least in the form of physical responses) but his knowledge, despite his will to the contrary, is extracted': 8 Wigmore, note 153, above, § 2265.
[186] See Allen and Mace, note 51, above, 276. But see H. R. Uviller, 'Foreword: *Fisher* Goes on the Quintessential Fishing Expedition and *Hubbell* is Off the Hook' 91 (2001) *Journal Criminal Law and Criminology* 311, 320–1 (observing that derivative protection in *United States v Hubbell* 'comes perilously close to treating the contents of a document as the indirect product of its production').

evidence-gathering over the product of information thereby revealed. On Daniel Seidman and Alex Stein's anti-pooling account of the privilege, for example, the lies of a guilty suspect should not be immune from censure because, by masquerading as the truths of an innocent suspect, these lies compromise the credibility of all innocent suspects in defending themselves.[187] Focusing on the process of evidentiary disclosure captures this ability of guilty suspects to manipulate incriminating information and thereby hurts the innocent by rendering their truthful appeals unreliable.[188] Giving attention instead to the product of incriminating information would fail to capture this reliability element of compelled evidence that serves to help innocent suspects.[189]

On this understanding of the physical/testimonial distinction, the important fact in *Schmerber* is that the accused did not play a conscious or purposeful role in the transfer of incriminating information to the government, since he was an unwilling donor for bodily material that was examined by a chemical analysis likewise beyond his control.[190] Since Schmerber's blood did not 'belong' to him in the way that his thoughts do, Louis Seidman explains, sticking a needle in his arm deprived Schmerber of any intentional participation that could qualify his blood test as 'testimonial'.[191] The results of the blood test in *Schmerber* counted as physical evidence 'not simply because the evidence concerned the suspect's physical body', the Court would later write, but also because of Schmerber's passive role in providing the evidence.[192] *Schmerber*'s progeny followed this process-based (as opposed to product-based) approach to defining testimonial evidence for purposes of the protection under the Fifth Amendment.

The Court's subpoena cases are instructive. The self-incrimination clause has been held to protect not the *product*, or informational content, of whatever is produced in response to the subpoena, but rather the *process* of providing a specified entity. Compare the application of the physical/testimonial distinction when faced with a subpoena to produce known tax forms in *Fisher v United States*,[193] as opposed to a subpoena to produce underspecified documents in *United States v Hubbell*.[194] In *Fisher*, the privilege did not apply because the product of the subpoena papers was already known by the government, so the process by which the suspect disclosed the information involved an effectively mechanical and therefore non-testimonial response.[195] In *Hubbell*, by contrast, the Court held that Fifth Amendment prevented the government's request for documents whose existence was unknown and whose contents could not be described with any

---

[187] Seidman and Stein, note 42, above, 453, note 79.
[188] Ibid.
[189] Michael Pardo tries to save the physical/testimonial distinction on innovative, epistemological grounds. Pardo distinguishes testimonial evidence as that which requires reliance by the fact-finder on the defendant as a source of epistemic authority. See M.S. Pardo, 'Self-Incrimination and the Epistemology of Testimony' (2008) 30 *Cardozo Law Review* 1023, 1037.
[190] Ibid. at 475–6.   [191] Seidman, note 148, above, 131–2.
[192] *Pennsylvania v Muniz* 496 US 582, 593 (1990).   [193] 425 US 391 (1976).
[194] 530 US 27 (2000).
[195] See *Fisher* 425 US at 410 (acknowledging that '[t]he act of producing evidence in response to a subpoena... has communicative aspects of its own').

specificity.[196] The privilege applied in *Hubbell* because providing the documents required the suspect's active participation, and thereby made use of the 'contents of his own mind'.[197] Although the testimonial *product* was essentially identical to that in *Fisher*, because the existence and location of the documents were not a 'foregone conclusion',[198] the testimonial *process* revealed to the government the suspect's knowledge that the documents both existed and also were within his own possession.[199] It was the process and not the product of subpoenaed information that made the evidence count as testimonial.

The Court also took this approach in *Pennsylvania v Muniz*, which involved a right to silence appeal to an answer to a question about the date of the defendant's sixth birthday at a sobriety check.[200] At a drunk driving stop, an officer asked Inocencio Muniz to perform 'three standard field sobriety tests: a "horizontal gaze nystagmus" test, a "walk and turn" test, and a "one leg stand" test'.[201] Muniz failed all three.[202] He was arrested and driven to the police station, where the officer videotaped his slurred responses to questions including his 'name, address, height, weight, eye color, date of birth, and current age'.[203] After 'stumbling over his address and age', Muniz was asked whether he knew 'the date of [his] sixth birthday'[204] he replied, 'No, I don't.'[205]

Writing for the majority, Justice Brennan distinguished the physical qualities of Muniz's slurred speech from the testimonial qualities of the sixth birthday question.[206] The 'slurred nature of his speech' involved Muniz's physical inability to articulate words in a clear manner due to 'the lack of muscular coordination of his tongue and mouth.'[207] The videotape evidence of the slurring, in isolation, constituted a physical process, which was 'not itself testimonial'.[208] By contrast, the sixth birthday question was testimonial because it required Muniz to make active use of his confused mental state.[209] Inferring Muniz's level of intoxication from his inability to answer a simple question turned on the process of his transmitting information about his drunkenness. Or so Justice Brennan contended:

> The Commonwealth ... argue[s] that this incriminating inference does not trigger the protections of the Fifth Amendment privilege because the inference concerns 'the physiological functioning of [Muniz's] brain,' ... which is asserted to be every bit as 'real or physical' as the physiological makeup of his blood and the timbre of his voice. But this characterization addresses the wrong question; that the 'fact' to be inferred might be said to concern the physical status of Muniz's brain merely describes the way in which the inference is incriminating. The correct question for present purposes is whether the incriminating inference of mental confusion is drawn from a testimonial act or from physical evidence.[210]

For evidence to count as 'testimonial', Justice Brennan argued, it was not enough for the accused simply to impart the information from his body or mind; he must

---

[196] See *Hubbell*, 530 US at 44–5. [197] Ibid. at 43. [198] *Fisher* 425 US at 411.
[199] *Hubbell* 530 US at 44–5. [200] 496 US 582, 587–8 (1990). [201] Ibid. at 585.
[202] Ibid. [203] Ibid. at 585–6. [204] Ibid. at 586.
[205] Ibid. [206] *Pennsylvania v Muniz* 496 US 582, 590–1 (1990).
[207] Ibid. at 590 (citation omitted). [208] Ibid.
[209] Ibid. at 598–9. [210] Ibid. at 593.

also take an active and deliberate part in transferring that information to the government. A 'testimonial act' requires that the suspect exercise control over the transmission of information; mere submission to a state's drawing out of information is not enough.[211]

17.5 The Mind–Body Distinction in Fifth Amendment Jurisprudence

How does the court's distinction between testimonial and physical evidence apply to the use of involuntary and incriminating neuroscience evidence in a criminal trial? Consider the reflections of Hans A. Linde, Justice of Oregon Supreme Court from 1977 to 1990.[212] Writing in concurrence in the otherwise unremarkable case of *State v Lyon*,[213] which held that polygraph tests were inadmissible as evidence

---

[211] In the next section, I challenge the jurisprudential distinction between testimonial and physical evidence. I am hardly the first to do so. Four scholarly critiques are prominent. For excellent discussion of this literature, see Pardo, note 189, above, 1037. First, Richard Nagareda has argued that the physical/testimonial distinction is at odds with the historical understanding of what it means 'to be a witness' under the Fifth Amendment, which he says is to produce compelled evidence, whether it is physical or testimonial in nature. R. A. Nagareda, 'Compulsion "to be a Witness" and the Resurrection of *Boyd*' (1999) 74 *New York University Law Review* 1575, 1658–9. See also *United States v Hubbell* 530 US 27, 49–56 (2000) (Thomas J concurring). Second, Mike Redmayne has argued that 'the most compelling rationale for the privilege is that it serves as a distancing mechanism, allowing defendants to disassociate themselves from prosecutions, [which]...suggests that no distinction should be drawn between requirements to speak and requirements to provide the authorities with documents, blood samples and the like': M. Redmayne, 'Rethinking the Privilege Against Self-Incrimination' (2007) 27 *Oxford Journal of Legal Studies* 209, 209. Third, Susan Easton has argued that there are no principled grounds on the basis of which to distinguish the communicative qualities of evidence that the court has classified as 'physical' (e.g. bodily samples or markings like scars and tattoos) as opposed to 'testimonial' (e.g. words or gestures intended to impart information): Susan Easton, *The Case for the Right to Silence*, 2nd edn (Aldershot, Ashgate, 1998) 217–8; '[W]hile the distinction might be justified on grounds of policy or expediency', Easton concludes, 'nonetheless it is artificial and problematic, because both samples and speech are subject to similar considerations and arguments' (ibid. at 220). Finally, Daniel Seidman and Alex Stein argue that 'a suspect's ability to tell uncontradicted lies can impose externalities because guilty suspects might harm innocent suspects by pooling with them through fabrications, lies, or omissions': Seidman and Stein, note 42, above, 480; see also Stein, note 102, above, 158–64, 200–4. The question for Seidman and Stein is not whether evidence is 'physical' or 'testimonial' in nature, but instead whether the evidence can be manipulated in a way that harms innocent defendants by making it possible for guilty defendants to pass off lies as truths, thereby rendering the latter less credible and less capable of protecting the innocent (ibid. at 476). Seidman and Stein argue that courts should replace the physical/testimonial distinction as a test for right-to-silence protection with a substitute distinction between externality-laden and externality-free evidence, according to which 'only the existence of a meaningful fabrication alternative should...activate the privilege' (ibid. at 480). My account is different because mine is the only one that can explain why the Fifth Amendment should prohibit the compelled brain imaging technology to extort incriminating thoughts from an unwilling suspect.

[212] See generally Robert F. Nagel (ed.), *Intellect and Craft, The Contributions of Hans A. Linde to American Constitutionalism* (Boulder, Westview Press, 1995); Symposium on the Work of Justice Hans Linde (1991) 70 *Oregon Law Review* 679; Sanford Levinson, 'Tiers of Scrutiny—From Strict Through Rational Basis—and the Future of Interests: Commentary on Fiss and Linde' (1992) 55 *Albany Law Review* 745, 746 (referring to Linde as 'easily one of the three most important state court judges in this century').

[213] 744 P2d 231 (Or 1987).

due to insufficient reliability,[214] Justice Linde expressed his widely shared but difficult-to-articulate reluctance to embrace even demonstrably reliable polygraph evidence:[215]

> I doubt the uneasiness about electrical lie detectors would disappear even if they were refined to place their accuracy beyond question. Indeed, I would not be surprised if such a development would only heighten the sense of unease and the search for plausible legal objections.[216]

Justice Linde said he would hesitate to admit into evidence any interrogation method, no matter how reliable, that purports to verify veracity in the thoughts of criminal suspects.[217] Citing a pair of obscure 1956 law review articles[218] discussing a German case[219] that denied admission of inculpatory polygraphy results on grounds of 'personal dignity',[220] Justice Linde argued that polygraph testing threatened to undermine certain 'fundamental tenets about human personhood'.[221] Justice Linde's concern about personhood had two parts. First, he worried that admitting polygraph testing into the courtroom would compromise the performative character of criminal trials, which he regarded as a constitutive element 'underlying our legal and social institutions'.[222]

In my argument that follows, I take a cue from Justice Linde's second argument that the use of polygraph testing threatened to reduce subjects to 'electrochemical systems to be certified as truthful or mendacious by a machine.'[223] Justice Linde conceded that the 'function of detecting a lie' among suspects and witnesses is undoubtedly a 'valuable' one in the American criminal justice system,[224] and that the polygraph is not altogether unique among interrogation techniques that seek to achieve this worthy objective by 'turn[ing] its subject into an object'.[225] But

---

[214] See ibid. at 232.
[215] Ibid. at 240 (Linde J concurring).
[216] Ibid. at 238 (Linde J concurring).
[217] Ibid. at 238. See also Kittay, note 124, above, 1390 ('[E]ven if fMRI technology could fully deliver on its potential, hopes of being welcomed in the courtroom are exceptionally hamstrung.').
[218] H. J. Kaganiec, 'Lie Detector Tests and "Freedom of the Will" in Germany' (1956) 51 *Northwestern University Law Review* 446, 449 (arguing that coerced polygraph tests to 'disclose the otherwise concealed psychic structure of the accused ... are an ... attempt to discover what might be present only in the unconscious of the accused'); H. Silving, 'Testing of the Unconscious in Criminal Cases' (1956) 69 *Harvard Law Review* 683, 693 (arguing that concern about 'human dignity' weigh against the admission of incriminating polygraph tests).
[219] *Lyon* 744 P2d at 240 (Linde J concurring); see 5 Entscheidungen des Bundesgerichtshofes in Strafsachen [Judgment of Bundesgerichtshof, West Germany Supreme Court] (1954), cited in Silving, note 218, above, 688–9, *and* Kaganiec, note 218, above, 446.
[220] *Lyon* 744 P2d at 240 (Linde J concurring).
[221] Ibid.
[222] Ibid. at 240–1 ('The cherished courtroom drama of confrontation, oral testimony and cross examination is designed to let a jury pass judgment on [the] truthfulness [of witnesses] and on the accuracy of their testimony.').
[223] Ibid. at 240.
[224] Ibid. at 239 (Linde J concurring) ('[T]he lie detector only purports to detect whether a person is uttering a lie', which '[b]eyond doubt ... is often a useful thing to know'.).
[225] Ibid. (acknowledging that the basic relationship between the conclusions about truthfulness and the physiological facts that polygraph testing attempts to 'independently establish' are similar in

polygraph testing is different in kind, Justice Linde argued, because of the '[i]nconsistency of physiological lie detection with fundamental tenets about human personhood [that] has been important in European objections to the polygraph, reflecting Christian and Kantian philosophical traditions as much as doubts of its accuracy'.[226] Though he did not elaborate as to what precisely those 'tenets about human personhood' consisted of,[227] Justice Linde overstated such concerns as they apply to polygraph testing,[228] since polygraphy yields unreliable results that are vulnerable to manipulation by subjects, interrogators, and evaluators.[229]

Nor does *Schmerber* resolve whether brain fingerprinting is more like forms of evidence such as speaking, writing, and nodding, which are protected by the Fifth Amendment, or more like fingerprinting, breathalyzer tests, and handwriting samples, which are not protected. While Justice Brennan's reflections on the use of involuntary testing to incriminate a criminal suspect extended beyond the facts of Schmerber's blood testing, whatever application the 'spirit and history'[230] *dicta* might have had for forensic neuroscience, it is not binding precedent for any future cases. Whether the fitting analogy for neuroforensics is to traditionally 'testimonial' evidence, or instead to traditionally 'physical' evidence, is far from clear.[231]

Until recently, reliable information about a subject's substantive knowledge required that a suspect actively communicate that information, whether by choosing to speak, write, or nod. But now, forensic neuroscience enables law enforcement officials to access communicative phenomena in a way that leaves the suspect 'no more control over the firing of neurons than Schmerber had in the way his blood responded to the test'.[232] Once a suspect's blood has been drawn, or EEG sensors affixed to his scalp, the results of the blood alcohol tests or lie-detection results are, for both, altogether involuntary. As Sarah Stoller and Paul Root Wolpe have observed, 'any active participation on the part of the subject [of neuroscientific

---

principle to accepted forensic techniques such as fingerprinting and blood testing for alcohol content levels).

[226] *State v Lyon* 744 P2d 231, 239–40 (Or 1987).

[227] See ibid. at 240.

[228] See J. R. McCall, 'The Personhood Argument against Polygraph Evidence, Or "Even if the Polygraph Really Works, Will Courts Admit the Results?"' (1998) 49 *Hastings Law Journal* 925, 941–4.

[229] As Department of Justice lawyer Michael R. Dreeben testified during oral argument in *United States v Scheffer* 523 US 303 (1998), '[t]he fundamental unreliability of polygraph evidence is underscored ... [by] the possibility that countermeasures can defeat any test': Oral Argument Transcript of Michael R. Dreeben on behalf of the Petitioner, available at <http://www.oyez.org/cases/1990-1999/1997/1997_96_1133/argument/>. See also C. M. Polizzi, 'A New View into the Truth: Impact of a Reliable Deception Detection Technology on the Legal System' (1995) 21 *Rutgers Computer and Technology Law Journal* 395, 398 note 9 ('We are familiar with the polygraph, the most widely used and most controversial method of lie detection.'); T. B. Henseler, Comment, 'A Critical Look at the Admissibility of Polygraph Evidence in the Wake of *Daubert*: The Lie Detector Fails the Test' (1997) 46 *Catholic University Law Review* 1247, 1247 ('The polygraph and other deception detection instruments have suffered through a tumultuous seventy years in the American legal system.').

[230] *Schmerber v California* 384 US 757, 764 (1966).

[231] A. J. Roberts, Comment, 'Everything Old is New Again: Brain Fingerprinting and Evidentiary Analogy' (2006) 9 *Yale of Journal Law and Technology* 234, 268.

[232] Stoller and Wolpe, note 28, above, 368–9.

testing] would be, like Schmerber's participation in the blood extraction and analysis, irrelevant to the results of the test'.[233]

Brain imaging is difficult to classify because it promises distinctly testimonial-like information about the content of a person's mind that is packaged in demonstrably physical-like form, either as blood flows in the case of fMRI, or as brainwaves in the case of EEG.[234] Forensic neuroscience measures physiological changes in the brain—chemical reactions to an outside stimulus—that reveal communicative processes such as recall and deception. When a subject is forced to undergo brain fingerprinting, the information that the test elicits is not precisely the subject's thoughts, at least in the way that we ordinarily tend to experience our thoughts as the deliberate transmission of ideas such as 'I am familiar with the scene of the murder', or 'I recognize the murder weapon.' In the India murder case, for example, the defendant Sharma did not speak, write, nod, or take any other active measure to communicate her thoughts in response to targeted stimuli. Instead, EEG sensors detected patterns of electrical activity in her brain that corresponded to a physical code for her stored knowledge, the incriminating content of which was used to prosecute her for murder.

What makes brain imaging unique is not only 'that the physical characteristics the scanner "observes" are imperceptible to the average observer'.[235] The more important reason that neuroscientific evidence is different is that it measures direct and involuntary brain activity that cannot, like polygraph testing, be effectively controlled by the subject undergoing interrogation.[236] And because EEG and fMRI test results appear in the form of high resolution computer-generated images, they are less likely than the polygraph to be misinterpreted by biased or imperfect evaluators.[237]

What Justice Brennan failed to appreciate is that the distinction between physical and testimonial evidence presupposes a flawed dualism between body and mind.[238] The physical/testimonial distinction requires that exclusively mental (and not bodily) processes comprise communicative meaning.[239] A division of

---

[233] Ibid. at 369.
[234] See Haynes et al., note 67, above, 324.
[235] Thompson, note 50, above, 348. But see ibid. at 357.
[236] See Lee et al., note 119, above, 163 ('[C]ontrolling one's cerebral activity to avoid detection is unfeasible.').
[237] See Giannelli, note 126, above, 905 ('The examiner's role cannot be overstated, because it is the examiner who decides whether there is sufficient indication of deception.'); Silberman, note 128, above, 142.
[238] For first-rate review of the philosophical literature that examines the meaning of testimonial knowledge, see M. S. Pardo, 'Testimony' (2007) 82 *Tulane Law Review* 119, 125–44.
[239] Susan Easton suggests that the communicative meaning we attach to body language in itself reveals the conceptual bankruptcy of Cartesian dualism. Easton states: '[I]t is generally accepted that non-verbal gestures, whether voluntary, subconscious or involuntary, may communicate information, even where the individual does not "intend" that behavior... In everyday life people constantly look for visual clues and make judgments according to demeanor, or the shape or position of the body, and often treat them as better guides to mental attitudes than the spoken word.': Easton, note 211, above, 217; 'While the distinction [between physical and testimonial evidence] might be justified on grounds of policy or expediency', Easton concludes, 'nonetheless it is artificial and problematic...' (ibid. at 220). Easton concludes that the rigid physical/testimonial distinction should be replaced with a

mind and body therefore underlies the doctrinal distinction between physical and testimonial evidence. The so-called 'mind–body problem is the problem of understanding how... the water of the physical brain is turned into the wine of consciousness'.[240] Mind/body dualism conceives of the sphere of human psychology as distinct from the sphere of human biology.[241] The contemporary version of mind/body dualism was articulated by René Descartes.[242] Following Plato and Aristotle's speculations that the faculties of intellect or soul are distinct from the physical organism,[243] Descartes argued for a divorce of mind from body, distinguishing mental phenomena such as consciousness and self-awareness from material organs such as even the brain.[244] On this account, the mind directs the body like a 'ghost in the machine'.[245]

Advances in cognitive imaging reveal the folly in the conceptual schism of mind and body.[246] Neuroscientists agree that the complex phenomena of thought and behaviour can be explained in terms of the neural activity of the brain.[247] Recent research into the human nervous system has uncovered the existence of discernible correlates in our brain chemistry for what were once thought of as the purely philosophical and psychological, including thoughts, emotions, and behaviours like decision-making,[248] free will,[249] moral

---

continuum that extends from oral communication at one end of the continuum; with body language (e.g. nodding) intended to communicate at the next point on the continuum; unintentional non-verbal communications (e.g. sweating) after that; followed by external bodily features (e.g. tattoos); then bodily materials voluntarily or involuntarily transmitted (e.g. blood or hair); and finally materials from a dead person (e.g. DNA) at the other end of the continuum: see ibid. at 218.

[240] C. McGinn, 'Can We Solve the Mind-Body Problem?' (1989) 98 *Mind* 349, 349.

[241] See W. D. Hart, 'Dualism' in Samuel Guttenplan (ed.), *A Companion to the Philosophy of Mind* (Oxford, Oxford University Press, 1996) 265–7.

[242] See Howard Robinson, Dualism, in Edward note Zalta (ed.), *The Stanford Encyclopedia of Philosophy* (Stanford, Stanford University Press, 2003), available at <http://plato.stanford.edu/archives/fall2003/entries/dualism/>.

[243] See, for example, 1 Plato, *Platonis Opera* (E. A. Duke et al., 1995); 2 Aristotle, *Metaphysics* (Metaphysica) (W. D. Ross (ed.), 1924).

[244] Gordon Baker and Katherine J. Morris, *Descartes' Dualism* (New York, Routledge, 1996), citing René Descartes, *Meditations on First Philosophy* (1641), in John Cottingham et al. (eds.), 2 *The Philosophical Writings of Descartes* (Cambridge, Cambridge University Press, 1984) 1–62).

[245] Brent Garland, *Neuroscience and the Law: Brain, Mind, and the Scales of Justice* (New York, Dana Press, 2004) 66.

[246] See Nemeroff et al., note 12, above, 672 ('Neuroimaging offers a powerful probe of brain state, but we are now faced with metaphysical questions; i.e., what is a brain state, and how is it related to the outward manifestations of behavior? This has the potential for degenerating into the old mind-body duality of Descartes...').

[247] See J. Greene and J. Cohen, 'For the Law, Neuroscience Changes Nothing and Everything' (2004) 359 *Philosophy Transactions Royal Society of London* 1775, 1775 (describing the very enterprise of cognitive neuroscience as the 'understanding of the mind as brain').

[248] See, for example, L. K. Fellows and M. J. Farah, 'Different Underlying Impairments in Decision-making Following Ventromedial and Dorsolateral Frontal Lobe Damage in Humans' (2005) 15 *Cerebral Cortex* 58; A. R. Damasio, 'How the Brain Creates the Mind' (2002) *Scientific America* 4 (originally appearing in December 1999 issue).

[249] See, for example, R. F. Wagner and H. Reinecker, 'Problems and Solutions: Two Concepts of Mankind in Cognitive–Behavior Therapy' (2003) 57 *American Journal of Psychotherapy* 401.

judgement,[250] personality,[251] consciousness,[252] and the self.[253] Among these neuroscientific insights is the discovery that even the most sophisticated operations of mind are deeply integrated with the mechanical operations of biological organisms.[254] Brain imaging technologies can localize complex psychological and behavioural functions to specific neural regions through spatial and temporal resolution of functional markers in the brain.[255]

That the mystery we associate with our thoughts and actions can be reduced to quantifiable networks of neural activity might be difficult to accept.[256] In one sense, new insight into the neurological basis of behaviour merely confirms what doctors suspected in 1848, when Phineas Gage, a Vermont railroad worker, had a thirteen-pound iron rod driven through his frontal lobe with such dramatic effects on his personality and behaviour that his friends declared that 'Gage was no longer Gage'.[257] In another sense, research into the human nervous system significantly advances our understanding of and confidence in the synthesis between mind and body.[258] Neuroimaging studies provide strong evidence that there is no freestanding metaphysical being that tells our brains what to think and do.[259] Instead, the mind is intimately bound up with the whole of the human organism, while the organism works together with both body and brain. This neuroscientifically informed picture of the integrated body and mind powerfully undermines the dualistic premise upon which the physical/testimonial distinction is built.

Recognizing the incoherence of mind/body dualism prompts us to acknowledge that what we really care about is the individual's control over his thoughts—those mental activities that comprise 'the sheer chaotic tropical luxuriance of the inner life.'[260] The relevant difference between a person's thoughts and his blood is

[250] See, for example, J. D. Greene et al., 'An fMRI Investigation of Emotional Engagement in Moral Judgment' (2001) 293 *Science* 2105; J. Moll et al., 'Functional Networks in Emotional Moral and Nonmoral Social Judgments' (2002) 16 *NeuroImage* 696.

[251] See, for example, T. Canli and Z. Amin, 'Neuroimaging of Emotion and Personality: Scientific Evidence and Ethical Considerations' (2002) 50 *Brain Cognition* 414.

[252] See, for example, G. Tononi and G. M. Edelman, 'Schizophrenia and the Mechanisms of Conscious Integration' (2000) 31 *Brain Research Reviews* 391.

[253] See, for example, S. J. Gillihan and M. J. Farah, 'Is Self Special? A Critical Review of Evidence from Experimental Psychology and Cognitive Neuroscience' (2005) 131 *Psychological Bulletin* 76.

[254] See Damasio, note 248, above, 9 ('[B]y 2050, sufficient knowledge of biological phenomena will have wiped out the traditional dualistic separations of body/brain, body/mind, and brain/mind.').

[255] See Daniel C. Dennett, *Consciousness Explained* (Boston, Back Bay Books, 1991).

[256] See Stoller and Wolpe, note 28, above, 369 ('[O]ur everyday conception of humanity still reflects dualistic notions of body and non-physical mind or soul'). 'Perhaps we think our minds are too opaque to ever be laid bare by a machine', another commentator speculates, 'or perhaps we so desperately cling to this last frontier of privacy that we are willing it to be impossible': Kittay, note 124, above, 1394.

[257] Malcolm Macmillan, *An Odd Kind of Fame: Stories of Phineas Gage* (Cambridge, MA, Harvard University Press, 2000) 11–13; John Fleischman, *Phineas Gage: A Gruesome But True Story About Brain Science* (New York, Houghton Mifflin, 2004), 2; M. Macmillan, 'A Wonderful Journey through Skull and Brains: The Travels of Mr. Gage's Tamping Iron' (1986) 5 *Brain and Cognition* 67, 85.

[258] S. R. Quartz and T. J. Sejnowski, *Liars, Lovers, and Heroes: What New Brain Science Reveals About How We Become Who We Are* (New York, HarperCollins, 2002) 138–41.

[259] See studies discussed in Antonio Damasio, *Descartes' Error: Emotion, Reason, and the Human Brain* (New York, Avon Books, 2005) 24–30, 258–60.

[260] T. Nagel, 'Concealment and Exposure' (1998) 27 *Philosophy and Public Affairs* 4.

neither that his thoughts communicate information that his blood does not, nor that thinking requires the use of communicative processes that bleeding does not. What matters is instead that our blood is readily separable from what we think important about us. Not so with our thoughts.[261] Conventional forms of criminal evidence can disclose all sorts of personal and private information about me—where I was, what I did, or why I did it. But when the state seizes my photographs, handwriting samples, or DNA, it does not deprive me of control over my mental life.[262] It does not compromise my ability to be in command of the use and disclosure of my thoughts; compelled brain imaging does.

This notion of mental control has normative significance because our thoughts are what anchor each of us as an individual person with an uninterrupted autobiographical narrative.[263] It is not just that physical manifestations of brain function correspond to the exercise of subjective awareness; it is, moreover, that those neural transmissions are, as the organ of mind and consciousness, what we tend to think of as most important about who we are.[264] At the core, that is, we think of ourselves as beings capable of self-reflection. The Fifth Amendment protects a suspect's control over his thoughts from unwanted government access and use.[265]

---

[261] Neuroscientist Donald Kennedy, editor of the journal *Science*, suggests that he seeks to keep his 'brainome' private because '[i]t is way too close to who I am'. 'Major Talks on Addiction, Neuroethics, and Depression Highlight Neuroscience 2003' *Neuroscience Quarterly* (Winter 2004) <http://web.sfn.org/index.cfm?pagename=neuroscienceQuarterly_04winter_neuroscience>; New York City Bar Association, Report from the Committee on Science and Law, 'Are Your Thoughts Your Own? 'Neuroprivacy' and the Legal Implications of Brain Imaging' (2005) <http://www.abcny.org/pdf/report/Neuroprivacy-revisions.pdf> (citing *Harrington v Iowa* 659 NW2d 509 (Iowa 2003)).

[262] John Fischer has suggested it would be interesting to consider whether we could ever consider our mental lives under our exclusive control if there exists an omniscient God who has complete knowledge of people's brain states.

[263] Neuroscientist Antonio Damasio refers to these features of selfhood as 'the individuality and continuity of a living organism': A. R. Damasio, 'Mental Self: The Person Within' (2003) 423 *Nature* 227, 229. See also S. A. Tovino, 'Functional Neuroimaging Information: A Case for Neuro Exceptionalism?' (2007) 34 *Florida State University Law Review* 415, 470 (exploring the 'possibility that fMRI will reveal back to the individual... stories that are inconsistent with the individual's dominant life narrative').

[264] See Garland, note 245, above, 34; see also H. T. Greely, *Neuroethics and ELSI: Some Comparisons and Considerations* (2004) <http://perpich.com/neuroed/archive/139.pdf>. Greely states (ibid.): '[I] am more than my genes. The genes are an important part of me, but I can be certain that they are not my essence; they are not my soul. When we shift that notion to the neuroscience area, though, I am not so confident. Is my consciousness—is my brain—me? I am tempted to think it is.'

[265] The notion of protecting the individual's control over the use of his thoughts is not altogether foreign to Supreme Court reflections on the constitutional right to silence. In *United States v Nobles*, the majority wrote that the Fifth Amendment prohibits 'state intrusion to extract self-condemnation' by 'protect[ing] a "private inner sanctum of individual feeling and thought"' 422 US 225, 233 (1975) (quoting *Couch v United States* 409 US 322, 327 (1973)); see also *Braswell v United States* 487 US 99, 126 (1988) (Kennedy J dissenting) ('Physical acts will constitute testimony if they probe the state of mind, memory, perception, or cognition of the witness.'). Professor Richard Uviller also argued as early as 1987 that 'personal control over the production of cognitive evidence, free of official coercion, is guaranteed by the self-incrimination clause of the fifth amendment': see H. R. Uviller, 'Evidence from the Mind of the Criminal Suspect: A Reconsideration of the Current Rules of Access and Restraint' (1987) 87 *Columbia Law Review* 1137, 1137. In light of *Hubbell*'s language emphasizing protection of the 'contents of [a suspect's] own mind', *United States v Hubbell* 530 US 27, 43 (2000), Professor Uviller took Brennan J's opinion in *Schmerber* to mean that 'no one can be forced to divulge cerebral evidence, to speak the contents and products of the mind': Uviller, note 186, above, 313. Ronald Allen

At least three objections may be raised against this account of the right to silence. The first objection is that advances in forensic neurotechnology could shift the landscape of moral perceptions to eliminate the mismatch I identified between existing right-to-silence rationales and intuitions about the privilege's application to brain imaging techniques. This objection underestimates the human faculty of impulse and imagination. Recall the Indian murder case, in which police were able to use brain imaging techniques to extort incriminating thoughts from the unwilling suspect. This vision is not so different, and less radical in fact, than similar possibilities portrayed in contemporary film and literature such as George Orwell's *1984*[266] and Steven Spielberg's *Minority Report*.[267] Popular media prompt us to reflect on or at least acknowledge the possibility of

and Kristin Mace have since built on Uviller's insights to argue that the privilege applies to 'the substantive content of cognition', which they define as 'the intellectual processes of acquiring, storing, retrieving, and using knowledge ... that allow one to gain and make use of substantive knowledge and to compare one's "inner world" (previous knowledge) with the "outside world" (stimuli such as questions from an interrogator)': Allen and Mace, note 51, above, 246–7. Allen and Mace suggest that the court has implicitly taken 'testimony' to mean 'substantive cognition', that is, the 'contents of [one's] own mind', or 'thoughts' that 'results in holding or asserting propositions with truth value', for example, that a person was present at the scene of a crime, or that he recognizes the victim, took part in the act in question. See ibid. at 266–7, 267 note 107 (citing *Doe v United States* (*Doe II*) 487 US 201, 211 (1988)) (quoting *Couch v United States* 409 US 322, 328 (1973); *Curcio v United States* 354 US 118, 128 (1957)); *United States v Hubbell* 530 US 27, 43 (2000) (also quoting *Doe II* and *Curcio*); see also *Braswell v United States* 487 US 99, 126 (1988) (Kennedy J dissenting). In proscribing the compulsion of incriminating 'testimony', they argue, the Court is actually saying that state action cannot compel the revelation of these sorts of thoughts from the criminal suspect: Allen and Mace, note 51, above, 266–7. On this account, the right to silence guarantees 'that the government may not compel revelation of the incriminating substantive results of compelled cognition': ibid. at 268.

However, Allen and Mace define protected cognition too narrowly: see ibid. at 266–7, 267 note 107. Because their theory of 'compelled cognition' incorporates the narrow precedential focus on those 'propositions with truth-value' that are prompted by state action, ibid. at 268, their account would deny Fifth Amendment protection in a hypothetical case in which police reliably scanned a suspect's mind for incriminating thoughts, provided that state officials did nothing to provoke or induce those thoughts, such as posing a question, or presenting the suspect with an image or recording that was relevant to the victim or crime scene. See W. Federspiel, Comment, '1984 Arrives: Thought (crime), Technology, and the Constitution' (2008) 16 *William and Mary Bill of Rights Journal* 865, 896–7.

Michael Pardo tries to resolve the gaps left by Allen and Mace to argue that the self-incrimination clause 'appl[ies]' when the government compels the tests in order to obtain evidence of the incriminating informational content of subjects' propositional attitudes', including 'inductive evidence of their beliefs, knowledge, and other mental states'. Pardo, note 46, above, 331–2. But Pardo's account of the Fifth Amendment suffers from similar limitations. He concedes that his version of the privilege 'would not preclude compelled tests when used for any purpose other than those that rely on incriminating propositional content. For example, if the tests could be used to determine mental capacity, intent, bias, voluntariness, etc., without relying on incriminating propositional content, then the privilege would not preclude such uses': ibid. at 332 note 205.

My theory can be distinguished in two ways. The first difference is methodological. Allen and Mace and Pardo use cognition as a conceptual tool to vindicate the right to silence. I go other way around in using the right to silence to protect a person's brainwork from state use. Second, Allen and Mace and Pardo conceive of the right to silence as preventing the state from eliciting self-incriminating testimony because it is a product of cognition. They perceive compelled extraction of this product as a revelational harm. My theory of mental control, by contrast, protects a person's thought processes directly, rather than via products, by prohibiting the state from seizing control over those processes, both self-incriminating and non-self-incriminating. See note 273, below, and accompanying text.

[266] George Orwell, *1984* (1949).
[267] *Minority Report* (DreamWorks, 2002).

mind-reading devices and of government access to the thoughts of individuals. The premise of the Indian case should not be so unfamiliar to most people that their present intuitions could not be taken as a reliable guide to their intuitions in a world in which such technologies were no longer imagined, but real, and their use by police interrogators were no longer impossible but routine. If present intuitions are a reliable guide, any divergence with existing principles is unlikely to disappear if the hypothetical were to become reality.

The second objection is that protection of a suspect's control over his thoughts cannot explain why the state can compel even confessions that are self-incriminating with a grant of immunity,[268] or why the privilege traditionally applies only to *incriminating* evidence or to *self-incriminating* evidence.[269] The immunity exception is overstated, since 'use immunity' applies to the act of production only.[270] While subpoenaed documents are not protected from government use under 18 USC §6002, the state cannot use whatever information is obtained from that act of production, whether directly or indirectly, to criminally prosecute the suspect.[271] The second point is that compelled evidence has not traditionally been protected under the Fifth Amendment unless it counts as both *incriminating*—in the sense that it could 'furnish a link in the chain of evidence needed to prosecute'[272]—and also *self*-incriminating—in the sense that the information is sought to incriminate the subject himself, as opposed to someone else, even family or friends.[273] This restraint on the privilege implies that, provided prosecutors did not use the information derived from compelled brain imaging, either directly or indirectly, as evidence against the suspect himself in a criminal proceeding, then the Fifth Amendment would not prohibit the state from gaining unwanted access to his mental life.

I reply by proposing that judges refine these limits on the right to silence, as they apply to state action that would deny a suspect control over the use of her thoughts. Whether in the course of interrogation,[274] trial,[275] or sentencing,[276] when state officials seek to acquire information that would deprive individuals of control over their thoughts, traditional limitations on the privilege against self-incrimination

---

[268] See *Kastigar v United States* 406 US 441, 445–7 (1972); see also R. Allen, 'Theorizing about Self-Incrimination' (2008) 30 *Cardozo Law Review* 729, 732.

[269] See Stuntz, note 37, above, 1232–4 ('The privilege applies only to testimony that is incriminating.').

[270] See 18 USC. § 6002 (1994).

[271] Exceptions to this rule include prosecution for perjury, for giving a false statement, or for failing to comply with the subpoena order. See ibid.

[272] *Hiibel v Sixth Jud. Dist. Ct. of Nev., Humboldt County* 542 US 177, 190 (2004) (quoting *Hoffman v United States* 341 US 479, 486 (1951)).

[273] See *Ullmann v United States* 350 US 422, 430–1 (1956).

[274] *Doyle v Ohio* 426 US 610 (1976) (upholding a constitutional due process right to remain silent after arrest without later penalty); see also *Miranda v Arizona* 384 US 436 (1966) (holding that the Fifth Amendment requires that a suspect be warned of his right to silence before any custodial interrogation).

[275] *Griffin v California* 380 US 609 (1965) (holding that the Fifth Amendment prohibits prosecutors and judges from commenting adversely on a defendant's failure to testify in a criminal proceeding).

[276] *Mitchell v United States* 526 US 314, 321, 330 (1999) (extending application of the Fifth Amendment privilege against compelled self-incrimination to criminal sentencing proceedings).

should not apply. This exception is sufficiently narrow in scope and modest in practice and does not, however, change what it means for a person to be a witness against himself.[277]

The third objection is that concerns about mental control, even if valid, should not take priority over the social goals of criminal justice, including the punishment of the guilty and the prevention of crime.[278] 'There is nothing intrinsic in privacy and other individualist values', some Fifth Amendment scholars have argued, 'that should allow them to trump such objectives of the criminal justice system as deterrence and retribution.'[279] Yet much in modern political theory has devoted itself to the proposition that each person possesses rights that carry greater weight than even considerations of the common good. Respect for the individual lies in conceiving of her both *analytically* as more than just one among a collective, whose group interests may not be fused together into a single sum total; and also *normatively* as an end in herself, who may not be used exclusively for purposes incompatible with her own.[280] Such respect demands deference to certain liberties that even the well-being of society as a whole cannot override. Worthy and serious though the goals of the criminal justice system are, they fail to outweigh the injury to the individual that is done when the state deprives a suspect of control over her mental life.

17.6 Conclusion

The law provides principles to guide research and technology of the human brain; so too can advances in cognitive neuroscience shape and inform the law.[281] Emerging brain imaging technology forces us to reckon with the prospect of evidence-gathering techniques that would enable the state to extract information directly from a suspect's brain, in a way that affords them no opportunity for control over the transmission of that information. In so doing, forensic

---

[277] Nita Farahany has proposed a more useful way of thinking about the testimonial dimension of self-incrimination doctrine: a *spectrum* ranging from structural evidence—like blood or bullets in the body—to functional evidence, like the criminal confessions. Farahany argues that brain imaging reveals several intermediate categories along the spectrum between the archetypically structural and functional. She calls these 1) autonomic, 2) visceral, 3) intention-based, and 4) evoked. Farahany characterizes autonomic evidence as involuntary or unconscious processes including brain function that control, for example, heart or respiration rate, salivation, perspiration, and diameter of the pupils, and perhaps even sexual arousal or foreign language processing. Visceral evidence constitutes feelings (as distinct from intellect); the examples Farahany gives are pedophilia and pathological gambling. Intention-based evidence, Farahany tells us, relates to a person's aim, plan, or design. Finally, evoked evidence, which is elicited or remembered in the conscious mind, includes knowledge, familiarity, and recall of complex images, thoughts, and words. See Nita A. Farahany, unpublished presentation, 'Incriminating Thoughts', delivered at the Fourteenth Annual Current Legal Issues: Law & Neuroscience, Inter-Disciplinary Colloquium at University College London, 6–7 July 2009.
[278] See Seidman and Stein, note 42, above, 455.
[279] Ibid.
[280] See Martha C. Nussbaum, *Sex and Social Justice* (Oxford, Oxford University Press, 1999) 62.
[281] See Adina Roskies, 'A Case Study of Neuroethics: The Nature of Moral Judgment' in Judy Illes (ed.), *Neuroethics: Defining the Issues in Theory, Practice, and Policy* (2006) 17, 18.

neuroscience reveals a tension between existing accounts of the right to silence and the plausible idea that individuals have a right of control over their thoughts from unwanted use by the government.

Some scholars have tried to resolve this tension by comparing brain imaging to more familiar types of criminal evidence. They argue that if we set aside reliability concerns, then whatever suspicion with which we might greet the introduction of neuroscience to the courtroom is in principle indistinguishable from misguided resistance to forensic advances of the past such as DNA, X-rays, and photography.[282] Yet this analogical approach is incomplete.

That we balk at the prospect of reliable brain imaging, ostensibly similar though it appears to accepted forms of evidence, suggests that the dominant right-to-silence framework—with its arguments about reliability, excuse, and psychological cruelty—cannot capture the values at stake in matters of involuntary lie detection. The physical/testimonial distinction at the core of self-incrimination doctrine is unlikely to protect a criminal suspect from the compelled use of fMRI or EEG. But this distinction presupposes a flawed conception of mind/body dualism. Brain imaging techniques that deprive individuals of control over their thoughts violate the 'spirit and history of the Fifth Amendment'.[283]

---

[282] See Pardo, note 46, above, 311 ('Photography, it was thought, potentially could usurp the power of courts to determine facts by shifting power to photography experts, and away from courts, to determine the true nature of reality. None of this happened, of course, because the evidence was eventually assimilated....'). Pardo ultimately gives a provocative epistemic justification for why brain imaging evidence should be treated as testimony for Fifth Amendment purposes. See Pardo, note 189, above, 1046 ('The results of polygraphs and other lie-detection tests, whether they call for a voluntary response or not, are testimonial because the tests are just inductive evidence of the defendant's epistemic state.' (citation omitted)).

[283] *Schmerber* 384 US at 764.

18

Minds Apart: Severe Brain Injury, Citizenship, and Civil Rights

*Joseph J. Fins**

18.1 Introduction

New insights from neuroscience about disorders of consciousness make it necessary to revisit presumptions about medical futility and severe brain which date to *Quinlan*, a landmark ruling allowing for the withdrawal of life-sustaining therapy in a young woman in the vegetative state.[1] Although *Quinlan* was laudable for enfranchizing patients and families with a right to die, the case's legacy has also had the unintended consequence of marginalizing a whole class of patients with disorders of brain injury, presumed to be beyond any of hope of recovery and thus undeserving of care. This implicit bias has led to discriminatory practices and a segregation of patients in the chronic care sector where they are viewed nihilistically, often receiving what is euphemistically described as 'custodial care'.

---

* MD, FACP, Chief of the Division of Medical Ethics at Weill Cornell Medical College where he serves as Professor of Medicine, Professor of Public Health and Professor of Medicine in Psychiatry; Attending Physician and Director of Medical Ethics at New York-Presbyterian Weill Cornell Medical Center and member of the Adjunct Faculty of Rockefeller University and Senior Attending Physician at The Rockefeller University Hospital. Correspondence should be addressed to: Joseph J. Fins, MD, FACP, Division of Medical Ethics, New York Presbyterian-Weill Cornell Medical Center, 435 East 70th Street, Suite 4-J, New York, NY USA 10021; tele: (212) 746-4246; fax: (212) 746-8738; email: jjfins@med.cornell.edu.

Dr Fins is the recipient of an Investigator Award in Health Policy Research (*Minds Apart: Severe Brain Injury and Health Policy*) from the Robert Wood Johnson Foundation. He also gratefully acknowledges grant support from: the Buster Foundation (*Neuroethics and Disorders of Consciousness*); the Richard Lounsbery Foundation (*Sustaining and Building Research Infrastructure for the Study of Disorders of Consciousness at Weill Cornell Medical College and Rockefeller University*); and the Jerold B Katz Foundation. The author also thanks: Professor Michael Freeman and the University College of London Faculty of Laws for the opportunity to contribute to this anthology and for the College's support; Dr Nicholas D. Schiff for his insights; Jennifer Hersh for her research assistance; and Amy B. Ehrlich for her counsel. Dr Fins is most grateful to the families who have shared their stories with us with such generosity.

[1] *Matter of Karen Quinlan* 70 NJ 10, 355 A2d 677 (1976); N. Cantor, 'Twenty-five years after Quinlan: a review of the jurisprudence of death and dying' (2001) 29 *Journal of Law, Medicine, and Ethics* 182.

Advances in our understanding of how brains are injured and recover, and more specifically neuroimaging findings in those who are severely injured, suggest that it is time to reconsider practice patterns for this population and ask the more fundamental question about what is owed to those with liminal states of consciousness. Neuroimaging will be particularly important to this recalibration of norms, practices, and expectations. By providing an alternative, or better yet, complementary perspective to the traditional clinical examination, these novel technologies will be able to point to discordances between the functional status of a patient's brain state as suggested by neuroimaging and by what is observed at the bedside.

Imaging research data suggestive of cognitive receptivity seen in a patient, otherwise inert by behavioural criteria, are already challenging assumptions about the presence or absence of consciousness. While these findings are not yet ready for clinical dissemination,[2] this nascent work anticipates future developments and clinical applications that are already necessitating a corrective of a heretofore unreflective nihilism.[3]

Despite the impressive progress that has been made in neuroscience, little has changed for patients with disorders of consciousness who struggle for access to even basic diagnostic and rehabilitative services as they remain sequestered from the scientific advances, which might make a difference in their lives. Although this is certainly a question of access to care, it is more than an entitlement issue. In my view, it is fundamentally a question of human rights in which a class of patients have been denied voice and been victimized by segregation and disinterest.

In this paper I consider how current practices and ethical norms should evolve to reflect progress in the neurosciences for this neglected population. In making these recommendations I draw parallels to the US Civil Rights Movement, a struggle worthy of emulation as we consider ways to achieve integration and parity in civic life for a population society has forgotten.

18.2 Quinlan's Other Legacy

Before we seek to reform current ethical norms and legal frameworks, it would be helpful first to understand how prevailing customs and practices evolved. How we view those with brain injury is closely linked to the landmark ruling in the case of Karen Ann Quinlan, a young woman left in the persistent vegetative state after a drug overdose.

The persistent vegetative state was first described in *The Lancet* by the Scottish neurosurgeon Bryan Jennett and American neurologist Fred Plum in 1972 as a state of 'wakeful unresponsiveness'.[4] Paradoxically, the eyes are open but there is no

---

[2] J. J. Fins and J. Illes, 'Neuroimaging and disorders of consciousness: envisioning an ethical research agenda' (2008) 8 *American Journal of Bioethics* 3.

[3] J. J. Fins, 'Constructing an ethical stereotaxy for severe brain injury: balancing risks, benefits and access' (2003) 4 *Nature Reviews Neuroscience* 323.

[4] B. Jennett and F. Plum, 'Persistent vegetative state after brain damage: a syndrome in search of a name' (1972) 1 *Lancet* 734.

awareness of self or the environment. Purposeless eye movements, along with sleep-wake cycles and autonomic regulation, are all controlled by a functioning brain stem devoid of higher cortical function.

Appreciating the futility of her condition, her parents sought to have her ventilator removed. The case reached the New Jersey Supreme Court and the request was permitted. Judge C. J. Hughes opined that withdrawal of life-sustaining therapy was to be allowed because it would be futile to provide additional care because of Ms Quinlan's irretrievable loss of a cognitive sapient state.

The court justified the removal of Ms Quinlan's ventilator based on the testimony of Dr Plum, who served as the court-appointed neurologist. Dr Plum described the elements of cognitive function as a mix of primitive or vegetative capabilities and higher order ones:

We have an internal vegetative regulation which controls body temperature which controls breathing, which controls to a considerable degree blood pressure, which controls to some degree heart rate, which controls chewing swallowing and which controls sleeping and waking. We have a more highly developed brain which is uniquely human which controls our relation to the outside world, our capacity to talk, to see, to feel, to sing, to think. Brain death necessarily must mean the death of both of these functions of the brain, vegetative and sapient.[5]

Drawing upon Dr Plum's testimony, Judge Hughes justified the removal of Ms Quinlan's ventilator as ethically warranted because there was no rationale to compel someone to be in a vital state devoid of sapience and cognition. Under such circumstances, all measures to preserve life could be seen as futile. Judge Hughes wrote:

We have no hesitancy in deciding... that no external compelling interest of the State should compel Karen to endure the unendurable, only to vegetate a few more measurable months with no realistic possibility of returning to any semblance of *cognitive or sapient life*.[6]

And thus was ensconced the legal basis for the right-to-die movement in the US, predicated as it was upon the utter futility of Ms. Quinlan's vegetative state. The *Quinlan* case—and the right to die—became an important cultural force, much like the contemporaneous US Supreme Court ruling in *Roe v Wade* granting women autonomy over their bodies and their reproductive rights.[7] In *Quinlan*, self-determination played out as a negative right to be left alone and to a refuse unwanted interventions.

This right would expand to non-vegetative terminally ill patients and lead to the acculturation of generations of physicians who have come to see requests to withhold or withdraw life-sustaining therapies as normative.[8] But with this transformation of

---

[5] See above, note 1.   [6] Ibid.
[7] *Roe v Wade* 410 US 113, 93 SCt 705, 35 LEd 2nd 147 (1973).
[8] Joseph J. Fins, *A Palliative Ethic of Care: Clinical Wisdom at Life's End* (Sudbury, MA Jones & Bartlett, 2006); J. J. Fins and F. Plum, 'Neurological diagnosis is more than a state of mind: diagnostic clarity and impaired consciousness' (2004) 61 *Archives of Neurology* 1354.

attitudes towards the right to die came entrenched views on the vegetative state, and by extension to others with severe brain injury.

Because this historic legacy was predicated upon the futility of further intervention in Ms Quinlan's case, clinicians have come to view the severely brain injured as a whole class beyond hope. This has led severe brain injury to be viewed nihilistically, that intervening in such states would be ethically disproportionate.[9] This perspective was amplified in the legal and public policy debate of subsequent high visibility right-to-die cases like Nancy Beth Cruzan[10] and Terri Schiavo.[11]

But in rereading the *Quinlan* decision it is interesting to note that history has recalled only part of Judge Hughes' admonition. In addition to asserting that there was no compelling reason to force her to endure the vegetative state, his approval for the withdrawal of life-sustaining therapy was contingent upon analogizing the vegetative state to that of a terminal illness: 'We perceive no thread of logic distinguishing between such a choice on Karen's part and a similar choice which, under the evidence of this case, could be made by a competent patient terminally ill, riddled by cancer and suffering great pain...'[12]

Although it was convenient to view the vegetative state as a terminal condition to allow for the withdrawal of life-sustaining therapy in a young woman like Ms Quinlan, time has shown this to be a flawed analogy.[13] The vegetative state need not be terminal absent a withdrawal of life-sustaining therapy.

Judge Hughes' analogy was also flawed in his mention of 'suffering great pain'. It was appreciated then, and demonstrated more recently, that patients in the vegetative state do not suffer pain. Recent work by Steven Laureys and colleagues have shown that vegetative patients cannot perceive pain because they lack the integrated pain network necessary for perception and processing of these sensations. In their study, when vegetative patients were exposed to painful stimuli only the primary sensory area is activated on functional MRI studies.[14] Network propagation of the percept stopped there: the sensory input did not activate the more fully integrated pain network necessary for the perception and experience of pain that is seen in normals.

Judge Hughes also conditioned his permission to remove life-sustaining therapy upon Ms Quinlan's doctors' ability to make an accurate diagnosis. He wrote:

Upon concurrence of the guardian and the family of Karen, should the responsible attending conclude that there is no reasonable possibility of Karen's ever emerging from

---

[9] See above, note 3.
[10] *Cruzan v Director, Missouri Department of Health* 497 US 261 (1990).
[11] J. J. Fins, 'Affirming the right to care, preserving the right to die: disorders of consciousness and neuroethics after Schiavo' (2006) 4 *Supportive & Palliative Care* 169.
[12] See above, note 1.
[13] J. J. Fins, 'Rethinking disorders of consciousness: new research and its implications' (2005) 35 *Hastings Center Report* 22.
[14] M. Boly, M. E. Faymonville, P. Peigneux, B. Lambermont, P. Damas, A. Luxen, M. Lamy, G. Moonen, P. Maquet, and S. Laureys, 'Cerebral processing of auditory and noxious stimuli in severely brain injured patients: differences between VS and MCS' (2005) 15 *Neuropsychological Rehabilitation* 283.

her present comatose condition to a cognitive, sapient state and that the life-support apparatus now being administered to Karen should be discontinued, they shall consult with the hospital 'Ethics Committee' or like body of the institution in which Karen is then hospitalized. If that consultative body agrees that there is no reasonable possibility of Karen's ever emerging form her present comatose condition to a cognitive, sapient state, the present life-support system may be withdrawn...[15]

While there was no doubt that Ms Quinlan was in the vegetative state, as confirmed by Dr Plum *in vivo* and by autopsy findings consistent with that state,[16] prognostication for patients with severe brain injury is not as clear-cut as the ideology surrounding these cases as those on the left and right might suggest. Diagnostic assessment should never reflect an ideological 'state of mind' but be an objective assessment of a patient's condition.[17]

Outcomes were never so simple or as clearly dichotomous with a miraculous recovery on the one hand and a dire outcome on the other. Instead there were shades of grey with outcomes occurring on a biological continuum.[18] We now appreciate that 23 per cent of comas resulting from anoxic brain injury will result in death or the permanent vegetative state but that 50 per cent for those who have sustained a traumatic injury will recover to some level of consciousness.[19]

Dr Plum understood that there was middle ground early on. In papers from his archives at Weill Cornell Medical College, he wrote about both the utter futility of some brain states vis-à-vis recovery to higher function and the indeterminacy and possibility of others. In a document, likely contemporaneous with the Quinlan era or late 1970s, Dr Plum noted his experience with acutely injured patients:

We have now studied over a 100 patients and without fail can identify patients within 24 hrs by their neurological signs alone who can not recover above a vegetative level no matter what treatment we undertake. Conversely, we can identify patients who will do well unless and intercurrent complication confounds all best medical efforts. This leaves a middle group for whom more information is needed but where presenting every effort at treatment must be made to know their maximal potential and how to judge their early signs...[20]

This cautionary note from decades past, written for the acute phase of injury, is just as applicable today for patients in chronic care with disorders of consciousness. Regrettably, many patients today diagnosed as being vegetative state may in fact be in the minimally conscious state (MCS), a condition in which there is definite,

---

[15] See above, note 1.
[16] H. C. Kinney, J. Korein, A. Panigrahy, P. Dikkes, and R. Goode, 'Neuropathological findings in the brain of Karen Ann Quinlan: the role of the thalamus in the persistent vegetative state' (1994) 330 *New England Journal of Medicine* 1469.
[17] J. J. Fins and F. Plum, 'Neurological diagnosis is more than a state of mind: diagnostic clarity and impaired consciousness' (2004) 61(9) *Arch Neurol* 1354–5.
[18] J. J. Fins and N. D. Schiff, 'Shades of gray: new insights into the vegetative state' (2006) 36 *Hastings Center Report* 8.
[19] J. Posner, C. Saper, N. D. Schiff and F. Plum, *Plum and Posner's Diagnosis of Stupor and Coma*, 4th edn (New York, Oxford University Press, 2007).
[20] New York Presbyterian Hospital Weill-Cornell Medical College: Fred Plum Archives, Box 2, undated ms, circa late 1970s.

albeit intermittent and episodic, evidence of awareness of self, others, or the environment.[21] Patients in MCS may say words or phrases and gesture or show evidence of memory, attention, and intention.

A recent paper surveying the accuracy of diagnoses of forty-one nursing home patients labelled as vegetative found that eighteen, or 41 per cent, were in fact in MCS,[22] helping to validate earlier smaller studies that found disturbing rates of misdiagnosis.[23] If we were to invoke the diagnostic standards articulated by Judge Hughes, such a diagnostic error rate would be impeachable. That too is a legacy of *Quinlan*.

18.3 Diagnostic Assessment and Societal Neglect

Diagnostic errors are even more grievous in the current era. In the years since *Quinlan*, the importance of diagnostic accuracy has become all the more critical now that the *persistent vegetative state has* been joined by related brain states from which it must be differentiated. This diagnostic differentiation is further complicated by the fact that these diagnoses may not be static. It is now appreciated that while some vegetative states are permanent, others are transient and the pathway to higher levels of functioning depending upon the nature of the injury.[24]

After receiving an injury severe enough to cause a coma, a self-limited and an eyes-closed state of unresponsiveness, a patient may: recover; sustain whole brain death; or pass into the vegetative state. In contrast to coma, the vegetative state is an eyes-open state of unresponsiveness in which there is only autonomic function supplied by an intact brain stem. These functions include the regulation of cardio-pulmonary function and maintenance of the sleep-wake cycle. By some classifications, the vegetative state becomes *persistent* after it has lasted thirty days. A vegetative state is considered permanent if has lasted three months following an anoxic injury. In contrast to the global insult of oxygen deprivation, the duration of a persistent vegetative state can be up to twelve months after a more selective traumatic injury. But once either threshold is crossed the vegetative state is considered irreversible.[25]

---

[21] J. T. Giacino, S. Ashwal, N. Childs, R. Cranford, B. Jennett, D. I. Katz, J. P. Kelly, J. H. Rosenberg, J. Whyte, R. D. Zafonte, and N. D. Zasler, 'The minimally conscious state: definition and diagnostic criteria' (2002) 58 *Neurology* 349.

[22] C. Schnakers, A. Vanhaudenhuyse, J. T. Giacino, M. Ventura, M. Boly, S. Majerus, G. Moonen, and S. Laureys, 'Diagnostic accuracy of the vegetative and minimally conscious state: clinical consensus versus standardized neurobehavioral assessment' (2009) 9 *BMC Neurology* 35.

[23] F. C. Wilson, J. Harpur, T. Watson, and J. I. Morrow, 'Vegetative state and minimally responsive patients: regional survey, long-term case outcomes and service recommendations' (2002) 17 *NeuroRehabilitation* 231; K. Andrews, L. Murphy, R. Munday, and C. Littlewood, 'Misdiagnosis of the vegetative state: retrospective study in a rehabilitation unit' (1996) 313 *British Medical Journal* 13; N. L. Childs, W. N. Mercer, and H. W. Childs, 'Accuracy of diagnosis of persistent vegetative state' (1993) 43 *Neurology* 1465.

[24] J. J. Fins, 'Neuroethics and neuroimaging: moving toward transparency' (2008) 8 *American Journal of Bioethics* 46.

[25] The Multi-Society Task Force on PVS, 'Medical aspects of the persistent vegetative state (1 and 2)' (1994) 330 *New England Journal of Medicine* 1499 and 1572.

Before these temporal milestones are reached patients may move into the minimally conscious state. Movement into MCS may occur surreptitiously and go unnoticed. Behaviours that distinguish VS from MCS are episodic and thus easily missed. There are additional challenges to diagnostic accuracy because of when the transition from VS to MCS occurs and where it takes place. At that juncture in their recovery, patients will likely have been transferred to chronic or long-term care and be in a setting of lesser medical acuity than the acute care hospital where definitive medical and surgical care was rendered. Because of the relative prestige and standing of these centres, physicians in chronic care may be hesitant to question the discharge diagnosis, even though a diagnostic revision would only reflect the ongoing biology of recovery and not represent an assessment error upstream in the provision of care.[26]

For whatever reason diagnostic errors are made, mistaking a MCS patient as vegetative comes at a cost because achieving the MCS threshold may be the herald of additional recovery.[27] Unlike patients who become permanently vegetative, patients who become MCS have a rather open-ended, and poorly understood, possibility for additional recovery.[28] These recoveries can take years or decades and can be quite dramatic as in the case of the Arkansan Terry Wallis who began to speak spontaneously nineteen years after traumatic injury.[29] Wallis who resided in a nursing home had been diagnosed as vegetative when he had actually been in the minimally conscious state. His parents' requests for imaging studies and an evaluation by a neurologist were denied.

Paradoxically, his mind continues to recover but his body still struggles with contractures that resulted from a lack of physical therapy while he was institutionalized. Functional neuroimaging studies, using diffusion tensor imaging (DTI) have revealed interval changes in his brain, specifically the sprouting of new axonal connections.[30] While these findings cannot explain his functional recovery, they do indicate that his brain remains in a dynamic and plastic state, *decades after injury*.

Although these neuroimaging findings may never be generalized to others in the minimally conscious state, his sequestration and decades-long neglect is not an isolated story but rather a public policy concern affecting a class of patients who are similarly situated. Since the advent of the minimally conscious state and reports like his the media has been ripe with other cases that speak to the recuperative potential

---

[26] J. J. Fins, M. G. Master, L. M. Gerber, and J. T. Giacino, 'The minimally conscious state: a diagnosis in search of an epidemiology' (2007) 64 *Archives of Neurology* 1400; see above, note 24.

[27] M. Lammi, V. H. Smith, R. L. Tate, and C. M. Taylor, 'The minimally conscious state and recovery potential: a follow-up study 2–5 years after traumatic brain injury' (2005) 86 *Archives of Physical Medicine and Rehabilitation* 746.

[28] Ibid.

[29] N. D. Schiff and J. J. Fins, 'Hope for "comatose" patients' (2003) 5 *Cerebrum: The Dana Forum on Brain Science* 7.

[30] H. Voss, A. Uluc, J. Dyke, R. Watts, E. J. Kobylarz, B. D. McCandliss, L. A. Heier, B. J. Beattie, K. A. Hamacher, S. Vallabhajosula, S. J. Goldsmith, D. Ballon, J. T. Giacino, N. D. Schiff, 'Possible axonal regrowth in late recovery from the minimally conscious state' (2006) 116 *Journal of Clinical Investigation* 2005.

of the severely injured brain.[31] The marginalization of this population has been the subject of an exploratory meeting at the Institute of Medicine, in which I took part. We considered the ethical and policy implications of late recovery from the minimally conscious state and urged a formal study of the problem space, from both a clinical and research perspective. We recommended that initial areas of focus should be defining the basic epidemiology of this population, elucidating mechanisms of recovery, and identifying clinical tools that could assist with diagnosis and prognosis in practice.[32]

18.4 On the Cusp of Progress

Ignoring these patients and sequestering them in a chronic care sector where their potential cannot be realized is especially concerning as we learn more about mechanisms of brain recovery and potential therapeutic interventions. We are just beginning to identify patients who demonstrate recuperative potential on neuroimaging studies that is not apparent on clinical examination. Other patients may be helped by pharmacologic or neuromodulatory interventions like deep brain stimulation.

Promising studies with the flu drug amantadine have helped to identify and foster prognostic potential and accelerate recovery.[33] Studies like these, and others investigating the sleeping pill, zolpidem,[34] while still equivocal seem to point to an effect in subsets of patients. This heterogeneity suggests that physiologically the causes of disordered consciousness are heterogeneous and that descriptive categories based on behavioural criteria will need to be replaced by a more highly refined diagnostic schema based on the underlying systems biology and circuit

---

[31] E. Racine and E. Bell, 'Clinical and public translation of neuroimaging research in disorders of consciousness challenges current diagnostic and public understanding paradigms' (2008) 8 *American Journal of Bioethics* 13.

[32] J. J. Fins, N. D. Schiff, and K. M. Foley, 'Late recovery from the minimally conscious state: Ethical and policy implications' (2007) 68 *Neurology* 304.

[33] R. D. Zafonte, T. Watanabe, and N. R. Mann, 'Amantadine: a potential treatment for the minimally conscious state' (1998) 12 *Brain Injury* 617; J. Whyte, D. Katz, D. Long, M. C. DiPasquale, M. Polansky, K. Kalmar, J. T. Giacino, N. Childs, W. Mercer, P. Novak, P. Maurer, and B. Eifert, 'Predictors of outcome in prolonged posttraumatic disorders of consciousness and assessment of medication effects: a multicenter study' (2005) 86 *Archives of Physical Medicine and Rehabilitation* 453; C. Schnakers, R. Hustinx, G. Vandewalle, S. Majerus, G. Moonen, M. Boly, A. Vanhaudenhuyse, and S. Laureys, 'Measuring the effect of amantadine in chronic anoxic minimally conscious state' (2008) 79 *Journal of Neurology, Neurosurgery, and Psychiatry* 225.

[34] S. I. Cohen and T. T. Duong, 'Review. Increased arousal in a patient with anoxic brain injury after administration of zolpidem' (2008) 87 *American Journal of Physical Medicine and Rehabilitation* 229; J. L. Shames and H. Ring, 'Transient reversal of anoxic brain injury-related minimally conscious state after zolpidem administration: a case report' (2008) 89 *Archives of Physical Medicine and Rehabilitation* 386; J. Whyte and R. Myers, 'Incidence of clinically significant responses to zolpidem among patients with disorders of consciousness: a preliminary placebo controlled trial' (2009) 88 *American Journal of Physical Medicine and Rehabilitation* 410; R. Singh, C. McDonald, K. Dawson, S. Lewis, A. M. Pringle, S. Smith, and B. Pentland, 'Zolpidem in a minimally conscious state' (2008) 22 *Brain Inj*ury 103.

mechanisms. With that additional knowledge, prediction of responders vs non-responders will become possible.

Our group's study of deep brain stimulation is also noteworthy.[35] Using a hypothesis-driven approach about circuit disorders in disordered consciousness,[36] we conducted central thalamic stimulation of a subject in MCS following traumatic brain injury six years earlier. There was objective behavioural evidence of improved attention, responsiveness, motor control, feeding, and communication using a validated and standardized assessment tool.[37] He regained the ability to take food by mouth, having been dependent upon a feeding, tube for nutrition prior to stimulation.

Most notable was a new-found ability to communicate verbally. Prior to stimulation the extent of his interaction with the outside world was inconsistent command following through eye movements. With stimulation he was able to achieve verbal output of six to seven words and respond to questions about his needs and wants. He is able to tell his mother that he loves her and respond when clinical staff asks about his willingness to participate in his rehabilitation activities.[38] Although these responses do not yet cross the threshold of decision-making capacity necessary for the return of legal competence and the ability to consent, they do indicate a return of some elements of personal agency at the level of assent.[39]

18.5 Consciousness Obscured

These efforts to restore functional communication take on all the more urgency, as a rights issue, because there is a class of individuals who are conscious who might be able to communicate—and given voice —with assistive devices or medication. This possibility was brought home in dramatic fashion by neuroimaging studies that show a profound discordance between what is observed on the clinical examination and what might be deemed possible after reviewing a functional neuroimage. Studies have shown that minimally conscious patients with rare responsiveness to the environment manifest widespread *integrated* language networks in the brain when exposed to taped narratives recorded by family members.[40] This integrative network response contrasts with the limited activation seen in vegetative patients

---

[35] N. D. Schiff, J. T. Giacino, K. Kalmar, J. D. Victor, K. Baker, M. Gerber, B. Fritz, B. Eisenberg, J. O'Connor, E. J. Kobylarz, S. Farris, A. Machado, C. McCagg, F. Plum, J. J. Fins, and A. Rezai, 'Behavioral improvements with thalamic stimulation after severe traumatic brain injury' (2007) 448 *Nature* 600.

[36] N. D. Schiff, 'Central thalamic deep-brain stimulation in the severely injured brain: rationale and proposed mechanisms of action' (2009) 1157 *Annals of the New York Academy of Sciences* 101.

[37] J. T. Giacino, K. Kalmar, and J. Whyte, 'The JFK coma recovery scale-revised: measurement characteristics and diagnostic utility' (2004) 85 *Archives of Physical Medicine and Rehabilitation* 2020.

[38] Charlie Rose Show, 'A discussion about brain stimulation' (2007), <http://www.charlierose.com/view/interview/8627>.

[39] N. D. Schiff, J. T. Giacino, and J. J. Fins, 'Deep brain stimulation, neuroethics and the minimally conscious state: moving beyond proof of principle' (2009) 66 *Archives of Neurology* 697.

[40] N. D. Schiff, D. Rodriguez-Moreno, A. Kamal, K. H. Kim, J. T. Giacino, F. Plum, and J. Hirsch, 'fMRI reveals large-scale network activation in minimally conscious patients' (2005) 64 *Neurology* 514.

whose only activation is at the level of the initial primary sensory area, the first way station into the brain.

These findings have been expanded to occasional vegetative patients who demonstrate widespread network activations but are vegetative by clinical criteria. One patient, reported by Owen and colleagues, was clinically in the vegetative state five months after traumatic brain injury. This patient activated the appropriate motor, spacial, and language networks when asked to imagine playing tennis, walking in her home or disambiguating a complex word.[41] My colleague and I described this as a state of *non-behavioural* MCS in which the response to one's environment is made evident by the integrated response seen on neuroimaging data, not overt behavioural criteria.[42]

Together these studies paint the disturbing possibility of consciousness obscured by a physical incapacity to communicate and make one's presence known. To come out of the shadows, these patients need assistive devices or diagnostic tests—like the aforementioned neuroimaging methods—that demonstrate that all is not as it seems on first glance.

The mother of a subject upon whom we conducted neuroimaging studies spoke of the power of these interventions to demonstrate the presence of consciousness, and validate efforts that might seem utterly futile if assessment were based solely upon behavioural criteria. Her twenty-four-year-old daughter had sustained a brain stem stroke just prior to college graduation two years earlier. She came to us carrying a vegetative diagnosis. The mother spoke of the importance of objective evidence indicating her child's needs and potential:

I want to be as aggressive as I can to help her communicate. But how can I? So I have to convince people that she could communicate. And the only, if I could convince people that she would be able to if we could just find a pathway and here's a speculative idea, so lets try it.[43]

Our clinical assessment and neuroimaging studies indicated that she was not vegetative but *at least* at the level of the minimally conscious state. This evaluation, and in particular the neuroimaging studies that indicated a degree of responsiveness that had not been seen behaviourally, was a potential corrective to the implicit nihilsim she had encountered. These objective data helped to legitimate goals of care, which fundamentally were geared towards the re-establishment of functional communication and giving voice to a conscious life.

18.6 Civil Rights Issues

There is a grave risk of error of omission if we fail to look for those who might be conscious or fail to augment those who are in liminal states. Without affirmative

---

[41] A. M. Owen, M. R. Coleman, M. Boly, M. H. Davis, S. Laureys, and J. D. Pickard, 'Detecting awareness in the vegetative state' (2006) 313 *Science* 1402.
[42] See above, note 18.
[43] Weill-Cornell Medical College, subject interview, 'Transcription: IN316W (NY visit)' (2008).

actions by others—upon whom these individuals must depend—conscious, sentient individuals may in fact be trapped in unresponsive bodies that cannot give voice to minds that might potentially become more active.

One can view the neglect of this population as an entitlement issue but we will fail this class of individuals if we leave it there and not assert that denying them potential voice is to deprive them of their basic civil rights. While their marginalization deprives them of appropriate clinical care, the basis for this concern is more than a concern about entitlements because such benefits are discretionary.

Instead, it is respect and regard for civil rights and the more fundamental issue that conscious individuals, who might be embraced more fully by our shared human community, are routinely ignored, sequestered, and segregated. This is an utter breach of their inalienable rights, rights that in this case may be dependent upon the provision of interventions, which remain experimental but will likely be expensive and scarce. Because of this it is important to view this technology as enabling a fundamental right to be enjoyed by all citizens, a class utterly dependent upon the goodwill of others, and their recognition of their reciprocal ethical obligation to provide access to technology that will restore their voice and give them more equal opportunities. If these responsibilities are not recognized by our common morality, they should be enforced by law.

Although I am advocating access to scientific interventions that only will become available in the future, I am worried that scientific advances will outpace society's embrace of individuals with disorders of consciousness as full members of our collective human community. Both science progress and progressive reform of entrenched attitudes will take decades. But while scientific progress proceeds, society has yet to view the marginalization of this population as a civil rights issue. It is time to start reflecting upon our biases so that when discovery is made, society will understand its moral obligation to make it available. We have to take a forward view and *expect* that early brain-computer interfaces will mature into technologies that will allow minds trapped inside unresponsive bodies to communicate, express love and fears, and perhaps *even sing*—to echo Dr Plum's testimony in *Quinlan*.[44]

Although such hopes may never be realized, this speculation is no less daring a prediction than suggesting at the time of *Plessy v Ferguson*,[45]—when separation of the races was opined as being equal—that an African-American would head one of the co-equal branches of government *and have the authority to nominate justices to the US Supreme Court*. The irony of the Obama presidency should not escape us as we imagine the future and strive to be on the right side of history. Reading the decision in *Plessy v Ferguson*, it is almost impossible for the modern reader to understand the logic of Justice Brown, who asserted for a majority of his brethren that:

Laws permitting, and even requiring, their separation in places where they are liable to be brought into contact do not necessarily imply the inferiority of either race to the other, and have been generally, if not universally, recognized as within the competency of the state

---

[44] See above, note 1.
[45] *Plessy v Ferguson* 163 US 537 (1896).

legislatures in the exercise of their police power. The most common instance of this is connected with the establishment of separate schools for white and colored children, which has been held to be a valid exercise of the legislative power even by courts of States where the political rights of the colored race have been longest and most earnestly enforced.[46]

The illogic of Justice Brown's argument is breathtaking. He makes a specious assertion that separation is not necessarily inferior. Then he justifies the precedent by citing other examples of beneficent segregation in the schools, further suggesting that other efforts separating the races were perhaps less intended.

Contrast this with the dissent of his contemporary, Justice Harlan, who wrote:

The white race deems itself to be the dominant race in this country. And so it is in prestige, in achievements, in education, in wealth and in power. So, I doubt not, it will continue to be for all time if it remains true to its great heritage and holds fast to the principles of constitutional liberty. But in view of the Constitution, in the eye of the law, there is in this country no superior, dominant, ruling class of citizens. There is no caste here. Our Constitution is colorblind, and neither knows nor tolerates classes among citizens. In respect of civil rights, all citizens are equal before the law. The humblest is the peer of the most powerful. The law regards man as man, and takes no account of his surroundings or of his color when his civil rights as guaranteed by the supreme law of the land are involved. It is therefore to be regretted that this high tribunal, the final expositor of the fundamental law of the land, has reached the conclusion that it is competent for a State to regulate the enjoyment by citizens of their civil rights solely upon the basis of race.[47]

Here, Justice Harlan is writing for the ages, if not for the 1896 decision, asserting the power of the Constitution against both race and class. With a clear sense of past decisions the Court would come to regret, Justice Harlan adds that he is in fact writing for a different day when separate and equal would be revealed as the moral fiction that it was. He makes an analogy to the *Dred Scott* decision which denied citizenship to all those of African descent:[48] 'In my opinion, the judgment this day rendered will, in time, prove to be quite as pernicious as the decision made by this tribunal in the *Dred Scott* Case.'[49]

Which side of history do we want to be on? It was fifty-eight years before *Brown v Board of Education*[50] overturned *Plessy v Ferguson*. Imagine the future promise of neurorehabilitation over a comparable period. It will be truly breathtaking. In a shorter span, I predict neuroscience will provide communication tools to a class of patients who heretofore could only hope for the isolation of a solitary consciousness devoid of the collective reciprocal exchange that makes for community. In this context, assistive devices will be *instrumental* in building community by allowing some individuals, who could not communicate with others, the opportunity to do so.

As we contemplate this progress, we might also imagine what the civil rights of our fellow citizens with disorders of consciousness ought to be. As the advertising

---

[46] From Justice Brown, Opinion of the Court, *Plessy v Ferguson* 163 US 537 (1896).
[47] From Justice Harlan, Dissenting Opinion, *Plessy v Ferguson* 163 US 537 (1896).
[48] *Dred Scott v Sandford* 60 US 393 (1856).
[49] See above, note 47.
[50] *Brown v Board of Education of Topeka* 347 US 483 (1954).

Minds Apart: Severe Brain Injury, Citizenship, and Civil Rights 379

campaign of the United Negro College Fund so aptly reminds us, *a mind is a terrible thing to waste*.[51] So informed, we should try to anticipate what will eventually become society's ethical and legal norms towards this population, much as Justice Harlan anticipated what was owed to young black school children six decades later in *Brown v Board of Education*. Although Justice Harlan was in the minority in *Plessy v Ferguson*, his dissent could have joined the majority in *Brown v Board of Education* when a unanimous court struck down the doctrine of separate but equal. Justice Harlan's opinion was as modern as that decision written decades later. With his eye to the future Justice Harlan confidently asserted:

> In respect of civil rights common to all citizens, the Constitution of the United States does not, I think, permit any public authority to know the race of those entitled to be protected in the enjoyment of such rights. Every true man has pride of race, and, under appropriate circumstances, when the rights of others, his equals before the law, are not to be affected, it is his privilege to express such pride and to take such action based upon it as to him seems proper. But I deny that any legislative body or judicial tribunal may have regard to the [p. 555] race of citizens when the civil rights of those citizens are involved. Indeed, such legislation as that here in question is inconsistent not only with that equality of rights which pertains to citizenship, National and State, but with the personal liberty enjoyed by everyone within the United States.[52]

Justice Harlan's view of citizenship, democracy, and equality of rights remain inspiring to this day and is instructive if we consider its application to those with severe brain injury, a similarly vulnerable class. The Constitutional predicate for Justice Harlan's dissent and the *Brown v Board of Education* court are both the Equal Protection Clause of the Fourteenth Amendment or, as noted by Chief Justice Warren, the 'Due Process Clause'.

This clause is only violated by acts of government, legislated or administered, that treat similarly situated individuals differently. While the discriminatory practices I have described towards those with severe brain injury are not directly the result of government action, careful study of Center for Medicaid and Medicare reimbursement regulations could result in challenges alleging that denial of services to this class of individuals has deprived them of equal protection under the law.

If these regulatory distinctions or denials were found too prejudicial, and not merely related to a diagnostic classification that could lead to benign outcomes, patients with severe brain injury could be construed as a suspect class based on the criteria outlined in *City of Cleburne, Texas v Cleburne Living Center, Inc.*[53] alleging discrimination over the placement of a group home for the mentally retarded. In that decision the US Supreme Court held that the mentally retarded were not a suspect class because this 'legitimate' diagnostic label had resulted in neither legislative neglect nor antipathy. Justice White wrote the majority opinion, noting:

---

[51] 'United Negro College Fund (1972-Present)', <http://www.adcouncil.org/default.aspx?id=134>.
[52] See above, note 45.
[53] *City of Cleburne v Cleburne Living Center, Inc.* 473 US 432 (1985).

Mentally retarded persons, who have a reduced ability to cope with and function in the everyday world, are thus different than other persons, and the States' interest in dealing and providing for them [p. 433] is plainly a legitimate one. The distinctive legislative response, both national and state, to the plight of those who are mentally retarded demonstrates not only that they have unique problems, but also that the lawmakers have been addressing their difficulties in a manner that belies a continuing antipathy or prejudice and a corresponding need for more intrusive oversight than is afforded under the normal equal protection standard. Moreover, the legislative response, which could hardly have occurred and survived without public support, negates any claim that the mentally retarded are politically powerless in the sense that they have no ability to attract the attention of the lawmakers.[54]

While we should be grateful for the political clout of those concerned about the mentally retarded in society, no comparable political prowess exists for advocates of those with disorders of consciousness. In contrast to the mentally retarded, which have well-established national organizations, such as the Association for Retarded Citizens (ARC), to represent their interests[55] and a long legacy of effective legislative initiatives, as typified by the Legislative Agenda for the 111th Congress,[56] the plight of those with severe brain injury is just beginning to garner national attention, prompted by media coverage of scientific developments and the epidemic of brain injury stemming from the wars in Iraq and Afghanistan.[57]

These efforts towards advocacy remain fledgling and fragmented; as yet there has neither been the coalescence of a need for concerted political action nor a broader public awareness to which it could be directed. Their still anaemic efforts geared towards brain injury cannot compare with established lobbies that represent the mentally retarded. No advocacy group specifically addresses the needs of patients with disorders of consciousness. And as a group this class of patients has been virtually ignored by brain injury advocacy groups. The mother of a young man struck down in a civilian motor vehicle accident days before deploying with the Marines for basic training put it this way:

Not really, because a lot of times these groups are so, like, fragmented. And especially for brain injury groups. There's not a lot of, you know, people who don't vote, I don't think there are a lot of groups out there to help them. I mean, even if you're paralyzed you can still vote, hopefully. But when you're brain injured, you don't, you don't, you don't want to, you know, there's no power there, none. And uh, and it is sad to find out like all these groups, like I was getting to the state, which is responsible for the health care, are you know they tell you, there's nothing they can do for you. There's no, nothing fits his needs. You know. It's you, if you have AIDS, they have a Medicaid waiver for that, if you're paralyzed there's a waiver for that. But brain injury, no.[58]

---

[54] Ibid.
[55] The Association for Retarded Citizens (The ARC), <http://www.thearc.org/NetCommunity/Page.aspx?pid=183>.
[56] Legislative Agenda for the 111th Congress, First Session 2009, Disability Policy Collaboration (DPC), *A Partnership of The Arc & United Cerebral Palsy* (Washington, DC, 2009).
[57] S. Okie, 'Traumatic brain injury in the war zone' (2005) 352 *New England Journal of Medicine* 2043; R. Glasser, 'A shock wave of brain injuries' (2007) <http://Washingtonpost.com>, B01.
[58] Weill-Cornell Medical College, subject interview, 'Transcription: IN314W advocacy' (2008).

If we accept that patients with disorders of consciousness are a suspect class then they would be entitled to the protections of the Equal Protection Clause. One such protection is integration. In the context of *Brown v Board of Education* and the Warren Court it was in education. For those with disorders of consciousness it would be, by analogy, inclusion in appropriate health care settings where proper care and rehabilitation can be rendered. With this argument in mind, it is especially fruitful to return to the Warren Court's emphasis on the instrumentality of a shared educational experience, in promoting equal protections for both black and white children. This line of argument is especially instructive as we imagine how to integrate brain-injured citizens into our shared society and give them the tools necessary to rebuild communal ties. Let us consider three key passages in *Brown v Board of Education* that are especially instructive by analogic reasoning.

In asserting their view, the Warren Court spoke of the primacy of public education as the key public policy concern of government and the pivotal role schooling plays in the development of children into mature, capable, and productive citizens in a democratic society:

Today, education is perhaps the most important function of state and local governments. Compulsory school attendance laws and the great expenditures for education both demonstrate our recognition of the importance of education to our democratic society. It is required in the performance of our most basic public responsibilities, even service in the armed forces. It is the very foundation of good citizenship. Today it is a principal instrument in awakening the child to cultural values, in preparing him for later professional training, and in helping him to adjust normally to his environment. In these days, it is doubtful that any child may reasonably be expected to succeed in life if he is denied the opportunity of an education. Such an opportunity, where the state has undertaken to provide it, is a right that must be made available to all on equal terms.[59]

The emphasis on education is particularly apt to how neurorehabilitation will be conducted in the future. Both childhood education and brain injury rehabilitation are linked to developmental processes. Education is coupled with the progressively maturing brain, with curricula geared to what is cognitively and socially possible for young and developing minds. Brain-injury rehabilitation can be also understood as hinging upon a recuperative process that often retraces earlier developmental milestones. This recapitulation of an earlier process is made necessary by the setback of injury. Like the education of a child, brain injury rehabilitation's purpose is to help an individual meet progressive functional goals and to maximize one's potential. If this premise is accepted, then rehabilitation is as important to the injured citizen as school is to the young child. In a Rawlsian frame, then, each provides the individual with the capabilities to maximize their participation in a shared democratic life.

As was the case with educational segregation of the races, sequestering severely brain-injured patients in chronic care facilities can prove clinically detrimental and juridically a violation of equal protection. Placement of these patients in venues

---

[59] See above, note 50.

outside of acute care and specialized rehabilitative centres can expose them to a therapeutic frame that accepts the immutability of their injuries and their inevitable decline. This segregation is no different than history's placement of young black children in inferior schools where their potential was wasted for lack of proper instruction, guidance and materials.

It is no different when a brain-injured patient with the potential recovery is placed in a nursing home or any facility that is ill equipped—or ill-disposed—to properly diagnosis their condition and unable to provide rehabilitation. It is important to remember that while a severely brain-injured patient may share a highly impaired functional status with a fellow nursing home patient with end-stage Alzheimer's, each are on distinct trajectories. It is important to remember that brain injury is an acquired insult, not a degenerative process, like the dementing illnesses. As such the outcome is not an inevitable process of decline and death. Proper diagnostic assessment, often unavailable in chronic care, can identify those patients who have the potential to improve with proper rehabilitation months and years after acute injury.

On an emotional level, we might also speculate that, sequestration in nursing homes with frail elderly cannot be good for a brain-injured patient's hopes for the future and sense of self. For brain-injured patients, a population disproportionately young, to be cohorted with patients who are decades older, separates them from their generational peers. It dislocates them from all the cultural markers that might be shared and which could be harnessed in the service of health and recovery.

Of course, one cannot know if substituting Big Bands for Bono will be therapeutic, but having the soundtrack of one's life as background noise might awaken an injured patient's sense of shared community. The mother of the patient who sustained a brain-stem stroke spoke of the importance of generational integration for her daughter. In commenting upon a facility, which she found favourable, she noted:

> Most of the patients are young. So she's in a place where they at least play appropriate music [p. l] for her age group. *It's like a dorm of young people, they just happen to all be brain injured.* [italics added] So, that, that makes, and then the families are sharing some of the same struggles.[60]

Even here, *Brown* speaks to the emotional consequences of segregation. Writing of the enduring psychological impact of this practice, the court observed that:

> To separate them from others of similar age and qualifications solely because of their race generates a feeling of inferiority as to their status in the community that may affect their hearts and minds in a way unlikely ever to be undone.[61]

While we cannot know if the segregation of brain-injured patients in nursing homes has a similarly enduring adverse effect on the psychology and outlook of

---

[60] Weill-Cornell Medical College, subject interview, 'Transcription: IN316W' (2007).
[61] See above, note 50.

individuals so sequestered, in our research activities bringing subjects to Weill Cornell Medical College for assessment and neuroimgaing studies, we have seen individuals become more interactive when they leave chronic care facilities and come to our Medical Center to be engaged by our investigative team.

We saw this in the young woman with the brain-stem stroke. Once we indicated that we appreciated that she was able to understand us, albeit absent easily discernible reciprocal communication on her part, she began to make a greater effort to let us know she was alert and aware. Although she could neither speak nor move her limbs, after we acknowledged her as part of our community and worthy of communication, she made vigorous use of her remaining ability to move her eyes to let us know that she was in fact 'in there'.

It is tragic to think that others like her could remain misdiagnosed, but they do, segregated in nursing homes away from the clinical and rehabilitative programme that might be better able to distinguish the unconsciousness of the vegetative state from the minimally conscious and locked-in states. This is a systemic problem as indicated by Schnakers et al.[62]

The Warren Court famously noted that when it came to the question of public education that segregation was a violation of the equal protection clause of the Fourteenth Amendment. They noted:

> We conclude that, in the field of public education, the doctrine of 'separate but equal' has no place. Separate educational facilities are inherently unequal. Therefore, we hold that the plaintiffs and others similarly situated for whom the actions have been brought are, by reason of the segregation complained of, deprived of the equal protection of the laws guaranteed by the Fourteenth Amendment.[63]

Could not the same arguments be made for the separate but truly unequal medical care received by those with disorders of consciousness, given the analogous role that proper diagnostic assessment and rehabilitation plays for members of this class?

And so, like those before us we have our chance, to write the contrarian view. We can make the argument for equal protection under the law for another class of citizens whose experience has the potential to be different because of the lessons learned from America's march towards civil rights for its African-American citizens. Though that journey remains incomplete it has progressed across the Selma Bridge to Pennsylvania Avenue. It is a reminder that long journeys begin with tentative steps. And so now, it is our turn to make some forward progress and echo *Brown* by asserting that:

> with respect to how we rehabilitate patients with severe brain injury . . . it is doubtful that any may reasonably be expected to succeed in life if denied that opportunity . . . an opportunity which must be made available to all on equal terms.

If the scientific difference between the vegetative and minimally conscious states is cortical integration, then the societal corollary necessary to facilitate functional

---

[62] See above, note 22.   [63] See above, note 50.

recovery is societal integration. It is only be ending the segregation of those with severe brain injury that we will be able to repatriate them into our shared community and ensure that they have proper access to coming advances in the neurosciences. Such scientific progress will come to naught absent a collective societal will to accept its responsibility to provide equal protections to this marginalized population.

19
Reciprocity and Neuroscience in Public Health Law

*A. M. Viens**

There is an underdeveloped potential for using neuroscience as a particular input in the process of law-making. This paper examines one such instance in the area of public health law. Neuroscience could play an important role in elucidating and strengthening the relevance of the conditions underlying and re-enforcing our ability to cooperate in balancing the benefits and burdens necessary to achieve particular goods; for instance, the protection of public health in an outbreak of pandemic influenza. In particular, I shall focus on how a better understanding of the neurobiological basis of reciprocity could be used to help increase support and compliance with public health laws—especially those involving restrictive measures (such as quarantine and isolation).

19.1 Neuroscience as an Input to Making Public Health Law

Law, at its most general level, aims at guiding and regulating human conduct. A large part of law is directed at providing reasons for individuals and groups to change undesirable dispositions, behaviours, and environments. Within the area of public health law, in particular, this view broadly pervades. Law is considered and utilized as an effective tool to protect and promote public health.[1] Public health laws and regulations function to advance the public's health as a population through influencing how individuals, as constituents of the population, make judgements and act.

An examination of the history of public health law within the nineteenth and twentieth centuries reveals how the use of legal rules, mechanisms, and institutions

---

* School of Law, Queen Mary, University of London and Joint Centre for Bioethics, University of Toronto. The author would like to thank Catherine R. McGowan for useful comments, and Michael Freeman.

[1] See, for instance, Lawrence O. Gostin, *Public Health Law: Power, Duty, Restraint* (Berkeley, University of California Press, 2000); Richard A. Goodman et al. (eds), *Law in Public Health Practice* (Oxford, Oxford University Press, 2007); Wendy E. Parmet, *Populations, Public Health and Law* (Washington, DC, Georgetown University Press, 2009).

were crucial tools in influencing individual judgement and action in relation to the advancement of public health. Laws concerning sanitation and immunization contributed massively to reducing the incidence of infectious diseases (e.g. cholera) and its associated mortality and morbidity. Laws mandating food safety (e.g. water fluoridation, food inspection), motor-vehicle safety (e.g. seatbelts, helmets, drunk driving), and safer workplace standards further promoted the health of the public. Most recently, legal barriers and restrictions to tobacco consumption (e.g. minimum age limits, taxes, public use restrictions) have been credited with contributing to the significant reduction of smoking-related chronic diseases. These have all been heralded as success stories in which the law played an indispensable part in helping to secure the public's health by providing individuals with the right kind of motivational and normative reasons to make choices and act in particular ways.

While perhaps not immediately obvious, neuroscience can and will be used within the formulation and enforcement of public health law. Knowledge and techniques derived from neuroscience could provide us with invaluable information about how we can guide and regulate individual behaviour more effectively in relation to advancing public health goals. This, however, is largely anticipatory. The issue of using neuroscience within law is still largely an academic exercise. The majority of the research being done at the interface of law and neuroscience, with a few exceptions, has focused on the theoretical role neuroscience could play in delivering particular outputs within the legal process, such as the use of neuroscientific data for evidentiary purposes—in particular tort law[2] and criminal law[3] being prime areas of debate.

One area of great interest has been the neuroscientific explanation of action. For example, there is a growing literature within cognitive science, and increasingly within law and philosophy, about the neural basis of violent, aggressive, and addictive behaviours and the extent to which the employment of neuroimaging technology may come to bear on questions of criminal guilt, responsibility, punishment, or parole.[4] It is indeed the case that these potential outputs within various

---

[2] A. M. Viens, 'The Use of Functional Neuroimaging Technology in the Assessment of Loss and Damages in Tort Law' (2007) 7 *American Journal of Bioethics-Neuroscience* 63; B. Grey, 'Neuroscience, Emotional Harm, and Emotional Distress Tort Claims' (2007) 7 *American Journal of Bioethics-Neuroscience* 65; A. Kolber, 'Pain Detection and the Privacy of Subjective Experience' (2007) *American Journal of Law & Medicine* 433.

[3] J. Kulynych, 'Psychiatric Neuroimaging Evidence: A High-Tech Crystal Ball? (1997) 49 *Stanford Law Review* 1249; J. A. Seiden, 'The Criminal Brain: Frontal Lobe Dysfunction in Capital Proceedings' (2004) 16 *Capital Defense Journal* 395; N. Eastman and C. Campbell, 'Neuroscience and Legal Determination of Criminal Responsibility' (2006) 7 *Nature Reviews: Neuroscience* 311; S. J. Morse, 'Brain Overclaim Syndrome and Criminal Responsibility: A Diagnostic Note' (2006) 3 *Ohio State Journal of Criminal Law* 397; Stephen J. Morse, 'Morse, Moral and Legal Responsibility and the New Neuroscience' in Judy Illes (ed.), *Neuroethics: Defining the Issues in Theory, Practice and Policy* (Oxford, Oxford University Press, 2006) 33; O. Carter Snead, 'Neuroimaging, Entrapment, and the Predisposition to Crime' (2007) 7 *American Journal of Bioethics-Neuroscience* 60; Y. Yang, A. L. Glenn, and A. Raine, 'Brain Abnormalities in Antisocial Individuals: Implications for the Law' (2008) 26 *Journal of Behavioral Science and the Law* 65.

[4] J. L. Bufkin and V. R. Luttrell, 'Neuroimaging Studies of Aggressive and Violent Behavior: Current Findings and Implications for Criminology and Criminal Justice' (2005) 6 *Trauma, Violence and Abuse* 176.

areas of the legal system raise important and interesting questions as to whether the implications of using neuroscience will have a fundamental effect on legal doctrine or practice—of which, the answers have been mixed.[5] While I maintain a healthy scepticism about the extent to which neuroscience will fundamentally affect law, instead, in this paper, I want to explore the potential benefit of using neuroscience as an input within the process of making, and enforcing existing, public health law. Indeed, this may be one area where the interface of law and neuroscience could soon have direct policy applications.

19.2 Neuroscience, Reciprocity, and Public Health Law

In this paper, I am only going to focus on one aspect of public heath law—that pertaining to pandemic preparedness and response. Amongst other functions, a large part of public health law in this realm seeks to regulate human conduct in an effort to prevent or mitigate the spread of contagion. The success of law concerning social distancing measures (such as closing public areas and enforcing quarantine and isolation), epidemiological surveillance, examination and treatment measures, travel and trade restrictions, etc. are all considerations that will, by and large, determine the government's ability to remove the threat of contagion and maintain public health. The best chance of public health laws fulfilling their function will be when these laws have wide public support and compliance.[6]

I want to suggest that neuroscience can play a part in promoting this support and compliance. In particular, I want to argue that work on the neurobiological basis of reciprocity may provide one way of increasing support and compliance with restrictive measures employed in circumstances like a pandemic. In the context of normative support for public health laws, which can promote the common good by restricting individual liberties, reciprocity plays an important role in legitimizing action in virtue of ensuring that justified infringements of rights only take place in a way that are not overly burdensome and provide the necessary support for those individuals being constrained by restrictive measures.[7] In the context of

---

[5] Cf. J. T. Cacioppo, et al., 'Just Because You're Imaging the Brain Doesn't Mean You Can Stop Using Your Head: A Primer and Set of First Principles' (2003) 85 *Journal of Personality and Social Psychology* 650; J. Bard, 'Learning from Law's Past: A Call for Caution in Incorporating New Innovations in Neuroscience' (2007) 7 *American Journal of Bioethics-Neuroscience* 73; D. S. Weisberg et al., 'The Seductive Allure of Neuroscience Explanations' 20 *Journal of Cognitive Neuroscience* 470.

[6] Most justificatory arguments for the use of restrictive measures, as well as empirical arguments for the likelihood with which individuals actually will be motivated to comply with such measures, rest on employing such restrictive measures on a voluntary basis. For example, self-administered quarantine or isolation when an individual feels they have been exposed or infected to pandemic influenza. Successful arguments for the justification and effectiveness of employing restrictive measures on a non-voluntary basis are more difficult to achieve. One reason why a neuroscientific understanding of what motivates human action will be of such interest to those working in public health law is because such information can be harnessed in a way to increase support and compliance with voluntary restrictive measures to a point where resorting to non-voluntary measures could be extremely rare.

[7] A. M. Viens, C. M. Bensimon and R. E. G. Upshur, 'Your Liberty or Your Life: Reciprocity in the Use of Restrictive Measures in Contexts of Contagion' (2009) 6 *Journal of Bioethical Inquiry* 207.

motivational compliance with public health laws, reciprocity also plays an important role in giving individuals incentives and disincentives for acting in ways to protect and promote their own and the public's health.

Reciprocity is already recognized as a value within public health law and ethics, and is especially relevant in the context of public health emergencies.[8] One problem is, however, there are different competing conceptions of reciprocity.[9] Some people understand reciprocity in terms of gratitude, in which individuals who accept some benefit from another person or institution have an obligation not to act against the interests of the provider of the benefit.[10] Others maintain that reciprocity should be understood in terms of mutuality or fair play, in which individuals who receive a particular benefit from participating in a mutually beneficial cooperative scheme or practice have an obligation to endure a fair share of the burden of such an arrangement.[11] Others still understand reciprocity in terms of the golden rule—in which reciprocity demands that I behave towards you as I would have you behave towards me.[12]

There are important philosophical questions about which conception should be adopted and related issues concerning the nature and scope of the concept; however, I shall put those normative considerations aside in this paper.[13] I am not going to defend a substantive theory of reciprocity, but instead restrict my claims to what may be required by reciprocity in the context of public health and

---

[8] R. E. G. Upshur, 'Principles for the justification of public health intervention' (2002) 93 *Canadian Journal of Public Health* 101; P. A. Singer, S. R. Benatar, M. Bernstein, A. S. Daar, B. M. Dickens, S. K. MacRae, R. E. G. Upshur, L. Wright, and R. Z. Shaul, 'Ethics and SARS: Lessons from Toronto' (2003) 327 *British Medical Journal* 1342; University of Toronto Joint Centre for Bioethics Pandemic Ethics Working Group, *Stand on Guard for Thee: Ethical Considerations in Preparedness Planning for Pandemic Influenza* (Toronto, Joint Centre for Bioethics, 2005), available at: <http:www.jointcentreforbioethics.ca/people/documents/upshur_stand_guard.pdf> (accessed 31 October 2009); F. Baylis, N. P. Kenny, and S. Sherwin, 'A Relational Account of Public Health Ethics' (2008) 1 *Public Health Ethics* 196. A. M. Viens, C. M. Bensimon, and R. E. G. Upshur, 'Your Liberty or Your Life: Reciprocity in the Use of Restrictive Measures in Contexts of Contagion' (2009) 6 *Journal of Bioethical Inquiry* 207.

[9] A majority of the conceptions are discussed in relation to political obligation; however, there have also been works examining the concept itself. See Lawrence C. Becker, *Reciprocity* (London, Routledge, 1986).

[10] Cf. Plato's *Crito*; H. Sidgwick, *The Methods of Ethics* (Indianapolis, Hackett Publishing Company, [1907] 1981); A. J. Simmons, *Moral Principles and Political Obligations* (Princeton, Princeton University Press, 1979), 162; J. English, 'What Do Grown Children Owe Their Parents?' in Onora O'Neill and William Ruddick (eds.), *Having Children* (Oxford, Oxford University Press, 1979) 351; A. D. M. Walker, 'Political Obligation and the Argument from Gratitude' (1988) 17 *Philosophy and Public Affairs* 191; N. S. Jecker, 'Are Filial Duties Unfounded?' (1989) 26 *American Philosophical Quarterly* 73.

[11] Cf. J. Rawls, 'Are There Any Natural Rights?' and 'Legal Obligation and the Duty of Fair Play' in S. Hook (ed.), *Law and Philosophy* (New York, New York University Press, 1964); J. Rawls, *A Theory of Justice* (Cambridge, MA, Harvard University Press, 1971) 97, 308; M. B. E. Smith, 'Is There a Prima Facie Obligation to Obey the Law?' (1972) 82 *Yale Law Journal* 950.

[12] Donald Pfaff, *The Neuroscience of Fair Play: Why We (Usually) Follow the Golden Rule* (New York, Dana Foundation, 2007). Also cf. D. Pfaff, M. Kavaliers, and E. Choleris, 'Mechanisms Underlying an Ability to Behave Ethically' (2008) 8 *American Journal of Bioethics—Neuroscience* 10.

[13] Cf. A. M. Viens and R. E. G. Upshur, 'The Concept of Reciprocity in Public Health' (in preparation).

how neuroscience may be used to help us understand what guides and motivates individuals to fulfil their reciprocal obligations in such contexts.[14]

19.3 Reciprocity in the Use of Restrictive Public Health Measures

Neuroscientific research has been able to shine a light on various factors underlying the internal regulation of human conduct. It may be possible to incorporate aspects of this internal regulation into the formulation of law as an external regulator of conduct. In better understanding the structures and mechanisms that lead us to reciprocate and cooperate with reciprocators, the use of neurobiological data within the construction and implementation of public health law may help us formulate more effective public health policy and practice by ensuring it is formulated in such as way that it most optimally promotes support and compliance. Both neurobiological and qualitative research relating to reciprocity appears to support the validity of this claim.

19.3.1 Neurobiological research

Donald Pfaff has recently argued, in his own work and with colleagues, that humans are wired for reciprocity and the widespread predisposition towards reciprocal behaviour depends, in part, on neurobiological mechanisms for fear and affiliative behaviours.[15] On this view, one aspect of acting ethically is dependent on an individual's refusal to perform other-regarding acts in a way that, if such an act were directed towards them, would cause a fear of harm. There are brain mechanisms that underlie the representation of an action and its envisioned consequences, according to Pfaff, in which a person blurs the differences between themselves and others, and, as a result, if the intended act under consideration is beneficial, the individual is more likely to perform the act and if harmful the individual is less likely to perform the act.

Extending this neurobiological explanation to the context of public health and the case of quarantine, one possibility is that individuals may have a tendency to avoid performing harmful acts—the consequences of which they themselves would fear. For non-infected individuals, this may mean motivation to undertake testing and adequate precautions, reporting exposed individuals breaking quarantine, among other considerations. For healthcare professionals, this may mean motivation to fulfill their duty to care in the face of increased risk and conflicting personal obligations to family. Even for infected individuals, who may have fallen victim to the consequence they feared, this still may mean motivation to cooperate with isolation measures, being compliant

---

[14] For more on these normative issues in relation to reciprocity and public health, see A. M. Viens, C. M. Bensimon, and R. E. G. Upshur, 'Your Liberty or Your Life: Reciprocity in the Use of Restrictive Measures in Contexts of Contagion' (2009) 6 *Journal of Bioethical Inquiry* 207.

[15] This section is based on A. M. Viens, 'Public Health, Ethical Behaviour and Reciprocity' (2008) 8 *American Journal of Bioethics–Neuroscience* 1.

with treatment, taking active steps not to expose or infect others. For government and public health officials, this may mean provision of the necessities of life, compensation and assurances of substantive and procedural protections for those subjected to restrictive measures (such as the right to counsel, the right to a hearing, a right to appeal quarantine and detention decisions, and relief from punishment for non-compliance or violation of restrictive measures on judicial review). If such an account is correct—and I think Pfaff's account is at least plausible—public health laws involving restrictive measures could make use of this neurobiological predisposition to reciprocate by ensuring that individuals doing their part are adequately supported and assisted in fulfilling their obligations, and concrete estimations of the benefits and burdens are accurately communicated to assure everyone that the balance of benefit over harm is highly favourable.

If Pfaff is correct that humans are wired to reciprocate under the right conditions, making use of the neurobiological mechanisms underlying behaviour as part of the input towards constructing and enforcing public health laws surrounding quarantine and isolation could result in increased support and compliance with such measures. This increased support and compliance will mean that the promotion and protection of public health will be more effective.

19.3.2 Qualitative research

The relevance of incorporating neuroscientific studies of reciprocity as a way of seeking to understand what mechanisms need to be focused on in order to help gain support and compliance for public health measures may not be such a speculative or fanciful idea.[16] In the past couple of years, there has been an increase in qualitative research on the use of social distancing and restrictive measures in public health law to deal with infectious diseases in countries like Canada, Australia, and the US.[17] In the form of town halls, citizen juries, questionnaires, and surveys, it has been repeatedly found that individuals cite reciprocity as a central issue within pandemic preparedness and response planning and that there is a relationship between support and compliance with public health laws involving restrictive measures and the extent to which these measures are shown to be reciprocal in nature and practice.

---

[16] This section is based on A. M. Viens, C. M. Bensimon, and R. E. G. Upshur, 'Your Liberty or Your Life: Reciprocity in the Use of Restrictive Measures in Contexts of Contagion' (2009) 6 *Journal of Bioethical Inquiry* 207.

[17] R. J. Blendon, J. M. Benson, and K. J. Weldon, 'Pandemic influenza and the Public: Survey Findings' Cambridge, MA: Harvard School of Public Health Project on the Public and Biological Security, available at <http://www.hsph.harvard.edu/panflu/IOM_Avian_flu.ppt> (accessed 31 October 2009); R. J. Blendon, C. M. DesRoches, M. S. Cetron, J. M. Benson, T. Meinhardt, and W. Pollard, 'Attitudes Toward the Use of Quarantine in a Public Health Emergency in Four Countries' (2006) 25 *Health Affairs* 15; N. M. Baum, S. D. Goold, and P. D. Jacobson, '"Listen to the People" Public Deliberation about Social Distancing Measures in a Pandemic' (2009) 9 *American Journal of Bioethics* 4; Joint Centre for Bioethics Pandemic Ethics Working Group, 'Public Engagement on Social Distancing in a Pandemic: A Canadian Perspective' (2009) 9 *American Journal of Bioethics* 15.

Most participants in these public engagement exercises were broadly supportive of the use of restrictive measures within public health law, so long as they were applied equitably and with appropriate reciprocal support mechanisms in place for those subjected to the measures. The research revealed that while citizens demonstrate self-interested concerns about issues such as job security, when it came to complying with restrictive measures, a significant amount of the justificatory support and motivational compliance expressed involved appeals to reciprocity.

Participants repeatedly emphasized that those complying with quarantine orders must be assured that they will not be disproportionately burdened or unnecessarily penalized. They also articulated the view that there was a collective obligation to provide social and material support to persons subjected to restrictive measures. For instance, in one Canadian study,[18] 85 per cent of respondents supported government-ordered quarantines and the suspension of individuals' rights (e.g. travel, assembly) during outbreaks. Ninety-five per cent of respondents supported ensuring that the basic needs (e.g. food, shelter, social support) of those quarantined were adequately met. Seventy-nine per cent of respondents supported the provision of support services after the quarantine ended. This advocacy of governmental power also extended to international authorities—97 per cent of participants supported travel advisories and restrictions to outbreak areas in order to stop a pandemic spread, even when the consequences included serious economic loss. One participant noted, 'For me this raises a question of interactive societal responsibility, so if society deems it necessary for [a person] to stay at home to protect society from the spread of infection by advising him of quarantine, then society must, in turn, be responsible to [that person] to ensure he is well provided for and will not suffer the results of his patriotic duty.'

This qualitative research supports the idea that when reciprocity, as a value, is reflected in public health laws and regulations, individuals will be more likely to support and comply with those laws. The evidence and information that we receive from neurobiological studies of the basis of reciprocity and the conditions under which reciprocal behaviour is more apt to be exhibited can be incorporated as an input within the law-making and enforcement process for public health law. Doing so has the prospect of helping public health law be more effective in achieving its aim, *viz.* protecting the public's health.

19.4 Objections

There are, of course, possible objections to using neuroscience as an input into public health law-making.

One objection may be that the use of neuroscience to increase or enhance compliance with the law is somehow unduly manipulative. Most of the suggested uses of neuroscientific studies and technology have been directed at their potential

---

[18] Joint Centre for Bioethics Pandemic Ethics Working Group, 'Public Engagement on Social Distancing in a Pandemic: A Canadian Perspective' (2009) 9 *American Journal of Bioethics* 15.

contribution in obtaining evidence that can be used in seeking to answer a legal question about guilt, responsibility or liability. However, in using neuroscience within the design, implementation or enforcement of law, it may be thought that such uses begin to cross the threshold from highlighting the usual merits of obeying the law to surreptitiously taking advantage of underlying motivational structures to get individuals to do what you want them to do—even if it is in our common interest. In one sense, this is indeed a worry—and one that is not foreign to those working in public health law. Public health interventions aimed at encouraging individuals to change their behaviour in individually and collectively beneficial ways regularly employ educational campaigns, incentives and deterrence mechanisms that have sometimes been criticized as being coercive or paternalistic, and possibly undermining of personal autonomy and responsibility.

To be sure, I am not advocating that in using neuroscience as an input of law-making that we employ some form of mind control or forced behavioural manipulation. The suggestion is merely that a greater understanding of the neurobiological basis of why individuals are disposed to reciprocate and why the presence of fulfilled reciprocal obligations can further enhance the reasons citizens already have to obey public health laws. In order to prevent any insidious or illegitimate use, it could perhaps be mandated that neuroscientific studies only be used to enhance *voluntary* measures and that any use of neuroscientific techniques or technologies that would render the measures involuntary would be prohibited.[19]

Another objection might be that neurobiological considerations only form a small part of the complex and diverse factors that regulate human behaviour. So, even if we could use our knowledge of how brain structures and systems underpin reciprocal behaviour and incorporate it within public health law, there is no guarantee that it will increase compliance in practice.[20] This is indeed the case. There is no guarantee of success; however, it is also an objection that can be equally levied against any measure we currently or could potentially employ to seek adherence with the law. The suggestion is merely that it is an additional promising measure to add within the process of law-making that already coheres with existing expressed values and preferences and shows promise of motivational efficacy. That being said, there is anecdotal evidence that we can point to that may lead us to think ensuring the requirements of reciprocity are respected within public health law could be beneficial. In the 2003 SARS outbreak in Toronto, for example, there was wide support and compliance with restrictive public health measures, in which reciprocity was seen to be promoted. Toronto public health officials reported that only twenty-two orders for mandatory detainment were necessary amongst the approximately 30,000 people who were voluntarily quarantined.

---

[19] There are other possible benefits to employing neuroscience as an input to law-making. It is likely that a greater understanding of the neurobiological basis of behaviour will also make public health educational campaigns, and incentive and deterrence activities, more effective.

[20] Pfaff thinks this might not be such a strong worry in this case. He claims that '... the mechanisms behind [the golden rule] must... be rather simple and dependable, and capable of operating without our having to sit down like The Thinker and cogitate on it—the way, for example, sleep is': note 11, above, 22.

One further objection might be that the use of neuroscience to increase or enhance compliance with public health law is potentially self-defeating. Understanding the brain mechanisms that predispose us to reciprocate and attempting to formulate public health laws that attempt to take advantage of such mechanisms will only have a chance of working if the government is indeed able to fulfil its reciprocal obligations and goals. We know from historical uses of restrictive measures,[21] and even from assessments of current capabilities in some jurisdictions for dealing with public health emergencies involving highly dangerous and contagious diseases, that the employment of restrictive measures cannot be guaranteed always to occur with adequate resources and facilities in place or administered in an appropriate way—as reciprocity would demand. If the expectation or promise of reciprocity that has been incorporated into law goes unfilled, it may actually result in reinforcing dissent and disobedience. This is also a potential problem—and one that would have to be taken into account. Indeed, it is a concern both for the use of neuroscience in the process of public health law-making and with current legal measures that already aim at being reciprocal in nature. I am inclined to believe that any significant dissent or disobedience would depend on the reasons for the government's failure to reciprocate. For instance, a government's failure to deliver on a promise to have enough anti-viral drugs stockpiled for 50 per cent of the population may not be met with dissent and disobedience, but if stockpiled anti-virals were distributed in an arbitrary or unfair way then that might be enough to trigger such dissent and disobedience. Be that as it may, the force of this objection may not be best directed at the use of neuroscience itself but in ensuring that whatever mechanisms are used to help increase support and compliance with law that it is done with prudence and assiduousness.

19.5 Conclusion

The law seeks to regulate behaviour in cooperative and other pro-social ways and the neurobiological mechanisms underlying such behaviours may provide promising additional material that could be used in the formulation and implementation of laws in order to increase support and compliance. In this paper, I have explored the area of public health law as one instance where this could be a possibility. The suggestion is still one of academic exercise, but as more and more public health researchers, practitioners, and lawyers are looking towards ways of using new tools to help promote and protect public health, the incorporation of neuroscience as a further explanatory component to understanding the motivation of human action related to public health presents itself as a possible option. It is an option that is viable and worth exploring further as an input to law-making.

---

[21] See *Jew Ho v Williamson* (1900) 103 F 10 (CCND Cal); *Williamson v Wong Wai* (1900) 103 F Rep 10; N. Shah, *Contagious Divides: Epidemics and Race in San Francisco's Chinatown* (Berkeley, University of California Press, 2001).

20

Pathways to Persuasion: How Neuroscience Can Inform the Study and Practice of Law[†]

Cheryl Boudreau, *Seana Coulson,** *and Mathew D. McCubbins****

Under what conditions can lawyers persuade jurors? During criminal and civil trials, lawyers seek to persuade jurors that their interpretation of the facts is the 'correct' one and that jurors should, therefore, reach a verdict that favours their client. Although lawyers have many persuasive tools at their disposal (e.g. rhetorical skill, detailed knowledge of the facts, colourful exhibits), one of the biggest challenges they face is convincing uninformed jurors to trust their statements, as opposed to those of opposing counsel. To succeed, lawyers must often determine how to present information and witnesses in the most persuasive way possible. Conversely, the challenge for jurors is to determine the truthfulness of the information that lawyers and witnesses present and reach a just verdict even when they are not fully informed. Indeed, because jurors do not witness for themselves the events surrounding particular crimes or disputes, they must rely upon the statements of lawyers and witnesses whom they do not personally know when making their decisions.[1] Further, because jurors typically lack scientific and mathematical

---

[†] The authors thank the National Science Foundation (grant #SES-0616904) and the Kavli Institute for Brain and Mind for providing financial support for their experiments. They are also grateful to participants in the University College London's colloquium on law and neuroscience, the University of California, Riverside's conference on experiments, and the University of Southern California School of Law's symposium on 'Neuroscience, Law, Economics, and Politics' for helpful comments on an earlier draft of this paper. They also thank Michael Freeman for his generous invitation to present their work at his colloquium at the University College London.

* Assistant Professor, University of California, Davis, Department of Political Science, One Shields Avenue, Davis, CA 95616; email: clboudreau@ucdavis.edu.
** Associate Professor, University of California, San Diego, Department of Cognitive Science, 9500 Gilman Drive, La Jolla, CA 92093; email: coulson@cogsci.ucsd.edu.
*** Provost Professor of Business, Law and Political Economy, University of Southern California, Marshall School of Business, Gould School of Law, and Department of Political Science, 699 Exposition Boulevard, Los Angeles, CA 90089; email: mmccubbins@marshall.usc.edu.

[1] A. Lupia and M. D. McCubbins, *The Democratic Dilemma: Can Citizens Learn What They Need to Know?* (Cambridge, Cambridge University Press, 1998); C. Boudreau, 'Jurors are Competent Cue-Takers: How Institutions Substitute for Legal Sophistication' (2006) 2 *International Journal of Law in Context* 293; C. Boudreau and M. D. McCubbins, 'Nothing But the Truth? Experiments on Adversarial Competition, Expert Testimony, and Decision Making' (2008) 5 *Journal of Empirical Legal Studies* 751.

expertise,[2] they are often unable to evaluate for themselves the quality of the expert testimony and evidence they receive.

Given that lawyers seek to persuade jurors and that jurors often lack the ability to evaluate critically lawyers' and witnesses' statements, it is important to understand the conditions under which lawyers can persuade uninformed jurors to trust and base their decisions upon the statements they make during a trial.[3] In an effort to identify these conditions, Lupia and McCubbins,[4] as well as Crawford and Sobel,[5] developed game theoretic models that yield the following predictions about when a knowledgeable speaker (be it a lawyer or an expert witness) will be able to successfully persuade jurors: 1) common interests between a knowledgeable speaker and jurors should induce jurors to trust the speaker's statements and base their choices upon them; 2) conflicting interests between a knowledgeable speaker and jurors should induce jurors to ignore the speaker's statements and make their decisions on their own; 3) institutions that are common in legal settings should sometimes induce jurors to trust a speaker's statements, even when the speaker's interests conflict with those of jurors. For example, a sufficiently large penalty for lying (which in legal settings may be a penalty for perjury or a loss of reputation) can remove a speaker's incentive to lie, and therefore lead jurors to trust and base their decisions on the speaker's statements.

Lupia and McCubbins[6] tested these predictions about persuasion in a series of behavioural experiments. Consistent with their predictions, they found that subjects' decisions to trust an unknown individual's statements depended upon the perceived trustworthiness of that other individual (dubbed 'the reporter' in their experiments and throughout this paper). For example, Lupia and McCubbins showed that subjects who perceived that the reporter shared common interests with them were significantly more likely to trust and base their decisions upon the reporter's statements than were subjects who perceived that the reporter's interests conflicted with their own. When the reporter had conflicting interests with subjects, but was made trustworthy by an institution (such as a penalty for lying that was large enough to ensure that the reporter had a dominant strategy to tell the

---

[2] G. C. Lilly (2001) 'The Decline of the American Jury,' (2001) 72 *University of Colorado Law Review* 53; M. A. Fisher, 'Going for the Blue Ribbon: The Legality of Expert Juries in Patent Litigation,' (2000-2001) 2 *Columbia Science and Technical Law Review* 1.

[3] Scholars in other disciplines also seek to understand trust. See, e.g., M. R. Delgado, R. H. Frank, and E. A. Phelps, 'Perceptions of Moral Character Modulate the Neural Systems of Reward During the Trust Game' (2005) 8 *Nature Neuroscience* 1611; B. King-Casas, D. Tomlin, C. Anen, C. F. Camerer, S. R. Quartz, and P. R. Montague, 'Getting to Know You: Reputation and Trust in a Two-Person Economic Exchange' (2005) 308 *Science* 78; M. Kosfeld, M. Heinrichs, P. J. Zak, U. Fischbacher, and E. Fehr, 'Oxytocin Increases Trust in Humans' (2005) 435 *Nature* 673; K. McCabe, D. Houser, L. Ryan, V. Smith, and T. Trouard, 'A Functional Imaging Study of Cooperation in Two-Person Reciprocal Exchange' (2001) 98 *Proceedings of the National Academy of Sciences* 11832; P. J. Zak, R. Kurzban, and W. T. Matzner, 'The Neurobiology of Trust' (2004) 1032 *Ann. NY Acad. Sci.* 224; P. J. Zak, R. Kurzban, and W. T. Matzner, 'Oxytocin is Associated with Human Trustworthiness.' (2005) 48 *Hormones and Behavior* 522.

[4] A. Lupia and M. D. McCubbins, *The Democratic Dilemma: Can Citizens Learn What They Need to Know?* (Cambridge, Cambridge University Press, 1998).

[5] V. Crawford and J. Sobel, 'Strategic Information Transmission' (1982) 50 *Econometrica* 1431.

[6] Note 4, above.

truth), subjects trusted the reporter's statements at a rate that was similar to the rate at which they trusted the reporter's statements when they knew that the reporter shared common interests with them. From these results, Lupia and McCubbins concluded that institutions can substitute for common interests because they, too, induced subjects to trust and base their decisions upon the reporter's statements.

Although Lupia and McCubbins's[7] theory and experiments help clarify the conditions for successful persuasion, they leave open the following question: do measures of brain activity and reaction time also indicate that subjects view in the same way information from a trustworthy individual (i.e. one who shares common interests with them) versus an individual who is otherwise untrustworthy (i.e. has conflicting interests with them), but is made trustworthy by an external institution? This question is an important one, since it has implications for both the study and practice of law. Indeed, if subjects' brain activity is different when they receive information from these two types of trustworthy individuals (even though their decisions and reaction times are similar), then this suggests that legal scholars and practitioners who seek to understand persuasion may not necessarily get the whole story if they only observe subjects' behaviour. More broadly, such a finding would also have implications for lawyers and witnesses, since it would indicate that the manner in which a speaker is made trustworthy (and not just trustworthiness itself) affects how jurors process information from that speaker. On the other hand, if subjects' decisions, reaction times, and brain activity are similar when they receive information from these two types of trustworthy individuals, then this indicates that institutions induce not only the same behaviour as common interests, but also the same cognitive processing of information.

To address this open question and illustrate how neuroscience can inform the study and practice of law, we describe our previous study that replicated Lupia and McCubbins's[8] experiments with electroencephalograph (EEG) and timed-response technology.[9] Although our behavioural results were consistent with Lupia and McCubbins's conclusion that institutions can substitute for common interests, our EEG results revealed something about trust and persuasion that we did not learn from observing subjects' decisions and reaction times. That is, they demonstrated that even though the reporter was, theoretically and behaviourally, equally trustworthy in the Common Interests and Penalty for Lying conditions, subjects processed information differently when it came from a reporter who was trustworthy by virtue of sharing common interests with them versus a reporter who was made trustworthy by an external institution. Indeed, across a range of cognitive responses, subjects' brain activity was different in the Common Interests condition, relative to both the Penalty for Lying and Conflicting Interests conditions. Interestingly, this processing difference existed even though subjects were equally likely

---

[7] Note 4, above.
[8] Ibid.
[9] C. Boudreau, M. D. McCubbins, and S. Coulson, 'Knowing When To Trust Others: An ERP Study of Decision Making After Receiving Information from Unknown People' (2009) 4 *Social, Cognitive, and Affective Neuroscience* 23.

to base their decisions upon the reporters' statements in the Common Interests and Penalty for Lying conditions and even though they took the same amount of time to make their decisions in these conditions. Given this difference between subjects' behaviour and brain activity, it appears that recording subjects' brain activity adds a new dimension to our understanding of jurors' decisions to trust and base their decisions on the statements of others.

This paper proceeds as follows. We begin with a brief overview of the EEG experiments that we used to study the behavioural and neural correlates of persuasion. We then describe our hypotheses, as well as the data and methods that we used to test them. Next, we summarize our experimental results on subjects' decisions, reaction times, and brain activity. We conclude with a discussion of the substantive and methodological implications that our research has for debates about persuasion in courtroom settings. Indeed, we summarize our previous research to illustrate the ways that neuroscience can contribute to our understanding of legal phenomena, such as persuasion and trust. Specifically, we emphasize that our experiments show the value of tying together both behavioural results and brain data in analyses of persuasion and trust. Although our study represents only a first step in this endeavour, we emphasize that future research on persuasion (and other topics of interest to legal scholars) can potentially benefit from simultaneously assessing behaviour and brain activity.

20.1 Using Neuroscience to Study Persuasion

In order to analyse the conditions under which a lawyer can successfully persuade a juror, we conducted laboratory experiments at two large public universities. When recruiting subjects, we posted flyers at various locations on campus, and we also sent out campus-wide emails to advertise the experiments. A total of fifty-nine healthy adults from these two university communities (thirty-seven men), aged eighteen to twenty-eight, participated in our experiments. We recorded behavioural responses and reaction times from all fifty-nine subjects, and we recorded the EEGs of twelve of these subjects.[10]

During the experiment, we asked subjects to predict the outcomes of coin tosses that they did not observe. We told subjects that they would earn 50 cents for each correct prediction that they made, and nothing when they made an incorrect prediction or failed to make a prediction. We also informed subjects that another subject in another room (dubbed 'the reporter') would observe each coin toss outcome and then send a report to them via computer about whether the coin landed on heads or tails. Importantly, we told subjects that the reporter could either lie about the coin toss outcome or tell the truth. Thus, before subjects made a prediction about each coin toss, they observed the reporter's report of whether the coin landed on heads or tails, but they did not know whether the report was

---

[10] Note 9, above.

truthful. As in Lupia and McCubbins's study,[11] the key factor that we manipulated was the perceived trustworthiness of the reporter, and we did this by varying the financial interests of the reporter, as well as the institutional context in which the reporter sent his or her report.

Specifically, we began the experiment by reading the instructions for the Common Interests condition to subjects. That is, we asked subjects to predict the outcome of an unseen coin toss after receiving a message from the reporter. We informed subjects that, in this condition, *both they and the reporter* earned 50 cents every time they, the subjects, correctly predicted the coin toss outcome, and nothing if they predicted incorrectly or failed to respond before the onset of the next coin toss. We reminded subjects that it was entirely the reporter's decision as to whether he or she sent a true or a false report via the computer. To ensure that subjects fully understood the instructions for the Common Interests condition, we gave them a quiz that asked them to say how much money the reporter earned under various circumstances. To motivate performance on the quiz, we paid subjects 25 cents for each quiz question they answered correctly. When we were sure that subjects understood how the reporter earned money in the Common Interests condition, ten experimental trials began.

Following the initial Common Interests trials, we read the instructions for the Conflicting Interests condition to subjects. Specifically, we told subjects that their task was the same as in the previous condition—to predict the outcome of an unseen coin toss after receiving a message from the reporter. We told subjects that while they themselves still earned 50 cents for each correctly predicted coin toss and nothing for incorrect predictions, the reporter now earned 50 cents for each *incorrect* prediction that subjects made. We then gave subjects a brief quiz on how much money the reporter earned under various circumstances, and we paid them 25 cents for each correctly answered quiz question. When we were sure that subjects understood how the reporter earned money in this condition, ten Conflicting Interests trials began.

Following the initial Conflicting Interests trials, we read the instructions for the Penalty for Lying condition to subjects. We told subjects that as in the previous (Conflicting Interests) condition, the reporter earned 50 cents for each of the subject's incorrect predictions, while the subject earned 50 cents for each correct prediction. We also told subjects that every time the reporter sent a false report, we deducted $1 from the reporter's experimental earnings. We then gave subjects a brief quiz on how much money the reporter earned under various circumstances, and we paid them 25 cents for each correctly answered quiz question. Because we quizzed subjects on how the reporter earned money and corrected their quizzes in front of them, they knew that the $1 penalty was large enough to ensure that the reporter always had an incentive to tell the truth about the coin toss outcome. When we were sure that subjects understood how the reporter earned money in this condition, ten Penalty for Lying trials began.

---

[11] Note 4, above.

Once subjects completed ten trials for all three conditions, we collected data for additional coin tosses in each of our three conditions. In order to control for learning and arousal effects, half of the subjects completed the second block of trials in order 1 (Common Interests, Conflicting Interests, Penalty for Lying), while the other half completed the second block of trials in order 2 (Penalty for Lying, Conflicting Interests, and Common Interests). On each coin toss, we recorded the amount of time that elapsed between the presentation of the reporter's report and subjects' predictions. We did not tell subjects that their predictions were being timed, and we did not give them any feedback until the end of the experiment. We also told subjects that the reporter did not observe their predictions about the coin toss outcomes.

20.2 Predictions

Lupia and McCubbins's[12] theory and experiments suggest that particular institutions (such as a sufficiently large penalty for lying) can substitute for common interests. Based on their results, we predicted that subjects would be equally likely to trust the reporter's statements (and, thus, base their predictions upon them) when the reporter shared common interests with them and when the reporter was made trustworthy by a penalty for lying. Following Lupia and McCubbins, we also predicted that when the reporter's interests conflicted with those of subjects, subjects would not trust the reporter's statements and, thus, not base their predictions upon them.

As for subjects' reaction times and brain activity in the three conditions, Lupia and McCubbins[13] do not offer predictions for these other measures. That said, based on their theoretical and experimental results suggesting that institutions can substitute for common interests, we expected to observe similar reaction times in the Common Interests and Penalty for Lying conditions, as well as similar brain activity in these two conditions. We also expected subjects' reaction times and brain activity to be different in the Conflicting Interests condition (relative to the Common Interests and Penalty for Lying conditions), as this was the one condition where the reporter was not trustworthy.

If our results supported these expectations, then our study would suggest several substantive and methodological conclusions. First, it would suggest that particular institutions substitute for common interests; that is, they induce not only the same decisions, but also the same reaction times and cognitive processing of information. Second, it would suggest that observing subjects' brain activity does not add much to our understanding of subjects' decisions to trust and base their decisions on the statements of others. Indeed, if subjects' brain activity simply mirrored their behaviour, then one might question whether there is any value added to using this technology. Stated differently, one might ask why legal scholars should record

[12] Ibid. [13] Ibid.

and interpret subjects' brain activity if it simply tells us the same thing that subjects' decisions and reaction times tell us.

Alternatively, if subjects' brain activity differed from their behaviour, then this would indicate that legal scholars and practitioners who seek to understand persuasion and trust may not necessarily get the whole story if they only observe subjects' decisions and reaction times. For example, it was possible that subjects in our experiments would process information *differently* when it came from a reporter who was trustworthy by virtue of sharing common interests with them versus a reporter who was made trustworthy by an external institution. This difference in the way that subjects process information from these two types of trustworthy reporters could exist even if they were equally likely to base their decisions upon these reporters' reports and even if they took the same amount of time to make their decisions with both reporters. Indeed, research in cognitive neuroscience shows that similar behavioural outcomes can be subserved by different neural mechanisms.[14]

20.3 Measuring Persuasion

To assess whether subjects were persuaded by the reporter's reports, we first examined the extent to which their predictions were the same as what the reporter reported in each experimental condition (i.e. how likely were subjects to predict 'heads' when the reporter reported 'heads' and predict 'tails' when the reporter reported 'tails' in each condition). Specifically, we analysed whether subjects' predictions matched what the reporter reported more than 50 per cent of the time. We used a 50 per cent baseline because we tossed a fair coin; thus, if subjects were simply choosing heads or tails randomly, then we would expect their predictions to match the reporter's reports 50 per cent of the time. If subjects were persuaded by the reporter's reports, then we should observe their predictions matching the reporter's reports more than 50 per cent of the time. We also recorded the amount of time that elapsed between each presentation of the reporter's report and subjects' predictions.

We recorded subjects' brain activity from twenty-nine tin electrodes that were arranged in an expanded version of the 10–20 system atop subjects' scalps.[15] In a nutshell, these electrodes recorded electrical activity in the brain due to postsynaptic potentials (i.e. graded voltage changes in the cell membrane of neurons) occurring in the cortex. Because electrical activity in response to a particular

---

[14] C. L. Grady, J. V. Haxby, B. Horwitz, M. B. Schapiro, S. I. Rapoport, L. G. Ungerleider, M. Mishkin, R. E. Carson, and P. Herscovitch, 'Dissociation of Object and Spatial Vision in Human Extrastriate Cortex: Age-Related Changes in Activation of Regional Cerebral Blood Flow Measured with [15O] Water and Positron Emission Tomography' (1992) 4 *Journal of Cognitive Neuroscience* 23; P. A. Reuter-Lorenz, J. Jonides, E. E. Smith, A. Hartley, A. Miller, C. Marshuetz, and R. A. Koeppe, 'Age Differences in the Frontal Lateralization of Verbal and Spatial Working Memory Revealed by PET' (2000) 12 *Journal of Cognitive Neuroscience* 174.

[15] M. Nuwer, G. Comi, R. Emerson, A. Fuglsang-Frederiksen, J.-M. Guerit, H. Hinrichs, and P. Rappelsburger, 'IFCN Standards for Digital Recording of Clinical EEG' (1998) 106 *Electroencephalography and Clinical Neurophysiology* 259.

event (i.e. stimulus) is quite small, the signal must be enhanced by averaging over a large number of trials. Specifically, for each subject in an EEG experiment, it is necessary to: 1) repeat the event of interest (which in our experiment is the reporter's report) many times in each experimental condition; and 2) time-lock segments of the EEG to that event so that those segments can be averaged together for each experimental condition. This averaging process reveals the subset of each subject's brain's electrical response that is temporally correlated with the onset of the event, and is known as an event-related potential (ERP). Once this averaging process is completed for each subject, all subjects' ERPs are averaged together to produce what is known as a grand average ERP. It is this grand average ERP that is used in statistical analyses.[16]

In our study, we time-locked subjects' EEGs to the onset of the reporter's report in each experimental condition. We assessed subjects' ERPs by measuring the mean amplitude of the waveform in intervals that captured various cognitive components of interest. Specifically, we analysed the mean amplitudes that we observed in each experimental condition by using three sorts of repeated measures ANOVAs: 1) midline analyses involving measurements taken from channels FPz, FCz, Cz, CPz, Pz, and Oz; 2) medial analyses involving measurements taken from channels FP1, F3, FC3, C3, CP3, P3, O1, and their left hemisphere counterparts; and 3) lateral analyses involving measurements from channels F7, FT7, TP7, T5, and their left hemisphere counterparts.

20.4 Results

20.4.1 Subjects' decisions

When subjects knew that the reporter shared common interests with them, they were significantly more likely to predict 'heads' when the reporter reported 'heads' and predict 'tails' when the reporter reported 'tails' than they would by chance. Specifically, when the reporter reported 'heads' in the Common Interests condition, there was a 94 per cent chance that subjects predicted 'heads'—a figure that was significantly greater than our 50 per cent baseline. Similarly, when the reporter reported 'tails' in the Common Interests condition, there was a 94 per cent chance that subjects predicted 'tails,' which was also significantly greater than our 50 per cent baseline. In the Penalty for Lying condition, subjects' predictions were also significantly more likely to match the reporter's reports than they would by chance. That is, when the reporter reported 'heads' in the Penalty for Lying condition, there was a 94 per cent chance that subjects predicted 'heads.' When the reporter reported 'tails' in the Penalty for Lying condition, there was a 94 per cent chance that subjects predicted 'tails'. These figures were both significantly greater than 50 per cent and were the same as in the Common Interests condition.

---

[16] Steven J. Luck, *An Introduction to the Event-Related Potential Technique* (Cambridge, MIT, 2005).

In the Conflicting Interests condition, however, subjects' predictions matched what the reporter reported only 50 per cent of the time. That is, regardless of whether the reporter reported 'heads' or 'tails', subjects were equally likely to predict 'heads' versus 'tails'. Taken together, these results were consistent with those of Lupia and McCubbins[17] and indicated that subjects were equally likely to trust the statements of a reporter who shared common interests with them and a reporter who was made trustworthy by a penalty for lying. They also demonstrated that subjects' behaviour in both the Common Interests and Penalty for Lying conditions was significantly different from their behaviour in the Conflicting Interests condition, where subjects did not trust the reporter's statements.

20.4.2 Reaction times

Our reaction time results showed a similar pattern. That is, subjects in the Common Interests and Penalty for Lying conditions took similar amounts of time to make their predictions after receiving the reporter's reports. Further, subjects in the Conflicting Interests condition were slower to make their predictions than were subjects in the other two conditions. Specifically, subjects in the Common Interests condition took, on average, 1,191 milliseconds to make their predictions of 'heads' or 'tails', while subjects in the Penalty for Lying condition took, on average, 1,157 milliseconds to make their predictions. This difference was not statistically significant. Subjects in the Conflicting Interests condition, however, took, on average, 1,318 milliseconds to make their predictions, which was significantly slower than subjects in Penalty for Lying and Common Interests conditions.

20.4.3 ERP results

Unlike our behavioural results, our ERP results demonstrated that subjects' brain activity was different when they received information from a reporter who shared common interests with them versus a reporter who was made trustworthy by a penalty for lying. Indeed, across a range of cognitive responses to the reporter's reports, we consistently found that subjects' brain activity was more similar in the Penalty for Lying and Conflicting Interests conditions than it was in the Penalty for Lying and Common Interests conditions. Specifically, prominent portions of the waveform included a negativity peaking approximately 100 ms after the onset of the reporter's report (the AN1), a positivity peaking approximately 200 ms after the onset of the reporter's report (the P2), a more broadly distributed positivity peaking at approximately 500 ms (the P3), a negative-going peak at 600 ms (the medial negativity), and subsequent slow wave activity we refer to as the late positive complex (LPC). With only one exception (i.e. the medial negativity[18]), each of

---

[17] Note 4, above.
[18] This is the only ERP component that showed a similar response in the Common Interests and Penalty for Lying conditions.

these cognitive responses indicated that subjects' brain activity was different in the Common Interests condition, relative to both the Penalty for Lying and Conflicting Interests conditions. Our results for the P3 and the LPC also demonstrated that there was not a significant difference in subjects' brain activity in the Conflicting Interests and Penalty for Lying conditions.

At a minimum, our ERP results indicate that subjects processed information differently when it came from an individual who shared common interests with them, relative to when it came from an individual whose interests conflicted with their own, but who was made trustworthy by an external institution. More broadly, these results may suggest that subjects' brains treated reports as more informative when the reporter shared common interests with them.[19] Further, that we observed differences in subjects' brain activity across these three conditions is remarkable because the visually presented stimuli were identical in each condition. It is also interesting that subjects' brain activity was different in the Common Interests and Penalty for Lying conditions even though both reporters were trustworthy (albeit for different reasons) in these two conditions.

20.5 Conclusion

In this paper, we provided an overview of our previous EEG experiments[20] to illustrate the potential that neuroscience has to inform the study and practice of law. Specifically, in our experiments, we analysed subjects' behaviour and brain activity in response to information from reporters whose trustworthiness stemmed from either the reporter's interests (vis-à-vis the subjects) or from an institution, such as a penalty for lying. We did so by recording the decisions, reaction times, and EEGs of subjects who guessed the outcome of an unseen coin toss after they received information from an anonymous reporter who knew the outcome of the coin toss, but was under no obligation to communicate it truthfully. Based upon Lupia and McCubbins's[21] theory and experiments, we predicted that subjects would be equally likely to base their decisions upon the statements of a reporter who was trustworthy by virtue of sharing common interests with them and a reporter whose interests conflicted with their own, but who was made trustworthy by a penalty for lying. Because Lupia and McCubbins's theory does not make predictions about subjects' reaction times and brain activity when receiving information from reporters in the Common Interests, Conflicting Interests, and Penalty for Lying conditions, we asked whether these two other measures would also yield results that are consistent with their conclusion that institutions can substitute for common interests.

---

[19] E. Donchin and M. G. H. Coles, 'Is the P300 Component a Manifestation of Context Updating?' (1988) 11 *The Behavioral and Brain Sciences* 357.
[20] Note 9 above.
[21] Note 4 above.

Our results indicated that although subjects behaved as if reporters in the Common Interests and Penalty for Lying conditions were equally trustworthy, their brain activity suggested that they processed information differently in the Common Interests and Penalty for Lying conditions. Subjects in both the Common Interests and Penalty for Lying conditions almost always based their predictions on the reporter's reports, while subjects apparently ignored the reporter's reports in the Conflicting Interests condition. Further, subjects' reaction times were similar in the Common Interests and Penalty for Lying conditions and were significantly faster than subjects' reaction times in the Conflicting Interests condition. Based on these behavioural responses, it appeared that subjects were equally likely to be persuaded by a reporter who shared common interests with them and a reporter who was made trustworthy by an institution, namely a penalty for lying.

In contrast, subjects' brain activity in response to the reporter's reports in the Common Interests condition tended to differ significantly from both the Conflicting Interests condition (as expected) and from the Penalty for Lying condition (contrary to our predictions). Thus, even though the reporter was, theoretically and behaviourally, equally trustworthy in the Common Interests and Penalty for Lying conditions, subjects processed information differently when it came from a reporter who was trustworthy by virtue of sharing common interests with them versus a reporter who was made trustworthy by an external institution. In this way, our results suggest that even though institutions can substitute for common interests in a behavioural sense, they do not necessarily induce the same cognitive processing of information.

As for the implications of our results, they indicate, at a minimum, that legal scholars who seek to understand persuasion and trust may not get the whole story if they only observe subjects' decisions and reaction times. Specifically, in our experiments, subjects processed information differently when it came from reporters who were trustworthy for different reasons, and this processing difference existed even though subjects were equally likely to base their decisions upon these reporters' reports and even though they took the same amount of time to make their decisions with both reporters. Given this difference between subjects' behaviour and brain activity, it is clear that recording subjects' brain activity adds a new dimension to our understanding of subjects' decisions to trust and base their decisions on the statements of others.

More broadly, our results have implications for lawyers who seek to present arguments, evidence, and witnesses in the most persuasive way possible during trials. Specifically, our results suggest that the manner in which a speaker is made trustworthy (and not just trustworthiness itself) affects how jurors process information from that speaker. Thus, lawyers who seek to persuade jurors should not necessarily assume that all perceptions of trustworthiness are created equal. Specifically, if the broader interpretations of our EEG results are correct (i.e. that subjects' brains treat reports as more informative in the Common Interests condition, relative to the Penalty for Lying condition), then lawyers (and witnesses) who seek to persuade jurors may benefit from conveying that they share common interests with jurors, as opposed to emphasizing their trustworthiness by appealing to institutional constraints. Of course, the question of whether and when the cognitive differences that we observed lead to changes in jurors' behaviour is an empirical question that we will explore in future research.

Finally, we emphasize an important methodological conclusion: namely, that EEG technology has much to offer legal scholars who seek to understand how jurors process information. First, because electricity travels at nearly the speed of light, the voltages that scalp electrodes record reflect the brain's activity at the same point in time; thus, EEG has excellent temporal resolution (approximately 1 millisecond) and provides a continuous measure of the online cognitive processing of information.[22] Given the many behavioural studies of the online processing model and other theories of cognition, it is clear that the direct real-time processing measure that EEG provides would be beneficial to many scholars. Indeed, in their study of the 'hot cognition' hypothesis that underlies the online processing model, Morris, Squires, Taber, and Lodge[23] take advantage of EEG technology to test this hypothesis, arguing that EEG allows for a better understanding of sensory and cognitive processing, as well as the activation of implicit attitudes. We could not agree more.

Second, EEG directly reflects the activity of neurons that are involved in the processing of information; therefore, EEG provides a direct measure of brain activity, in contrast to other neuroimaging techniques, such as fMRI, that provide more indirect measures that are based on blood oxygenation levels or blood flow.[24] Further, unlike other neuroimaging techniques, EEG is much less expensive (the supplies needed to test each subject cost between $1 and $3) and much less invasive (i.e. subjects simply wear a cap atop their heads that contains small electrodes). Thus, EEG provides legal scholars with a unique, practical way of simultaneously observing decision-making and the cognitive processing of information. Further, given the differences that we observed between subjects' behaviour and brain activity in our study, it appears that recording subjects' brain activity via EEG can potentially add a new dimension to our understanding of persuasion, trust and other legal phenomena—a dimension that we cannot necessarily tap if we only record behavioural responses.

---

[22] Note 16 above.
[23] J. P. Morris, N. K. Squires, C. S. Taber, and M. Lodge, 'Activation of Political Attitudes: A Psychophysiological Examination of the Hot Cognition Hypothesis' (2003) 24 *Political Psychology* 727.
[24] Note 16 above.

21

The Juridical Role of Emotions in the Decisional Process of Popular Juries

*Laura Capraro**

The study of law is, in general, the result of a scientific approach that tends to privilege 'rationale' over the influence of 'emotions' and 'intuitions' within the context of juridical reasoning.

Emotions hold a fundamental role even in instances such as criminal cases, in which results—as supported by research and findings furnished by cognitive neuroscience—are strongly conditioned by 'emotions'. The latter, while belonging to the realm of 'reason', are not an effective impediment to its functionality but, rather, greatly contribute to the reasoning process.

21.1 Background

Even if most trials in common law jurisdictions (e.g. US, UK, Canada, Australia, New Zealand) are conducted—always bearing in mind the difference in the provisions of law of the different countries—without the jury's contribution, it is undeniable that those conducted by jury are the most interesting, since, even if less significant considering its frequency, are those having as main object the most serious crimes.

The jury is one of those topics where one can observe extremely contrasting opinions. There are those who radically oppose it[1] and there are those who embrace its presence within the system.[2] These 'split' views essentially derive from the fact

---

* Assistant Professor in Criminal Procedure at the University of Rome 'Tor Vergata'.

[1] B. S. Oppenheimer, 'Trial by Jury' (1937) 11 *University of Cincinnati Law Review* 142: 'We commonly strive to assemble 12 persons colossally ignorant of all practical matters, fill their vacuous heads with law which they cannot comprehend, obfuscate their seldom intellects with testimony they are incompetent to analyze or unable to remember, permit partisan lawyers to bewilder them with their meaningless sophistry, then lock them up until the most obstinate of their number coerce the others into submission or drive them into open revolt.'

[2] Lord Devlin, *Trial by Jury* (London, Stevens & Sons, 1966) 164, according to whom '... No tyrant could afford to leave a subject's freedom in the hands of twelve of his country-men. So that trial by jury is more than an instrument of justice and more than one wheel of the constitution: it is the lamp that shows that freedom lives.'

that jury members are not subjects with a formal preparation within a juridical environment, and therefore, in almost all cases, have no specific competence or skill in the field in which they are called to operate. Therefore, according to the opposing view, very often they are not even able to understand fully the 'instructions' received from the judges, and, above all, make their decision and deliver their verdict more on the basis of their prejudices or, at best, by common sense than on the basis of the evidence given by the parties during the trial, according to which they should actually decide. On the other hand, it is interesting to note that several empirical researches—top of which that famous one conducted by Kalven and Zeisel and published in 1966[3]—proved a high level of trust put by the same juridical professionals in this institution: most judges in fact concordantly affirm that, despite the undeniable lack of preparation of the jury members, they perfectly play their role of 'fact-finders', and that themselves—professional judges—would have issued, in three-quarters of the cases, a judgment coinciding with that expressed by the jury's unmotivated verdict. These results attest that jurors not having knowledge of the law do not represent an obstacle to the delivery of a verdict substantially and significantly shared by professional judges. According to Feigenson, this could be explained by the concept of 'Total Justice',[4] whose aspiration guides the jurors' actions and therefore compensates for their objective lack of preparation.

So, if is true that the judges have declared—in these several studies conducted on the matter—that they substantially share the decisions taken by the jury, it is also true that one of the most ruthless critics of the jury system was nonetheless a judge, J. Frank, whose theories are a major source of, as it is well known, American Legal Realism. Frank, starting from empirical observation that the decision seems to be a product of emotion and not of reason, comes to the conclusion that a decision represents an act of will of the subject (judge) or the subjects (jury) summoned to express a judgment, and that therefore the jury's verdict is substantially an act of pure will.[5]

Beyond these extreme positions, the juries' decision-making process traditionally represents a reason for discussion between scholars belonging to different disciplines such as law and mind philosophers, juridical psychologists and jurists and, most recently, neurologists and neuroscientists. Although I examined the whole

---

[3] H. Kalven and H. Zeisel, *The American Jury* (Boston, Little Brown, 1967). Such work is considered still today a classic on the topic and, even if during the years have been highlighted the numerous methodological mistakes made by the research group, still represents an important landmark on jury research.

[4] N. Feigenson, *Legal Blame, How Jurors think and talk about Accidents* (Washington, DC, American Psychological Association, 2000) 5, 104.

[5] Naturally, reference is to be made to J. Frank, *Law and the Modern Mind* (New Brunswick and London, Transaction Publishers, 2009) (1930): 'The Jury (...) determine the rights of the respective parties and the jury's determination of these rights is guided by no real regard for "rules", abstract or otherwise. The decision of many case are products of irresponsible jury caprice and prejudice' (191), since '... Usually the jury are neither able to, nor do they attempt to, apply the instructions of the court' (185). In the final analysis, the problem for Frank derives from the fact that '[adequate fact-finding] requires devoted attention, skill in analysis, and, above all, high power of resistance to a multitude of personal biases. But these qualities are obviously not possessed by juries' (192).

process of juridical decision-making, I intend to focus on a notorious controversial argument: determining the nature of juridical reasoning. Whether it is directly a product either of reason or of emotion, and, whether it is a result of the interaction of both, and, in that case, how such interaction works.

Beside the strong influence exercised by the rationalistic approach of Immanuel Kant's thought, which still at present significantly influences the most traditional conception of law throughout the western world (at least on the Continent), we are all aware of the increasingly significant presence of a completely different point of view in the light of which emotion is given back the role it should play. From this point of view Neuroscience seems today to put itself on the same line of thought initiated in the mid 1930s by the American Legal Realism whose deeply innovative positions we have previously mentioned.

The Realist School brings to extreme consequences the approach where emotion—or intuition—and not reason governs juridical judgment. This reaches disruptive assertions,[6] which in any case can be credited for giving law scholars food for thought, since, even today we rely too often on rigorous distinctions between what is 'rational' and what is not, and on simplistic juxtapositions between what belongs to the emotional world and expressions of pure rationality, or—as it is preferable referred to nowadays—'logic'.

Today we know that the development of cognitive function requires an extremely complex mental process. Nevertheless, what we are interested in verifying is how jurors reach the decisions they make, and more specifically—in the light of the outcome of the research conducted by cognitive neurosciences—whether it is possible to state that emotion influences the decisional process and, should this be the case, whether such contribution either facilitates or hinders the decision itself. Particular attention will be paid to the problem of the so-called Emotional Evidence, whose solution can be greatly helped by research and techniques developed by neuroscience.

We already anticipated that the most recent neuroscientific studies produced a significant result in regards to the understanding of the nature of decisional processes: they in fact acknowledge an important function to emotions, which work alongside reason as an integrated whole when called upon processing a rational thought inherent to a decision.[7] Therefore, in this sense, we may almost consider definitely overcome—at least in reference to Anglo-Saxon scientific culture—the approach aimed at disowning, or rather denying, the contribution given to decisions by emotions. From this perspective, what I will focus on is how the decisional processes characterize criminal proceedings.

This point of view is considered an outcome of the approaches that are still prevalent in many countries, such as Italy for example, within which rationality is premise. Therefore, in this case, the aim of juridical reasoning and of judicial proceedings is to adopt a mechanistic view of the judicial decision. According to

---

[6] Note 5, above.
[7] M. Koenigs, L. L. Young, R. Adolphs, D. Tranel, F. Cushman, M. Hauser, and A. Damasio, 'Damage to the Prefrontal Cortex Increases Utilitarian Model Judgements' (2007) 446 *Nature* 908–11.

this traditional approach, based on the criteria that the carrying out of a trial is deemed a product of rational logic, the subject summoned to decide (judge or jury), should restrict himself to applying an abstract provision to an ascertained fact, accurately avoiding, as if it were actually possible, any conditioning by emotions.[8]

21.2 Jury Instructions and the Persistent 'Demonization' of Emotions

It is a generally shared opinion that emotions play an important role in decision-making, even though it is a strongly rooted belief that emotions hinder rational reasoning and moral judgements.[9] The guidelines that the judge[10] usually gives to the jury ('Jury Instructions') confirm such statements. In fact, in US law it is deemed appropriate to recommend to the jury panel to avoid being influenced by their emotions in any way, because there is clearly the conviction—this is the implicit conceptual assumption—that the pursuit for a balanced judgement on a subject's conduct could be diverted by the contribution that emotions may possibly give to the decisional process. Therefore, the belief that emotions are deceitful, once again emerges. In fact, according to such an idea, emotions deeply undermine the possibility of rational thought. Throughout trials leading towards judgment (sentence, verdict) everything must be based on 'pure reason', under the unrealistic assumption that cognitive processes are closer to reality when formed without the negative conditioning produced by emotions.

Jury Instructions, while postulating the possibility of isolating the rational component of cognition from the emotional one, clearly admit what contextually they would like to deny: whenever a decision needs to be taken, emotion certainly exists. Therefore the complication arises when we attempt to erase consequences deriving from such 'interference'. The problem rests entirely in the following question: why are Jury Instructions focused on asking jurors to avoid being conditioned by emotions and not suggest a valid thought process to facilitate their decision-making? Furthermore, *how* can they *concretely* achieve this result, *how can they assure that their judgement does not depend on emotion?*

---

[8] Within the Italian system, the principle that must govern the judge at the moment of 'evaluation' of evidence is 'libero convincimento', which means, in an extremely brief manner, the absence of predetermined criteria chosen in advance by the legislator. According to the prevailing literal and systematic reconstruction of the meaning of such principle '[In its correct meaning,] "libero convincimento" requires the judge to evaluate the evidence on the basis of *rational* criteria such are those derived from logic, experience and laws of science'. So M. Nigro and P. Tonini, 'Libero convincimento' in G. Spangher (ed.), *Procedura penale*, *Systematic Dictionaries series* (Milan, Il Sole 24 Ore, 2008) 333.

[9] Kant was surely not the first thinker to deem that the nature of emotions was irrationality (for example, just think about Plato's *Phaedrus*), but surely was the one to stress the supposed irrational nature of emotions, as far as coming to the statement according to which these latter corrupt the moral judgement.

[10] For example, in New Jersey it is recommended to the members of the jury: 'It is your duty to weigh the evidence calmly and without bias, passion, prejudice, or sympathy, and to decide the issues upon the merits' (*New Jersey Criminal* (2007)). In North Dakota jurors are instructed according to the following: 'Your decision must not be influenced by sympathy or emotion' (*State Bar Association of North Dakota* (2008)).

The answer to this is as simple as it is intuitive: the reason why jurors are not guided through the difficult task of 'ignoring' emotion appearing on the path to knowledge they are about to go through, is that the request contained in the 'Instructions' is unrealistic, since, as already acknowledged by almost all fields, from economy[11] to strict finance[12] to broader social studies,[13] emotions play a fundamental and insuppressible role in the decision-making process. It would be, in any case, very difficult, if not totally impossible to demand from the subject summoned to take a decision to separate the two components, emotional and rational, which both contribute integrally although in different ways to the decision-making process and to the expression of a final judgment.

What we would like herewith to highlight is that emotions not only represent a relevant and insuppressible part of the cognitive process, but that, most of all, they considerably contribute to the success of the process, as demonstrated by the studies—herewith below reported—conducted by neuroscientists and neurobiologists on the decisional capability of subjects who have suffered certain kinds of brain damage.

21.3 The Contribution Given by Neuroscience to the Understanding of the Relationship Between Emotion and the Decision-Making Process

According to the rationalistic approach, as well as to common sense, it is always necessary to exclude emotions, which are by definition a corrupting, or at least a contaminating element for thought in order for reasoning to be considered satisfying and be capable of achieving optimal results. On this basis—as appropriately underlined by Antonio Damasio[14]—it is possible to find Descartes' dualistic elaboration, according to which the mind, or thought, is separate from the brain and the

---

[11] M. Lindstrom, *Buyology: Truth and Lies about Why We Buy* (New York, Doubleday, 2008); *Brand Sense: Build Powerful Brands Through Touch, Taste, Smell, Sight, and Sound* (New York, Free Press, 2005).

[12] D. Salzman and E. Trifan, *Emotions, Bayesian Inference, and Financial Decision Making* (unpublished manuscript, 2005).

[13] Nowadays there is no field immune from the consideration of the relevance to be given to the human brain's emotional component: as one example, we would like herewith to remind that today public social spaces in general, as well as the most commercial ones are to the public's liking—as demonstrated by the studies—when the architecture of the environment manages to stimulate the user's *sensitivity* and therefore to conquer his emotional side (in this regard, we talk about 'emotional' architecture or design or 'neuro–design'); in the same way, in order to sell a product it is now accepted that more than the cost–quality ratio, it counts the marketing strategies' capability to move the client's emotions ('neuro-marketing' or 'emotional marketing'). These last years' focus on the importance that—on the basis of neuroscientific studies—must be acknowledged to emotion, is so evident that it took the features of some kind of revenge of emotion on characteristics that were once preeminent, so that someone even talks about a sort of '*Neuro-mania*': P. Legrenzi and C. Umiltà (Bologna, Il Mulino, 2009).

[14] A. Damasio, *Descartes' Error: Emotion, Reason, and the Human Brain* [*L'errore di Cartesio*] (Milan, Adelphi, 2008) 336.

body. Theoretically, therefore, the mind that traces back to all human skill and abilities relevant to logic and reason (*res cogitans*)—is actually located outside the body, which in fact represents the so-called *res extensa*.[15] In this perspective, reasoning and moral judgement, as well as all information processing conceived by the human mind, are presumed to be distinctly separate from the structure and function of a biological organism, from which, on the contrary, emotion and passion derive.

Neuroscience, having the need to rely on neurobiology, neuroanatomy, neurophysiology, neurochemistry, together with those disciplines relating to the mind as a specialized subject, in order to understand the mechanisms governing its function, questions the following postulate.[16]

The hypothesis proposed by Antonio Damasio, in particular, is that, of the existence of the so-called somatic marker, which, 'by making more efficient and precise...the decision-making process',[17] represents the link between cognitive and emotional processes. The somatic marker is a hypothesis referring to the phenomenon theorized by neurologists. When given a stimulus, emotionally associated to a given circumstance, the brain produces (*marks*) a signal of activity, either conscious or unconscious, positive or negative; hence the definition: 'somatic marker'. Neurological patients suffering brain damage of a specific region of the frontal lobe—a topic of research from Damasio's group—presented a lack of decisional capacity associated with considerable alteration in the capability to have feelings. From clinical and experimental observation of several cases in a twenty-year span, the group was able to obtain elements supporting the hypothesis that reason is perhaps less 'pure' than we think or hope it would be.[18] The conclusions reached do not surely confirm that feelings and emotions guide our thought and actions, or that humans are not rational beings. Nevertheless, what has been confirmed once and for all is that 'certain aspects of emotions and feelings are indispensable for rationality'.[19]

---

[15] Reference is made to what is considered as the heart of the Cartesian thought, '*Cogito, ergo sum*': 'I thence concluded that I was a substance whose whole essence or nature consists only in thinking, and which, that it may exist, has need of no place, nor is dependent on any material thing; so that "I," that is to say, the mind by which I am what I am, is wholly distinct from the body, and is even more easily known than the latter, and is such, that although the latter were not, it would still continue to be all that it is': R. Descartes, *Discours de la méthode* [*Discorso sul metodo*] (Milan, Oscar Mondadori,1993) .

[16] '...a full understanding of the human mind requires an integrated perspective: the mind should not only start from a "cogito" not belonging to the realm of biological tissues, but should also be related to an entire organism, having a brain and an integrated body and fully interacting with a physical and social environment. But the mind really soaked into the body as I see it, does not abandon the most refined levels of activity, those constituting its soul and spirit. From my point of view, soul and spirit, in all their dignity and human measure, are now statuses, complex and unique of an organism'; and more, '...I am not affirming that the mind is into the body. I am affirming that the body's contribution to the brain is not limited to regulatory effects or to the support of all vital functions, but also includes *contents* which are an integral part of the functioning of the normal mind': A. Damasio, note 14, above, 341, 309.

[17] Ibid. 245.
[18] Ibid. 18.
[19] Ibid. 19.

Through the hypothesis of the somatic marker, Damasio not only relates cognitive contents[20] to emotions, but acknowledges their determining role as a cognitive *guide*.[21]

According to Damasio, if it were realistically possible to accept Decartes' approach, and if formal logic, product of pure rational thought uncontaminated by passions, which represents the best possible hypothesis of mental elaboration, then subjects suffering prefrontal brain damage, presenting a reasoning deficiency related to areas of the brain governing emotions, should represent examples of perfection regarding logical reasoning and decision-making.

Unfortunately, the studies performed demonstrate that it is not so, and those subjects, whose function of the brain area governing emotions is impaired presented on the contrary a disability in their decisional capability. In other words, pure reason presents disabling limits,[22] contrary to Kant and those sustaining the absolute necessity that the basis for the pursuit of justice judgements should not be contaminated by the bias of emotions.

David Pizarro, when imagining the existence of a rational Kantian mind, recalls a very suggestive analogy: 'the Vulcan' of the *Star Trek* series. Vulcans are completely rational humanoids; nevertheless, they are incapable of feeling emotions. With regard to moral judgements, Vulcans—states Pizarro—by applying the general theory of principles, would be surely able to formulate very accurate ones. However, the dilemma of the insufficiency of having a purely rational mind would emerge in the insuperable incapacity of having such a thought or the awareness of a morally significant fact. In other words, Vulcans would not be able to emotionally conceive nor contextualize a situation and this impossibility to feel emotions[23] would paralyse thought and therefore impair action. This is what actually happens to patients suffering prefrontal brain damage, which from a neurological point of view might seem normal, because they are not affected by any speech impediments, impaired movements, odd sensations, memory loss, logical and mathematical disabilities, but are nevertheless incapable to feel, and therefore *to reason*.

The aspiration for a universal behavioural standard, which emotions would hinder,[24] is what seems to be conditioning moral philosophers in the elaboration of a justice judgement model. However, they forget that for human beings, as well as for all other animals, well before the pursuit of a standardized behavioural code comes the certainty of their undeniable limits. The concreteness of subjectivity prevails on the abstractness of universality. In this sense, the significant discovery is that:

---

[20] Also the philosopher R. De Sousa (*The rationality of emotion* (Cambridge, MA, MIT, 1987) amongst others, deems and maintain that emotions have contents and nature intrinsically rational.
[21] 'Emotions are not a luxury; they play a role in communicating meanings to others, and may also fulfil the function of cognitive guide': A. Damasio, note 14, above, 191.
[22] Ibid. 271; D. Pizarro, 'Nothing More than Feelings? The Role of Emotions in Moral Judgments' (2000) *Journal for the Theory of Social Behavior* 371.
[23] Pizarro, in this regard, talks more specifically of 'empathy'.
[24] J. B. Rawls, whose theory of distributive justice is more than well-known, hypothesized the necessity for the so-called '*veil of ignorance*', beyond which justice statements should be formulated, with the aim to annul the influence exercised on judgements by the bias of emotions.

the rationality apparatus, traditionally deemed *neo*cortical, does not operate without the apparatus of biologic regulation, traditionally deemed *sub*cortical. It seems that nature built the first not only on top of the second, but also *with* this and *from this*... Neocortex results engaged *together* with the most ancient brain nucleus, and rationality is the effect of their joint activity.[25]

In other words, the systems governing the normal processes of emotion, feelings, reason, and decisions-making are deeply interconnected,[26] and, as matter of fact, reduction of the emotional experience may lead to a significant reason for irrational behaviour.

21.4 'Emotional Evidence'

Studies conducted by juridical and forensic psychology[27] demonstrate that a given category of evidence, usually admitted and considered usable for decision-making purposes,[28] influences the jury towards a bias decision. Contemporary North American doctrine, considering such a hypothesis, talks about 'Emotional Evidence', in order to indicate evidence that, regardless from being presented by either the prosecution or the defence, is characterized by a high emotional content.

In order radically to summarize the results obtained by such research, we may confirm that it has been substantially demonstrated that the so-called Emotional Evidence is potentially suitable to lead the same jury to issue *more severe punitive* judgments, being irrelevant from where such evidence was introduced (whether presented by the defence or the prosecution: 'indiscriminant punitiviness'[29]). All existing literature on the effects on the jury's decisions, for example the showing of particularly gruesome or post-mortem photographs related to the crime they must judge, or of the hearing of victim impact statements, especially in death penalty trials, demonstrate that whenever such evidence having high emotional content (Emotional Evidence) has been admitted in mock trials, the verdicts have ended in

---

[25] A. Damasio, note 14, above, 188–9.
[26] Ibid. 97.
[27] For example B. Myers, D. Godwin, R. Latter, and S. Winstanley, 'Victim Impact Statements and Mock Juror Sentencing: The Impact of Dehumanizing Language on a Death Qualified Sample' (2004) *American Journal of Forensic Psychology* 40. For a rich and interesting reference bibliography, reference should be made to J. Salerno and B. Bottoms, 'Emotional Evidence and Jurors' Judgments: The Promise of Neuroscience for Informing Psychology and Law' (2009) *Behavioral Sciences and the Law* 273.
[28] The Supreme Court in *Payne v Tennessee* (1991) stated that in the sentencing phase of a death penalty trial, the victim impact statements—representing a classic example of Emotional Evidence—can in no way damage the defendant, since they are only functional to determining the level of the offence perpetrated (thus, without straying from the so-called probative effect). In a more recent sentence (*Kelly v California* (2008)), however, the same court, expressed a negative opinion on the opportunity to admit video tributes to victims, since they would not be useful from a probative point of view, being only 'emotionally evocative'.
[29] J. H. Goldberg, J. S. Lerner, and P. E. Tetlock, 'Rage and Reason: The Psychology of the Intuitive Prosecutor' (1999) 29 *European Journal of Social Psychology* 781–5.

conviction or, in any case, resulted with a more punitive sentence in a significantly higher number of cases.

Therefore, under certain circumstances there are cases in which emotions, vital to the decision-making process, may actually corrupt judgement. In this way the issue becomes that of determining as accurately as possible, the fine boundaries of Rule 403 of American Federal Rules of Evidence[30] which lie between the 'probative' and the 'prejudicial' use of evidence.

Even if no doubt exists in the fact that it is inevitable that during a death penalty trial the jury may examine evidence that may also be very 'emotionally disturbing', the question that we must ask is: should the use of Emotional Evidence be regulated or is such regulation unnecessary? Of course, it is important for the jury to know how the victim was killed, because we have to know whether one deadly blow was inflicted or the perpetrator of the crime tormented the victim's body with 100 stabs; therefore, the admission of photographs aimed at giving ground to a reconstruction instead of another can be considered probative. But the question is: is it necessary to view fifty photographs, or are three photographs to be considered sufficient? From our point of view it would be appropriate to take provisions concentrating on regulating, in a more rigourous manner, the admission of evidence falling under such category.

We were talking about *probative* and *prejudicial* use of evidence. In other words, this means to verify when and under which conditions the emotional aspect of a decision-making process prevails. It is evident, in the case of Emotional Evidence, that we no longer deal with the physiological interaction between rationality and emotion, but we are clearly entering a different field: that of the pathology of such relationship.

Once again neuroscience can assist law. Its contribution plays a determining role in the acknowledgment of the necessity of emotions within the decision-making process. In the same way its contribution can be illuminating[31] towards verifying whether an increased or prevailing influence of emotions exists. These processes must be necessarily linked to a decreased 'cognitive effectiveness' and, therefore, ultimately provide an aid to judges in the interpretation of Rule 403.

Through the fMRI technology, which substantially measures blood flow in areas of the brain during neurological activity, it is possible to relate brain activity recorded in a given area to the cognitive activity the subject is carrying out. In this way, it is possible to record different brain activity in the examined subjects, according to their being exposed to less or more intense emotional stimulation. The results obtained so far by neuroimaging studies[32] demonstrate that a more intense

---

[30] 'Although relevant, evidence may be excluded if its probative value is substantially outweighed by the danger of unfair prejudice, confusion of the issues, or misleading the jury, or by considerations of undue delay, waste of time, or needless of cumulative evidence.'
[31] Mostly through the use of the functional magnetic resonance imaging (fMRI). For a deeper analysis of the issues related to the problem of Emotional Evidence, see J. Salerno, B. Bottoms, note 27, above, 273.
[32] Amongst others, we highlight those from J. D. Greene, B. R. B. Sommerville, L. E. Nystrom, J. M. Darley, and J. D. Cohen, 'An fMRI Investigation of Emotional Engagement in Moral Judgement'

emotional stimulation produces brain activity that is more significant in those brain areas responsible for emotional reaction and less activity is detected in areas related to cognitive activity in general. For this reason it is probably time to conduct new research and empirical experiments on the matter with a specific focus on legal decision-making, which until now has not been the subject of specialized study, with the goal to either confirm or prove false the results obtained so far.[33]

21.5 Conclusions

As a result of the previous account, the growing potential of neuroscience's effectiveness in the redetermination of numerous juridical categories emerges with great strength, although it will probably, in the near future, be revised in order to accommodate new discoveries.

The importance of the studies conducted in these fields is therefore seemingly taken for granted. Nevertheless, it seems appropriate briefly to highlight at least two aspects related to this matter: on one hand, the limits of neuroscience; on the other, the risks that may arise.

In the first place, even if the strong bond between the mind and the brain is now clear, it would be in any case arbitrary to confine the mind to brain function. In particular, the main problem, which researchers themselves are aware of,[34] derives from the fact that to verify, thanks to technologies like the fMRI, that a certain area of the brain is activated by a given activity performed by the subject, is one thing. But, it is a very different task that of deducing brain function and linking it to brain activity of its relative area: 'Just because we can see activation of particular areas of the brain with fMRI technology does not mean we understand specifically the function of those brain areas'.[35]

Secondly, it would be wise not to underestimate the effects of the use of neuroimaging evidence that may influence the jury's decision-making process; such effects may lead the jury to lose its primary and essential duty: to be the judge of the fact.

Many of the tools that may be made available in the near future to the judicial system (e.g. neuroscientific lie detection; studies aimed at demonstrating the different *brain* configuration and predisposition of certain subjects that develop antisocial behaviour—see the neuroimaging and scientific evidence in general) are considered to be trustworthy by the jury, and therefore present a risk of actually depriving them of their actual role.

(2001) 293 *Science* 2105–8 and H. R. Heekeren, I. Wartenburger, H. Smidt, K. Prehn, H. Schwintowski, and A. Villringer, 'Influence of Bodily Harm on Neural Correlates of Semantic and Moral Decision Making' (2005) 24 *NeuroImage* 887–97.

[33] J. Salerno and B. Bottoms, note 27, above, 273 ss.; O. Goodenough, K. Prehn, 'A Neuroscientific Approach to Normative Judgment in Law and Justice' in S. Zeki and O. Goodenough (eds.), *Law and the Brain* (Oxford, Oxford University Press, 2006) 97–8.

[34] Very recently, P. Legrenzi, and C. Umiltà, note 13, above.

[35] J. Salerno and B. Bottoms, note 27, above.

In fact, the studies carried out demonstrate that when neuroimaging evidence is introduced, for example by the defence, in order to demonstrate the insanity of the defendant, the jury tends to support the relevant request with significantly increased willingness: the jury basically trusts the 'expert witness' presenting 'evidence regarding the defendants' brain abnormalities in murder trials' to the point that their role as 'finder of the fact'[36] is affected. Therefore one can only imagine the direction[37] this may lead us to, not only relative to juries, but finally toward a shift in our conception of the nature of free will in general.

---

[36] B. Garland (ed.), 'Monitoring and Imaging the Brain' in B. Garland, *Neuroscience and the Law* (New York—Washington D.C., Dana Press, 2004) 21: 'The evaluation of witnesses, and the credibility of and weight given to their testimony, are matters for the "finder of fact", the body charged with determining the facts in the matter before the court. In jury trials this is the jury; in bench trials, it is the judge. In allowing scientific testimony regarding truthfulness into evidence, the court may well be invading the purview of the jury... Members of the jury may weight "scientific evidence" more heavily than their opinion as formed by their own sense, and may do so specifically on the matter of truth.'

[37] M. Gazzaniga, *The Ethical Brain* [*La Mente Etica*] (Turin, Codice Edizioni, 2005) 86.

22

Possible Legal Implications of Neural Mechanisms Underlying Ethical Behaviour

*Donald Pfaff and Sandra Sherman**

This paper will suggest that recent work in neuroscience pointing towards a physical/hormonal basis for moral reciprocity—the 'do unto others' dictum commonly called the Golden Rule—may have implications for how legal concepts have developed and should be applied. We start from the assumption, however, that while neuroscience can now perhaps demonstrate that moral reciprocity is the product of how human brains have evolved, it would be facile to argue that the law simply reflects this evolution, and incorporates (or should incorporate) a 'do unto others' ideology into its basic, jurisprudential structure. As Portia so cunningly proves in *The Merchant of Venice*, 'the quality of mercy' is just one factor among many that is weighed by judges before reaching a legally acceptable decision.

In modern American jurisprudence, the Golden Rule has no place in a courtroom. Indeed, courts will routinely reverse a decision where one party was permitted to argue that members of a jury should put themselves in the plaintiff's place when deciding damages. In one important case, *Whitehead v Kmart Corp.*,[1] the court rejected the linguistically archaic formulation of moral reciprocity—the 'do unto others' dictum—so as to underscore how alien the notion is to a modern civil proceeding: 'The Fifth Circuit has forbidden plaintiff's counsel from explicitly requesting a jury to place itself in the plaintiff's position and do unto him as they have them do unto them.'[2] Thus, while *Whitehead* acknowledges the phenomenon that we propose here, i.e. that humans tend towards empathy, the case utterly banishes it as inimical to the formal administration of justice.

---

  * Donald Pfaff is Professor and Head of the Laboratory of Neurobiology and Behavior at The Rockefeller University, a Fellow of the New York Academy of Sciences, a member of the Advisory Board of the National Academy of Sciences, a Fellow of the American Academy of Arts and Sciences, and serves on the editorial boards of numerous scholarly journals. Sandra Sherman is Assistant Director of the Fordham Intellectual Property Law Institute, and formerly Professor at the University of Arkansas and Georgia State University, fellow of the National Endowment for the Humanities and the Andrew Mellon Foundation, Visiting Fellow of Lucy Cavendish College, Cambridge University, and the Institute for Research in the Humanities at the University of Wisconsin-Madison, Visiting Scholar at Harvard.

  [1] 163 F 3d 275 (5th Cir 1998).
  [2] Ibid. at 278.

We plan to provide evidence suggesting that the court's position in *Whitehead* contravenes how we now understand the brain, i.e. it flies in the face of how humans are wired. Consider the facts. In *Whitehead*, a woman and her teenage daughter sued a department store after two assailants abducted and robbed them in the store's parking lot. After the abduction, the assailants took turns sodomizing the woman outside her vehicle while the other held the daughter inside. At trial, plaintiff's counsel argued to the jury that:

The incident took approximately two hours from when they were abducted to when they were released. And I calculated it, and that's 7,200 seconds. And I want for you to just for a couple of seconds to see—when I say start, that's ten seconds.
Ten seconds.
And can you imagine how it would feel to have a knife in your side or a knife on your leg or a pistol on your neck for ten seconds?[3]

In overruling the lower court, the Fifth Circuit held that 'even assuming he was not explicitly invoking the Golden Rule, counsel was clearly inviting the members of the jury to put themselves in the place of the plaintiffs when deciding damages'.[4] The fact that the jury was asked to empathize—to do what juries often do without being encouraged—was grounds for reversal. It is as if, in putting a price on what those women endured, the jury was supposed not even to think about how they themselves might have felt during the ordeal.

What we think the law should consider is whether, given our current understanding of the brain, such wholesale dismissal of human empathy is still appropriate. Is there room for empathy in a courtroom dealing with issues that are usually left to actuaries? Should neuroscience be dismissed as 'fuzzy' science when stacked up against hard financial calculation? At present, most courts would seem to say yes. In another damages case, this time involving the sexual harassment of firemen, the court relied on *Whitehead* to throw out a decision in which financial awards calculated down to the exact penny were found tainted by the jury's response to a Golden Rule argument.[5]

What scares courts about relying on emotion in damages cases? As the court said in *Moody v Ford Motor Co.*,[6] 'A Golden Rule argument is "universally recognized as improper, because it encourages the jury to depart from neutrality and to decide the case on the basis of personal interest and bias rather than evidence."'[7] The court sets up a dichotomy: empathy on one side (bad) vs neutrality/rationality on the other (good). From a neuroscientific perspective, it would have been interesting to ask jurors whether they felt comfortable being told that their feelings were out of

---

[3] Ibid. at 278.
[4] Ibid.
[5] See *Alexander v City of Jackson* 2008 US D Lexis 29879 (S D Miss 2008).
[6] 506 F Supp 2d 823, 836 (N D Okla 2007).
[7] The court cited *Blevins v Cessna Aircraft Co.* 728 F 2d 1576, 1580 (10th Cir 1984). See also *US v Moreno* 947 F 2d 7, 8 (1st Cir 1991), observing that a Golden Rule argument 'improperly 'encourages the jury to depart from neutrality and to decide the case on the basis of personal interest and bias rather than on the evidence' (quoting *Forrestal v Magendantz* 848 F 2d 303, 309 (1st Cir 1988). See also *Dole v U.S.A. Waste Services, Inc.* 100 F 3d 1384, 1388 (8th Cir 1996).

bounds in this case, in which a young man was killed in a rollover accident, and his parents' counsel appealed to the jury's own sense of personal fragility:

> It's going to happen. It might happen when you're on your way to school. It might happen when your mom takes your kids to day care. It might happen when you're in a rush to work. It might happen because a child runs in front of you, and you try to avoid it. It's going to happen.[8]

In the dock, is this nightmare not what everyone was quietly confronting? How could they not have been? What we want to present is evidence that courts might at least consider when examining how juries are motivated and whether those motivations are fair to all parties.

According to the *Moody* court, empathic feelings have no place in awarding damages because comparing oneself to the victim is *ipso facto* unfair: 'Plaintiff's counsel's remarks in this case went to the likelihood that jurors would suffer a personal injury similar to that suffered by plaintiff's decedent..., and he asked the jurors to award damages as if they had been personally harmed.' Why should the law diverge so much from how people naturally react, especially since we now *know* that the brain pushes us in the direction of empathy? Is there something special about the law such that it *should* require us to forgo empathy? The *Moody* court starts from an assumption, which is suspect from a neuroscientific point of view, that reason and disinterest are corrupted by identification with another human being—at least in so far as we are called upon to award that person monetary damages: 'Because of the prejudicial nature of these arguments and the likelihood that these statements aroused the passions of the jurors, plaintiffs' counsel's Golden Rule arguments should be considered as a factor in ruling on the fairness of the trial as a whole.'[9]

We take the position that normal people—the kind that normally get onto juries—naturally shrink from doing harm. In a legal context, where harm is ritualized and can include failure to compensates adequately someone for a wrong that he has suffered, we think it is *natural*—and potentially desirable—that jurors be permitted to at least register feelings of personal identification. An appropriate degree of leeway in this regard could mitigate harm that might otherwise be done through a failure to empathize.[10] The neuroscientific discussion in this article is meant to justify this recommendation, and to demonstrate that splitting legal processes off from natural human empathy may take too harsh a view of empathy. Empathy is not mere, uncontrolled passion, but rather an instinct for avoiding another's harm. Viewed in this context, is there harm in seeking to avoid harm?

In some connections, legal institutions already take account of our instinct to identify with others, and to trust that because we extend ourselves they will reciprocate. Every society since the Phoenicians has had some form of contract

---

[8] Ibid.
[9] Ibid. at 837.
[10] An appeal to jurors' empathy is not always reversible error, since other factors can mitigate that appeal. See *Lovett v Union Pacific Railroad Company* 201 F 3d 1074 (8th Cir 2000) and *Doorbal v McNeil* 2008 US Dist Lexis 74939 (2008).

law, in which individuals undertake to satisfy each other's expectations. While the law has built up extensive mechanisms to compensate for non-performance, and in some cases to ensure that the parties operate in good faith, the fact is that no procedures for reciprocal exchange could operate in the absence of trust. Contract law is built on the assumption that I would *prefer* your performance to any form of compensation, and that you *understand* my preference sufficiently so as to endeavour to meet it. In the law, a contract is 'the meeting of minds', not just intellectually but also in terms of a shared will to perform.[11] This sharing is antecedent to any formal legal arrangement, and makes such arrangement plausible, indeed possible. In other words, the legal institution of contract is a constructive example of how mutual identification can operate constructively within a legal framework.

There is a substantial body of scholarship on the role of good faith in contract law—what it means, where it should be applied, whether it can be enforced.[12] Our evidence suggests, however, that the very notion of contract, that is of mutual reciprocal undertaking, could not have developed (and could not operate practically) unless people believed in the basic responsiveness of others. That responsiveness, we argue, emanates from an identification among contractual partners sufficient so that each partner will be motivated to avoid harming the others. In practical terms, that means that if you hire me to perform some task, I will actually try to do it. Just like you, I will want to maintain my reputation in the community, and I will want to earn money rather than owe it. If legal institutions can take on board these basic notions, now reinforced by brain research, then perhaps this type of understanding is possible in other areas of the law.

* * *

When individuals behave toward each other in a civil and thoughtful manner—as for example, when they perform the mutual obligations entailed in a contract—what brain mechanisms regulate their behaviour? Here we propose a mechanism operating in the Central Nervous System (CNS) in such a manner as to produce thoughtful, cooperative, and even altruistic behaviours. The first part of this paper proposes that the widespread ethical principle called the Golden Rule depends in part on neurobiological mechanisms for fear. A theoretically efficient possible neuroscientific explanation of reciprocal altruism posits that one person's envisioned fates and fears are merged with another's, producing a *loss* of personality-defining information and consequent blurring of identities.[13] To that set of

---

[11] See *Roland v Transamerica Life Ins. Co.* 2009 US App Lexis 14179 (5th Cir 2009) at 5, holding that for a contract to exist there must be a shared intent: 'Roland further argues that he, Transamerica, and a physician actually agreed on an alternate plan of care. Roland bases this argument on the fact that he submitted a plan to Transamerica, and Transamerica countered with a proposed plan of its own. But that was it. A proposal and a counter-proposal do not a meeting of the minds make.'

[12] See Jack Beatson and Daniel Friedman (eds.), *Good Faith and Fault in Contract Law* (Oxford, Clarendon, 1995), Roger Brownsword, Norma J. Hird, and Geraint Howells (eds.), *Good Faith in Contract: Concept and Context* (Dartmouth, Ashgate, 1999), A. D. M. Foote (ed.), *Good Faith in Contract and Property* (Oxford, Hart Publishing, 1999); Reinhard Zimmerman and Simon Whittaker (eds.), *Good Faith in European Contract Law* (Cambridge, Cambridge University Press, 2000). See also J. F. O'Connor, *Good Faith in International Law* (Dartmouth, Ashgate, 1991).

[13] Donald Pfaff, *The Neuroscience of Fair Play* (Washington, Dana Press, 2007).

mechanisms are added the mechanisms for affiliative behaviours. Finally, we confront the problem of aggressive, violent behaviour. We suggest that the capacity of a person to behave according to the Golden Rule depends on a balance among social behaviour mechanisms in which those producing prosocial behaviours, that is behaviours beneficial to others, outweigh those producing antisocial behaviours, that is, behaviours that harm others. In this paper we refashion a summary of the theory to render it relevant to the relation between neuroscience and the law.[14]

This neuroscientific theory is not intended to deny the existence of evil behaviour or to solve courtroom disputes. Instead, it provides a parsimonious idea of neural mechanisms underlying one essential step in the production of reciprocally altruistic responses that potentially can aid courts and juries in sorting out complex questions in at least some areas of legal dispute.

22.1 Nervous system mechanisms producing behaviour that obeys the negative form of the Golden Rule

One aspect of ethical behaviour is to refuse to act towards other humans in a manner which, if such an act were directed toward you, you would be afraid.[15]

Fear mechanisms have been comprehensively reviewed and therefore will be described here very briefly.[16] Both unlearned and learned fear depend on groups of neurons collectively named the amygdala. Its neuroanatomical connections are well understood. Fear signals reach the amygdala both directly and indirectly, through the thalamus and cortex. Efferents from the amygdala reach the frontal cortex, septum, hippocampus, hypothalamus and central grey.

In terms of neurochemical inputs, one of the most important influences on fear signalling is the excitatory, rapidly acting neurotransmitter glutamate. When glutamate binds to its receptors, calcium channels are opened, triggering calcium-dependent kinase cascades in which intracellular signalling is accomplished by chains of reactions that operate by adding phosphate groups to proteins in such a manner as to alter their shapes and electrical charge. A second neurochemical example uses the neuropeptide Brain Derived Neurotrophic Factor (BDNF), known to participate in stress and fear. BDNF binds to its specific receptor on an amygdaloid neuron to trigger a different set of phosphate-adding reactions from the glutamate-triggered reactions summarized above. One of the most important neurochemical outputs from the amygdala is the neuropeptide Corticotrophin Releasing Hormone (CRH). CRH not only has stress and fear-related actions in the CNS but also, through the hypothalamic/pituitary/adrenal axis, causes the secretion of steroidal stress hormones such as cortisol (in a rat, a similar hormone, corticosterone) that help the entire body to deal with stress and fear.

---

[14] Donald Pfaff, M. Kavaliers, and E. Choleris, 'Biology of the Ability to Behave Ethically' (2008) 8 *American Journal of Bioethics-Neuroscience* 10–19.
[15] Ibid.
[16] J. LeDoux, 'Emotion Circuits in the Brain' (2000) 23 *Annual Review of Neuroscience* 155–84.

The effects of cortisol on behaviour depend upon whether or not adrenal steroid hormone levels have been elevated chronically and repeatedly. If not, the effect of cortisol effect is restorative. It brings the body's systems from an emergency state back toward a normal state. However, crucially, if the animal has been subjected to chronic fear and stress such that adrenal hormones would have been called on again and again, then cortisol steroidal stress hormones can actually amplify subsequent behavioural responses to fearful stimuli.

Note that these fear mechanisms in the amygdala do not work effectively if the CNS has not been aroused. Norepinephrine and dopamine-containing axons coming from the brainstem reach the amygdala and constitute two parts of the ascending systems supporting generalized CNS arousal.[17] In turn, Roozendaal and McGaugh have reported that the proper operations of amygdaloid mechanisms related to fear depend on these synaptic inputs that release norepinephrine and dopamine.

In addition to these fear mechanisms to which we will refer in the section below, new support for an important role of the amygdala in the generation of altruistic behaviours comes from the observation that children who grow up having suffered bilateral damage to the amygdala have difficulty in behaving in a socially appropriate manner.

With this brief review in mind, the stage is set to theorize how fear mechanisms come into play in ways that foster ethical behaviour by avoiding acts of violence.

22.2 A Four-Step Neuroscientific Theory: Shared Fears Underlie a Universally Ethical Behaviour Pattern

The best explanations of complex phenomena are the most parsimonious. It is possible to achieve this type of scientific explanation of reciprocal altruism, the Golden Rule behaviour described above, by invoking the primitive brain mechanisms governing fear that were summarized. The explanation will not require fancy tricks supporting learning and memory but instead will invoke the easiest step of all: the *loss* of information. This hypothetical explanation proceeds in four steps.

First, consider a person's, M's, action toward another, N. Before this act occurs, it is represented in M's brain, as every act must be. Motor acts being represented in one's own brain, so-called 'corollary discharges', were conceived first in 'reafferenz theory' and supported by a large body of experimental data.[18] Action representation to one's own brain remains of current interest in neuroscience.

Second, this act will have consequences for N individual which M can predict and envision. Then comes the crucial step.

Third, to achieve a feeling consistent with 'Golden Rule' behaviour, M *blurs* the difference between the other individual and himself to an abstract intermediate image. For example, in terms of face recognition, neurons and inferotemporal cortical regions specialized for that function are well-documented. Mechanisms for blurring,

---

[17] Reviewed in Donald Pfaff, *Brain Arousal and Information Theory* (Cambridge, MA, Harvard University Press, 2006).

[18] R. Held and S. Freedman, 'Plasticity in Human Sensorimotor Control' (1963) 142 *Science* 455–62.

besides simply reducing cortical neuron reliability, include adding noise to the mechanism or altering temporal phases of inputs. Adding markedly to excitability in the cerebral cortex allows for a *cross-excitation* from one represented image to another. As a result, instead of seeing the consequences of his act solely for the other individual, M sees them for himself. As an example posed for absolute clarity, if M had been planning on knifing N in the stomach, he *loses the difference* between N's body and his own. This loss of information is easy to posit because any one of the many steps required for the neurobiology of fear, by its loss would provide the loss of information this theory supposes. As a result, the knifing is less likely to occur because he shares the other person's fear.

Additionally, the mirror neuron system (MNS) in the cerebral cortex may provide still another mechanism that permits a blurring of the difference between the person beginning an act and the target of that act. The MNS allows the connection between one's planned actions towards another and the intended recipient's predicted feelings that would arise as the consequence of the planned action. Once this identification with the recipient has been achieved, it can be transmitted to the amygdala through a recently documented MNS-limbic system. In fact there is an activation (as measured with fMRI) of this MNS-limbic connection while observing and imitating facial expressions of emotions in normal children, that is absent in autistic children. In summary, we have proposed several ways in which the perceived difference between the actor and the intended victim can be reduced. This blurring of personal difference would make it easier for the actor to avoid harming the intended victim.

Fourth, and finally, if the consequences of M's intended act are good for N, he does it; if the consequences for N are bad, then M does not do it.

This explanation of an ethical decision by the would-be knifer has an attractive feature. Usually we have to recognize and remember differences between ourselves and others. However, the explanation of an ethical decision given here involves only the *loss* of information, not its acquisition or storage. The learning of complex information and its storage in memory are very hard to understand. However, the *loss* of information is easy to understand, because it only requires the breakdown of any single part of the complex memory-storage processes, whether they be intricate biochemical adaptations, subtle synaptic modifications or precise temporal patterns of electrical activity. Damping any one of the many mechanisms involved in memory can explain the blurring of identity required by this explanation of Golden Rule-related behaviour.

Leaving out any one of the mechanisms involved in social recognition or memory allows us to identify with the person toward whom we are about to act. Moreover, among the theoretical mechanisms described above, the individual mechanism left out could differ from person to person and from occasion to occasion. In mechanistic terms, therefore, it is incredibly easy to achieve a sense of shared fate with another. All individuals who forget things appreciate that the loss of information occurs with ease. In the knifing example imagined above, as a result of a blurring of identity—a loss of individuality—the knifer temporarily puts himself in the other person's place. Because that person would be afraid, so will he. He avoids an unethical act because of shared fear.

As a side point, we note that the more the other person looks or sounds like us, the easier it is to blur the difference between us. A theoretical consequence of this fact is that altruistic behaviours toward kin should be easier to achieve than toward other, completely unrelated, individuals.

22.3 Mechanisms for Behaviour that Obey the Positive Form of the Golden Rule

One statement of the Golden Rule instructs us 'to love one another even as we love ourselves'.

In order to explain peoples' friendly, helpful behaviours toward each other one need not rely exclusively on their avoidance of harmful acts as has been proposed above. Positive, affiliative behaviours support ethical responses as well. Using the same logic as spelled out in the four steps described above, if we *blur the difference* between the other person and our self, then we behave well because we would want to be the recipients of positive, friendly acts.

Mechanisms underlying affiliative behaviours have received much attention.[19] Did social behaviours spring from the lap of evolution spontaneously? Most ethologists would say not. Some believe that sexual behaviours lay down the *bauplan* for a wide range of social behaviours. Hormonal, neural, and genomic mechanisms of male and female sexual behaviours are relatively well understood. Briefly, in female laboratory animals, estrogens, having bound to Estrogen Receptor-alpha (ER-alpha) in neurons in the ventrolateral subdivision of the ventromedial nucleus of the hypothalamus (VMH), facilitate the transcription of several genes and elevate the frequency of action potentials in those neurons. Outputs from VMH activate a spinal-midbrain-spinal circuit that executes lordosis behaviour. In male laboratory animals, testosterone and its metabolites bind, most importantly, to androgen or estrogen receptors in neurons of the medial preoptic area. With their activity augmented, as well, by dopaminergic inputs, these neurons are essential for both the behavioural (mounting and thrusting) and the autonomic (erection and ejaculation) functions of male reproductive performance.

Branching out from mating behaviours themselves, another obvious foundation for social affiliation lies in familial relationships. For brain mechanisms, the most heavily studied among these are parental behaviours, with an emphasis on maternal behaviour.[20] Briefly, maternal behaviour in laboratory animals depends on high levels of estrogens coupled with a sudden decline in progesterone levels. Neurons that are particularly sensitive to these hormones and whose activity is essential for the performance of maternal behaviours are found in the dorsolateral portion of the preoptic area, near the bed nucleus of the stria terminalis and the anterior commissure. Axons from these neurons travel through the medial forebrain bundle to the

---

[19] L. J. Young, 'Cellular Mechanisms of Attachment' (2001) 40 *Hormones and Behavior* 133–8.
[20] Reviewed in M. Numan and T. Insel, *The Neurobiology of Parental Behavior* (Heidelberg, Springer Verlag, 2003).

midbrain, thus to activate circuitry whose exact nature remains to be discovered. Genes involved include not only ER-alpha but also prolactin and its receptor.

A neuropeptide that is important for maternal behaviour and that also links mechanisms for sex behaviour to maternal behaviour to a wider range of affiliative behaviours is oxytocin (OT).[21] Expressed by both magnocellular and parvocellular neurons in the paraventricular nucleus(PVN) of the hypothalamus, OT serves as a circulating hormone responsible for controlling the contractility of smooth muscles (e.g. in the uterus) and also as a neuropeptide distributed in the brain through axons emanating from PVN. Some of these fibres innervate the amygdala. There, deposition of OT facilitates social recognition and interfering with OT action reduces social recognition.[22] Actions of OT in the forebrain remain within the domain of sex hormone action. Both the ER-alpha and the ER-beta gene are required for normal social recognition.

In humans, OT increases trust. Ernst Fehr and his colleagues set up a game in which volunteers played a game for money. In this game two subjects play either the role of 'investor' or 'trustee'. If the investor decides to trust the anonymous trustee and give money to him, the total amount available goes up. The trustee can either share the monetary increase caused by the investor's transfer—in which case both players end up with a greater payoff—or he can keep the proceeds. Results showed that administration of OT to the investor via a nasal spray significantly increased investors' trust, more than doubling the amount of money transferred to the investor.

The reverse is true as well. If one perceives trustworthy behaviour directed toward us, one's oxytocin levels go up. Zak and his colleagues used a game similar to that used above by Fehr. Most importantly, the index of trust shown by player 1 toward player 2 is the amount of money player 1 is willing to send to player 2. Likewise, trustworthiness of player 2 is shown by the amount of money player 2 transfers back to player 1. When player 2 received intentional signals of trust from player 1 together with the money, player 2's blood OT levels were significantly higher that player 2 controls without signals of trust. Furthermore, trustworthy behaviour by player 2, measured by the amount of money returned by player 2 to player 1, was higher in player 2s whose OT was higher. Thus, higher OT levels were linked both to more trusting and more trustworthiness.

In sum, results from neuroendocrine and neurobiological studies appear to support the inference that some aspects of humans' positive, affiliative behaviours toward each other have roots in mechanisms for more primitive maternal and even sexual behaviours. In turn, it can be argued that humans can behave in this friendly, ethical way, in part, because they *blur* the distinction between the other's identity and their own and treat the other in the way they themselves would like to be treated.

---

[21] C. S. Carter and E. B. Keverne, 'The Neurobiology of Social Affiliation and Pair Bonding' in D. Pfaff et al. (eds.), *Hormones and Brain Behavior*, vol. I (San Diego, Academic Press, 2002) 299–339.
[22] E. Choleris, S. R. Little, J. A. Mong, S. V. Puram, R. Langer, and D. W. Pfaff, 'Microparticle-based Delivery of Oxytocin Receptor Antisense DNA in the Medial Amygdala Blocks Social Recognition in Female Mice' (2007) 104(11) *Proc. Natl. Acad. Sci. U S A.* 4670–5.

22.4 Aggression and Violence

Despite the arguments above, it is impossible to ignore the transgressions of ethical principles manifest in acts of aggression and violence, both individual and at the level of organized violence. The neurobiology of aggression has been studied intensively and well-reviewed and thus will be treated here only briefly.

One prominent fact is that a disproportionate percentage of violent acts are committed by young males. Surveying statistics on an international basis, Daly and Wilson demonstrated that the lifetime curve of the murders of unrelated males by males closely followed the lifetime curve of testosterone levels in blood. Testosterone, acting both on the developing brain and in the adult, increases aggression through three biochemical routes: 1) as testosterone itself; and 2) following conversion to dihydrotestosterone, in both cases involving binding to the androgen receptor; and 3) following aromatization to estradiol, thus involving binding to estrogen receptors. In humans and laboratory animals, certain anabolic androgenic steroids, chemically modified to delay chemical breakdown and thus prolong muscle-building effects, can heighten aggression and even violent behaviour.

In the brain, important sites for the facilitation of aggression by testosterone and its metabolites include the bed nucleus of the stria terminalis, the amygdala and the septum. One of the neurochemical systems involved is the production and release of vasopressin, but the full range of the neurochemical effects of testosterone remain to be discovered.

Among the neurochemical systems important for the regulation of aggression, serotonin stands out.[23] Reducing serotonergic transmission heightens aggression; and increasing serotonergic signalling reduces aggression. Dramatically, Caspi et al. reported that young men with a modified allele of the gene encoding an enzyme, monoamine oxidase-A, that chemically breaks down serotonin, had higher levels of conduct disorder, greater dispositions toward violence, higher frequencies of antisocial personality disorder, and more convictions for violent criminal offences. These results only appeared among young men who had been severely maltreated as children. If this finding can be replicated, then it could be inferred that two forces for violence, early maltreatment and a genetic alteration, 'multiplied' each others' effects on antisocial behaviour.

A new and active field of work has begun to link specific genes to aggression in experimental animals. Deletion of the gene encoding the enzyme neuronal nitric oxide synthase strongly promotes aggression by male mice.[24] Genes in a small region of the Y chromosome termed the non-pseudoautosomal region are required for normal aggression. More complex are the contributions from genes of estrogen receptors. In male mice, deleting the gene for estrogen receptor-alpha (ER-alpha) abolishes aggressive behaviour. Clearly, this field is just beginning to develop.

---

[23] B. Olivier, 'Serotonergic Mechanisms in Aggression' in G. Bock and J. Goode (eds.), *Molecular Mechanisms Influencing Aggressive Behaviors* (London, Wiley, 2005).

[24] R. J. Nelson (ed.), *Biology of Aggression* (New York, Oxford University Press, 2006).

For all of the neurobiological work on biologically adaptive, stereotyped aggressive behaviours in laboratory animals, our understanding of how regulated, low-level aggression turns into violence among humans is sorely lacking, according to reports of the US National Research Council. Neither for individual violent acts nor for organized violence among groups is there a simple 'technical fix'. Only one thing is clear. Scientific approaches to the reduction of violence need not be limited to the endocrine and neurobiological mechanisms that have received much experimental attention. Indeed, reducing testosterone levels, increasing serotonergic transmission, and avoiding alcohol all will help. However, it also seems apparent that psychological and socioeconomic alterations are also necessary. That is, avoiding the humiliation of young men by subjecting them to early experiences in families with extremely low incomes (maximizing income disparities in wealthy societies) as well as providing adolescent males with positive visions of their roles in society (e.g. through initiation rites) are non-neuroscientific tactics likely to help in reducing violence by young men.

22.5 Balance of Neural Mechanisms Bearing on Ethical Behaviours

Above, two lines of evidence have been presented that support a theory of how ethical behaviours are produced, behaviours that the initiator himself would like to receive. One line of thought emphasized avoiding harmful behaviours that the person himself would fear. The other line of thought documented mechanisms for loving, affiliative behaviours. However, it was also recognized that aggressive behaviours frequently become violent, at both the personal and the organized, social level, yielding behaviours that are exactly the opposite of the ethical responses described in the larger part of this review.

How does one explain what any individual will do at any moment, among these competing forces? Currently, there is no convincing answer to this question, and the explanations certainly will not be simple. We envision particular types of interactions between genes and environment, with an emphasis on certain critical periods during development.

One approach to the question will surely have to do with genetic influences on an individual's temperament. Long-lasting dispositions towards certain types of behaviours have been documented, first with respect to introversion vs extraversion and then with respect to a variety of personality dimensions. Psychiatrists with training in genetics have begun to report influences of specific genes on behavioural tendencies such as risk seeking, harm avoidance, and others. This field of psychiatric genetics is still in its infancy.

Whatever the genetically influenced disposition of an individual, his ethical choices will also be influenced by his social history. One particularly sensitive time is during the neonatal period. Vasopressin and oxytocin neuropeptide systems are affected by early social experience with neglected children displaying less functional OT systems and lower levels of OT and trust. Harlow's early work with maternally deprived monkeys revealed long-lasting deficits in social behaviour. Further, in human society, this can become a vicious circle: a mother who had been abused as an infant may go on to abuse her own children. In the work of Meany and his colleagues, one of the mechanisms for the long-lasting effects of neonatal stress can be tied to the activity of CRH neurons in the PVN and the amygdala. Doubtless, other mechanisms remain to be discovered. In CRH and other neurons, methylation of specific portions of DNA could provide the long-lasting biochemical changes that influence social behaviours.[25] Again, the enduring changes within the nerve cell nucleus are not likely to be limited to methylation. This field of work is just beginning.

Another particularly sensitive time for the development (or obfuscation) of ethical principles is during adolescence, during which the brain is still developing. This is true, for example, with respect to exacerbating or preventing violence among young males and with respect to avoiding the effects of stress in young women. Again, there exists the potential for a vicious circle. High degrees of stress can produce abnormalities of social behaviour that subsequently cause that individual to become even more socially isolated. For these phenomena, no brain mechanisms are yet discovered. In lower animals such as laboratory mice, it is known that loss of the gene for ER-beta leads to higher aggression in pubertal but not in older mice. However, this is just one example—a large number of similar genomic and neurochemical influences is to be expected. It is hypothesized that environmental influences particularly during these critical periods interact with genetic dispositions in complex ways to produce ethical behaviours or failures thereof.

All of these data point to interactions of genes and environment in the regulation of neuronal mechanisms for social behaviours. Some of these behaviours are positive, friendly, and affiliative. Others are antisocial. To the extent that we understand such gene/environmental interactions, how do we view their implications for the law?

22.6 Perspective and Outlook

We began our discussion examining torts and contracts. In closing, we would like to say a word about criminal law. If it is true that humans are naturally motivated to shrink from harming other people, then what are the implications for someone who, for example, has committed first degree murder, and deliberately undertaken egregious harm? Does neuroscience suggest that this person is somehow not wired like most of us, and that his 'fault' is more anatomical than moral? Arguments about

---

[25] M. Szyf. et al., 'Maternal Programming of Steroid Receptor Expression and Phenotype through DNA Methylation in the Rat' (2005) 26 *Frontiers in Neuroscience* 139–62; F. A. Champagne et al., 'Maternal Behavior Associated with Methylation of Estrogen Reception Promoter and Estrogen Receptor-alpha Expression in the Medial Preoptic Area of Female Offspring' (2006) 17 *Endocrinology* 2909–15.

mental capability have been around for a long time in establishing *mens rea*, and courts continue to wrestle with the issues. We are uncertain as to how the research discussed here should affect criminal law, but we think it should be examined as the law moves forward.

We recognize that there is serious, principled opposition to any such enterprise. Stephen Morse, for example, argues that even such high-profile cases as *Roper v Simmons*,[26] holding that 'juvenile offenders cannot with reliability be classified among the worst offenders', do not constitute a green light. Rather, he notes that while 'perhaps the neuroscience evidence [offered by numerous *amici*] actually played a role in the decision... there is no evidence in the opinion to support this speculation'.[27] In adolescents, he claims that:

> at most, the neuroscientific evidence provides a partial causal explanation of why the observed behavioral differences exist and thus some further evidence of the validity of the behavioral differences. It is only of limited and indirect relevance to responsibility assessment, which is based on behavioral criteria.[28]

In effect, Morse reads *Roper* as an implicit rejection of neuroscientific evidence for the purpose of determining *mens rea*, since the court failed explicitly to rely on what amounted to copious legal arguments in favour of such evidence.

Morse is in fact a categorical opponent of the use of neuroscientific evidence for the purpose of establishing responsibility, setting the bar so high that we think only a categorical shift in the *nature* of such evidence (which we modestly claim to offer here) would stand even a chance of his seriously shifting his ground:

> Even if there were a perfect correlation between brain states and the behavioral criteria for responsibility, the brain states would be nothing more than evidence of the behavioral states. Such a correlation is a fantasy based on present knowledge and probably always will be when we are considering complex human actions. If the person meets the behavioral criteria for responsibility, the person should be held responsible, whatever the brain evidence may indicate, such as the presence of an abnormality. If the person does not meet the behavioral criteria, the person should be held not responsible, however normal the brain may look. Brains are not held responsible. Acting people are.[29]

For Morse, neuroscientists demonstrate a reductive confusion concerning the relation between the brain and complex, intentional action. Our response is that, at least so far, no one has raised the issue of empathy—that is, our apparently built-in capacity to resist doing harm—which we believe could constitute a game-changer in so far as how neuroscientific evidence might be of use in determining criminal responsibility. If in fact a court were not simply to confront a particular neuroscientific 'state', but rather a fundamental failure to act on impulses basic to human nature, then perhaps the type of categorical opposition that Morse displays

---

[26] 543 US 551, 569 (2005).
[27] See 'Brain Overclaim Syndrome and Criminal Responsibility: A Diagnostic Note' (2006) 3 *Ohio StateL. Rev.* 397, 410.
[28] Ibid. at 409.
[29] Ibid. at 405.

would not (at least immediately) be justifiable. We think such approach is worth a shot.

We are encouraged by studies such as those of Henry T. Greely, who writes about the potential utility of neuroscience to the law. In a recent article, Greely provides a measured assessment of how neuroscience could affect the evaluation of *mens rea*:

> Neuroscience seems unlikely to lead to major changes in our view of criminal responsibility, but it will make a difference in some individual cases where it convinces us that the defendant truly and convincingly could not control his actions. Whether that means we treat him more leniently or more harshly is not clear, but we are likely, on occasion, to treat some defendants differently.[30]

Like virtually all analysts, Greely is postulating that a complex calculus may develop whereby the acts of a particular individual, the circumstances of his action, and the potential of neuroscience to shed light on the underlying relevant motivations, may yield guidance in criminal cases.[31] We think that the virtue of our own approach—postulating an inborn proclivity towards empathy—is that it reinforces Greely's relative optimism towards the potential contributions of neuroscience by eliminating the need for *ad hoc* dependence on technology. If neuroscience can actually demonstrate that we are hardwired for empathy, then virtually every case involving *mens rea* will, at some point and in some way, have to take account of why the defendant fell away from an empathetic norm. Thus while Greely suggests that on account of neuroscientific evidence we may 'on occasion... treat some defendants differently', we claim that we will always have to treat all defendants differently, taking into consideration a new array of variables.

What remains (apart from hoping that courts will acknowledge our position) is to work out what new tests should be introduced in order to make any departure from empathy (i.e. any apparent imbalance between prosocial and antisocial behaviours) an element in determining *mens rea*. Would this require relying on technology yet again? Possibly. But we think that establishing first principles must now be our first priority.

---

[30] 'Law and Revolution in Neuroscience: An Early Look at the Field' (2009) 42 *Akron Law Review* 687, 700.

[31] Greely assumes that technology could play a key factor in any neuroscientific analysis that a court might undertake: 'People studying the ethical, legal, and social implications of neuroscience have to walk a tightrope. We have to worry about the implications if the technology does work, but we also have to remember that there cannot be any good implications of using an ineffective technology' (ibid. at 707).

23

What Hobbes Left Out: The Neuroscience of Compassion and its Implications for a New Common-wealth

James D. Duffy[*]

> The only way to construct such a Common Power (commonwealth), as may be able to defend them form the invasion of foreigners, and the injuries of one another, and thereby to secure them in such sort, as that by their own industry, and by the fruits of the earth, they may nourish themselves and live contentedly; is to confer all their power and strength upon one man, or upon one assembly of men.[1]

Much has changed, and much has happened, in the four centuries since Hobbes wrote *Leviathan*. However, his jaundiced view of human nature and the necessity of enforced moral codes remains the dominant organizing principle of our social and legal systems. Our judicial restraints have multiplied and the number of our incarcerated citizens continues to escalate dramatically. Given this apparent paradox, one can be forced into one of two conclusions, i.e. human beings are incapable of being morally constrained—or, our current model of societal morality is inadequate (or simply wrong). In this paper, I will argue for the latter. Recent advances in the social neurosciences are providing us with insights into ourselves that require us to re-evaluate and evolve our models of social morality. Rather than replacing our current models, these insights provide us with an opportunity to support the emergence of social systems that not only inhibit anti-social behaviours, but actually support pro-social behaviours by our citizens. Recent advances in social neuroscience are providing the theoretical scaffolding to support radical shifts in our understanding of personal identity and social relationships. In particular, the recent discovery of mirror neuron systems challenges our illusion of absolute autonomy and suggests that we are the product of a co-creative emergent process. These new insights suggest specific strategies that our educational and legal systems can employ in fostering the emergence of a new common-wealth driven by compassionate intention—and not by the politics of fear and control.

---

[*] Professor of Psychiatry, University of Texas, M. D. Anderson Cancer Center, Fellow, McGovern Center for Health, Healing, and the Human Spirit.

[1] Thomas Hobbes, *Leviathan, or the Matter, Forme, and Power of a Commonwealth, Ecclesiasticall and Civil* (1651) 118.

This paper provides a broad overview of:

1. current concepts such as empathy, compassion, altruism;
2. recent findings in the neuroscience of social cognition;
3. recent theories on the neural correlates of conventional and care-based morality;
4. findings on the neurology of psychopathy as a breakdown in compassion capacity;
5. research on the impact of contemplative practices on neural systems sub serving empathy;
6. the implications of these findings for developing interventions that can enhance empathy and thereby diminish antisocial behaviours.

23.1 Definitions—What Hobbes Left Out

> The first cause of absurd conclusions I ascribe to the want of method; in that they begin not their ratiocination from definitions.[2]

As Hobbes stated, any rational and scientific discourse must be built upon a foundation of shared definitions and a common lexicon. Constructs such as empathy, sympathy, compassion, and altruism have long been the purview of the humanities and considered to be too 'subjective' to lend themselves to neuroscientific study. Although no universal definitions for these terms currently exists, there has however been sufficient consensus developed to provide the basis for research in this arena.

Empathy is a modern term that is most likely derived from the German word *Einfühlüng*, described as 'feeling into' another individual's experience. Although there is no current consensus definition of empathy, there is general agreement that empathy involves two core components, i.e. a cognitive and an emotional perspective. Cognitive empathy requires that the subject has the capacity to construct a cognitive framework that reflects the internal mental state of the other person. This mind-reading capacity is referred to as 'theory of mind' and involves the medial prefrontal cortex (anterior cingulate cortex), the temporoparietal junction, and the temporal poles.[3] Individuals with autism exhibit diminished capacity for mind-reading and have been reported to exhibit reduced activation in these same brain regions to social cues (defined by the American Psychiatric Association DSM-IV as 'the presence of markedly abnormal or impaired development in social interaction and communication and a markedly restricted repertoire of activities').[4]

Sympathy is a term that is frequently used as synonymous with empathy. However, it describes the subject's extrapolation (distortion) of the other's experience. It therefore contains a heavily biased perspective shaped by the observer's attitudes.

Emotional contagion is the process of vicariously experiencing the other's experience. In essence, emotional contagion results from the observer's inability to construct clear

---

[2] Ibid. 24.
[3] C. D. Frith and U. Frith, 'The Neural Basis of Mentalizing' (2006) 50(4) *Neuron* 531–4.
[4] *American Psychiatric Association Diagnostic and Statistical Manual*, 4th edn, Revised (Place, APA Press, 1996) 66.

Table 1

	Key Definitions
Empathy:	'The process that results in the subject acquiring a cognitive and/or an emotional appreciation of the other's (object's) experiential state.'
Sympathy:	'The state in which the subject feels 'sorry for' the other as a result of perceiving distress in that other.'
Emotional contagion:	'An emotional state in an observer as a direct result of perceiving the emotional state of another.'
Compassion:	'Describes the intention to be of benefit to other(s).'
Altruism:	Describes actions that are motivated by the intention of being of benefit to other(s).'

(ego) boundaries between self and the other person. The prefrontal cortex appears to play the key role in enabling the observer not to become personally distressed by witnessing the emotional state of the other. This ability to modulate one's emotional response is crucial when one encounters potentially overwhelming emotionally charged situations. This imperturbability provides the non-reactive flexibility necessary to support non-reactive and reasoned responses i.e. *equanimity*.

The term *compassion* has become a part of everyday conversations and is frequently invoked by such diverse groups as politicians, marketing firms and religious leaders. Despite this widespread usage, or perhaps because of it, the precise definition of compassion has remained opaque and open to an interpretation biased by the agent who employs the term. Compassion does not describe 'suffering with' (which would be akin to emotional contagion) but rather describes the *intention* to be of benefit to others. As such, compassion represents the moral context within which an individual relates to others. Compassion therefore provides the foundation for the individual's character and relationship to others. An individual who acts with compassionate intention will adhere to the 'Golden' and 'Silver' rules of ethical conduct and will manifest *altruism* as actions that benefit others (see Table 1).

23.2 Empathy and Compassion as the Bootstraps of Evolving Social Systems

> In regard to moral qualities, some elimination of the worst dispositions is always in progress even in the most civilized nations. Malefactors are executed, or imprisoned for long periods, so they cannot freely transmit their bad qualities.[5]

It is fair to state that empathy has been the social bootstrap that has supported the emergence of increasingly complex societies. Empathy is essential for maternal nurturance and the establishment of affiliative bonds. Paul McLean's triune brain model suggests that the emergence of a limbic system in the paleomammalian brain

---

[5] C. Darwin, *The Descent of Man* (1871) 166.

of the earliest birds and therapsids enabled the nurturance of offspring and the establishment of familial nests as the progenitor of the complex social clusterings that characterize homo sapiens. Building upon this simple but elegant model, it is reasonable to suggest that the limbic system, as the mediator of both positive and negative emotions, was co-opted to support the emergence of moral systems that are intended to support the short-term and long-term viability of the group.

Several recent paleontological findings indicate that Neanderthals and middle Pleistocene hominins provided nurturant and compassionate care to disabled members of their clan. Furthermore, primates and even rodents and birds have been reported to exhibit altruistic behaviours to con-specifics.[6] Byrne and Whiten in the 'social brain hypothesis' suggested that increasingly complex social environments are the primary selective pressure for the explosive growth in human brain over the past few millennia. It is reasonable to suggest the evolution of human beings has occurred as a consequence of our neural capacity to create and support trusting and affiliative relationships across broader and broader domains of social connectivity. Whilst the social groups of the earliest humans included their immediate clan, post-modern humans must continually create these affiliative relationships with individuals with whom they may never have significant indirect physical contact.[7] Boyd and Richerson suggest that:

> in such culturally evolved cooperative social environments, social selection favoured genes that gave rise to new, more pro-social motives. Moral systems enforced by systems of sanctions and rewards increased the reproductive success of individuals who functioned well in such environments, and this in turn led to the evolution of other regarding motives such like empathy and social emotions like shame.[8]

This hypothesis suggests that the emergence of empathy along with emotions such as shame and guilt provided the evolutionary impulse to promote the development of more egalitarian social systems where decisions are not driven by fear but by loyalty and avoiding pain towards others. In essence this describes the emergence of a care-based morality that complements, rather than challenges, earlier fear-based social systems. These mutually supportive social systems enabled homo sapiens to rapidly acquire collaborative problem solving that allowed him to survive the rapidly changing climate variations that have occurred since the mid-Miocene era starting approximately 400,000 years ago. The emergence of social cognition allowed man to generate novel and flexible social responses to ever-changing environmental changes with slowly evolving rule-based moral conventions providing the cultural boundaries within which man could explore and rapidly generate creative solutions. Those

---

[6] A. Gracia, J. L. Arsuaga, I. Martínez, C. Lorenzo, J. M. Carretero, J. M. Bermúdez de Castro, E. Carbonell, 'Craniosynostosis in the Middle Pleistocene Human Cranium 14 from the Sima de los Huesos, Atapuerca, Spain' (2009) 106(16) *Proc. Natl. Acad. Sci. U S A* 6429–30.

[7] R. W. Byrne, A. Whiten (eds.), *Machiavellian Intelligence: Social Expertise and the Evolution of Intellect in Monkeys, Apes, and Humans* (New York, Oxford University Press, 1988).

[8] R. Boyd and P. J. Richerson, 'An Evolutionary Model of Social Learning: the Effect of Spatial and Temporal Variation' in R. R. Zentall and B. J. Galef (eds.), *Social Learning: Psychological and Biological Perspectives* (Hillsdale, NJ, Erlbaum, 1988) 29–48.

Table 2

Conventional Morality	Care-Based Morality
Rule Based	Pragmatic
Fear Driven	Empathy Driven
Prefrontal Cortex Cortex(Ventrolateral frontal and ventromedial frontal cortices)	Paralimbic Cortex (Insula, Cingulate, Orbito-frontal cortices)
Static andpredictable	Evolving and creative
Top-Down	Egalitarian
Enforced	Engaged
Reactive	Enactive
May constrain antisocial behaviours (depending upon the motivation of dominant authority).	Vulnerable to manipulation by predatory (psychopathic) agents.

groups (tribes) that provided less restrictive conventional moral boundaries also provided their members with more behavioural spaciousness within which to generate more innovative and effective responses to a challenge. As Boyd and Richerson point out, only a few other taxa have made cooperation and collaborative behaviours the foundation of their social architecture. However, those that have, i.e. social insects and humans, have become remarkably successful and have become the majority of the planet's biomass. However, as Boyd and Richardson also point out, it is difficult for these systems to evolve because, at least initially, because they are vulnerable to exploitation by non-conforming members who exploit the altruism of others in the social group.[9] However, the conventional morality of the group can restrain the behaviours of such individuals and provide safe space within which a care-based morality can flourish. In this context, Hobbes was correct when he called for the imposition of moral constraints; however, Hobbes failed to recognize that these alone were insufficient to supporting the emergence of a self-learning evolving and viable social system. Darwin recognized the evolutionary advantage conferred by empathic social systems when he wrote: 'At all times throughout the world tribes have supplanted other tribes; and as morality is one important element in their success, the standard of morality and the number of well-endowed men will thus everywhere tend to rise and increase.'[10]

23.3 Social Factors as a Determinant of Empathic Connection

A number of social factors have been identified as inhibiting empathic communion between individuals. These include:

---

[9] R. Boyd and P. J. Richerson, 'Culture and the Evolution of Human Cooperation' (2009) 364 *Philosophical Transactions of the Royal Society* 3281–8.
[10] C. Darwin, *The Descent of Man* (1871) 159.

1. Physical hierarchies—any situation where one individual has significantly more physical dominance over the other will impede empathic communication between the parties. Furthermore, even constructed differential physical hierarchies such as one party standing while the other sits, will negatively influence the empathic communication.
2. Power hierarchies—when one party has more social authority than the other.
3. Expectation of brief relationship—where the parties do not expect to maintain their social linkage beyond the current encounter.
4. Expectation of poor outcome—if one or both parties expect that the outcome of their relationship will be unfavourable.
5. Inadequate communication between parties—verbal and/or non-verbal communication between parties facilitates empathic communication. Furthermore, Ginsberg and Buck have proposed that these behaviours serve to facilitate a third channel of communication that they describe as 'spontaneous communication'. This third non-literal, non-propositional channel serves to convey experiential data that binds participants at an energetic level (e.g. birds that maintain complex flock patterns without evidence for any overt communication patterns).

The implications of these social determinants of empathic communication is clear. Social systems that alienate, dehumanize, or limit direct communication between individuals will inevitably degrade care-based moral systems. This is graphically illustrated as the so-called 'Lucifer Effect' and can explain many of man's inhumanities to man where entire social groups (and not just individual psychopaths) have acted in apparent disregard of the suffering of their victims.

These factors also suggest that modern society, with the decay of extended family systems, migratory populations, and increasing reliance of impersonal communication technologies is at risk for creating decay in care-based moral behaviours.

23.4 Morality as an Emergent Property of Complex Systems

Simply describing the neural architecture supporting the generation of moral behaviours is not sufficient to understanding the emergence of complex social behaviours. What is necessary is a theoretical framework for describing the dynamics of these constantly shifting relationships, a framework that extends beyond the strange loop of linear mechanics and allows for the constantly emerging complexities of evolving social organizations. Growing out of autopoietic models of life, an enactive approach offers such a framework and suggests:

1. Living beings are autonomous agents that generate information through reciprocal interactions with the environment.
2. The central nervous system does not function in a sequential pattern but rather generates its own coherent patterns (meaning-based) according to its reciprocal experiences with other information sources (including other brains).

3. Cognition is a form of embodied action with the goal of creating self-referential meaning.
4. The outside environment is not experienced by the brain as a fixed external reality but is a relational realm that is uniquely experienced subjectively by each person based upon how that person relates to that environment.
5. 'Subjective' experience (consciousness) is not an ephiphenomenon of brain function but is central to its construction.[11]

To quote Colombetti and Thompson: 'According to the enactive approach, the human mind is embodied in our entire organism and embedded in the world, and hence is not reducible to the structures in our head. Meaning and experience are created by, or enacted through, the continuous reciprocal interactions of the brain, body, and the world.'[12]

In the context of moral cognition, the enactive approach suggests that behaviours consistent with the 'Golden Rule' are the consequence of:

1. the empathic communion that is inherent to any enactive inter-subjective experience;
2. the meaning that the organism generates from being immersed in a culture that is configured by cognitive symbols and laws.

According to the enactive model of consciousness and life, rather than losing information through our shared experience, each person is continuously generating information based on the symbiotic transfer of information at social, cognitive, emotional, chemical, and energetic levels. This model suggests that fear represents just one of many bits of information that construct both an individual and a shared experience of present reality (consciousness).

This approach also implies that the Golden Rule is the necessary and ultimate (albeit sometimes long-delayed) consequence of a universe in which each sentient being exists within an energetic embrace where the destiny of each is inextricably linked to the other. Contrary to Thomas Hobbes, the enactive model suggests that love (compassionate intent) is the central impulse of any community that is to survive beyond one generation. According to this view, compassionate intention is the primary atavistic impulse and should be fostered if the group is to continue to flourish.

23.5 Mirror Neurons and the Neural Substrate of Empathy

Mirror neurons represent a class of neurons that reflect both the actions *and the intentions* in the observed behaviours of others. Mirror neurons were discovered by

---

[11] Giovanna Colombetti and Evan Thompson, 'The Feeling Body: Towards an Enactive Approach to Emotion', in Willis F. Overton, Ulrich Miller, and Judith Newman (eds), *Body in Mind, Mind in Body: Developmental Perspectives on Embodiment and Consciousness* (Place, Lawrence Erlbaum, 2007).
[12] Ibid.

Giacomo Rizzolatti in the 1980s when he coincidentally observed that some motor neurons in a monkey responded in the same way regardless of whether the monkey was watching a person pick up a piece of food or performing the action himself.[13] Although ethical and practical limitations have precluded the direct single-cell electrical study of these mirror neurons in humans, functional neuroimaging does strongly support their presence in the human brain. Following on Rizzolatti's discovery, mirror neurons have been identified not only in the motor cortex, but also in areas mediating affective and interpersonal processes including empathy (i.e. anterior cingulate cortex and insula cortex). In other words, mirror neuron systems exist for not only motor behaviours but also emotional and sensory experiences (e.g. disgust, pain, happiness, sadness etc.). Furthermore, mirror neurons have been demonstrated selectively to respond to the intentions of the observed other (e.g. responding differentially to whether the other person is attempting to open or close the door).

Mirror neurons exhibit congruent selectivity, i.e. they respond selectively to the specific details of a particular action e.g. mirror neurons will respond selectively to apparently similar observed actions (e.g. watching another person clench their fist versus squeeze a ball). These attributes strongly support a role for the mirror neuron system (MNS) in empathizing with others. This is supported by the study by Gazzola et al. that found a positive correlation between subjects' scores on a measure of empathy with activation of the MNS.[14]

The above findings indicate that our brains are not insulated, or isolated, from one another but are constantly engaged in a dynamic and precise dialogue of intersubjective communication. This provides all parties with a 'shared experience' that spans both first and third-party perspectives.

23.6 The Neural Correlates of Empathy and Morality

Rather than a unitary model of moral socialization, Blair[15] has proposed that moral transgressions can be subsumed under two distinct categories i.e.:

Care-based morality: This describes moral (i.e. non-harming) behaviour that is motivated by vicariously experiencing the distress of others—and therefore avoiding such behaviours as a means of avoiding vicarious distress. In essence, care-based morality describes behaviour that is directed by the 'Golden' and 'Silver' rules. The shared experience generated by the mirror neuron systems of socially engaged individuals extends beyond the sensorimotor mirror neuron system and includes neural modifiers sub-serving contextual event representations (the prefrontal cortex), social colouring (the superior temporal cortex), and basic emotional states

---

[13] L. Cattaneo and G. Rizzolatti, 'The Mirror Neuron System' (2009) 66(5) *Archives of Neurol* 557–60, Review.

[14] V. Gazzola, L. Aziz-Zadeh, and C. Keysers, 'Empathy and the somatotopic auditory mirror system in humans' (2006) 16(18) *Current Biology* 1824–9.

[15] R. J. R. Blair, 'The Amygdala and Ventromedial Prefrontal Cortex in Morality and Psychopathy' (2007) 11(9) *Trends in Cognitive Sciences* 387–92.

including not just fear, but also positive emotional states such as reward, attachment, and sadness (limbic-paralimbic regions). Care-based morality evolves out an autopoietic relationship between the engaged parties and therefore is capable of manifesting novel and atavistic consequences that extend beyond linear expectations (i.e. so-called punctuated equilibrium).

Moral social convention: This is defined by Blair as the code of conduct imposed by a societal group, e.g. traffic rules, codes of professional ethics. Transgressions in this area are typically viewed by others as being 'wrong' but not necessarily 'bad'. An example of a moral social transgression would be a motorist who injures a pedestrian because he failed to notice a stop street sign versus a man who deliberately runs down his philandering spouse (care-based moral transgression). Social conventions are enacted to maintain social order and allow higher status individuals to constrain the behaviour of lower ranking individuals in the social group. Rather than empathy, the primary emotional reinforcer of social conventions is the fear of retribution elicited in the individual who transgresses the rules. Based on findings with patients with lesions to the ventrolateral prefrontal cortex (e.g. Phineas Gage), Blair has proposed that aversive social cues (such as angry expressions by authority figures) activate the ventrolateral prefrontal cortex that constructs social engrams that serve to inhibit the individual from similar behaviours in the future. This hypothesis is supported by a recent finding that subjects who transgress a social convention activate their prefrontal cortex only when an audience is present.[16]

23.7 Psychopathy as a Disorder of Emotional Empathy

Psychopaths exhibit a marked deficit in emotional empathy, i.e. they do not vicariously experience the emotional distress of others but do have a cognitive appreciation of the other's experience. This empathic emotional indifference means that the psychopathic person comes to view others as inanimate objects that can be manipulated in the same way one might move a chess piece on a board. Given this 'cold-blooded' disposition, it is understandable that psychopathy is a clinical construct that describes a personality style that includes callousness, a lack of guilt, lack of remorse, sensation-seeking, and impulsiveness. Specifically, psychopathic individuals exhibit very profound deficits in their capacity for care-based morality and make less distinction between the care-based versus conventional morality.[17] In addition, they do not show deficits in executive cognition unless there is an affective component to the task. This finding suggests that the primary deficit in psychopathy is one of affective, rather than cognitive, empathy. Consistent with this hypothesis is the finding that psychopathic individuals perform poorly in recognizing sad and fearful expressions and exhibit reduced autonomic

---

[16] Ibid.
[17] R. J. R. Blair, 'A Cognitive Developmental Approach to Morality: Investigating the Psychopath' (1995) 57 *Cognition* 1–29.

response when witnessing another person in distress.[18] Socially healthy individuals typically exhibit increased amygdale activation in response to witnessing distress in others, and therefore find this experience aversive. They therefore learn (mediated *via* projections form the amygdala to the prefrontal cortex) to avoid behaviours that produce this distress in others (and vicariously) in themselves. Functional neuroimaging reveals that psychopathic individuals exhibit less amygdala, anterior cingulated cortex and ventromedial prefrontal cortex response to negatively laden words and fearful images. Further support for amygdaloid dysfunction as the core dysfunction in psychopathy is provided by the recent report by Yang et al. that individuals with psychopathy show very significant bilateral reductions in amygdale volume with a positive correlation between amount of volume loss and psychopathy score.[19]

This data indicates that the psychopath does not have a reduced ability to understand cognitively the emotions of others but does not experience subjective arousal to their distress. In essence, psychopaths appear to exhibit a deficit in the linkage between emotion-oriented limbic systems and heteromodal association cortex involved in the development and extrapolation of novel cognitive strategies. This linkage is provided by transitional paralimbic neural structures that include the insula and anterior cingulate cortex. The psychopath is therefore unable to 'learn' how to behave in a manner that is supportive of wellbeing in *both* themselves *and* in others. This developmental (learning-disability) model of psychopathy is supported by research that shows that the inverse relationship between empathy and antisocial behaviour actually increases as the child grows older.[20]

23.8 Meditation as the Technology of Empathy

Some traditions (e.g. Tibetan) have developed practices that have been specifically designed to counteract self-centeredness and support concern for the welfare of others. These techniques have employed contemplative approaches that train attention, stabilize emotion, and enhance a motivation to be of benefit to others. One can immediately see that such techniques, if effective, would be of tremendous atavistic value to the collective society that incorporated them into their educational system. Recent research has demonstrated that specific contemplative practices that are intended to enhance compassion and empathy (e.g. Tibetan tonglen meditation practice) produce robust and predictable alterations in neural function involving neural systems sub-serving empathy. These findings include:

---

[18] E. C. Finger, A. A. Marsh, N. Kamel, D. G. Mitchell, J. R. Blair, 'Caught in the Act: the Impact of Audience on the Neural Response to Morally and Socially Inappropriate Behavior (2006) 33(1) *Neuroimage* 414–21.

[19] Y. Yang, A. Raine, K. L. Narr, P. Colletti, A. W. Toga, 'Localization of Deformations within the Amygdala in Individuals with Psychopathy' (2009) 66(9) *Arch Gen Psychiatry* 986–94.

[20] P. D. Hastings, C. Zahn-Waxler, J. Robinson, B. Usher, D. Bridges, 'The Development of Concern for Others in Children with Behavior Problems' (2000) 36(5) *Dev. Psychol.* 531–46.

1. Tibetan monks practising compassion meditation elicit robust state and trait increases in the power and synchronization of gamma-frequency (i.e. > 30mHz) electrophysiological activity. Gamma-wave synchrony has been postulated to reflect cross-cortical cerebral hemispheric synchronization that may provide the electrophysiological dipole for stable self-representational function (e.g. ego identity). It is interesting to note that both schizophrenia and borderline personality disorder, two psychiatric conditions that are associated with poor ego identity, have been reported to be associated with unstable gamma synchrony.

2. Tibetan monks with extensive training (>10,000 hours) in compassion meditation exhibited stronger neural responses to negative emotional sounds in the anterior cingulated and anterior insula (both structures that are intimately involved in empathy). This finding suggests that cultivating compassionate intention can enhance empathic stimuli associated with distress in others.[21] The monks also exhibited greater activation in the neural circuitry implicated in the cognitive appraisal of other's mental states, i.e. temporoparietal junction, superior temporal sulcus, medial prefrontal cortex, and posterior cingulated cortex. This activation was most pronounced to the right hemisphere—a finding that is consonant with previous findings of a right-sided dominance for social cognition.

3. Long-term meditation appears to produce structural changes in areas associated with empathy. In this regard, Lazar et al. reported that long-term meditators have been reported to exhibit increase in the cortical thickness of the anterior insula cortex.[22]

4. Mindfulness meditation shifts the practitioner from a predominantly self-referential narrative perspective (NF) to a more flexible experiential focus (EF) in which decisions are less likely to driven by selfish (i.e. primarily self-oriented priorities) and more by a broader perspective that includes a wider range of priorities (including long-term perspectives that include the overall wellbeing of all those who might be affected by the decision at hand). Stated differently, mindfulness meditation enables the individual to 'think before acting' and thereby reduces impulsive stimulus-driven behaviours (that are characteristic of the psychopathic individual). An EF is also less likely to be shaped by (narrative-driven) stereotypes that may distort the individual's capacity for empathic concern for the other. The EF is associated with increased activation of 'empathy-related' pathways including the right insula cortex.[23]

[21] A. Lutz, J. Brefczynski-Lewis, T. Johnstone, R. J. Davidson, 'Regulation of the Neural Circuitry of Emotion by Compassion Meditation: Effects of Meditative Expertise' (2008) 3(3) *PLoS One* 1897.
[22] S. W. Lazar, C. E. Kerr, R. H. Wasserman, J. R. Gray, D. N. Greve, M. T. Treadway, M. McGarvey, B. T. Quinn, J. A. Dusek, H. Benson, S. L. Rauch, C. I. Moore, B. Fischl, 'Meditation Experience is Associated with Increased Cortical Thickness' (2005) 16(17) *Neuroreport* 1893–7.
[23] N. A. Farb, Z. V. Segal, H. Mayberg, J. Bean, D. McKeon, Z. Fatima, A. K. Anderson, 'Attending to the Present: Mindfulness Meditation Reveals Distinct Neural Modes of Self-reference' (2007) 2(4) *Soc Cogn Affect Neurosci.* 313–22.

5. Meditation produces increased vagal tone as measured by heart-rate variability. Porges et al. have proposed the *polyvagal* theory that posits the evolution of the vagus in support of affiliative behaviours mediated i.e. the antithesis of 'flight versus fight' behaviour mediated through the sympathetic nervous system. Increased vagal tone would be consistent with the individual experiencing others as less threatening and therefore being more empathically available to experience their distress.

6. Brief loving-kindness meditation practice by inexperienced practitioners has been reported to produce very significant increases in feelings of social connection and more positive implicit and explicit reactions towards others and self. This finding has obvious implications for supporting trusting affiliative connections that will support empathy between the parties.

The implications of these recent findings are profound. Simply stated, neuroscience has demonstrated that simple but specific contemplative practices produce sustained changes in neural functioning that enhance the practitioner's capacity for empathy. Although sages have been preaching the benefit of these practices for millennia, we now have scientific proof of the efficacy and mode of action of these practices. In effect, we are in the process of developing a technology that is capable of enhancing our moral capability—most particularly in the realm of care-based morality.

23.9 Towards a New Common-wealth of Compassion

> Unfortunately, we find that no matter how sophisticated and well administered our legal systems, and no matter how advanced our methods of external control, by themselves these cannot eradicate wrongdoing. Observe that nowadays our police forces have at their disposal technology that could barely have been imagined fifty years ago. They have methods of surveillance which enable them to see what formerly was hidden; they have DNA matching, forensic laboratories, sniffer dogs, and, of course, highly trained personnel. Yet criminal methods are correspondingly advanced so that really we are no better off. Where ethical restraint is lacking, there can be no hope of overcoming problems like those of rising crime. In fact, without such inner discipline, we find that the very means we use to solve them becomes a source of difficulty itself. The increasing sophistication of criminal and police methods is a vicious and mutually reinforcing cycle.[24]

As stated by His Holiness the Dalai Lama above, modern societies run the risk of creating social environments that impede empathy, and therefore undermine care-based morality. One societal response to this challenge might be to create more stringent and intrusive legal prohibitions that can maintain convention-based morality. Unfortunately, the motivations of those who construct these conventions is not guaranteed. Such a response would be harkening back to a tribal consciousness where individuals are ultimately subjugated to the will of a 'higher' authority

---

[24] The Dalai Lama, *Ethics for a New Millennium* (New York, Riverhead Books, 1991).

(i.e. the re-emergence of fascist and totalitarian states). Although this might have temporary utility, in the longer term such a response would deny one irrefutable Darwinian fact, i.e.: evolution always moves forward.

Given the above, if the human species is to continue to evolve, and move beyond fear-based consciousness it becomes apparent that we (society) must develop a means for actively supporting and enhancing care-based morality.

In this paper, and building upon recent neuroscientific findings, I have laid the foundations for the argument that:

1. Care-based morality is dependent upon empathic communication.
2. If human societies are to continue to evolve, even remain viable, they must move beyond fear-based conventional morality.
3. As an autopoietic system, care-based morality provides the necessary foundation for the future evolution and survival of human societies.
4. Viable societies require both conventional and care-based moral systems.
5. The breakdown of family units, migratory workforces, and the mechanization of communication systems threatens to impede empathic connections and encourages the emergence of fear-based conventional morality (and the imposition of an increasing number of punitive laws).
6. Contemplative practices enhance the neural systems that support empathy, and therefore care-based morality.

23.10 Contemplation as the Next Technological Revolution

> The more we develop compassion, the more genuinely ethical our conduct will be.[25]

Based on the above, it is reasonable to propose that contemplative practices should be employed by modern society as a necessary means of supporting our survival and evolution. This proposal is built on a reasoned analysis of current neuroscientific data, and does not need to be a matter of intuitive faith or religious dogma.

Stated succinctly, contemplative practices could, and should, become the *technology of morality*. Perhaps these practices represent the next great technological advance. Just as fire, tools, mechanization, the printing press, industrialization, and information technologies spurred mankind through successive punctuated evolutionary leaps, contemplative practices, as the technology of empathy, will spur the emergence of an egalitarian society motivated by compassion, and not by fear. An autopoietic society organized around the intention of compassion would manifest a society that we cannot even begin to comprehend. Just as our caveman forebears would have been incapable of imagining our twenty-first century society, so we cannot even begin to imagine the bounty generated by this emerging consciousness. The alternative future is predictable. Any future driven by fear, according to the

---

[25] Dalai Lama, ibid.

simple logic of Newtonian cause and effect, will dictate a closed-loop of diminishing resources, escalating conflict, and eventual demise.

The legal system should assume a more positivist stance and begin to employ contemplative practices as a means of both preventing and remediating antisocial behaviours.

As described above, although psychopathy appears to have a strong genetic determinant, the expression of the condition is strongly influenced by developmental factors. As William James, the father of modern psychology stated, 'An education in attention (i.e. contemplation) would be an education par excellence.'

Until recently, Western approaches to bringing about a decrease in antisocial behaviours have focused on counteracting these behaviours by raising awareness of individuals from the disliked 'other' group and through desensitization to social cues that support violence or prejudice. More recently however, more positive psychological practices that foster prosocial emotions and behaviours are beginning to undergo scientific evaluation. In what may be the first study of its kind, Hutcherson et al. reported that a brief seven-minute loving-kindness meditation practice produces statistically significant increases in the subjects' implicit and explicit emotional positivity towards neutral strangers.[26] Sumter et al. have reported that female prisoners participating in seven consecutive weekly meditation sessions of two and a half hours exhibited less anger, less anxiety, and less physical assaultiveness towards other prisoners who were not engaged in the programme.[27] These very early and limited findings suggest that more funding should be allocated to rigorously researching the impact of contemplative approaches in criminal populations. Furthermore, recent research in real-time neurofeedback indicates that individuals can train themselves to activate more robustly brain regions that mediate empathic behaviours (such as the anterior insula cortex). These findings suggest that prosocial digital technologies could be developed—much the same way a person with heart disease might participate in physical rehabilitation.

Most importantly, contemplative educational strategies could be developed to enhance the moral development of all children, and specifically to address the 'prosocial learning disability' of children who are at risk for developing psychopathic characteristics. This contemplative education would foster the emergence of care-based moral systems and provide the necessary balance for an educational process that has become over-burdened with fact-based learning.

The character of such an education is described by Judith Lief (past president of Naropa University):

The point is not to abandon scholarship but to ground it, to personalize it and to balance it with the fundamentals of mind training, especially the practice of sitting meditation so that inner development and outer knowledge go hand in hand...A balanced education

---

[26] C. A. Hutcherson, E. M. Seppala, J. J. Gross, 'Loving-kindness Meditation Increases Social Connectedness' (2008) 8(5) *Emotion* 720–4.
[27] M. T. Sumter, E. Monk-Turner, and C. Turner, 'The benefits of meditation practice in the correctional setting' (2009) 15(1) *J Correct Health Care* 47–57.

cultivates abilities beyond the verbal and conceptual to include matters of heart, character, creativity, self-knowledge, concentration, openness and mental flexibility.

An educational system that balanced the genius of our science with the wisdom of our collective experience would become the foundation for a new common-wealth based on compassion—and not fear.

I have attempted in this paper to provide a rationale, based on neuroscientific data, for developing educational and legal systems that take a proactive role in supporting and enhancing prosocial behaviours. These new insights provide a vision of a future that is driven by compassion and offer the potential for creating a common-wealth that lies far beyond anything Thomas Hobbes could even have imagined.

24

Neuroscience and the Free Exercise of Religion

Steven Goldberg*

24.1 Introduction

Dr Michael Persinger, a prominent neuroscientist and author, can stimulate your right temporal lobe and quite possibly cause you to experience a sense that something God-like is in the room.[1] Persinger draws no small conclusions from this. 'What is the last illusion that we must overcome as a species?' he asks. And he answers: 'That illusion is that God is an absolute that exists independent of the human brain—that somehow we are in his or her care.'[2]

Professor Michael Gazzaniga, head of a leading academic Center for the Study of the Mind, relies on studies showing that specific parts of the brain are activated when people make moral judgements to conclude that religious teachings must give way to what he calls a 'universal ethics'.[3] 'It is not a good idea to kill', according to Gazzaniga, 'because it is not a good idea to kill, not because God or Allah or Buddha said it was not a good idea to kill.'[4]

Other scholars and commentators have supported variations on these themes, concluding, for example, that modern neuroscience undermines belief in a monotheistic God or in the Bible.[5] Still others maintain that brain-imaging studies fit perfectly with support for 'militant atheism'.[6]

---

* Until his death in 2010, James and Catherine Denny Professor of Law, Georgetown University Law Center, Washington, DC. The author would like to thank Miriam Goldberg, Louis Michael Seidman, and Girardeau Spann for their comments and suggestions.

[1] Professor Michael A. Persinger, the author of *Neuropsychological Bases of God Beliefs* (New York, Praeger, 1987), discussed this aspect of his work on National Public Radio on 19 May, 2009. See <http://www.npr.org/templates/story/story.php?storyId=104291534> (accessed 16 July 2009).

[2] Ibid.

[3] Michael S. Gazzaniga, *The Ethical Brain* (New York, Dana Publisher, (2005) 163–78.

[4] Ibid 165.

[5] Bennett Gordon, 'Neuroscience Could Be Killing God' (4 June 2008) *Utne Reader* <http://www.utne.com/2008-06-04/SpiritualityNeuroscience-Could-Be-Killing-God.aspx> (accessed 16 July 2009).

[6] Kelly Bulkeley, 'The Cognitive Revolution and the Decline of Monotheism' (30 May 2008), *The Immanent Frame* <http://www.ssrc.org/blogs/immanent_frame/2008/05/30/the-cognitive-revolution-and-the-decline-of-monotheism/> (accessed 16 July 2009).

If these beliefs take hold in popular culture, they could, over time, have substantial implications for the legal status of religion. If you look across the US and Europe you will see variation in how free exercise of religion claims are handled, yet you will nonetheless see a broad consensus that religious practices should be respected and tolerated unless they disrupt public order.[7] But that consensus might erode if religion came to be seen as simply a by-product of biology and not a very attractive by-product at that. And, needless to say, the establishments of religion that linger in some European countries[8] would also have difficulty surviving a transformed view of religion as a branch of neurology.

Neuroscience will not, however, fundamentally undermine the social and legal status of religion. We know this is so for two reasons. First, those who believe brain science erases supernatural religion have logical and philosophical hurdles they have not begun to surmount. Secondly, and more importantly, the challenges that are being raised today have been with us for over 100 years and have had little to no impact.

24.2 Neuroscience, Metaphysics, and Ethics

As to the first point, there simply is a lot more intellectual work to be done to make the case that neuroscience decisively undermines religious belief. Persinger's apparent belief that empirical testing can falsify supernatural claims is baffling. Karl Popper himself made clear that science simply cannot speak to the truth of metaphysics. That is the whole point of saying that scientific claims must be falsifiable. Progress in science has come from the recognition that the statement 'God exists' is not scientific because it is not testable. Thus alleged scientific proofs of God's existence are, in fact, undermined by the modern approach to science. But that is completely different from saying that 'God exists' is false. Science cannot tell us that.[9]

As Popper himself put it, falsifiability is:

A criterion of demarcation between empirical science on the one hand and pure mathematics, logic, metaphysics, and pseudo-science on the other... The broad line of demarcation between empirical science... and... metaphysics... has to be drawn right through the very

---

[7] See e.g. J. Tomasso, 'Separation of the Conjoined Twins: A Comparative Analysis of the Rights to Privacy and Religious Freedom in Great Britain and the United States' (2002) 54 *Rutgers Law Review* 771.

[8] See e.g. C. Evans and C. Thomas, 'Church-State Relations in the European Court of Human Rights' (2006) 2006 *Brigham Young University Law Review* 699.

[9] For an account of Popper's thinking on metaphysics, see e.g. Stephen Mumford, 'Metaphysics' in Stathis Psillos and Martin Curd (eds.), *The Routledge Companion to Philosophy of Science* (London, Routledge, 2008) 26–35. Mumford makes clear that Popper's views are disputed by many philosophers, but it remains true that Popper has the 'largest methodological influence on actual scientists' in modern times. S. O'Connor, 'The Supreme Court's Philosophy of Science: Will the Real Karl Popper Please Stand Up?' (1995) 35 *Jurimetrics Journal* 276. For further support for the proposition that 'neuroscience ultimately cannot disprove a metaphysical soul', see Paul Root Wolpe, 'Religious Responses to Neuroscientific Questions' in Judy Illes (ed.), *Neuroethics: Defining the Issues in Theory, Practice, and Policy* (Oxford, Oxford University Press, 2006) 289, 292–3.

heart of the region of sense—with meaningful theories on both sides of the dividing line—rather than between the region of sense and nonsense.[10]

But does Persinger level at least a practical attack on religion? Here, too, his position is wanting. Suppose stimulation of the brain by a clinician triggers a religious experience. That might undermine the faith of someone who thought only God could trigger such an experience. But many people of faith would have no difficulty in accepting that God can also do what a clinician can do. After all, neuroscientific studies have shown that placebos operate on the pathways of pain in much the same way as opiates.[11] If a placebo can trigger the brain, it would be odd to suppose God could not. And of course for many it is God who made sure in the first place that people have brains that can experience the spiritual.

It is also hard to give great weight to Gazzaniga's notion that our knowledge of the brain will lead to a universal ethics that will supplant traditional religions. First, there is his apparent commission of the naturalistic fallacy. He seems to assume that what is tells us what ought to be. But any serious study of the brain, just like any serious study of the Bible, would turn up substantial evidence that some people are predisposed to enslave others.[12] Yet that does not make slavery morally acceptable.

Even when brain science allegedly uncovers a helpful moral teaching, Gazzaniga's hope for a universal ethics is not very promising. He finds that, in certain settings, people believe that we should not kill. But a cursory glance around the globe suggests that we do, as a species, differ quite a bit on issues like pacifism and capital punishment. In a world in which abortion ranges from a crime to a constitutional right, we may have to put the emergence of a universal ethics based on science on hold for a little while.

24.3 William James and the Lessons of History

Yet the biggest problem with the neuroscientific attack on religion is that we have been through this before, and religion has emerged essentially unscathed. By the early days of the twentieth century, medicine and science had progressed to the point where the ecstatic experiences of great religious leaders were often attributed to physical abnormalities: Saint Paul was viewed as an epileptic, Saint Teresa as an hysteric, and so on.[13]

---

[10] Karl Popper, *Realism and the Aim of Science* (La Salle, IL, Open Court Press, 1982) 175–6 (emphasis in the original).
[11] A. J. Kolber, 'A Limited Defense of Clinical Placebo Deception' (2007) 26 *Yale Law & Policy Review* 89–90.
[12] The acceptance of slavery in the Old and New Testament and in other religious texts along with the persistence of slavery in modern times are among the topics discussed in B. K. Freemon, 'Slavery, Freedom, and the Doctrine of Consensus in Islamic Jurisprudence' (1998) 11 *Harvard Human Rights Journal* 1.
[13] William James, *The Varieties of Religious Experience: A Study in Human Nature* (London, Longmans, Green & Co, 1902) 16.

It was against this backdrop that William James delivered the Gifford Lectures on Natural Religion at the University of Edinburgh. James's lectures, published in 1902 as *The Varieties of Religious Experience*, took as their subject precisely the subjective experiences of those religious visionaries who often fell into trances, heard voices, saw visions, and the like.[14] James had no interest in the 'ordinary religious believer who follows the conventional observances' of Christianity or any other faith.[15] He said that it 'would profit us little to study this second-hand religious life' marked as it was by 'fixed forms of imitation and retained by habit'.[16] James was interested precisely in people like Saint Paul.

As a philosopher and a psychologist who also had a medical degree, James was well situated to address religion and the mind. And, indeed, he began his talk by arguing that to 'plead the organic causation of a religious state of mind... in refutation of its claim to possess superior spiritual value is quite illogical and arbitrary.'[17] We cannot improve upon the reason he gave:

[T]here is not a single one of our states of mind, high or low, healthy or morbid, that has not some organic process as its condition. Scientific theories are organically conditioned just as much as religious emotions are; and if we only knew the facts intimately enough, we should doubtless see 'the liver' determining the dicta of the sturdy atheist as decisively as it does those of the Methodist...[18]

Today it is popular to give functional MRIs to see how the brain is behaving when religious or moral sentiments are under consideration. In the future it may be equally popular to scan the brains of scientists while they think about dark matter or superstring theory or neuroscience itself. The results may be fascinating, but they will tell us nothing about the merits of the topics on the thinkers' minds.

Since William James, the subject of the connection between our physical selves and our religious beliefs has come up time and again. Just fifteen years ago, when the Human Genome Project, rather than brain scanning, was the reductionist theory of the day, the cover of *Time* magazine announced the discovery of The God Gene and asked 'Does our DNA compel us to seek a higher power?'[19]

The God Gene turned out to be hard to pin down. But the bigger point is that it scared very few people of faith. Just as with the results of MRIs, a gene for religion

---

[14] William James, *The Varieties of Religious Experience: A Study in Human Nature* (London, Longmans, Green & Co., 1902) 16.
[15] Ibid 10.   [16] Ibid.
[17] Ibid 17. James concludes this sentence by saying, 'unless one has already worked out in advance some psycho-physical theory connecting spiritual values in general with determinate sorts of physiological change': ibid. James here is rejecting the idea that an experience produced by a certain part of the brain is morally superior because that part of the brain is somehow privileged. He writes that 'medical materialism... has no physiological theory of the production of... its favorite states, by which it may accredit them; and its attempt to discredit the states which it dislikes, by vaguely associating them with nerves and liver, and connecting them with names connoting bodily affliction, is altogether illogical and inconsistent': ibid.
[18] Ibid.
[19] 'The God Gene: Does our DNA Compel Us to Seek a Higher Power? Believe It or Not, Some Scientists Say Yes' (2004) *Time*, 25 October (Cover).

that is seen by some as an evolutionary adaptation that undermines God is taken by others as evidence that God himself has hardwired us for faith.[20]

Indeed in recent years, religious organizations, far from being cowed by developments in brain science, have sponsored conferences that have featured presentations on the compatibility of neuroscience and faith. At the Woodstock Theological Center, Dr Malcolm Jeeves argued that 'all the evidence in neuroscience points to us as a...unity...[T]here are these intimate relations between mental life, emotional life and neural structures, including the spiritual dimension'.[21] By the spiritual dimension, Jeeves said, 'I mean essentially a relationship with God however you formulate that.'[22] At the Center for Theology and the Natural Sciences' conference on Neuroscience, Religious Experience and the Self, Professor David Ray Griffin concluded that 'the existence and persistence of religion', including 'a direct experience...of a Holy Actuality', is consistent with the scientific point of view.[23]

It is reasonable to wonder why people of faith are comfortable with neuroscience while many remain deeply troubled by the theory of evolution. Neuroscience, after all, is completely embedded in and dependent on the evolutionary framework: working neuroscientists are not likely to be creationists.

While it is, of course, possible to reconcile evolution with belief in God, there is no denying that Darwin's work caused a crisis in traditional religion.[24] Why is neuroscience not doing the same?

The reason is that evolution is not simply a set of facts, or a model, or an increase in our knowledge about the material nature of human beings. Evolution is a narrative. It is a story about who we are and where we are going. From a human point of view, it is not a particularly uplifting narrative, indeed it is not even centred on humans. And evolution is a counter-narrative to that told by traditional religions. They tell a story in which we are specially created and have a special destiny. Giving up a central narrative that gives meaning to our existence is not easy.

Consider the argument put forward by Ron Carlson and Ed Decker in their book *Fast Facts on False Teachings*, in which they defend evangelical Christianity and attack evolution. Carlson and Decker write that according to evolution, 'You are the descendant of a tiny cell of primordial protoplasm washed up on an ocean beach 3½ billion years ago...You are a mere grab-bag of atomic particles, a conglomeration of genetic substance...[Y]ou came from nothing, you are going nowhere.'[25] The Christian view, in contrast, says, 'You are the special creation of a

---

[20] See e.g. Dean Hamer, *The God Gene: How Faith is Hardwired into our Genes* (New York, Doubleday, 2004).
[21] Malcolm Jeeves, 'Neuroscience and the Soul' (1998) *Woodstock Forum* 19.
[22] Ibid.
[23] David Ray Griffin, 'Scientific Naturalism, The Mind-Body Relation, and Religious Experience' (2002) 37 *Zygon* 376.
[24] See e.g. Steven Goldberg, *Seduced By Science: How American Religion Has Lost Its Way* (New York, New York University Press, 1999) 25–39.
[25] Ron Carlson and Ed Decker, *Fast Facts on False Teachings* (Eugene, Oregon, Harvest House, 2003) 62.

good and all-powerful God. You are the climax of His creation... [Y]]ou are unique.... Your Creator love[s] you.'²⁶

Unlike evolution, neuroscience can be seen as simply providing a deeper understanding of the wonderfully complex special brain that God has given us. And those neuroscientists who think that the discipline's deterministic framework is somehow inconsistent with religion need to meet some Calvinists.

24.4 Incremental Legal Change As Law Reacts to Neuroscience

So neuroscience will not fatally undermine religion or fundamentally erode its legal status. But that is only half of the story.

Recent research in brain science will play a role in the law in a manner that is typical of scientific advances. The role will be modest rather than grandiose, incremental rather than sweeping. Advocates for a party in a dispute will seek to use current research findings as part of their effort to advance their cause. If the courts judge the proffered evidence to be sufficiently reliable the evidence will be admitted. It may then play a part in the tribunal's final decision. An individual litigated matter will not necessarily have broad implications, but over time the underlying science may become a feature of certain disputes.

Under this scenario, neuroscience could play a role in law and religion disputes in the courtroom, just as it can play a role in other legal settings. Let me give a hypothetical example to clarify my point before turning to a few actual cases that have already arisen.

Under current law in many American states, parents can be excused from their ordinary obligation to give their children traditional medical treatment when the parents object on religious grounds.²⁷ Christian Scientists are the main beneficiaries of this religious exemption. This statutory exemption, where it exists, is not unlimited. Parents, regardless of their religious beliefs, are never free to prevent their children from receiving life-saving care. But when a treatment is related to health rather than life, and a parent chooses to use prayer rather than conventional medicine, that parent will in certain states be immune from child neglect charges.

As with all legal doctrines, hard cases can arise under these statutes. How dangerous is the child's illness? Are the parents' objections really based on religion, as required by the statute? These kinds of issues are determined on a case-by-case basis.

So consider the following hypothetical case. A single father has a young daughter with a serious arm injury. The doctor wants to intervene with antibiotics. The father, who is not affiliated with any organized religious group and has not been to church for years, suddenly announces that he will pray over his daughter rather

---

²⁶ Ibid 63.
²⁷ For a discussion, with references, of the material in this paragraph, see Kei Robert Hirasawa, 'Are Parents Acting in the Best Interests of their Children when they Make Medical Decisions Based on their Religious Beliefs?' (2006) 44 *Family Court Review* 318.

than bring her back to the doctor's office. He says that he is hearing voices from God telling him to keep his daughter at home.

When the matter comes before a judge, she is quite troubled. She is concerned that the child could have an infection that will become very serious. She knows that under the statutory exception for religious practice it should not matter whether the father is a member of an organized church, but she is concerned that his statements are better described as erratic or troubled rather than religious.

Suppose now that the state, in an effort to force treatment, introduces neuroscientific evidence that the father's experience of hearing voices is typical of a particular brain abnormality usually viewed as a type of mental illness. Perhaps the state has even obtained brain scans of the father that support this view.

I believe it is quite plausible that this neuroscientific evidence could play a role in the judge deciding that treatment should be given to the daughter over the father's objection. Technically, the judge might find that the father's objections are not religious under the statute, although the decision could also involve a stated or unstated concern that the injury could be potentially life threatening. My point is simply that the neuroscience might matter to this judge.

The immediate objection to this prediction is that it appears flatly to contradict our earlier observation that neuroscience will not undermine religion. What happened to the arguments we heard a moment ago? Attributing the voice of God that the father is hearing to a particular part of his brain should not undermine that voice any more than attributing a doctor's brilliant diagnosis to a part of the doctor's brain should undermine the diagnosis.

The answer is that there is a contradiction, but it is one that we see in the law every day. Consider the insanity defence, which negates criminal liability if because of mental illness an individual cannot tell right from wrong.[28] Why do we regard an individual as not guilty by reason of insanity if his moral compass has been thrown off by a mental defect, but not if it has been impaired by extreme poverty or drug addition? Surely it is the brain that is implicated in all of these cases.

Or consider the doctrine in some jurisdictions that excuses you from a contract if you entered into the deal under the compulsion of a mental disease.[29] To a philosopher or a scientist any contract you made might be traced back to your genetic endowment, your upbringing, and the latest advertisements you saw, but that sort of compulsion argument will never work.

Obviously the courtroom plays fast and loose with deep philosophical ideas like determinism. The legal system's treatment of these concepts is rarely logically consistent. Instead the law treats new scientific theories on a rather *ad hoc* basis until some sort of broad societal consensus arises.

There are two primary reasons judicial decision-makers do not seek scientific and philosophical consistency above all else.[30] First, in resolving a case, a judge is

---

[28] See e.g. *R v M'Naghten* (1843) 10 C and F 200; 8 ER 718, 722 (HL).
[29] See e.g. *Faber v Sweet Style Mfg. Corp.* 242 NYS 2d 763 (Sup Ct NY 1963).
[30] I have discussed the material in this and the succeeding paragraph in Steven Goldberg, *Culture Clash: Law and Science in America* (New York, New York University Press, 1994) 13–20.

seeking a peaceful and socially acceptable resolution of a dispute. People need a chance to be heard, values like religious liberty and child welfare need to be weighed, public understanding needs to be considered. The accurate resolution of a factual dispute is not the only, or even the primary, goal.

Secondly, every judge understands that not to decide is a decision. If a judge said, 'I won't order a young child to receive treatment until there is complete consensus on the neuroscience underlying the voices the father is hearing, and full philosophical agreement on whether the father had a real choice in making his decision', then no child would ever be treated over the parent's objection. And if the judge put the burden on the parent to prove beyond any doubt *his* scientific and philosophical assumptions then religious liberty would never be respected. The law does not have the luxury of waiting, so it moves along, case by case, making imperfect decisions in the best way it can.

It is not that there are no relevant legal presumptions here. Whether it be religious belief, criminal responsibility, or the making of contracts, the law, as Justice Cardozo put it, begins 'with a robust common sense which assumes the freedom of the will as a working hypothesis'.[31] But, as he well knew, there are places where we put this assumption aside.

24.5 Recent Cases

Although it did not involve religion, there has already been one major US Supreme Court case where modern brain science played a role. In *Roper v Simmons* the court, in 2005, ruled that it violated the Constitution for the state of Missouri to execute individuals who were under the age of eighteen when their crimes were committed.[32] In finding that the death penalty in those instances constituted cruel and unusual punishment, the court relied in part on an *amicus* brief filed on behalf of scientists and doctors that argued that adolescent brain development is immature because the frontal lobe is not fully developed, and it is this part of the brain that is responsible for impulse control and good judgement.[33]

One can already hear the shouts of protests. When the frontal lobe is fully developed, does that mean that free will magically appears? And does this maturing of the lobe happily take place precisely when one turns eighteen? And so on.

But the court knew all this and more. Its decision drew strength from the emerging field of neuroscience—and science does lend some prestige to a judicial ruling—but it relied as well on precedent, on the unwillingness to execute children in most American states and in most countries around the world, and on a broader discomfort with the death penalty itself.[34] Even when the court referenced the

---

[31] *Steward Machine Co. v Davis* 301 US 548, 590 (1937).
[32] *Roper v Simmons* 543 US 551 (2005).
[33] Ibid 569–70. See also A. Haider, '*Roper v. Simmons*: The Role of the Science Brief' (2006) 3 *Ohio State Journal of Criminal Law* 369.
[34] See e.g. E. F. Emens, 'Aggravating Youth: *Roper v. Simmons* and Age Discrimination' (2005) 2005 *Supreme Court Review* 51.

scientific studies, it did so in a cautious way, embedding them in common sense and legislative judgments:

> [A]s any parent knows and as the scientific and sociological studies respondent and his *amici* cite tend to confirm, '[a] lack of maturity and an underdeveloped sense of responsibility are found in youth more often than in adults and are more understandable among the young. These qualities often result in impetuous and ill-considered actions and decisions.'... In recognition of the comparative immaturity and irresponsibility of juveniles, almost every State prohibits those under 18 years of age from voting, serving on juries, or marrying without parental consent...[35]

So in the field of law and religion, I would expect neuroscience to make some limited but important appearances in the years ahead. There already are some cases of this type. In 2008, the Minnesota Court of Appeals affirmed a decision by a district court civilly committing Allison Fischer to a mental institution and ordering the involuntary administration of neuroleptic medication by that institution.[36]

At the time of the case, Fischer was a homeless twenty-two-year-old with no income. She was asking strangers if she could live with them, was expressing grandiose ideas such as winning the World Cup, and was not sleeping regularly. A psychologist testified that she exhibited grossly disturbed behaviour and faulty perception. Although Fischer insisted that she was not mentally ill, it is not surprising that the Court of Appeals affirmed the district court holding that she was a danger to self and others and could be committed and treated.

What gives the case particular interest is that much of Fischer's behaviours and perceptions involved religion. Her grandiose ideas included planning to convert China, Russia, and Africa to Catholicism, she refused to have blood drawn and to remove her clothes on religious grounds, and she was described by her mother as being religiously preoccupied. Moreover, the court did not rely solely on the psychologist presented by the government. The district court appointed a 'certified neuropsychiatrist' who found among other things that Fischer 'is in a manic phase of bipolar disorder' and that '[s]he does not have the ability to manipulate information rationally'.[37] In the end both the district and appellate courts brushed aside Fischer's explicit freedom of religion claims.

It is not clear from the record what tests this 'certified neuropsychiatrist' used, and it is far from clear that the case would have come out differently if the neuropsychiatrist had not testified. But this is a typical example of using science to bolster a case that turns on other factors as well, and it is clear that neuroscience is available to play that role.

Another straw in the wind is the recent decision in *Boone v Missouri*, in which an inmate resisted participating in a therapy programme on the ground that 'the Word of God only has the answers' to his problems and that 'only the Lord Jesus Christ

---

[35] *Roper v Simmons* 543 US 551, 569 (2005).
[36] *In the Matter of the Civil Commitment of Allison Fischer*, State of Minnesota Court of Appeals A07-1531 (2008).
[37] Ibid 3–4.

and His Holy Spirit can make any necessary changes in my life'.[38] Failure to participate in the programme lengthened the amount of time Boone had to spend incarcerated. The court's rejection of Boone's position turned on many factors, but it is worth noting that the court quoted without disapproval the finding of a therapist that Boone's religious convictions were evidence of 'internalized distortions' in his mind.[39]

Fischer and *Boone* are not leading precedents, and they certainly do not mean that the medical profession is going to begin to view religion generally as an illness or that courts would have any patience with such an approach.[40] But they do suggest that in close cases, emerging findings relating neuroscience to religion may find a role in some judicial contexts.

Given the way the law necessarily works, there is one final paradox that is worth noting here. Consider the hypothetical case I gave concerning a father who suddenly hears the voice of God and therefore does not want to give his daughter traditional medical treatment. Among the factors that a court would consider in the real world is the fact that the father in my hypothetical was not a member of an organized religious group. If he were, by contrast, a life-long Christian Scientist who had come to his views, not because of sudden voices from above, but because of long-standing study of standard Christian Science texts, a court would be more reluctant to view his decision as evidence of mental illness and thus would be more likely to rule in his favour. And this is independent of whatever a brain scan of the father would show. The court would be quite understandably reacting to the fact that the religious exemption to child abuse statutes under consideration were largely supported by and largely aimed at protecting Christian Science and similar established groups such as Jehovah's Witnesses. The fact is that no court is going to medicalize and undermine the religious beliefs of millions of people.

The paradox becomes clear if we look at this from William James's point of view. His defence of the authenticity of religious experience was aimed precisely at the ecstatic and extreme experiences of religious visionaries. He was quite uninterested in the 'second-hand' religious life of ordinary parishioners. The law views the matter just the other way around, giving weight to the preferences of well-established churches while tending to devalue the spiritual claims of unorthodox individuals. Odd perhaps, but an inevitable by-product of a system in which science is just one input in a broad effort to resolve cases in a socially acceptable manner.

---

[38] *Boone v Missouri* 147 SW 801, 804 (Court of Appeals of Missouri, Eastern District, 2004).
[39] Ibid 805.
[40] With the narrow, controversial exception of the 'deific decree' doctrine, courts and psychiatrists have long agreed that religious belief should not be viewed as an illness. See e.g. C. Hawthorne, '"Deific Decree": The Short, Happy Life of a Pseudo-Doctrine' (2000) 33 *Loyola of Los Angeles Law Review* 1755; G. H. Morris and A. Haroun, '"God Told Me to Kill": Religion or Delusion?' (2001) 38 *San Diego Law Review* 973.

25

Steps toward a Constructivist and Coherentist Theory of Judicial Reasoning in Civil Law Tradition*

*Enrique Cáceres***

This paper aims at presenting a theoretical model of judicial reasoning that satisfactorily integrates partially provided explanations by three different theoretical research paradigms: Philosophy of Law, Legal Epistemology and Artificial Intelligence and Law.

The model emerges from the application of knowledge elicitation and knowledge representation methods and uses the theory of neural networks as a theoretical metaphor to generate explanations and visual representations.

The epistemological status of the model is of constructivist stripe: it is in line with the contemporary research tendencies within cognitive psychology that propose that judicial reasoning may be better understood if a coherentist and a connectionist approach is taken.

25.1 Introduction

For Professor Hart, recurrent questions in jurisprudence may be adequately answered if law is conceived as the union of primary and secondary rules.[1] Among the

---

* In this paper, I present part of the research I am conducting within the framework of a post-doctoral programme on law and cognition under the tutorage of Dr Sandra Castañeda at the Laboratory of Cognitive Development and Complex Learning of the Faculty of Psychology of the National Autonomous University of Mexico. The research is being sponsored by the National Council for Science and Technology of Mexico. I am very grateful to Edinburgh University's School of Law's SCRIPT, where I concluded the final stage of this paper. There I enjoyed an excellent academic environment during the month prior to this Congress as a visiting professor to the university. I particularly wish to thank Professor Graeme Laurie, Director of SCRIPT and Professor Burkhard Schafer, co-director of SCRIPT and the co-founder of the Joseph Bell Center for Legal Reasoning and Forensic Statistics.

** Institute for Legal Research, National Autonomous University of Mexico; email: encacer@hotmail.com.

[1] H. L. A. Hart, *The Concept of Law*, 2nd edn (P. A. Bulloch and J. Raz (eds.)) (Oxford, Clarendon Press, 1994).

latter, we find the adjudicatory rules used by judges to decide the legal cases under dispute.

This implies that providing a satisfactory explanation of the cognitive processes behind judicial reasoning is a key issue for achieving a clear understanding of one of the most important legal institutions (the judiciary).

Nonetheless, explanations of judicial reasoning are not homogeneous. Different theoretical communities offer different explanations. Each focuses on certain aspects of this very complex phenomenon based on certain theoretical and methodological assumptions and commitments.

The communities I consider more important and that will serve as my point of departure are the following: philosophy of law, legal epistemology, and indirectly artificial intelligence and law.

Each of these communities may be characterized as follows:

There are two basic positions within the philosophy of law: those who claim that judicial reasoning is syllogistic (civil law tradition philosophers) and those who claim that judges solve disputed legal cases based on subjective psychological processes (American legal realism).

Within legal epistemology, we can distinguish two positions: Larry Laudan's, who directs his analysis towards the epistemology of US criminal law. According to Laudan, the aim of legal epistemology is to evaluate whether or not legal procedures and legal evidentiary practices are or are not justified—as systems that are engineered to obtaining true beliefs—compared to the same parameters as other investigative systems, namely scientific ones.[2]

The other position in legal epistemology assumes that certain vague phrases used by adjudicatory rules—such as 'beyond all reasonable doubt' or the 'firm conviction of the judge'—are themselves the cause of arbitrary and irrational judicial decision-making when disputed legal cases are being solved by the judiciary. The main authors of this version of legal epistemology assume that the problem of arbitrariness and irrationality might be overcome if we import to judicial practices methods for empirical inquiry that have been proven efficient in science.[3]

---

[2] Larry Laudan, *Truth, Error, and The Criminal Law, an Essay in Legal Epistemology* (Cambridge, Cambridge University Press, 2006); see also Larry Laudan, 'The Social Contract and the Rules of Trial: Re-Thinking Procedural Rules'; 'Deadly Dilemmas; Is Reasonable Doubt Reasonable?'; 'The Elementary Epistemic Arithmetic of Criminal Justice'; 'Strange Bedfellows: Inference to the Best Explanation and the Criminal Standard of Proof'; 'Legal Epistemology: The Anomaly of Affirmative Defenses'; 'The Presumption of Innocence: Material or Probatory?'; 'Deadly Dilemmas II: Bail and Crime'; 'Need Verdicts Come in Pairs?'; 'Taking the Ratio of Differences Seriously: The Multiple Offender and the Standard of Proof, or, Different Strokes for Serial Folks'; 'Re-Thinking the Criminal Standard of Proof: Seeking Consensus About the Utilities of Trial Outcomes'; and 'Aliados extraños: la inferencia a la mejor explicación y el estándar de prueba penal' (*PROBLEMA, Anuario de Filosofía y Teoría del Derecho*, 1, 2007, at: <http://www.juridicas.unam.mx/publica/librev/rev/filotder/cont/1/pr/pr10.pdf>.

[3] Michele Taruffo, *La Prueba de los Hechos*, 2nd edn (Madrid, Trotta, 2006); Michele Taruffo, *La Prueba* (Madrid, Marcial Pons, 2008); Jordi Ferrer, *Prueba y Verdad en el Derecho*, 2nd edn (Madrid, Marcial Pons, 2005); Jordi Ferrer, *La Valoración Racional de la Prueba* (Madrid, Marcial Pons, 2007); Marina Gascón, *Los Hechos en el Derecho. Bases Argumentales de la Prueba*, 2nd edn (Madrid, Marcial Pons, 2004).

For its part, the International Association for Artificial Intelligence and Law (AIL) has centred its attention on non-monotonic logic, on dialogical and defeasible argumentative processes within a legal procedure, and on the relationship of these processes with narratives, abductive reasoning, and inference to the best explanation.[4]

A comparative analysis of the different *explanans* of these communities yields the following considerations:

Philosophy of law focuses on the relationship between judicial reasoning and legal norms, whether these norms are considered premises or conceived as useless, but not when dealing with problems related to evidence evaluation and dialogical argumentation. Legal epistemology concentrates on factual propositions and evidence, but not on dialogical argumentation and narratives. The AIL community pays little attention to normative problems and focuses on factual propositions, dialogical argumentation and narratives, but does not address fundamental conceptual issues of legal epistemology. Some of this community's work on abduction and inference to the best explanation can be criticized in that they ignore the role of the standard of proof as a means to distribute potential errors (such as false acquittals and false convictions). In addition to this critique, AIL's analyses of multiple narratives for the same disputed legal case are based on events or states of affairs that presuppose static legal norms. This idea does not take into account that given the constitutive function of legal norms, legal facts are institutional facts that depend on those selfsame norms, and that the definition of these institutional facts arises from the reflexive equilibrium of the different norms that could possibly apply to the case, factual assertions by the parties, evidence, and doctrine. The atomistic conception of evidence on which these analyses are based conceals some deep and very difficult problems regarding the corroboration of evidence (holistic evaluation).

---

[4] Some of the major works on this trend within AI and Law are the following: Douglas Walton, Chris Reed, and Fabrizio Macagno, *Argumentation Schemes* (Cambridge, Cambridge University Press, 2008); Douglas Walton, *Argumentation Methods for Artificial Intelligence in Law* (Lecture Notes in Artificial Intelligence Series) (Berlin, Springer, 2005); Douglas Walton, *Legal Argumentation and Evidence* (University Park, Penn State Press, 2002); Douglas Walton, 'Argument Visualization Tools for Corroborative Evidence' (Proceedings of the 2nd International Conference on Evidence Law and Forensic Science, Institute of Evidence Law and Forensic Science, Beijing, 2009, 32–49); Douglas Walton and Thomas F. Gordon, 'Legal Reasoning with Argumentation Schemes' (12th International Conference on Artificial Intelligence and Law) (Carole D. Hafner (ed.)) (New York, Association for Computing Machinery, 2009, 137–46); Douglas Walton, 'Visualization Tools, Argumentation Schemes and Expert Opinion Evidence in Law' (2007) 6 *Law, Probability and Risk* 119–40; Henry Prakken, 'Formalising ordinary legal disputes: a case study' (2008) 16 *Artificial Intelligence and Law* 333–59; Henry Prakken, 'A Formal Model of Adjudication Dialogues' (2008) 16 *Artificial Intelligence and Law* 305–28; Tom Gordon, Henry Prakken, and Doug Walton, 'The Carneades Model of Argument and Burden of Proof' (2007) 171 *Artificial Intelligence* 875–96; F. J. Bex, S. W. van den Braak, H. van Oostendorp, H. Prakken, H. B. Verheij, and G. A. W. Vreeswijk, 'Sense-making Software for Crime Investigation: How to Combine Stories and Arguments?' (2007) 6 *Law, Probability & Risk* 145–68; Henry Prakken, 'Analysing Reasoning about Evidence with Formal Models of Argumentation' (2004) 3:1 *Law, Probability & Risk* 33–50; Floris Bex, Henry Prakken, Chris Reed, and Douglas Walton, 'Towards a Formal Account of Reasoning about Evidence: Argumentation Schemes and Generalisations' (2003) 11 *Artificial Intelligence and Law* 125–65.

25.2 Problem

How can we incorporate such differing explanations about what would have to be assumed into the same *explanans*? What is the structure implied by these theories, if any exists? What still needs to be developed in order to have a better understanding of judicial reasoning?

25.3 Theoretical Proposal

The central idea of this paper is that a descriptive and naturalized reflection on judicial reasoning has not been carried out. To date, these discussions have been speculative (philosophy of law), evaluative (Laudan's legal epistemology), prescriptive (Ferrer et al.) and focused on logical aspects (AIL).

In this paper, I present a preliminary outline of the descriptive and naturalized model that I believe should be also assumed by the work carried out by the referred communities. My claim is based on the following: an epistemic evaluation is more effective if it considers the pragmatic use of the system of investigation and not solely on a set of rules out of touch with the community practices of its users. It is not acceptable to disqualify judicial reasoning as irrational on the grounds of an inadequate analysis of vague terms without having a model of this reasoning on which to base such criticism. Lastly, an explanation of arguments and narratives that ignores the dynamics of the cognitive schemes judges use leads to an oversimplification.

25.4 The Constructivist-Connectionist Approach: A New Point of Departure

25.4.1 Developing EXPERTIUS

EXPERTIUS is the name given to a judicial decision-support system developed at the Institute for Legal Research of the National Autonomous University of Mexico. Its underlying theoretical model adopted an atomistic conception of evidence and the cascaded model, which implies that the probative weight of each evidentiary item is transferred to the factual propositions sought to be proven.[5]

---

[5] For works on Expertius, the Mexican expert system in family law, see: E. Cáceres, 'EXPERTIUS Technical Report' (rendered to the National Council on Scientific and Technological Research of Mexico); E. Cáceres, 'EXPERTIUS: A Mexican Judicial Decision-Support System in the Field of Family Law' (Proceedings of the 21st International Conference on Legal Knowledge and Information Systems, JURIX, 2008); E. Cáceres, 'Inteligencia artificial, derecho E-Justice (el Proyecto IIJ-Conacyt' (2006) 116 *Boletín Mexicano de Derecho Comparado* at <http://www.juridicas.unam.mx/publica/rev/boletin/cont/116/inf/inf12.htm>. For studies relating to the relationship between legal constructivism and artificial intelligence and law, see: E. Cáceres, 'Cognitive Legal Constructivism and Artificial Intelligence Applied to Law (CONACYT-IIJ Project: Expert Systems for the Assistance in Judicial Rulings)' Enrique Cáceres, 'Constructivismo Jurídico Fáctico y Elicitación del Conocimiento en el Proyecto Conacyt-IIJ-CCADET-STJT' (*La ciencia del derecho procesal constitucional. Estudios en homenaje a Héctor Fix-Zamudio en sus*

Nonetheless, a series of cognitive phenomena of great importance emerged from the knowledge elicitation activities. These phenomena had been overlooked by the communities referred to above.

Some of these phenomena are conceptualized as follows:

Determinant tendency: Judges decide based on previously formed heuristic structures that are stored in long-term memory. In each specific case, these structures are reactivated in the work-memory. It is not true that each and every case is solved individually, based on its own merits.

Self-fulfilled prophecy: When triggered, heuristic schemes that are strengthened by practice leads judges to 'pre-decide' the case. In other words, based on previous schema, judges often know what to look for and what they are going to find beforehand. The coherence of the final decision is usually more a result of the global condition of the judge's heuristic schema than of the case itself.

Interdependent evaluation of evidence: The probative weight assigned to each piece of evidence and the way each one is connected to the other usually conforms to predetermined decisions. In most cases, these predetermined decisions include common-sense beliefs that affect the neutral application of the law.

Cognitive refraction: When schema have been systematically reinforced, they become 'automatic cognitions' that can come to hinder relevant information from being filtered to identify and deal with a certain case differently than others of the same kind.

Cognitive individualization: When a certain type of cases does not motivate sufficiently reinforced schema or when community practices have not been properly shared, similar cases are usually solved in different ways or based on arguments that despite their plausibility, are different from one another. This happens, for instance, when legal reforms have recently been introduced into the legal system or there is a change in precedents.

25.4.2 The experimental research of Dan Simon and his Coherentist-Connectionist Model[6]

Upon finishing EXPERTIUS, Simon was found to have carried out experimental research on juries, the results of which are very similar to those of EXPERTIUS. Some of them include:

---

cincuenta años como investigador del derecho, t. XI, *Justicia, federalismo y derecho constitucional*. 2008. at: <http://www.bibliojuridica.org/libros/6/2563/21 .pdf>); E. Cáceres, 'A Constructivist Model for the Judicial Determination of Facts' (Proceedings of the Eleventh International Conference on Artificial Intelligence and Law, 2007); Enrique Cáceres, 'Juridical Constructivism, Artificial Intelligence, and E-Justice (The Mexican IIJ-Conacyt Project)' (Workshop on The Role of Legal Knowledge in E-Government, Proceedings of the Eleventh International Conference on Artificial Intelligence and Law, 2005). For works relating to cognition and applied legal epistemology, see: E. Cáceres, 'Cognition, Epistemology, and Reasoning about Evidence within the Legal Domain' (2008) 2 *PROBLEMA, Anuaro de Teoría y Filosofía del Derecho* at <http://www.juridicas.unam.mx/publica/librev/rev/filotder/cont/2/pr/pr11.pdf>.

[6] Dan Simon, 'A Third View of the Black Box: Cognitive Coherence in Legal Decision Making' (2004) *The University of Chicago Law Review*.

- The two test groups adjusted the same set of evidence in such a way so it would be coherent with their initial classification of the facts in terms of innocence or guilt.
- The coherentist displacements (changes in the general state of the system) carried out during a decision-making process are determined by the previous states of the system.
- The coherentist displacements operate at an unconscious level, which induces us to revise in depth the supposed rational and conscious control of judicial decisions.
- Each coherentist displacement has an impact on the structure of the heuristic system at a global level.
- Elements that could be viewed as peripheral can have a significant influence on the final decision, as happens with subjective beliefs.
- The subject's motivations and attitudes play a very important role in the final decision. A group of individuals in favour of the death penalty displayed a propensity to producing guilty verdicts, while individuals against the death penalty tended to produce innocent verdicts.
- Coherent structures are volatile; they tend to disappear over time if they are not systematically reinforced.

25.4.3 Connectionism, jurors (common law tradition), and judges (civil law tradition)

Although Simon's research was centred on jurors, his results can be extended to judges in the civil law tradition. However, this tradition implies a series of relevant cognitive differences that arise from its positivistic assumptions and the central role legislation and doctrine (differences at the belief level) play in this tradition, as do dogmatic theories. On the other hand, given the aspects that elude Simon's theoretical objectives, his work does not deal with the problem of argumentative schema visualization that AIL deems significant. Presented as a theoretical metaphor, I present a descriptive model of the cognitive processes carried out by civil law tradition judges. I call it the 'Constructivist–Connectionist Mental Model of Judicial Reasoning' (CCMMJR) or, to simplify, the 'Judicial Connectionist Mental Model' (JCMM). I use criminal law as our point of reference; nonetheless I believe the general theoretical assumptions of the model can apply to any other legal domain within the same tradition. Though yet to be demonstrated, criminal law doctrine assumes that this model can be instantiated by any type of crime.

25.5 The Constructivist–Connectionist Mental Model of Judicial Reasoning of the Civil Law Tradition

25.5.1 Mental models

The term 'mental model' was first used by Kenneth Craik in 1943.[7]

Some definitions of the term in question include: 'mental representations of the real world or imaginary situations'; 'representations of reality that people use to

---

[7] Kenneth Craik, *The Nature of Explanation* (Cambridge, Cambridge University Press, 1943).

understand a specific phenomenon'; 'psychological representations of the real world or imaginary situations'; 'an explanation of a person's thought processes on how real world operates'; and 'declarative models of the way the world is organized and that contain general or abstract knowledge and concrete instances'.[8]

Everybody has mental models about the world in which we live. For instance, we have models about our nation's history, its heroes and its eras; about what a gothic cathedral is or more specifically Notre Dame; about the distribution of space in our homes. Mental models are the navigation charts that determine the way we interact with the world and with others.

Scientific theories are also mental models of an abstract nature that allow us to understand, predict, and manipulate the world.

From the perspective of the connectionist model of memory, the symbolic elements that make up mental models (an image, for instance) are supervening structures. They supervene on neural connectivity patterns and brain functions that involve the amygdale, the hippocampus, and the cortex. This means that emotions also play a fundamental role in the development of these mental models.

As seen in the above definitions, mental models may contain both abstract knowledge and concrete instances. This implies the use of episodic memory (concrete experiences) and semantic memory (abstract structures), as well as long-term memory (heuristic knowledge) and working memory (decisions regarding concrete cases). These kinds of models are not states, but an ongoing process of establishing structural connections between abstract schema and experiences.

There are different types of mental models: conceptual models, structural models, causal or procedural models, and complex models (models that incorporate all three other types).[9]

For the purposes of this paper, I will use the term 'mental model' to allude to any complex system of mental representations, as well as their production rules.

Regarding the epistemological status of these mental representations, they can include mental images or propositions related to real, imaginary and performatively constituted events. These events can correspond to the 'external world', possible worlds, or socially constructed reality. They may be about past, present, or future phenomena; be explicit or implicit; or based on ordinary knowledge or expertise (specialized knowledge).

The type of mental model I propose involves propositional representations (normative constructs), mental images, and performatively constituted events that belong to a socially constructed reality rooted in legal discourse, as well as expert implicit or explicit knowledge.

---

[8] J. J. G. Van Merriënboer, R. E. Clark, and M. B. M. De Croock. 'Blueprints for Complex Learning: The 4C/ID-model' (2002) 50(2) *Educational Technology, Research and Development* 39–64.

[9] E. Peñalosa and S. Castañeda, 'El Análisis Cognitivo de Tareas, Base para el Diseño de Instrumentos de Evaluación en el Aprendizaje en Línea' (2009) 2(1) *Revista Iberoamericana de Evaluación Educativa* 162–85 at <http://www.rinace.net/riee/numeros/vol2-num1/art9.pdf>; S. Castañeda, 'A Cognitive Model for Learning Assessment' (2002) 12(1–4) *International Journal of Continuing Engineering Education and Life-long Learning* 94–106.

As to the question of whether the mental model is concrete, abstract, scientific, or common sense, it is a mixed model comprised of abstract concepts that correspond to the general nature of legal norms. It is also pseudo-scientific in the sense that it incorporates elements from legal theory and principles of common sense (given that judicial decisions presume these kinds of beliefs).

The architecture of the mental model is a complex one made up of diverse concepts, the relationships between them and causal links among them. This structure, which justifies the use of a neuronal metaphor, will be further explained below.

25.5.2 Connectionism

The two fundamental paradigms of cognitive sciences are known to have been built on theoretical metaphors. One of these metaphors is that of computers and the other is the neural brain structure.

For this paper, we will use an artificial connectionist model, which I define as follows:

Formal definition: A neural network is a directed graph with the following characteristics:

Each node i corresponds to a state variable xi.
Each connection (i,j) for nodes i and j corresponds to a weight $wij\ \varepsilon R$.
Each node i corresponds to a threshold θi.
Each node i corresponds to a function $fi\ (xj,\ wij,\ \theta i)$, which depends on the weight of its connections, the threshold, and the conditions of the j nodes connected to it. This function provides 'the new state of the node'.[10]

McClelland informally defines a neural or connectionist network as a system with the following components:

A set of elementary level processors or artificial neurons
A pattern of connectivity or structure
Dynamics of activation
Learning dynamics or rules
The operational environment.[11]

An example of each of these components is given below.
A graphical representation of an artificial neuron is the following:[12]

---

[10] Bonifacio Martínez del Brío and Alfredo Sanz Molina, *Redes neuronales y sistemas difusos* (México, Alfaomega Ra-Ma, 2002) 24.
[11] J. L. McClelland and D. E. Rumelhart (eds.), *Parallel Distributed Processing*, vol. 2, *Psychological and biological models* (Cambridge, MA, MIT Press, 1986).
[12] D. E. Rumelhart and J. L. McClelland (eds.), *Parallel Distributed Processing*, vol 1, *Foundations* (Cambridge, MA, MIT Press, 1986).

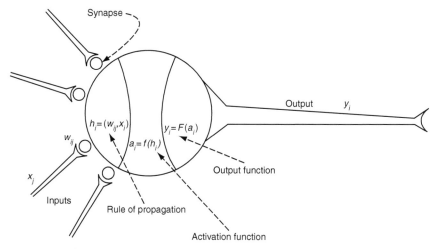

Figure 1 Artificial neuron.

The following charts illustrate the neural connectivity pattern.[13]
The connection between pre-synaptic (j) and post-synaptic (i) neurons:

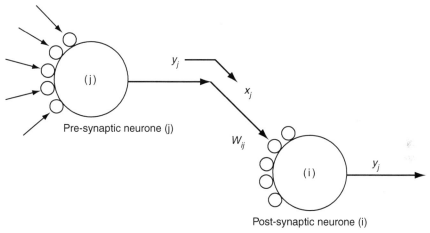

Figure 2 Neural activity pattern.

Figure 3 (below) presents a visual representation of a neuron with multiple pre-synaptic input and post-synaptic output.[14]

The structure of a neural network is necessarily composed of two layers: an input layer made up of input neurons without synapses and an output neuron that includes output neurons without synapses (vid., p. 24). Input neurons are those

---

[13] Bonifacio Martínez, 15.
[14] Ibid 19.

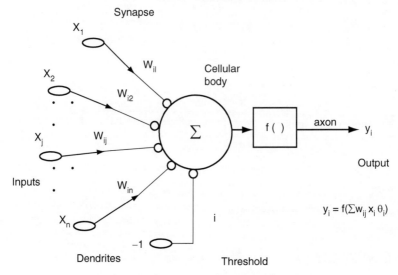

Figure 3 Neuron with multiple pre-synaptic input and post-synaptic output.

that receive the environmental stimuli while output neurons are those that provide the information capable of triggering motor behaviour. A visual representation is shown below:[15]

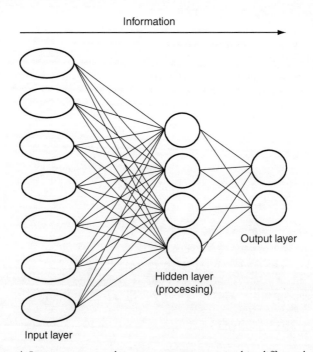

Figure 4 Input neurons and output neurons organized in different layers.

[15] Ibid. 24.

Connectivity patterns may operate both unidirectionally and bidirectionally on an intra and inter scale, and recurrently. See the following illustration:[16]

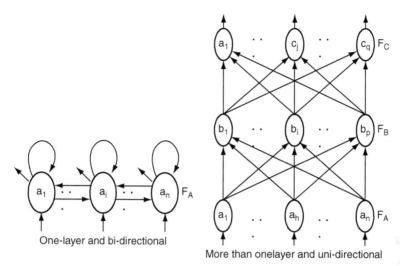

Figure 5 Connectivity dynamics.

The dynamics of activation are determined by the function $fi\ (xj,\ wij,\ \theta i)$.

The learning dynamics imply the changes in the state of the entire network and includes three possible dynamics that correspond to three temporal scales:

a) Dynamics of state (fast temporal scale) that corresponds to the activation of the network and is equivalent to a way of remembering.
b) Dynamics of pattern (intermediate temporal scale), which is slower than the previous one, and is equivalent to a way of learning that implies changing the parameters of activation between nodes or connection intensities.
c) Graph dynamics (slow temporal scale), which is the slowest of them all, and implies reshaping the structure of the network.[17]

25.5.3 Mental models, elicitation, and representation of judicial knowledge

One of the known features of an expert's heuristic knowledge is that it is amassed knowledge the specialist is not aware of.

Therefore, qualitative research on the elicitation of heuristic knowledge is needed to develop mental models about judicial reasoning.

For our purposes, we have worked with one of the few judges in Mexico who specializes in oral criminal procedures, which have been recently incorporated into Mexican legal system by way of a constitutional reform.

---

[16] Ibid. 23.   [17] Ibid. 34.

We used the following elicitation techniques:

- *Unstructured interview*: This was used at the beginning of the sessions to establish initial contact with the judge's reasoning processes.
- *Shadowing self*: With this technique, the judge analysed videotaped trials in which she participated and made observations, explaining the mental processes taking place in her mind at that moment.
- *Interruption analysis*: The researcher uses this technique to identify specific points of interest that arise during the shadowing self stage and asks in-depth questions for further clarification.
- *Mind maps*: Two kinds of mind maps are developed: those made by judges themselves, and those done by researchers. The latter reconstruct certain explanations in a theoretical format, which are then submitted to the judge for successive corrections until an accurate representation is attained.

Analyzing the collected data led to the idea of using an artificial neural network as a theoretical metaphor of the judicial mental model.

25.5.4 The mental model

As indicated above, the connectionist mental model of judicial reasoning constitutes a second-order theoretical metaphor in the sense that it is based (though not exclusively) on the theoretical metaphor of artificial neural networks, which is, in turn, based on the theory of biological neural networks. The reason behind this is that the theory of artificial neural networks is a simplified model of biological neural networks that is useful for the theoretical purposes of this paper.

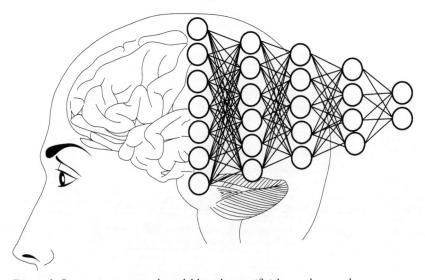

Figure 6 Connectionist mental model based on artificial neural networks.

Coherentist Theory of Judicial Reasoning in Civil Law Tradition 471

The characterization of the model is provided below in Figure 7.

The neural system corresponds to the mental space of an expert legal operator. Its sensors are made up of his sensory organs and its effectors, by his motor behaviour, basically those corresponding to the performative aspects of linguistic or verbal behaviour.

The definition of networks is then expressed as follows:

25.5.4.1 Neurons

Neurons are the concepts that belong to the following classes: narratives, evidence, the legal definition of the crime in question, factors that exclude criminal responsibility, general criminal law and the final sentence.

The correlation between this point and the connectionist theory of human memory lies in that these concepts supervene on neural connectivity patterns.

25.5.4.2 Pattern of connectivity or structure

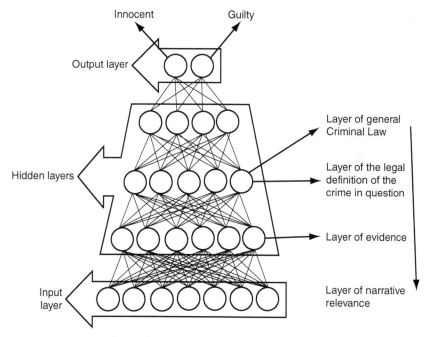

Figure 7 Structure of mental model of judicial reasoning.

a) A multi-layered network with hidden layers.
b) An input layer that corresponds to the terms contained in the case narratives the legal operator considers legally relevant.
c) A hidden layer immediately next to the input layer that corresponds to the evidence.

d) A hidden layer of a higher priority than the layer of evidence that corresponds to the definitional components (properties) of the statutory definition of the crime.
e) A hidden layer of a higher priority than the definition of the crime that corresponds to the factors that exclude criminal responsibility and are consistent with the statutory definition of the crime in question.
f) A hidden layer of a higher priority than that which excludes criminal responsibility that corresponds to the concepts contained in the general theory of criminal law.
g) These connections may be bi-directional, inter-layer, intra-layer, and feedback connections. The connectivity density is high and operates not only between adjacent layers, but among all the layers.

25.5.4.3 *Dynamics of activation*

It is assumed that different neurons are activated according to the fulfillment of an *active-state-threshold*, which is equivalent to a *certainty state* or a heuristic belief. It is also assumed that the intensity of neural connections does not operate deterministically, but in a fuzzy or plausibilistic manner. The model has three states of intensity which are represented by three different colours: yellow for 'Low', orange for 'Medium', and red (which correspond to the fulfillment of the threshold and of the respective neuron(s)). Synaptic connections may be excitative or inhibitory.[18]

Learning dynamics imply different changes in the arrangement of the network. This set of changes is linked to the timeline representing the sequence of events of a

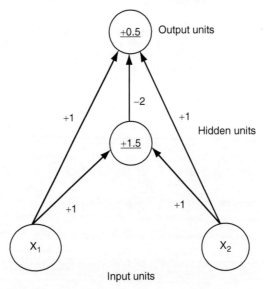

Figure 8 Learning dynamics.

---

[18] Antonio Crespo, *Cognición humana* (España, Editorial Centro de Estudios Ramón Areces, S.A, 2002) 125.

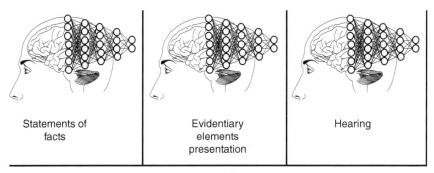

Figure 9 Time line of the judicial mental model.

given legal procedure. It also implies state dynamics that correspond to the activation of connectivity patterns in the memory of the system when in contact with the environment (the disputed legal case in question). Finally, it entails pattern dynamics derived from the modifications to the values assigned to the different concepts throughout the procedure.

We have deliberately excluded graph dynamics because we assume this would require legislative reforms to the procedure.

Operational environment, which is conformed of the courthouse and the inputs of the disputed legal cases brought before the judicial operator. These inputs are basically conceived in a narrative mode (the stories presented by the parties, witnesses, expert opinions, etc.).

Below is an example of how the Mental Model works:

25.5.4.4 The Hattori case

Hattori, a seventeen-year-old Japanese student who was a guest at the Haymaker family's home, and Haymaker junior (also seventeen) were invited to a Halloween party organized by their senior high classmates.

On the afternoon of Saturday, 17 October 1992, the two teenagers headed to Baton Rouge (where the party was going to take place). Hattori was dressed up as 'John Travolta' while Haymaker pretended to have an injured arm and wounds in his face.

They had not realized they had the wrong address until it was too late. So, they ended up at Peair family's home thinking the party was there.

The Peairs family consisted of Rodney (the husband), Bonnie (the wife), and their daughters Brittany and Stacey, eleven and seven years old respectively.

Haymaker junior rang the doorbell. When Bonnie opened the door, she could only see Haymaker, but when she turned and saw Hattori, she immediately shut the door in the boys' faces in a show of intense fear without allowing them to explain what had brought them there.

Once inside, Bonnie shouted frenetically to Rodney (who was upstairs) who grabbed a gun and came down at once. Bonnie then took her daughters and ran to hide in a backroom downstairs from where she was about to call the sheriff's office.

On coming downstairs Rodney could hear his wife shouting. When he reached the backroom, he saw that they were scared to death. Rodney and Bonnie looked at each other for a few seconds, and without a word, Rodney headed for the door.

Bonnie did not make that phone call as she intended to and told her daughters to stay in the room. She stood up and joined her husband on his way to the front door.

On his way out, Rodney took one of the many weapons found in the house, a Magnum 44 pistol that was already loaded. Before reaching the door, he had already taken the safety off and his finger was on the trigger ready to fire.

Without exchanging any sort of information as to what Bonnie had seen, both husband and wife finally opened the door.

Hattori and Haymaker junior were already leaving after what seemed to them a strange reaction from Bonnie.

When they heard the door opening, they turned around. Hattori started walking towards Bonnie and Rodney saying 'I've come to the party'. At that moment, Haymaker shouted to Hattori not to go further, but Hattori did not listen since he did not believe he was in any danger and simply wanted to ask if they knew about the party.

Rodney shouted 'Freeze!' Since Hattori was a foreign student who had just arrived to the US, he did not understand the instruction and kept walking toward the couple.

When Hattori was about five feet away, Rodney lifted his weapon to his shoulder and fired at Hattori's chest, killing him.

The operational environment is made up of various inputs that correspond to the various narratives found in the case: that of Rodney Peairs, of Bonnie Peairs, of Haymaker, of Haymaker's parents, and so on.

Each party declares the following:

The accused states: I was in my room when my wife, frenetically shouted that I ought to come downstairs quickly with a pistol. Then, I saw a strange-looking individual coming from the front yard to the door. I opened and shouted 'Freeze,' but he did not listen and kept coming towards me. So, I shot him in the chest.

The wife of the accused: The news said an extremely dangerous robber had killed a lot of people in the neighborhood. When I saw the strange-looking guy lying in my garden, I immediately thought it was him.

Statement of the victim's friend: Hattori was an exchange student from Japan. It was Halloween and we were dressed up for the occasion. We had the wrong address for the party. With his limited English, he didn't understand the colloquial expression 'freeze'.

The judge reconstructs the case based on the above testimonies to produce a coherent narrative.

It should be mentioned that in order to avoid unnecessary complications, we have used a simple example that does not include the problem of having two or more versions that could explain the same evidence. However, it should be noted

Coherentist Theory of Judicial Reasoning in Civil Law Tradition 475

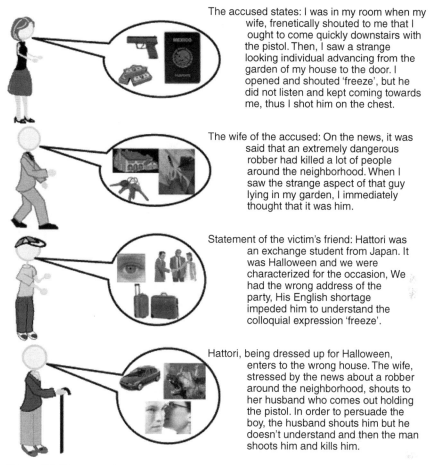

Figure 10 Case narratives.

that this problem, which is particularly relevant in the field of artificial intelligence, is found at this point of our model.

The legally relevant elements of the reconstructed narratives that activate the input layer are the following:

→Relevant Discoursive Elements
1) The accused shoots and causes the Hattori's death.
2) Hattori trespassed private property.
3) Hattori had a menacing aspect.
4) Hattori did not heed the instruction to stop.

As a result of the activation of the input layer (legally relevant facts), a pattern of connectivity is activated between the input layer and the layers that correspond to

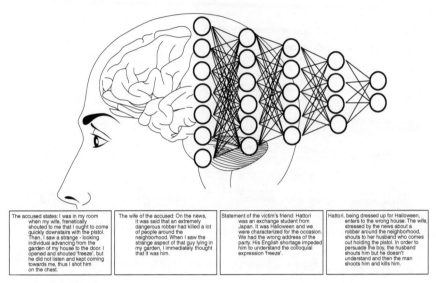

Figure 11 Judicial narrative reconstruction.

the statutory definition of the crime (possible homicide), and to the factors that exclude criminal responsibility (possible self-defence) related to legally relevant facts 2–4. The intensity of the connection has low plausibility value.

The statutory definition of homicide instantiates the theoretical concept of a crime, so the respective neuron is activated at the corresponding layer.

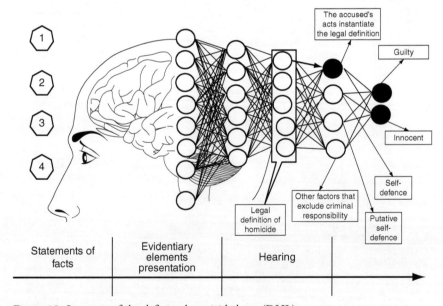

Figure 12 Structure of the defining hormicide layer (DHL).

Coherentist Theory of Judicial Reasoning in Civil Law Tradition 477

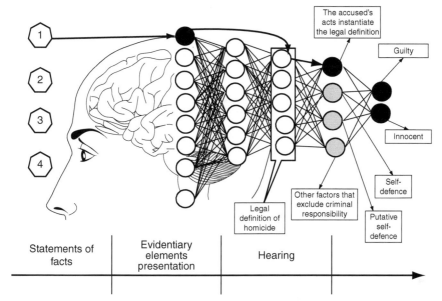

Figure 13 Activation of the DHL by discursive element 1.

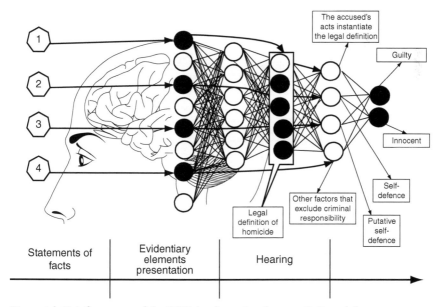

Figure 14 Reinforcement of the DHL by discursive elements 2, 3, and 4.

The following inputs are given during the probatory stage:

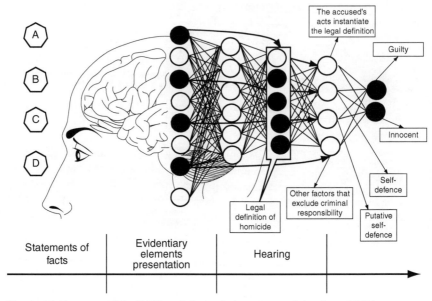

Figure 15 Structure of the DHL and the excluding responsibility layer (ERL).

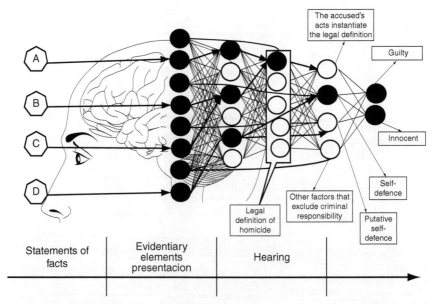

Figure 16 Reinforcement of the DHL by evidentiary elements (EE) 'A', 'B', 'C', and activation of the ERL by EE 'D'.

These pieces of evidence activate the second hidden layer and correspond to the confession ('A') and the two expert opinions ('B' and 'C').

Reinforcement of the pieces of evidence (excitative weight) increases the intensity of the connection in the second hidden layer, the corresponding threshold is fulfilled, and the neuron corresponding to the justified belief that Rodney committed homicide, is activated. However, it must be pointed out that this belief does not yet satisfy the required threshold to activate the layer that corresponds to the general theory of criminal law because the other neurons corresponding to the factors that exclude criminal responsibility such as self-defence are still activated, even though their weight may be 'low' (inhibiting).

Nevertheless, we should keep in mind that there is also evidence 'D' that corresponds to Hattori's friend. This person's testimony indicates first of all, that the victim was a Japanese student (which neutralizes the hypothesis of Hattori being dangerous); second, that he did not master the language (which explains why he kept walking toward Rodney despite the warning given); and third, that they had the wrong address (which explains why Hattori had entered Peairs' property with such confidence).

These elements create a pattern of inter-layer connectivity towards the neuron corresponding to self-defence. Since the pattern has an inhibitory weight, the belief in possible legitimate self-defence is deactivated.

Nonetheless the layer corresponding to the factors that exclude criminal responsibility remains active with a low plausibility value, due to a possible subjective legitimate self-defence (putative self-defence). Putative self-defence implies that the homicide had been the result of Bonnie and Rodney's experiencing an insurmountable subjective belief that they were facing a real threat with a plausibility value of 'Medium'.

But there are two other pieces of evidence.

'E' is a testimony that indicates that Hattori did not walk towards the Peairs in a threatening way but cheerfully. It also shows that Hattori's 'John Travolta' costume would not lead someone to think he was dangerous.

'F' is another testimony that indicates that Hattori had only taken two steps towards Bonnie and Rodney when he was shot.

The neurons activated by these elements generate a pattern of connectivity with inhibitory weight with respect to the neuron that corresponds to the factor that excludes criminal responsibility of subjective legitimate self-defence. This last neuron of subjective self-defence, along with the layer of all the factors that exclude criminal responsibility, is deactivated. In this state, an inhibitory weight is transferred to the layer that corresponds to unjustified crime, surmounting the threshold that becomes an active state. This state generates a last and new pattern of connectivity with the output layer that activates the neuron that corresponds to 'criminally responsible' that activates the system effectors and its corresponding motor behaviour, which in this case leads to pronouncing a sentence with the corresponding performative effects of changing the state of affairs of the legal operator's operational environment. As a result, a new legal reality is built.

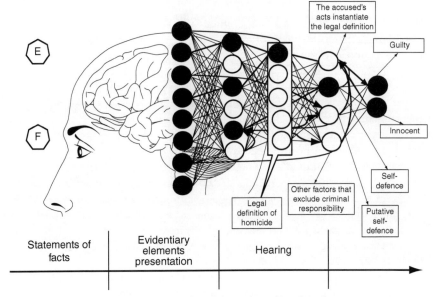

Figure 17 Activation of ERL with a plausibility value of 'medium'.

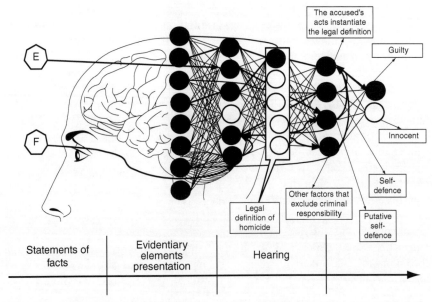

Figure 18 Deactivation of ERL by EE 'E' and 'F', and final state of the system (Guilty).

25.6 Discussion

If the connectionist mental model of judicial reasoning is accurate, it provides a descriptive explanation of the cognitive processes carried out by judges.

Based on this model, it is possible to evaluate the degree to which epistemological requirements applicable to any system of investigation leading to true, or at least justified, beliefs are satisfied. An evaluation of this kind seems more fruitful than those limited to pre-legislative abstract considerations on how, in theory, a criminal procedure can be implemented.

The model also draws attention to the fact that the vague expressions used in adjudicatory rules should be understood from a systematic perspective, and not as linguistic expressions without any context at all. These expressions seem to denote states of cognitive closure in mental models that presuppose a set of complex processes, the rationality of which lies in their holistic operation. Metaphorically speaking, these states of cognitive closure are equivalent to the effects of the expression 'checkmate' in chess, which has no meaning without the constitutive function of the rules it implies and the final state of certain mental models generated on the basis of these same constitutive rules.

Another positive aspect of the model is that it is possible to clearly see that discussions about conflicting narratives in law (a typical discussion in the AIL community) distort and hide highly complex processes that lie beyond the scope of the dialogical dimension between the narratives. Many of their generalizations only make sense within the framework of a complex mechanism of reflexive equilibrium occurring at different layers in the system.

The model also provides a broad holistic context in which discussions about opposed narratives have a place of their own, at least in the civil law tradition. The holistic nature of the model appears to overcome the simplifications associated with the atomistic model of evidence. The model is compatible with both the logic of evidentiary reasoning (which includes parameters, factors and critical questions) and non-monotonic logic (since a neural network can also operate as a Turing machine and may thus represent logical operations as well). The conceptual device of different thresholds makes it possible to think in fuzzy logic applications.

Based on the model developed in this paper, the idea that judicial reasoning may be better explained from a constructivist–coherentist–connectionist approach is reinforced. The states dynamics of the model's networks show the constant search for coherence characteristic of judicial reasoning. However, it also shows that, contrary to what can happen with jurors (for whom common beliefs or general social representations play a crucial role), in the case of the civil law tradition and of the system of beliefs that constitute it, different mental models with specific properties are constructed by the operators acting within the framework of this tradition.

25.7 Future research

In this paper, we have presented a simplified version of networks. We have not included a variety of aspects that are relevant for legal epistemology and procedural law. Nonetheless, it seems that within the topography of the model we can pinpoint elements like the standard of proof as an intermediate layer (just before the output layer) with a threshold equivalent to the degree of severity the standard of proof in question requires.

It is also possible to include other concepts that legislators use in the model to secure the standard of proof, such as presumptions and burdens of proof. These elements may be represented as default weights, etc.

In the computer field, the model may serve as a guide for developing three kinds of applications: a judicial knowledge elicitation application that would allow to build qualitative and quantitative models of cases and of problem-solving strategies (to build a database of EXPERTIUS-type scenarios); an expert system (EXPERTIUS II) that would support judicial decision-making; and a system for teaching judicial reasoning online.

Within the domain of teaching legal-reasoning skills, the model is compatible with the pedagogical theories oriented at complex learning. It is especially compatible with theories that use the paradigm of complex mental models as a starting point. Identifying different concepts and the different patterns of connectivity patterns linking them, as well as their operational dynamics, coincides with the classification of mental models suggested by Sandra Castañeda and the C4/ID model. Future research along this line could consist of identifying the cognitive tasks and skills required to perform them at each layer in the network. It could also consist of identifying adequate pedagogical strategies to induce those skills.

25.8 Social Justification

One of the reasons that makes developing the kind of research proposed in this paper very attractive in Mexico is the political will to put an end to a long, dark history of corruption within criminal law institutions. This political will has taken the form of a set of constitutional reforms that advocate the adoption of oral criminal procedures. At the moment, we have a handful of expert judges to work with. The plan is to elicit their expert knowledge and represent their mental models based on the general model outlined in this paper. We then could seize the advantages of online teaching to design courses that could be applied nationwide. We have eight years to incorporate the above-mentioned oral legal procedure in Mexico, but if we are not aware that law is built on psycho-cognitive processes, and not out of dead linguistic entities, history will repeat itself and we will let other (corrupt) mental models continue to perform the role of being the basis upon which a non-desired social reality is created by the activities of one of the most important institutions that justifies and allows us to understand the law: the judiciary branch.

26

Evolutionary Jurisprudence: The End of the Naturalistic Fallacy and the Beginning of Natural Reform?

*Morris B. Hoffman**

It is becoming increasingly clear that some human moral intuitions, and therefore great swathes of the law, are deeply embedded products of natural selection, not mere cultural constructs. But how much traction does that insight actually produce when it comes to evaluating particular laws against their evolved moral core? There are many roadblocks, but one of the most daunting is the difficulty known to philosophers as 'the naturalistic fallacy'. The central idea of the naturalistic fallacy, most famously articulated by the Scottish philosopher David Hume, is that the world of morals cannot be completely derived from the world of observed behaviours. That is, the 'is' is not the 'ought'. We may have evolved predispositions to kill in certain circumstances, for example, but that does not justify murder as a moral act. The naturalistic fallacy seems to be a giant stumbling block in the path of any sensible formulation of an evolution-based ethics, let alone a useful tool for legal evaluation.

But why cannot the 'is' at least *inform* the 'ought'? In fact, Hume himself never suggested that it could not, and neither, exactly, did the author of the modern articulation of the naturalistic fallacy, the English philosopher G. E. Moore. Instead, what both of them were complaining about was the apparent categorical difference between the 'is' and the 'ought', and the troubling tendency of early moral philosophers to skate back and forth between the two without any logical foundation to support them. But natural selection may be just the kind of connective tissue to support a limited kind of traverse across Hume's gap.

---

* District Judge, State of Colorado, Second Judicial District (Denver); Adjunct Professor of Law, University of Colorado; Member, MacArthur Foundation Law and Neuroscience Project; Research Fellow, Gruter Institute for Law and Behavioral Research. Of course, the author's views in this essay do not necessarily reflect the views of his judicial colleagues, the MacArthur Foundation or the Gruter Institute. The author thanks his law clerk, Cameron Munier, and intern, Robert Barlow, for their research and editorial help. He also thanks Theodore Blumoff, Fiery Cushman, William Pizzi, and Gideon Yaffe for their helpful remarks on earlier drafts.

In this essay, after surveying the naturalistic fallacy's philosophical landscape, and some of the scientific literature on the evolution of morality, I will propose a set of principles that legal decision-makers might use to determine what to do, if anything, when our moral intuitions conflict with existing legal doctrine. I will finish by applying those principles to the felony murder rule, arguing that the broadest Anglo-American forms of that rule are dissonant with our intuitions of moral blameworthiness and should be abolished.

26.1 The Fallacy of the Naturalistic Fallacy[1]

Hume started this all with the following passage from his *Treatise of Human Nature*:

In every system of morality with which I have hitherto met, I have always remark'd, that that the author proceeds for some time in the ordinary way of reasoning, and established the being of a God, or makes observations concerning human affairs; when of a sudden I am supriz'd to find, that instead of the usual copulations of propositions, *is*, and *is not*, I meet with no proposition that is not connected with an *ought*, or an *ought not*. This change is imperceptible; but is, however, of the last consequence. For as this *ought* or *ought not*, expresses some new relation or affirmation, 'tis necessary that it should be observ'd and explain'd; and at the same time that a reason should be given, for what seems altogether inconceivable, how this new relation can be a deduction from others, which are entirely different from it.[2]

This was the entirety of Hume's ruminations on what has since become known as the 'is–ought problem,' or sometimes 'Hume's gap'. Hume was complaining about the way in which moral philosophers and political writers moved from the 'is' to the 'ought' without so much as an ontological pause. He never suggested that the trip could not be made, that the 'is' must, by operation of some hidden and sacred axiom, never have anything to do with the 'ought'. He was, instead, making an observation that these ideas are canonically separate, and that to move from one to the other requires something more than mere logic. In modern philosophical parlance, moral truths cannot be derived solely from non-moral ones.

But notice that Hume was very much *not* saying that the two worlds of description and prescription were unbridgeable. Quite the contrary, Hume, like virtually every serious philosopher before the post-modern era, very much believed that the two worlds of nature and morals were connected.[3] He connected them not with Greek *a priori* notions like truth, beauty, or virtue, but with the rather astonishing pre-Freudian and pre-mirror neuron insight that we develop morals

---

[1] I borrow this section title from Francis Fukuyama, who titled a section of one of his books *Why the Naturalistic Fallacy is Fallacious* (Fukuyama (2002)).
[2] Hume (1740).
[3] MacIntrye (1959).

from our own desires to experience pleasure and avoid pain, and then externalize them to others via empathy:

> [A]nd as everything, which gives uneasiness in human actions, upon the general survey, is called Vice, and whatever produces satisfaction, in the same manner, is denominated Virtue; this is the reason why the sense of moral good and evil follows upon justice and injustice.[4]
>
> All mankind so far resemble the good principle, that, where interest or revenge or envy perverts not our disposition, we are always inclined, from our natural philanthropy, to give the preference to the happiness of society, and consequently to virtue above its opposite. Absolute, unprovoked, disinterested malice has never perhaps place in any human breast; or if it had, must there pervert all the sentiments of morals, as well as the feelings of humanity.[5]

I know it is wrong to hit you (even though people often hit each other) not because my family, culture, or God teaches me, but simply because I know it hurts when someone hits me.

Hume went on not only to formulate this theory of the natural origins of moral intuition, but to construct an explanation of the tension between selfishness and selflessness that produces in us a conflict about when to follow those moral intuitions (more modernly, when to cooperate) and when to disregard them (defect):

> After men have found by experience, that their selfishness and confined generosity, acting at their liberty, totally incapacitate them for society; and at the same time have observed, that society is necessary to the satisfaction of those very passions, they are naturally induced to lay themselves under the restraint of such rules, as may render their commerce more safe and commodious.[6]

These are remarkable pre-Darwinian observations, and put paid to any suggestion that Hume believed the 'is' of nature can never inform the 'ought' of morality, since he recognized that morality itself was somehow a product of that nature.

Hume's original is–ought complaint was named the 'naturalistic fallacy' by G. E. Moore, who expressly applied it to evolution by posing this single question: 'This conduct is more evolutionarily successful, but is it good?'[7] This is, of course, a perfectly legitimate question, precisely because the evolved 'is' is no more organically good than any other kind of 'is'. But just as with Hume's observations, Moore's most definitely do not mean that the two worlds of evolved behaviour and morality are unrelated. The answer to Moore's question is that evolved behaviours are not *necessarily* good, but that does not exclude the possibility that our deepest notions of which behaviours are good and which are bad might themselves be products of natural selection.

Critics of an evolutionarily-informed view of morality often overstate the nature of Hume's and Moore's observations, turning them into a kind of intellectual demilitarized zone, where nature and morality are forever forbidden to copulate, as Hume might say. The Social Darwinists, and other misunderstanders of natural selection, are no doubt largely responsible for this knee-jerk invocation of the

---

[4] Hume (1740). [5] Hume (1751).
[6] Hume (1740). [7] Moore (1948).

naturalistic fallacy. Indeed, Herbert Spencer was one of Moore's main targets.[8] But the Social Darwinists had a profoundly misguided and oversimplified idea of natural selection. In their world, evolution produces brutally selfish machines, red in tooth and claw. And although much of the animal kingdom has its share of red teeth and red claws, it is also full of cooperation, from vampire bats[9] to chimpanzees.[10] The very evolution of social animals is a paradigm of the nuanced complexity of natural selection. Living in groups gave us a net individual survival advantage, but at the short-term individual cost of cooperating.

These ideas have left us with a much richer model of human nature than Spencer's caricatures. We generally cooperate with in-group members, but are always opportunistically testing them for chances to defect, while at the same time vigilantly guarding against their defections. The complex web of interrelationships between in-group members left us, as I will discuss in more detail in the next section, with a core of moral intuitions. Part of the evolved 'is' *is* the 'ought'.

Yet long after the articulation of this new synthesis about human nature, critics of evolutionary psychology, and especially of any kind of evolutionarily based moral realism, continued to trot out the naturalistic fallacy. One of my favourite versions of this misuse is the argument that retribution cannot be a legitimate justification for criminal punishment because it is grounded on personal revenge, and just because revenge is natural does not mean that it is good. An even stronger, utilitarian, version of this profoundly mistaken critique of retribution is that revenge is bad, and that the law must overcome our instincts for revenge with rational, scientific sentencing models whose legitimacy depends entirely on whether the punishment results in measurable rehabilitation and/or deterrence.[11]

The post-Darwinian truth that these critiques miss is that universal human punishment institutions owe their very existence to deeply rooted notions of individual revenge,[12] or what evolutionary theorists more carefully call second-party punishment. Second-party punishment occurs when the victim retaliates against the wrongdoer; third-party punishment occurs when someone other than the wronged person punishes the wrongdoer. There is a rich literature on the evolution of second-party punishment in many different kinds of animals,[13] and a growing literature on the evolution of third-party punishment in non-human primates[14] and humans.[15] Some very recent fMRI work has even shown that there is substantial overlap between the neural correlates of second- and third-party punishment in humans, suggesting an evolutionary connection between them.[16]

One need only look at the ubiquitous practice of ostracism—present not only in all human societies but also in many non-human primates[17]—to see how impoverished is the view that the 'is' of revenge does not inform the 'ought' of punishment. Every time a primate mother swats her child for, say, poking the mother in the eye, she completes an act of second-party punishment. When she

---

[8] Snare (1992). [9] Wilkinson (1984). [10] Hockings et al. (2007).
[11] Marcus (2004). [12] Hoffman and Goldsmith (2004). [13] Jensen et al. (200.
[14] Flack et al. (200. [15] Fehr et al. (200; Henrich et al. (2006.
[16] Buckholtz et al. (2009). [17] Gruter and Masters (1986).

inflicts the same punishment when the child pokes a sibling in the eye is she acting as a second- or third-party punisher? The answer, of course, is a little bit of both, because the close connections between the family members mean that an injury to one is an injury to all. This was precisely the environment in which we evolved—in small groups of mostly related individuals. The social tensions inherent in that environment—when do I defect, when do I cooperate—turned our second-party revenge into third-party punishment. Jane Lancaster said it best in her essay on ostracism in chimps: 'The cold shoulder is only a step along the way to execution.'[18]

26.2 The Science of Moral Realism

It seems indisputable, once we recognize that behaviours can have evolutionary components, that behaviours about behaviours—that is, our judgements about whether some behaviours are better than others—can also have evolutionary components. There is in fact a vast and growing literature on the evolution and neuroscience of moral intuition, both human and non-human.[19] And although no contemporary non-human primates seem to have moral systems quite like ours, they have all the behavioural ingredients necessary—attachment, bonding, cooperation, defection, defection-detection, and empathy—for what Michael Shermer has called 'pre-moral sentiments'.[20] Reciprocal altruism is itself a kind of pre-moral sentiment, requiring the ability to give and accept benefits with an anticipation of a promised return.

In their simplest form, moral intuitions are those evolutionarily predisposed cooperating behaviours that stand in strategic tension with short-term defection. I may be better off in the short run stealing that food that everyone else worked so hard to gather, but I risk being expelled from the group and losing all the long-term benefits of group living. The very fact that we evolved in small groups suggests a net adaptive benefit to such groups, and thus suggests that evolution would have put a premium on presumptively cooperative behaviours, and least on net.

But do not misunderstand. 'Cooperative' behaviour is not some *a priori* good. It comes from the relentless selfishness of evolution, a selfishness, at least in humans, that was filtered through the lens of intense social living. We do not cooperate for the good of the group; we cooperate because, on net, the group is good for our individual fitness. The ultimatum game is perhaps the most elegant example of this idea of selfishness disguised as selflessness. The reason I do not offer you only one of my ten dollars is probably not because of any abstract duty I feel to be fair to you, but simply because I recognize that you will reject that offer and I will not get anything.

---

[18] Lancaster (1986).
[19] de Waal (2006); deWaal (1996); Flack and deWaal (2000); Greene and Paxton (2009); Greene et al. (2001); Hauser (2006); Joyce (2007); Wright (1995).
[20] Shermer (2004).

Sufficiently sentient animals have a built-in problem when it comes to navigating the social tension and making the defect/cooperate decision in the real and complex world beyond economic games. If left without any heuristics, homo economicus would try to calculate the risks and benefits in each situation. How much will my genes and I really benefit from stealing that food? What are the costs, that is, what are the risks of detection and what is the likely reaction of the group? These are, in fact, impossible calculations, and they themselves will consume precious time and brain power. It would be much better if I were equipped with built-in presumptions that were triggered in common situations.

I have argued elsewhere that we evolved three central heuristics for navigating the most common social problems we faced at our emergence: 1) exchanges must be voluntary; 2) promises must be kept; and 3) serious violations of 1 and 2 must be punished.[21] Like all heuristics, especially evolutionary ones, these rules are far from iron-clad, and humans regularly violate them at the margins. That's because the self v. others tension remains dynamic and delicate. Indeed, it is not at all inconceivable that we simultaneously evolved a predisposition toward a given behaviour *and* a moral sense that the evolved behaviour is wrong. This kind of asymmetry could have served an important buffering function against some antisocial behaviours that are highly beneficial to the defector but seriously destabilizing to the group if we engaged in them in excess. Moreover, we cannot forget about the out-group. Our evolved cooperation was cooperation aimed at members only of our own group. Members of other groups were for the most part existential threats, not objects of potential cooperation.

Rule 1—exchanges must be voluntary—might be restated as 'we cannot steal things from each other', and I suggest that Rule 1 forms the foundations of tort and criminal law. I cannot intentionally take your property or health or life without being punished, and cannot even accidentally do so without compensating you, unless I was as careful as I should be. We will revisit the distinction between intention and accident when we discuss the felony murder rule. Of course, the group itself (and, later, the state) is, by its very nature, exempt from Rule 1. It regularly takes members' property, and sometimes health and even lives, and indeed such taking is in many ways its *raison d'être*. But evolution may also have equipped us with intuitions about the limits beyond which even the group or state may not go, based on the idea that an overbearing group or dominant member risks destabilization if too many individuals decide to leave the group.[22] The Enlightenment idea that the state could not encroach on certain inalienable individual rights was a political and philosophical reformulation of this central notion that there were adaptive limits to the authority of the group or dominant members.

Rule 2—promises must be kept—is a sentient animal's version of reciprocal altruism, and it was essential to our emergence as a relentless trader. Our commerce could not really bloom until we were able to solve the problem of non-simultaneous exchange, and we did that by exchanging promises for goods, and eventually even

---

[21] Hoffman (2004).   [22] Rubin (2002).

promises for promises, instead of goods for goods. Rule 2 is the foundation of contract law, the institution that critically increased the reliability of promises.

These two rules are not really rules unless their violation is punishable in some fashion. Rule 3—serious violations of Rules 1 and 2 must be punished—reflects both the need for the deterrent impact of punishment but also (by punishing only serious violations) the adaptive value of forgiveness and proportionate punishment.[23]

These ruminations about the nature of our embedded moral intuitions are not just theoretical ramblings. There is an exploding body of behavioural and neurological evidence that is beginning to map our evolved moral intuitions. Let me briefly touch on two areas that are particularly relevant to the criminal dissonances addressed later in this paper: 1) the universality of intuitions of relative blameworthiness; and 2) psychopathy.

26.2.1 Relative blameworthiness

In 2007, Paul Robinson and his colleagues published a series of groundbreaking papers showing that people across many demographics were in remarkable, and finely tuned, agreement about the relative blameworthiness of different crimes.[24] Participants were asked to rank twenty-four different crime narratives from least 'blameworthy' to most (without being given any definition of 'blameworthy'). This was not a matter of asking whether robbery or speeding was more serious. The twenty-four narratives not only spanned a wide range of harm, they also required subjects to make a host of seemingly complex and nuanced judgements about crimes that appeared very similar. In in-person versions of the experiment (there were also web-based versions), participants were often seen changing their minds about their ordering. I took a sample of the web-based version, and found myself switching my initial ordering several times.[25]

To give readers a flavour of how finely tuned these judgements were, here are three so-called 'adjacent' narratives—ones Robinson and his co-authors believed participants would rank right next to each other—all at the highest end of anticipated blameworthiness:

John works out a plan to kill his sixty-year-old invalid mother for the inheritance. He drags her to her bed, puts her in, and lights her oxygen mask with a cigarette, hoping to make it look like an accident. The elderly woman screams as her clothes catch fire and she burns to death. John just watches her burn.

John kidnaps an eight-year-old girl for ransom, rapes her, then records the child's screams as he burns her with a cigarette lighter, sending the recording to her parents to induce them to pay his ransom demand. Even though they pay as directed, John strangles the child to death to avoid leaving a witness.

---

[23] Godfray (1992); Hoffman and Goldsmith (2004); McCullough (2008).
[24] Robinson and Darley (2007); Robinson and Kurzban (2007); Robinson, Kurzban, and Jones (2007).
[25] On the other hand, what people seem to do is make pair-wise comparisons, so what appears to be 'switching' may really just be inserting new narratives into the previous ordering.

A woman at work reveals John's misdeeds to his employer, thereby getting him fired. John devises a plan to get even with her. The next week he forces the woman into his car at knife point and drives her to a secluded area where he shoots her to death.[26]

All the narratives along the continuum involved close calls just like these. And yet subjects exhibited a remarkable degree of agreement. Virtually every participant ranked the twenty-four narratives in exactly the same way, and where there were differences they were typically just one ranking off in adjacent narratives. The pair-wise agreement was in excess of 90 per cent, and a more sophisticated measure was even more impressive.[27] Equally impressive, these results hold over several different demographic slices. It seems all people have a deeply embedded, uniform, and sophisticated sense of relative 'blameworthiness' that cuts across gender, age, race, income, and education.[28]

Just as interesting, Robinson and his colleagues constructed twelve additional narratives that they guessed would not recruit any deeply shared moral intuitions. These involved things like drug use, drunk driving with no harm to others, and bestiality. And, in fact, the results were close to random.[29] This lack of agreement for non-core crimes was likewise independent of all measured demographics. The difference between core and non-core crimes deserves much more thought and experimental attention. But in general, it appears core crimes touch on the things that humans all care about for our survival—being free from having our property stolen and our persons violated. That is, behaviours that violate Rule 1—exchanges must be voluntary.

These experiments demolish the conventional relativistic attack on retribution—that there is no such thing as 'just deserts' because what is 'just' in one kind of case, with one kind of defendant, with one kind of judge and in one kind of culture may not be 'just' in other kinds of circumstances. It also strongly suggests that what humans do when we punish, at least when we punish serious violations of Rule 1, is grounded on just a few deeply held, evolved, notions—notions about harm, intentionality, and kin.

26.2.2 Psychopathy

One of the best ways to see the deeply embedded moral intuitions with which most of us are equipped, and how those moral intuitions contribute to our human nature, is to look at a discrete group of people who lack them—psychopaths.

---

[26] Participants almost universally ranked these three scenarios as follows (from least blameworthy to most): 1) abduction shooting of co-worker; 2) burning mother for inheritance; and 3) ransom, rape, torture, and strangling of eight-year-old.

[27] Kendall's W ('K'), also known as the 'coefficient of concordance', is a statistical measure of the agreement between rankers of ranked lists, which takes into consideration not only differences in rankings but the degree of those differences. When K=1.00, all rankings are exactly the same. When K=0, the rankings are random. The in-person results from the Robinson et al. study (n=64) had an average K of 0.95 ($p < 0.001$). The web-based results (n=246) had an average K of 0.88 ($p < 0.001$).

[28] To my knowledge, the Robinson study has not yet been done in pre-industrial societies, as the experimental economists have done with some of their games (Henrich et al. (2001)).

[29] K=0.55 ($p < 0.001$).

To oversimplify, psychopaths have no empathy or conscience.[30] They are, in some ways, the real homo economicus. They act purely for short-term individual gain, unencumbered by the delicate social tensions that bedevil the rest of us. They are what humans would have been had we not evolved in small groups. As Martha Stout put it, what distinguishes the psychopath from the rest of us 'is an utterly empty hole in the psyche, where there should be the most evolved of all humanizing functions'.[31]

Psychopathy is particularly important for the legal system because psychopaths are responsible for a grossly disproportionate amount of crime, especially violent crime. Psychopathy occurs in the general male population at the rate of roughly 1 per cent, yet the rate in prisons is between 15 per cent and 25 per cent.[32] In other words, prisoners are up to twenty-five times more likely to be psychopaths than are non-incarcerated males. There is no other variable—substance abuse, race, poverty, childhood abuse, or anything else—more highly correlated to being in prison than psychopathy. In fact, of the estimated 700,000 males in the US aged 18–50 who are psychopaths, at any one time roughly one-third of them are in prison.[33] Here's another startling statistic: 50 per cent of all police officers killed in the line of duty are killed by psychopaths.[34]

Once they get to prison, psychopaths spin the revolving door faster than anyone else. They use their manipulative skills to dishonestly obtain early releases at about 2.5 times the rate of their non-psychopathic colleagues.[35] Once released, they recidivate early, often, and violently. In one Canadian longitudinal study, 60 per cent of psychopaths committed a new *violent sexual* crime within four years of release, compared to 15 per cent of non-psychopaths.[36] At the study's ten-year end, a whopping 77 per cent of psychopaths had committed a new violent sex offence, compared to 40 per cent of the non-psychopaths. This is what people do who have no moral intuition.

Behavioural and neuroimaging studies are making it clear, contrary to some early suppositions that psychopathy is caused by frontal lobe damage,[37] that what psychopaths are missing are contributions from the paralimbic regions of the brain, regions that have been well documented to be associated with moral reasoning.[38] Psychopaths have profoundly lowered activation across virtually all

---

[30] Robert Hare has developed what is still the clinical gold-standard for assessing psychopathy, consisting of two sets of criteria, one affective and one behavioural (Hare et al. (1990)). Psychopathy remains, controversially, unrecognized by the DSM, which continues to describe the overlapping diagnosis of anti-social personality disorder (Kiehl and Hoffman (in submission)).
[31] Stout (2005).
[32] Kiehl and Hoffman (in submission).
[33] Ibid.
[34] Rain and Sanmartin (2001).
[35] Porter (2009).
[36] Harris et al. (1991).
[37] Lezak (2004).
[38] The paralimbic regions are generally below the neocortex, adjacent to and including some limbic structures. They form a kind of girdle surrounding the medial and basal aspects of the two hemispheres. They contain many important structures known to be associated with moral reasoning, including the anterior temporal cortex, the ventromedial prefrontal cortex, the amygdala, the insula, the temporal pole, and the cingulate.

of these key moral reasoning areas.[39] They are completely capable of logical reasoning, but their brains lack the moral heuristics—Rules 1, 2, and 3—with which the rest of us come equipped.

So when we consider what complex and fortuitous forces keep most of us out of prison and put others of us in—our families, our personalities, our friends, our economic circumstances, our race—we should never forget what appears to be the most statistically important driver of all: whether we are in the 99 per cent who have moral intuitions or the 1 per cent who do not.

26.3 Legal Dissonances and What to Do About Them

As I have argued above, most law instantiates, and is an extension of, our neurologically embedded moral intuitions. That is, to turn Holmes's famous bad man quote on its head, law is largely a prediction of how most neurologically intact humans would agree we should behave towards one another. But of course some laws conflict with our intuitions.

For example, all American jurisdictions, as well as England, have evidentiary rules that generally recognize that a jury may not be told about a criminal defendant's prior felony convictions. I will call this rule the 404(b) rule, after the American Federal Rule of Evidence.[40] I suspect, based on my discussions over the years with hundreds of jurors after trials, not to mention my own human intuitions, that 404(b) violates our deep and strongly held sense that the best predictor of future (or otherwise unsubscribed) behaviour is past behaviour. It is not hard to imagine that such a strategy would have been keenly adaptive in a world where our principal predator was each other. Social scientists call these past behaviours 'priors'. There is no data more important to a strategic actor than the other actor's priors. He is probably trustworthy if he was trustworthy before, and he is probably dangerous if he was dangerous before. And yet few would argue that 404(b) rules should be abandoned because they 'conflict' with our natural predilection to want to know about, and act on, priors.

On the contrary, we tolerate and even encourage this particular dissonance because 404(b) is itself a recognition of the unacceptable dangers in the intuition. Every juror in every criminal case would love to know whether this defendant has been convicted of any other crimes, especially if those priors are similar to the charged crime. But the law has made a policy decision that our intuition about priors is simply too strong for the purposes of the criminal law. In a weak case—a case in which the ordinary principles of burden of proof and proof beyond a

---

[39] Kiehl (2006).
[40] The American federal rule is FRE 404(b), which has also been adopted in one form or another in all states. The English have the same general rule in § 101 of the Criminal Justice Act 2003 (c. 44), though its exceptions are generally more forgiving than the American version. One giant and unique American exception is that a defendant's prior felony convictions may be disclosed to the jury if the defendant chooses to testify, though the jury is then instructed that they may consider those priors only for the purpose of impeaching the defendant's testimony, and not for propensity.

reasonable doubt should result in an acquittal—the sheer power of the intuition about priors will, or so goes this argument, overpower the jurors and increase the chances that innocent prior felons will be convicted. We may not all agree with this policy assessment, but we can probably all agree that in making it the law is attempting to deal head on with the impact of our strong intuition about priors. That is, this is not the kind of dissonance that necessarily requires closure.

Indeed, 404(b) may be just one example of what Owen Jones has called 'time-shifted rationality'.[41] Embedded predilections that made evolutionary sense 100,000 years ago may not make any sense today, and one of culture's (and law's) tasks is to save us from those evolved predilections that simply no longer work in our current environment. Relying heavily on priors made perfect evolutionary sense in deciding whether that stranger coming over the hill was friend or foe, but today 404(b) is a recognition that our criminal justice systems need to fight against that intuition if they are to remain careful truth-finding institutions and avoid becoming parodies that rely on 'the usual suspects'.

But other legal principles seem to conflict with our intuitions in an entirely different, less satisfying, way. The harshness of American sentencing practices is one example. Contrary to some conventional wisdom, the levels of punishment that ordinary people think are appropriate are significantly lower than the levels that the law imposes, and in particular various doctrines within the law (three-strikes, mandatory minima for drug offences, the adult prosecution of juveniles, complicity, the felony murder rule, and, in a few states, the abolition of the insanity defence).[42] And although such work could be criticized as unrealistic—believe me, there is nothing more daunting than having to sentence a real rapist with a real family in a real courtroom in front of a real victim and her family—its artificiality is actually one of its strengths. Asking subjects with no axe to grind—unburdened by the emotions of revenge or the weight of real decision-making—to tell us how they would punish hypothetical wrongdoers gives us a powerful tool to gauge just desert. It is, to my mind, all the more shocking that experimental subjects, who could take this cost-free opportunity to be tough on crime, as our politicians think all of us would be but for those soft-hearted judges and their soft-hearted laws, are in fact substantially less harsh than the laws themselves.

These counterintuitive results are consistent with studies examining the differences between judge and jury sentencing. Contrary to predictions, in those American states that have moved from jury to judge sentencing in non-capital felonies, judges are substantially harsher sentencers than their jury predecessors.[43] This difference, of course, can be explained by many things, including the notion that judges get hardened and sceptical by being exposed to so much human maltreatment. And of course it does not touch on the point that the main sources of the dissonance are certain special legal rules, since even in jury-sentencing states juries and judges alike are bound by those special rules. But it does resonate with the broader point that

---

[41] Jones (2001).
[42] Robinson et al. (2009); Robinson and Darley (2007).
[43] Smith and Stevens (1984).

there is a huge gap between what most of us say to each other in general (throw away the key) and what we actually say and do when faced with a particular crime, a particular victim and a particular criminal, even hypothetically.

This general punishment dissonance, unlike the 404(b) dissonance, should trouble all of us, at several different levels. What does it say about the effectiveness of our political systems that elected officials believe the best way to stay elected is to vote to increase the penalties for crime, when present sentences already exceed what ordinary people think are appropriate? What does it say about those ordinary people as voters, who, in the privacy of a behavioural study and even in the bright lights of a jury sentence, will be less harsh than what they seem to demand of our elected representatives in the privacy of the voting booth? And, perhaps most importantly, what will this kind of chronic and gnawing gap do to the enduring legitimacy of the law?

Don't get me wrong. I am not one of those critics who think our justice system is inappropriately punitive as a general matter. Indeed, I am a card-carrying retributivist. But the moral force of every just sentence is put in peril by every unjust one, and as long as sentences in general exceed what ordinary citizens think is appropriate, the law risks losing its prescriptive powers.

But now we are back to the naturalistic fallacy. How can we distinguish, on any kind of principled basis, the difference between dissonances like the 404(b) rule against the admission of priors—which I have characterized as a conscious effort by the law to overcome a dangerous evolutionary intuition—and unnaturally long sentences—which I have characterized as the danger? Why is the real danger not that natural sentences are too short, and that the law in its wisdom recognizes that without draconian doctrines like mandatory minima and three strikes laws the sentences that human judges impose will be inadequate? Am I just picking and choosing between intuitions and legal doctrines I like and those I do not?

Let me suggest several different approaches that might be useful in distinguishing dissonances that seem distorting from those that seem curative. First, and maybe most importantly, we need to think carefully about the claimed 'natural' principle with which the legal doctrine conflicts. In the 404(b) example, there should really be little doubt that our intuition to place great weight on priors is real, is powerful, and is probably evolutionarily rooted. But long versus short sentences? Remember, the Robinson concordance studies were about *relative* blameworthiness not absolute punishment amounts. Other than the proportionality command that the punishment must fit the crime, I suspect there is no tangible, widely shared, natural 'core' against which we can reliably measure absolute sentence lengths.[44] Amounts

---

[44] Robinson et al. call this the 'problem of endpoint'. Even though we all have powerful and refined notions of which kinds of crimes are worse than other kinds of crimes, those notions do not inform actual sentence lengths, since we may well disagree about where this range of relative crimes fits onto a yardstick of actual punishment. And it is not just a problem of endpoint. Even if we all agreed that the worst crime deserves sentence X, and that the least serious deserved sentence Y, there is no obvious reason why our imagined list should result in a linear distribution of punishments. The very worst crime may be twice as bad as the second worst. In the end, ordinality tells us limited things about cardinality.

of punishment vary widely across cultures and over history. Abortion and adultery were both banishable offences, and therefore typically capital, among the Cheyenne;[45] now, the one is an American constitutional right (of sorts) and the other only a matter of ethical aspiration. As late as the seventeenth century, in a society as 'civilized' as England, virtually every serious crime was at least theoretically punishable by death, from stealing a horse to writing a bad cheque.[46]

In fact, some recent research suggests that humans use different brain systems, with different results, when we decide how 'wrong' a third party's conduct is versus how much punishment to impose.[47] Most importantly, as I have already mentioned, the empirical studies have shown that these 'unnaturally' long sentences are the products largely of the operation of discreet legal doctrines. Those doctrines are fair game for a naturalist critique because they may in their own right conflict with some important embedded intuitions, and indeed in the next section I will engage in just such a critique of the felony murder rule. But I suspect there is no general dissonance that demands closure when it comes to harsh sentencing because there is likely no identifiable core when it comes to sentence length.

Second, we might examine the legal doctrine to see if *it* advances some core intuitions. After all, if we are dealing with competing cores then maybe there is no dissonance at all, just a policy choice about how the clash of two core principles should be resolved in a particular cultural and legal context. An example of this might be the doctrine of criminal complicity, under which one who aids another person in the commission of a crime is treated equally guilty. It seems, on the one hand, clearly wrong to punish Joe for Sam's actions; but on the other hand it is easy to imagine that our ancestors were deeply aware of the increased dangers that *groups* of predators posed, especially out-groups. The law of conspiracy—which runs counter to the principle that most punishable crimes need to result in actual harm—may be another example of our deep suspicion of out-groups. A legal principle's aetiology and pedigree may tell us something about whether it advances a core belief.

Third, we might think of trying to gauge how far the dissonant legal rule has in fact strayed from the core principle. This will often be a difficult task, because of course it is typically a qualitative undertaking, though the case of punishment allows some quantitative inquiry. So, for example, legal doctrines drive sentences beyond the levels of punishment that ordinary people impose—but how *much* beyond?

Fourth, in some instances we might think of a directional kind of test. If an evolved core principle tends to result in judgments about a given set of circumstances that takes those judgments in a certain direction, does the dissonant legal rule take them in the same or a different direction? That is, is the dissonance a dissonance just in magnitude or in kind? Here, blameworthiness rather than punishment might be the kind of measure more amenable to a directional test.

---

[45] Hoffman and Goldsmith (2004).
[46] 4 Blackstone (1765–9).
[47] Cushman (2008).

If a given circumstance gives us the moral intuition that a wrongdoer is less (or more) blameworthy than without the circumstance, but operation of the legal rule has the opposite effect, this is the kind of directional dissonance that may well suggest closure. For example, those who propose a rule that would excuse psychopaths—that is, that would include psychopathy as a mental disease or defect that could come within the definition of insanity[48]—are probably proposing a directionally dissonant rule, if we assume an intuition to punish psychopaths even more than non-psychopaths.

Finally, we might think of whether the legal principle is riddled with exceptions, or with jurisdictions rejecting it outright, with the idea that it may be a dissonant core driving all or most of those exceptions and rejections. This seems especially fruitful given the paradigm of the common law operating over time. When Rule X drives a dissonant Result Y, individual judges will tend to refuse to apply the rule, either outright or because of some crafted exception, and if enough of them refuse or except, the common law will abolish the rule outright or develop recognized exceptions to it. A rule rejected by a significant number of jurisdictions, or riddled with swallowing exceptions, suggests a rule that may conflict with our juridical intuitions. But here again, caution is in order. There are no doubt many exception-riddled and minority rules that do not say anything about any deep-seated natural intuitions of justice. The hearsay rule and its gazillion exceptions may be an example. The exclusionary rule may be another.

Let us now apply these five tests to a famously dissonant legal rule—the felony murder rule—in an effort to ask whether that rule, at least in its most robust variation, should be abolished.

26.4 The Felony Murder Rule

Although jurisdictions vary in their labelling, the most serious kind of murder—triggering the most serious available punishment and typically called first degree murder—generally requires not only a purposeful killing but also that the killer acted after deliberation. The felony murder rule is an exception to this requirement of purposefulness and deliberation. As we shall see, there are many varieties of the felony murder rule, but the classic formulation renders a person guilty of first degree murder if that person commits a felony during the course of which someone is killed. The required predicate felonies are typically enumerated, and the most common are arson, rape, burglary, robbery, and kidnapping. But notice how broad this classic formulation is: its passive voice permits its operation when the person charged with felony murder does not commit the actual killing but, for example, a confederate does, and even when the killed person is the confederate. Its central principle, though, is that it makes a felon guilty of first degree murder even though the felon had no intention, let alone a deliberated one, that anyone would die.

---

[48] Morse (2008).

Thus, if John decides to rob a bank, and in the course the robbery he accidentally drops his gun and it discharges and kills a guard, John is, by operation of this most robust version of the felony murder rule, just as guilty of first degree murder as if he had killed the guard intentionally after planning to do so. It is this version of the felony murder rule that I will now subject to our evolutionary analysis.

26.4.1 The dissonance

The empirical evidence is clear that ordinary people simply do not view our fumbling robber as equally blameworthy, or even close to equally blameworthy, as if he planned to murder the guard. In a 1995 study, subjects judged a perpetrator of an accidental killing committed in the course of an aggravated robbery as being roughly as culpable as if he had committed manslaughter (requiring a reckless killing) rather than first degree murder.[49] A 2009 study had similar results.[50]

This intuition is entirely consonant with the criminal law's ancient and central axiom that intentional acts are more blameworthy than unintentional ones. *Mens rea* is indeed the golden road to responsibility, and this golden road seems to have served all cultures across all ages.[51]

Although there have been isolated strict liability exceptions, every civilization that has left a record—including the Babylonians, Jews, Egyptians, Greeks, and Romans—has adhered to the general rule that criminal responsibility requires both an unlawful act and some intention to commit it.[52] The root of the *mens rea* requirement is a deeply held, arguably evolved, intuition distinguishing intentional wrongs from accidents. As Holmes said, even a dog knows the difference between being kicked and tripped over.

Given this profound natural discomfort with punishing accidents as if they were intentional, it comes as no surprise that over the last seventy-five years the felony murder rule has been one of the criminal law's most frequent targets of reform, second perhaps only to the death penalty. The English abolished the rule in 1957,[53] and although forty-three American states still have some form of it, in all but fifteen it has been watered down to the point of non-existence, thanks in large part to the efforts of the American Law Institute, which effectively recommended its abolition in its 1962 version of the Model Penal Code.

There is no doubt that the felony murder rule conflicts with our deepest notions of criminal responsibility, but does that mean it is unintentionally distorting those notions or intentionally curing them? Let's consider the other four factors.

---

[49] Robinson and Darley (2007), at 36, study 3.
[50] Robinson et al. (2009).
[51] The phrase '*mens rea*' is Latin for 'guilty mind,' and comes from the English legal precept '*Actus non facit reum nisi mens rea sit*' ('An act is not guilty unless the mind is guilty'), whose wording dates at least as far back as Henry I in the early 1100s, and which is probably attributable to St Augustine (Sayre (1932)).
[52] Radin (1932).
[53] Homicide Act 1957, 5 & 6 Eliz. 2, c. 11, s. 1 (Eng.).

26.4.2 Does the rule itself have a core?

The answer seems to be no. There are very few reported examples, in any legal culture, of anything akin to broad and persistent rules about strict criminal liability, even, and maybe especially, for murder. We know that under Roman law for some period of time slave owners were strictly liable for the acts of their slaves, including homicide. But this was more akin to the strict tort liability imposed on owners of animals for damage done by the animals than any broad extension of responsibility beyond fault. There are also remnants in modern Anglo-American law of the ancient Norman doctrine of frankpledge, under which all members of a shire could be liable for the crimes of a single member if they failed to turn the criminal over. But the general liability imposed by frankpledge was always conditional on this failure, and it never found its way more broadly into well-settled notions of individual blameworthiness. There was, in theory, a strict liability view of homicide for a short period of time in medieval England, but this view did not persist.[54]

The felony murder rule is unknown in Continental legal systems. And despite a good deal of mythology surrounding its common law origins, recent scholarship has demonstrated that its robust form was not adopted by any court in England until the mid-1800s, and even then many courts required some level of culpability, and those that did not often invited jurors to ignore the rule.[55] Indeed, by the time the rule gained any traction at all, English commentators were already criticizing it harshly.

American states adopted it largely by misunderstanding. By the time of the Revolution, several English commentators[56] had incorrectly written that the robust rule had been established by the common law. Early American commentators, like Swift, copied the mistakes of their English brethren, and soon robust felony murder statutes were being adopted in all states on the mistaken belief that English courts had already done so.[57]

26.4.3 The magnitude of the dissonance

As mentioned above, empirical studies have shown that ordinary people view an accidental killing during the course of a felony roughly as serious as they view a reckless killing.[58] The results of a 2009 study were even more dramatic. When asked to rank scenarios for 'blameworthiness' on a scale of 1 to 24, with 24 being the most

---

[54] Once the royal courts took jurisdiction of all homicides in the middle of the twelfth century, and throughout the thirteenth century, accidental killings were fully prosecutable as if they were intentional, though defendants in such cases were eligible for royal pardon, for which they often had to pay (2 Pollock and Maitland (1898) 470–81).
[55] Binder (2004).
[56] Holt, Hawkins, Foster and, ambivalently, Blackstone.
[57] Ibid.
[58] Robinson and Darley (2007).

blameworthy, subjects ranked an intentional, planned ambush killing at a mean of 23.3, but ranked a burglary felony murder at 17.9, and a drug felony murder at 14.7, both even lower than reckless manslaughter (19.0).[59] That is, ordinary citizens view these kinds of felony murders as roughly 23 per cent and 37 per cent less serious than first degree murder.

This dissonance, already significant in the blameworthiness dimension, becomes substantially magnified by laws of punishment. For instance, in my state, Colorado, which is quite typical, reckless manslaughter is punishable by a prison sentence of between two and six years, with probation as an alternative. First degree murder carries either the death penalty or a mandatory life sentence without the possibility of parole.

26.4.4 Directional dissonance

There are two ways to think of this factor, depending on how we characterize the dissonance. The robust form of the felony murder rule considered here imposes liability for first degree murder when our intuitions do not. So if our yardstick for directional dissonance is whether equivalent liability is imposed, the felony murder rule is in this binary sense directionally opposed to our intuitions.

But of course in the end the choice is not whether these defendants should be treated as if they committed first degree murder or just the predicate felony. No jurisdiction of which I am aware ignores the fact that a death occurred during the commission of the predicate felony; all of them hold the felon in such circumstances liable to some degree for the death, just not as liable as they would if the felon killed the victim purposefully and after deliberation. Even in the absence of any version of felony murder, such defendants will typically be treated more harshly by way of other doctrines, including aggravation at sentencing.

The dissonance of the felony murder rule is the moral equivalence it forces. What rubs us the wrong way about it is not that John is punished more than if no one had died, but that he is punished just as much as if he purposefully planned the guard's murder. So, can we really say the felony murder rule is 'directionally dissonant' with our intuitions, in the same way, say, as excusing psychopaths might be? Probably not, but in considering this factor we do need to remember that the core of the dissonance is itself about relative blameworthiness.

26.4.5 Exceptions and rejections

The felony murder rule is so strewn with exceptions and rejections that it is no longer even clear what is meant by the phrase. As already mentioned, England expressly abolished it in 1957, and Continental European systems never recognized it. Two

---

[59] Robinson et al. (2009). The burglary hypothetical involved an unarmed defendant who agreed to burglarize a home with an unarmed friend, during which the friend shoots and kills the resisting homeowner with a gun the friend finds in the homeowner's nightstand. The drug hypothetical involved a drug dealer who brought drugs to a party at which one guest used too much and died from an overdose.

states, Hawaii and Kentucky, do not recognize it at all, and a third, Michigan, has abolished it by judicial decision.[60] Several states expressly reject its central equivalency outright, making felony murder some lesser degree of murder. As already mentioned, the Model Penal Code effectively abolished the rule by expressly requiring some level of culpability with respect to the killing (recklessness). Many jurisdictions have similarly abolished it by applying their default rules of culpability.

Even those dozen or so jurisdictions with a facially robust version typically riddle that version with innumerable exceptions. Some require that the person charged with felony murder be the proximate cause of the death; others permit it to be applied to one accomplice when the other accomplice does the killing. Some jurisdictions apply it when it is the accomplice who dies, but most do not. Some jurisdictions apply it when the death occurs during flight from the commission of the felony, others do not.

Some jurisdictions purport to make felony murder equivalent to first degree murder in name only, but then make an exception inside their sentencing statutes. In death penalty states, only a handful of states even purport to make felony murder a capital offence. And in fact the US Supreme Court has held that death is unconstitutionally cruel and unusual for pure strict liability forms of felony murder, and that to impose a capital sentence the state must prove at least that a defendant was reckless with respect to the killing.[61]

26.4.6 Abolish the rule?

So what do these lines of inquiry tell us about whether the felony murder rule should be abolished? The outcomes are somewhat mixed, but on the whole I think they suggest that the handful of remaining jurisdictions that retain a robust form of the rule should abolish it because it irreconcilably conflicts with our deepest notions of relative blameworthiness, proportional punishment and individual responsibility.

Those notions are central to our punishment intuitions. We blame and punish intended harms, not unintended ones. We are able to discriminate, and indeed insist on discriminating, between the act of inflicting intentional harm and the act of misjudging risk, even when that misjudgment is gross and made in the course of another criminal activity. And the amount of that discrimination is not nominal, it is massive. We agree felons should be punished more when someone dies in the course of the felony, but we most definitely do not agree that the punishment should be equal, or anything close to equal, to an intentional, planned killing.

The legal rule equalizing these acts, by contrast, has profoundly questionable pedigree. It was not, despite rumours to the contrary, well-recognized at common law, or really in any mature legal systems that we know about. That we have broadly rejected the rule, and created innumerable exceptions to it, further suggests that it must give way, and in large part has already given way, to overriding and deeper intuitions of justice.

---

[60] Dressler (2006).
[61] *Tison v Arizona* 481 US 137 (1987); *Enmund v Florida* 458 US 782 (1982).

26.5 Conclusion

We must be cautious about the utility, and wisdom, of trying to inform legal policy with truths about our deepest evolved natures. But we should never ignore those natures when they are palpably pertinent. The 'is' can sometimes inform the 'ought'. When we are confident that a human behaviour has an evolutionary core, and that a legal doctrine conflicts with that core, we need to ask ourselves whether the legal doctrine reflects a cultural judgement that the core is one of those time-shifted rationalities that once made evolutionary sense but no longer does. If the answer to that question is no, then we should try to examine whether the dissonant rule itself is based on some core principle, the amount of the dissonance, its direction, and how the legal rule seems to have fared over time, in deciding whether to abandon the rule.

These naturalistic inquiries will seldom be dispositive. But in deciding how we should behave towards one another, and how we should treat misbehaviour, we can no longer ignore that the human animal comes with some deeply preset notions about both. Whether and how much we punish is not some modern cultural invention; it is what we have been doing to and for each other since our emergence. When the law, through doctrines like the robust felony murder rule, forces us to deaden distinctions that cannot be deadened, it risks losing its legitimacy.

26.6 References

American Law Institute, Model Penal Code (ALI 1981).
G. Binder, 'The Origins of American Felony Murder Rules' (2004) 57 *Stan. L. Rev.* 59.
W. Blackstone, *Commentaries on the Laws of England* (Oxford, Clarendon, 1765–69).
J. Buckholtz et al., 'The Neural Correlates of Third-Party Punishment' (2009) 60 *Neuron* 930.
F. Cushman, 'Crime and Punishment: Distinguishing the Roles of Causal and Intentional Analyses in Moral Judgment' (2008) 108 *Cognition* 353.
F. de Waal, *Primates and Philosophers: How Morality Evolved* (Princeton, Princeton University Press, 2006).
F. deWaal, *Good Natured: The Origins of Right and Wrong in Humans and Other Animals* (Cambridge, MA, Harvard, 1996).
J. Dressler, *Understanding Criminal Law*, 4th edn (Newark, LexisNexis, 2006).
E. Fehr and U. Fischbacher, 'Third-Party Punishment and Social Norms' (2004) 25 *Evol. & Hum. Behav.* 63.
J. Flack et al., 'Policing Stabilizes Construction of Social Niches in Primates' (2006) 439 *Nature* 436.
J. Flack and F. deWaal, 'Any Animal Whatever: Darwinian Building Blocks in Morality of Monkeys and Apes' (2000) 7 *J. Conscious. Stud.* 1.
F. Fukuyama, *Our Post-Human Future: Consequences of the Biotechnology Revolution* (London, Picador, 2002).
H. Godfray, 'The Evolution of Forgiveness' (1992) 355 *Science* 206.

J. Greene and J. Paxton, 'The Patterns of Neural Activity Associated with Honest and Dishonest Moral Decisions' (2009) 106 *Proc. Nat. Acad. of Sciences* 12506.

J. Greene et al., 'An fMRI Investigation of Emotional Engagement in Moral Judgment' (2001) 293 *Science* 2105.

M. Gruter and R. Masters (eds.), *Ostracism: A Social and Biological Phenomenon* (New York, Elsevier Science Publishing, 1986).

R. Hare et al., 'The Revised Psychopathy Checklist: Reliability and Factor Structure' (1990) 2 *Psychol. Assessment* 338.

T. Harris et al., 'Psychopathy and Violent Recidivism' (1991) 15 *L. & Hum. Behav.* 625.

M. Hause, *Moral Minds: The Nature of Right and Wrong* (New York, Harper Perennial, 2006).

J. Henrich et al., 'Costly Punishment across Human Societies' (2006) 312 *Science* 1767.

J. Henrich et al., 'In Search of Homo Economicus: Behavioral Experiments in Fifteen Small-Scale Societies' (2001) 91 *Am. Econ. Rev.* 73.

K. Hockings et al., '*Chimps Share Forbidden Fruit*' (2007) 2 *PLoS One* 9.

M. Hoffman, 'The Neuroeconomic Path of the Law' in S. Zeki and O. Goodenough (eds.), *Law and the Brain* 3 (Oxford, Oxford University Press, 2004).

M. Hoffman and T. Goldsmith, 'The Biological Roots of Punishment' (2004) 1 *Ohio St. L. Crim. L.* 627.

D. Hume, *A Treatise of Human Nature*, vols. I & II (1740) (New York, Merchant, 2009).

D. Hume, *An Inquiry Concerning the Principles of Morals* (1751) (New York, Cosimo Inc., 2009).

K. Jensen et al., 'Chimpanzees are Vengeful but not Spiteful' (2007) 104 *Proc. Nat. Academy of Sciences* 13046.

O. Jones, 'Time-Shifted Rationality and the Law of Law's Leverage: Behavioral Economics Meets Behavioral Biology' (2001) 95 *Nw. U.L. Rev.* 1141.

R. Joyce, *The Evolution of Morality* (Cambridge, MA, MIT, 2007).

K. Kiehl, 'A Cognitive Perspective on Psychopathy: Evidence for Paralimbic System Dysfunction' (2006) 142 *Psych. Res.* 107.

K. Kiehl and M. Hoffman, '*Psychopathy and Crime*' *Jurimetrics: The Journal of Law, Science and Technology* (forthcoming).

J. Lancaster, 'Primate Social Behavior and Ostracism' in Gruter and Masters, Ostracism at 68.

M. Lezak et al., *Neurophysiological Assessment*, 4th edn (Oxford, Oxford University Press, 2004) 79.

A. MacIntyre, 'Hume on "Is" and "Ought"' (1959) 48 *J. Phil. Rev.* 45, reprinted in W. D. Hudson (ed.),*The Is-Ought Question* (New York, St. Martins Press, 1969) 35–50.

M. Marcus, 'Sentencing in the Temple of Denunciation: Criminal Justice's Weakest Link' (2004) 1 *Ohio St. J. Crim. L.* 671.

M. McCullough, *Beyond Revenge: The Evolution of the Forgiveness Instinct* (San Francisco, Jossey-Bass, 2008).

G. E. Moore, *Principia Ethica* (Cambridge, Cambridge University Press, 1948).

S. Morse, 'Psychopathy and Criminal Responsibility' (2008) 1 *Neuroethics* 205.

F. Pollock and F. Maitland, *The History of English Law*, 2nd edn (Cambridge, Cambridge University Press, 1898).

S. Porter et al., 'Crime Profiles and Conditional Release Performance on Psychopathic and Non-psychopathic Sexual Offenders' (2009) 14 *Legal & Crim. Psychology* 109–18.

M. Radin, 'Intent, Criminal' in E. Seligman and A. John (eds.), 8 *Encyclopdia of the Social Sciences* 126 (London, Macmillan & Co. Ltd., 1932).

A. Raine and J. Sanmartín (eds.), *Violence and Psychopathy* (New York, Kluwer Academic/Plenum Publishers, 2001).

P. Robinson, G. Goodwin, and M. Reisig, 'The Disutility of Injustice' (in submission) (2009), available at SSRN at <http://papers.ssrn.com/sol3/papers.cfm?abstract_id=1470905>.

P. Robinson and J. Darley, 'Intuitions of Justice: Implications for Criminal Law and Justice Policy' (2007) 81 *S. Cal. L. Rev.* 1.

P. Robinson and R. Kurzban, 'Concordance and Conflict in Intuitions of Justice' (2007) 91 *Minn. L. Rev.* 1829.

P. Robinson, R. Kurzban, and O. Jones, 'The Origins of Shared Intuitions of Justice' (2007) 60 *Vand. L. Rev.* 1633.

P. Robinson and J. Darley, *Justice, Liability and Blame: Community Views and the Criminal Law* (Boulder, Westview Press, 1995).

P. Rubin, *Darwinian Politics: The Evolutionary Origins of Freedom* (New Brunswick, Rutgers University Press, 2002).

M. Shermer, *The Science of Good and Evil* (New York, Henry Holt & Co., 2004).

F. Sayre, '*Mens Rea*' (1932) 45 *Harv. L. Rev.* 974.

F. Snare, *The Nature of Moral Thinking* (London, Routledge, 1992).

B. Smith and E. Stevens, *Sentencing Disparity and the Judge-Jury Sentencing Debate: An Analysis of Robbery Sentences in Six Southern States* (1984) 9 *Crim, J. Rev.* 1.

M. Stout, *The Sociopath Next Door* (New York, Broadway Books 2005).

G. Wilkinson, 'Reciprocal Food Sharing in the Vampire Bat' (1984) 308 *Nature* 181.

R. Wright, *The Moral Animal* (New York, Vintage Books, 1995).

27

The History of Scientific and Clinical Images in Mid-to-Late Nineteenth-Century American Legal Culture: Implications for Contemporary Law and Neuroscience

*Daniel S. Goldberg**

27.1 The Power of Scientific and Clinical Images in American Legal Culture

The central claim of this paper is that understanding the history of scientific and clinical imaging evidence is critical to informing analysis of the contemporary role of neuroscientific evidence in American legal culture. Justifying this thesis requires two principal tasks. First, I assess the epistemic and evidentiary status of scientific and clinical images in late nineteenth- to early twentieth-century American legal culture. I argue that these images became so powerful at this time that they catalysed the adoption of a distinct regime for managing and regulating their use as evidence. The nexus of this regime is the emphasis on the reliability of a particular scientific or clinical method as the central criterion for admitting evidence generated by use of that method. While the modern rise of scientific and medical imaging techniques during the mid-to-late nineteenth century did not create this regime *ex nihilo*, I argue that without these imaging techniques and, in particular, without the use of the x-ray in American courts, the foundational role that methodological reliability plays as a key criterion for admissibility might have been substantially delayed or even denied.

---

* JD, PhD, Assistant Professor in the Department of Bioethics & Interdisciplinary Studies at the Brody School of Medicine, East Carolina University. The author would like to thank Michael Freeman, the audience at the Fourteenth University College London Current Legal Issues Interdisciplinary Colloquium, and Jennifer L. Mnookin for their assistance, comments and criticisms.

Second, I argue that through a deeper understanding of the social and cultural power of the scientific image in shaping contemporary American law of evidence, one is in a better position to evaluate the scope and significance of neuroimaging evidence in American legal culture. More specifically, while many neurolaw scholars warn of the potentially prejudicial effects of neuroimages, there is less scholarship explaining why it is that such neuroimages are any more likely to pose significant risks of undue prejudice than many other forms of scientific and medical evidence. Given the extent to which science and medicine operate to legitimize all manner of epistemic claims within and without legal discourse, it is fair to question why neuroimages are deemed exceptional as to inspire the relative outpouring of academic and professional attention.

American science and technology studies provide ample evidence to support the idea that the role technology plays in American culture is itself a significant driver in the resources expended to analyse that role. Nevertheless, the power of the visible image in American legal culture provides a substantive reason for the apparently augmented concern directed to the evidentiary use of neuroimages. This concern is compounded by the fact that neuroimages are not simply scientific and clinical images of the inner body, but are images of the brain. This matters because there is evidence to suggest both that such images are central to dominant conceptions of (modern) Western science and medicine,[1] and to the ways in which individuals conceive of their own biomedical personhood.[2]

Accordingly, while much of the attention on the status of neuroscientific evidence and in particular neuroimaging evidence has focused on whether the evidence meets the requisite standards of reliability,[3] I submit that such a focus already assumes too much. Rather, the inquiry ought to begin with an understanding of the role scientific and medical images play in accelerating the regime by which methodological reliability becomes the evidentiary standard. Subsequently, greater attention may be directed to the question of whether the power of such images is so great that even if the process by which they are produced becomes sufficiently accepted and reliable, the risks of undue prejudice are so great as to typically outweigh the probative value of neuroimaging evidence. While the formulation embodied in this assessment, codified in both the Federal and Uniform Rules of Evidence,[4] is rightfully understood as a catch-all objection that will not

---

[1] Robert Martensen, *The Brain Takes Shape: An Early Modern History* (Oxford/New York, Oxford University Press, 2004).

[2] Joseph Dumit, *Picturing Personhood: Brain Scans and Biomedical Identity* (Princeton, Princeton University Press, 2003).

[3] See, e.g., Walter Sinnott-Armstrong, Adina Roskies, Teneille Brown, and Emily Murphy, 'Brain Images as Legal Evidence' (2008) 5(3) *Episteme* 359; Jane Campbell Moriarty, 'Visions of Deception: Neuroimages and the Search for Truth' (2009) 42 *Akron Law Review* 739; Joelle Anne Moreno, 'The Future of Neuroimaged Lie Detection and the Law' (2009) 42 *Akron Law Review* 717; Noel Shafi, 'Neuroscience and Law: The Evidentiary Value of Brain Imaging' (2009) 11 *Graduate Student Journal of Psychology* 27; Jane Campbell Moriarty, 'Flickering Admissibility: Neuroimaging Evidence in the U.S. Courts' (2008) 26 *Behavioral Sciences & Law* 29.

[4] Fed. R. Evid. 403; Unif. R. Evid. 403.

often succeed,[5] comprehending the power of neuroimages in American legal culture may provide some reason for thinking that such objections may or at least should have comparatively greater rhetorical force.

Why should a paper concerned with the contemporary role of neuroscientific evidence in the US spend any time at all discussing scientific and medical imaging evidence in *fin-de-siècle* American courts? The answer rests on a dialectic model of history, one in which prior events, ideas, and conditions continually operate to shape and inform contemporary discourse. The benefit of this framework is primarily that it facilitates comprehension of the powerful moulding role of social and historical context. I endeavour to show here that contemporary American discourse on neurolaw looks the way it does precisely because of past events, ideas, and conditions. More specifically, the historical pathways by which scientific and medical images became regulated and assimilated into the American law of evidence suggest some important implications for neuroimaging evidence. If, as I will argue here, such images profoundly reshaped the ways in which that law regulated all scientific and medical evidence, there is thus good reason for questioning the proper role and evidentiary status of contemporary neuroimaging evidence.

The use of visual images in Western legal culture is, like that culture itself, ancient in origin. Metzger notes the occasional use of pictures in litigation under Roman law,[6] and there is evidence that maps and diagrams in particular were relevant to property disputes since at least the high Middle Ages. However, in her history of the introduction of the photograph into nineteenth-century American legal culture, Jennifer Mnookin notes that even as late as the 'first two-thirds of the nineteenth century, there existed no well-developed general evidentiary category of "models, maps, and diagrams"'.[7] Rather, models, maps, and diagrams were understood as a 'narrowly defined grouping generally thought to apply only to specific categories of cases'.[8]

At risk of periodicity, the present historical analysis focuses on American legal culture in the late nineteenth- to early twentieth-century. This focus is justified because, as Mnookin notes, something quite distinctive happens in the US during the late nineteenth century regarding the use of images as evidence. The changes in the evidentiary status, applicability, and significance of these images are critical to understanding how the American law of evidence regulates scientific and clinical images both in the early twentieth century and how the law governs the use of such images today. Section III of this paper is devoted to assessing the closeness of the connection between these two discourses.

---

[5] E.g., Michael Teter, 'Acts of Emotion: Analyzing Congressional Involvement in the Federal Rules of Evidence' (2008) 58 *Catholic University Law Review* 153; Victor J. Gold, 'Federal Rule of Evidence 403: Observations on the Nature of Unfairly Prejudicial Evidence' (1983) 58 *Washington Law Review* 497.

[6] Ernest Metzger, *Litigation in Roman Law* (New York, Oxford University Press, 2002).

[7] Jennifer L. Mnookin, 'The Image of Truth: Photographic Evidence and the Power of Analogy' (1998) 10 *Yale Journal of Law and Humanities* 1, 59.

[8] Ibid. 60.

27.2 The Evidentiary Status of Scientific Images in Late Nineteenth- to Early Twentieth-Century American Legal Culture

27.2.1 Nineteenth-century changes: the photograph

On one level, the most obvious reason for the dramatic changes that occurred regarding the evidentiary status of images in late nineteenth-century America is the invention of the photograph. However, as important as the photograph is, it is a grievous error to analyse the effect of a technical innovation separate and apart from the social and cultural factors that shape the understandings, interpretations, and usages of the object.[9] The significance of the photograph is wholly a product of the ways in which the photograph was understood, interpreted, and used.[10] Accordingly, analysing how the photograph changed the American law of evidence requires due attention to some of the larger social and cultural movements, ideas, and conditions of the time, and how these latter factors facilitated the changes in the law of evidence. This will be of particular importance to my analysis here because, as important as the photograph undoubtedly was in bringing about changes in doctrine and in legal culture, I argue below that the x-ray is of equal or greater importance. Situating the role of the photograph in the larger social and historical context lays the appropriate groundwork for understanding how the *fin-de-siècle* discourse over the evidentiary status of the x-ray is related to the current concern over neuroimaging evidence.

Invented by Louis-Jacques Mande Daguerre in 1839, the photograph quickly penetrated American society. 'By the middle of the 1850s, Americans of all classes had sat for portraits...'[11] For a variety of reasons mostly related to technical limitations in the photographic process, photographs did not significantly diffuse into American courts until the early 1880s, when high quality dry plates became available for public use.[12] What is particularly salient in understanding the impact of the photograph is what Mnookin identifies as:

> two competing paradigms [that] governed the understanding of the photograph. One emphasized its ability to transcribe nature directly, while the other highlighted the ways in which it was a human representation. From the first perspective, the photograph was viewed as an especially privileged kind of evidence; from the second perspective, the photograph was seen as a potentially misleading form of proof.[13]

---

[9] This is a basic tenet of both science and technology studies, and the history of science, technology, and medicine. In the former, see Bruno Latour and Steve Woolgar, *Laboratory Life: The Construction of Scientific Facts* (Princeton, Princeton University Press, 1986); in the latter, see, e.g., Shai Lavi, *The Modern Art of Dying: A History of Euthanasia in the United States* (Princeton, Princeton University Press, 2006).
[10] Note 7, above, 7–14; Jennifer Tucker, *Nature Exposed: Photograpghy as Eyewitness in Victorian Science* (Baltimore, John Hopkins University Press, 2006).
[11] Note 7, above, 7.
[12] Ibid. 13.
[13] Ibid. 4.

Both Mnookin and Jennifer Tucker argue that despite the somewhat oppositional nature of these paradigms, neither prevails. Mnookin analyses in detail the arguments offered for and against the admission of photographic evidence in an 1869 case involving the well-known 'spirit' photographer William H. Mumler. Mumler was charged with fraud for producing photographs that 'appeared to show hazy, ghostlike figures looming in the background'.[14] Witnesses for the prosecution—which included noted ringmaster P. T. Barnum—challenged the authenticity of Mumler's photographs by way of supporting the prosecution's case for fraud.[15]

These witnesses emphasized the artificial nature of the photographs, that they were profoundly susceptible to human manipulation. However, what is most important for my purposes is that even those witnesses who challenged the validity of the photographs by 'invok[ing] photographic manipulability' did so 'not to dismantle photographic authority, but to preserve it'.[16] Mnookin traces how the witnesses attempted to affirm the authority of photographs in two ways. First, they emphasized that manipulation was easily detectable in part because the mechanical nature of photography meant that traces of unusual activity and processes would inevitably remain.[17] Second, they testified that photographic fakery required volition on the part of the hucksters, and hence that qualified photographic experts were able 'to delineate falsified photographs from authentic ones...'.[18] This second point is particularly important, and I shall return to it repeatedly in the remainder of this essay.

For my purposes, the central point here is that at least partially adversarial conceptions of the epistemic status of photographs seemed to coexist in late nineteenth-century American legal culture. Tucker observes that 'scientific photography, no less than other types of visual images, exhibited [in the nineteenth century] a multiplicity of meanings for different audiences in diverse viewing contexts'.[19] At times—as in the Mumler case—stakeholders simultaneously deemed photographs as faithful replications of the external reality of the world and as artifices easily susceptible to human fabrication. This is not to deny the real tension that existed between these different conceptions of the evidentiary and epistemic status of the photograph, but simply to note what Mnookin refers to as the 'indeterminacy of photographic meaning', the idea that photographs in late nineteenth-century American culture were construed in myriad and multivalent ways, including in ways that emphasized opposing if not contradictory interpretations. Though her account is not strictly a legal history, Tucker argues for a similar point, explaining that '[s]tudying the historical and disciplinary frameworks in which scientific photographs emerged as empirical proof yields a more complex understanding of the epistemologies associated with scientific photography than the "unshaking belief" theory of nineteenth-century photographic attitudes'.[20]

---

[14] Ibid. 27.
[15] Ibid. 36–7.
[16] Ibid. 37.
[17] Ibid. 38.
[18] Ibid. 39.
[19] Note 10, above, 235.
[20] Note 10, above, 235.

And yet, in an important sense, this role of the photography in late nineteenth-century American culture epitomizes the history of scientific objectivity in the West. In their recent analysis of this history, Lorraine Daston and Peter Galison emphasize that prevailing contemporary notions of scientific objectivity did not simply appear *ex nihilo*.[21] In fact, they argue, the concept of objectivity itself has gone through several important changes since the early modern era, perhaps most notably in the change from what they term 'truth-to-nature' objectivity to 'mechanical' objectivity.[22] The key difference between the two centres on the role of the subject in producing the relevant knowledge. In the older 'truth-to-nature' conception, it was widely understood that the subjective skills of the investigator and of the artist were critical to assessing the validity and authenticity of the scientific images that typically appeared in early modern compendia, *materia medica*, botanicals, and atlases.[23]

Moreover, in a truth-to-nature paradigm, faithfully representing a particular specimen's correspondence to external reality was never the prime objective. Rather, through the collaborative skills of the investigator and the artist, the typical aim was to present an image of the universal archetype of the *desideratum*. The images that appeared in the pages of the scientific atlas would not have been understood as accurate representations of any particular and imperfect specimen; the goal was rather to present the perfect archetype of the flower, plant, animal, etc.[24] Thus, far from attempting to excise human influence on the scientific process, under the truth-to-nature paradigm, such influence was a marker of accuracy and quality.

In contrast, Daston and Galison note, the concept of mechanical objectivity featured the ideal of excising human influence in the production of scientific images.[25] It is, I suggest, no coincidence that Daston and Galison (among others) trace the development of this form of objectivity to the middle decades of the nineteenth century. Nor is it an accident that Daston and Galison choose scientific images since the early modern era as the central object of their analysis. Why the concept of mechanical objectivity only begins to take shape in the middle of the nineteenth century is an important story, but one that is beyond the scope of this paper. The key is to understand: 1) that an ideal of scientific inquiry that highlighted the need to remove human influence from mechanical, scientific methods begins to emerge concurrent with the rise of the photograph; and 2) that a focus on images is at the very core of western discourse on objectivity for at least the last 400 years. Furthermore, it is a limited conception indeed that understands the reach and import of the concept of objectivity solely in terms of its impact on the cultures of science and medicine. What Daston and Galison demonstrate is that at least

---

[21] Lorraine Daston and Peter Galison, *Objectivity* (New York, Zone Books, 2007).
[22] Ibid. 55–105; 115–83.
[23] Ibid. 55–105.
[24] Humans *qua* scientific objects were no exception to this trend. Nancy Siraisi documents the influence of this conception in the framework of Vesalius's *De fabrica*, the most influential Western atlas of human anatomy in the last milennium. Nancy G. Siraisi, 'Vesalius and Human Diversity in *De Humani Corporis Fabrica*' (1994) 57 *Journal of the Warburg and Courtauld Institutes* 60.
[25] Note 21, above, 115–83.

since the early modern era in the West, the ways in which images are interpreted, the meanings attributed, and the role of the human agent in producing those images are part of a deep social and cultural discourse in which scientific and medical praxis is situated.

These two notions of objectivity ('truth-to-nature' and 'mechanical') elucidate the competing paradigms and reactions to photographs that Mnookin and Tucker identify with respect to the photograph. The concern with the prospect of human artifice and manipulation of photographs reflects the emerging concept of mechanical objectivity, in which the highest and most reliable forms of knowledge are produced with as little subjective influence as possible on the process of production. Accordingly, the possibility, exemplified in the Mumler hearing, that photographs could be easily manipulated and even faked through human artifice undermined the ideal of mechanical objectivity. Breidbach shows that a similar concern significantly impeded the acceptance and development of scientific microphotography during the nineteenth century.[26] Moreover, while the ideal of mechanical objectivity was at root a scientific ideal, the power of science in shaping public discourse is evident in the similarity of concern over the extent of human influence in producing photographs (whether the microphotography of the laboratory or the portraits of the home).

Yet it would be simplistic and erroneous to view the concern over human manipulation of lay and scientific photographs as a linear triumph of mechanical objectivity over truth-to-nature objectivity. Quite the contrary; as discussed above, even those witnesses in the Mumler hearing who testified as to the photographs fraudulence did so by emphasizing that various features related to the manipulability of the photographs provided indicia of reliability (i.e. that artifice would leave tell-tale signs that could be read by qualified experts). The fact that these witnesses invoked the qualifications of experts to read the signs of improper influence suggests the enduring power of truth-to-nature objectivity. This is because in the latter schema, the expertise of the investigator is critical to ensuring the veracity of the images; just as with the discourse on photographs in the Mumler hearing, the subjective influence of the scientific expert is a central reason for *ensuring* the accuracy and knowledge claims of images in truth-to-nature objectivity.

Thus, at least in some sense, the multivalent and at times paradoxical view of photographs in late nineteenth-century American legal culture reflects the changing frameworks of objectivity, from truth-to-nature to mechanical. That these two schema both exist in some tension and appear simultaneously in reaction to the use of photographic images in late nineteenth-century American courts is actually a feature of both a dialectic notion of history in general and of the Western history of objectivity in particular.

However, while these frameworks existed simultaneously in late nineteenth-century American legal culture, one should not infer that the coexistence was

[26] Olaf Breidbach, 'Representation of the Macrocosm: The Claim for Objectivity in 19th Century Scientific Microphotography' (2002) 35 *Journal of the History of Biology* 221.

peaceful and seamless. In emphasizing both the epistemic problems with excessive subjectivity in the production of photographs and the idea that subjective expertise was the key to shoring up the validity of photographs, American stakeholders experienced no small amount of cognitive dissonance. Given these difficulties, it is unsurprising that judges in particular sought to resolve the tension through the rhetorical mainstay of Anglo-American common law: the analogy.[27]

The problems posed by photographs could be minimized in scope and significance if the photograph was seen as similar in the salient aspects to other pieces of evidence. Even though models, maps, and diagrams were only occasionally used in the US as evidence prior to the nineteenth century, American stakeholders attempted to address the use of photographs as evidence and the competing epistemological frameworks for interpreting them as simply the latest variation of a type of evidence that had been in use, albeit limited, for centuries. Mnookin argues that this way of understanding the evidentiary role of photographs helped 'defuse its novelty', protected 'the reign of words' over images in legal culture, and preserved judicial (rather than expert) authority over the admissibility of images.[28]

In any case, the analogical treatment of photographs became more or less settled by the *fin-de-siècle*. In 1904, the Dean of the School of Law at Northwestern University, John Henry Wigmore, produced his magisterial treatise on the American law of evidence, which remains influential over a century later. In section 790, Wigmore addresses the evidentiary status of photographs, and summarizes the analogy:

> A document purporting to be a map, picture, or diagram, is, for evidential purposes simply nothing. Except so far as it has a human being's credit to support it. It is mere waste paper—a testimonial nonentity. It speaks to us no more than a stick or a stone. It can of itself tell us no more as to the existence of the thing portrayed upon it than can a tree or ox. It is somebody's testimony, or it is nothing.[29]

According to this framework, photographs, like models, maps, and diagrams, were simply a way of presenting a witness's verbal testimony in visual form. This meant, of course, that the photograph was not afforded the privilege of 'speaking for itself' as evidence in American courts. Rather, the witness whose testimony was considered to be represented in the photograph had to lay the appropriate evidentiary foundation before the photograph could be admitted into evidence. To establish such a foundation, the witness had to authenticate the image, which generally required testimony that the image fairly and accurately represented the substantive facts about the world it was offered to illustrate.[30] Of course, it typically took no special skill or qualifications for a witness to provide a founda-

---

[27] Note 7, above, 43–59.
[28] Ibid. 54–6.
[29] John H. Wigmore, *A Treatise on a System of Evidence in Trials at Common Law* (Boston, Little Brown & Co., 1904).
[30] The term of art in American law for this criterion is 'the truth of the matter asserted therein', such that a proponent of photographic evidence must generally provide testimony corroborating the truth of the matter asserted in proffering the photograph.

tion for a non-scientific photograph, but only sufficient familiarity with the state of the external world represented in the photograph. This requirement sustained the evidentiary authority of the judge in so far as '[j]udging the truth of the visual representation required exactly the same thing that judging the truth of any testimony required: an assessment of the credibility of the witness'.[31]

However, it is in this requirement of authentication, still codified in Federal Rule of Evidence 901(a), that this brief history of the photograph ends and the legal history of the x-ray begins. For, perhaps unlike the photograph, which, by the late nineteenth century had been assimilated into at least somewhat comfortable legal and evidentiary canons, the x-ray would submit to no such treatment. For the x-ray, the analogy to models, maps, and diagrams would fail.

Before turning to the x-ray, it is important to recapitulate the argument thus far, if for no other reason than that I have not, to this point, said anything of substance regarding law and neuroscience. This reticence is intentional; part of my argument is that discussions of neuroscientific evidence and the law have tended to be ahistorical, and have not contextualized the current discourse in the rich and instructive history of the use of images in American courts over the last century-plus. The first step in contributing to an historically informed understanding of the epistemic and evidentiary role of scientific and clinical images in American courts is briefly surveying the introduction and debate over the status of the photograph in late nineteenth-century American culture.

Though many features of this story are compelling, for my purposes, what matters most is the way that interlocutors emphasized subjectivity and human influence on the method of producing photographs as both a push and a pull on the epistemic authority of the photograph. While many detractors of the use of photographs as evidence pushed back against its ubiquity by stressing how deceptive they could be, ironically, both proponents of the evidentiary use of photographs and even some detractors relied on the notion of subjective knowledge (in the form of expertise) to pull up the reliability and veracity of the photograph. Photographs were both idealized as revealing correspondence with reality (and thereby producing truth) and degraded as an artifice susceptible to human manipulation, ripe for tricking the gullible.

Finally, the discourse on photographs in late nineteenth-century American legal culture demonstrates the effect schemas of objectivity had, with their implications for the production and arrangement of knowledge in American society. While the suggestion that (necessarily subjective) expertise could buttress knowledge claims based on photographic evidence reflected an older, truth-to-nature conception of objectivity, derogation of the evidentiary quality of photographs due to the ease of manipulation and chicanery marks the effect of an emerging conception of mechanical objectivity.

---

[31] Note 7, above, 44.

27.2.2 The x-ray

Why was the x-ray image less susceptible to the analogy by which photographs could be categorized with models, maps, and diagrams? This disanalogy was the result of the one salient aspect in which x-rays could not possibly be interpreted as relevantly similar to models, maps, diagrams, and photographs: a (lay) fact witness is unable to testify that the x-ray fairly and accurately represents the 'truth' of the 'inner body' captured via the image. While a lay witness could certainly authenticate a photograph of, for example, a tree said to mark the boundary of a property line by affirming their familiarity with the tree and the accuracy of the representation, the same could obviously not be said about an x-ray of a person's femur. This simple evidentiary impediment would come to have profound effects on the American law of evidence, effects that resonate to the present, and that suggest much of importance for the debate over the role of neuroscientific evidence.

Though the story of early American roentgenology is important for a number of reasons in the history of medicine,[32] my focus here is on the reception and regulation of x-rays as evidence in *fin-de-siècle* American courts. This history is itself intertwined with the early roentgenologists and their rapid experimentation and utilization of x-ray techniques, as concerns about the legal use of x-rays are evident relatively soon after x-ray work began in the US in early 1896.[33]

The single word that best captures the history of the x-ray in American culture is 'power'. In other work, I argue that it is critical to imagine, 400 years after Vesalius castigated his brethren for failing to perform anatomical studies manually, how extraordinary it must have been in 1896 for American roentgenologists to see remotely the inner body of a living person (rather than at post mortem).[34] Rebecca Herzig explains: 'Invisible, active at a distance, and powerful beyond any received understanding, the uncanny x-ray invited religious comparison.'[35] The early American roentgenologists and physicians were acutely aware of this power, and the resultant anxiety is evident not least in the pervasive fears of what scientists and physicians often termed 'x-ray litigation'.[36] Indeed, Tal Golan, the author of the seminal legal history of x-rays in American culture, notes that as early as 1897, concerns over 'x-ray litigation' were a major topic in almost every medical professional association meeting in that year.[37] Writing in the *Northwestern Lancet* in 1900, one O. C. Strickler recounts his belief that the power of the x-ray alone is

---

[32] Daniel S. Goldberg, 'Suffering and Death Among Early American Roentgenologists: The Power of Remotely Anatomizing the Living Body in Fin-de-Siècle America' (2010) submitted paper.
[33] Note 32, above.
[34] Ibid.
[35] Rebecca Herzig, 'In the Name of Science: Suffering, Sacrifice, and the Formation of American Roentgenology' (2001) 53(4) *American Quarterly* 563.
[36] It should be said that the *fin-de-siècle* featured a significant increase in American medical malpractice litigation. This fact provides some context for the fears of x-ray usage in the service of such claims.
[37] Tal Golan, *Laws of Men and Laws of Nature: The History of Scientific Expert Testimony in England and America* (Cambridge, MA, Harvard University Press, 2004) 198.

sufficient to prompt a mentally and emotionally vulnerable patient to initiate malpractice litigation:

> I made an examination of Colles' fracture some time ago, after removal of the splints, where, had I shown the shadowgraph to the patient a malpractice suit would surely have resulted... The shadowgraph showed imperfect apposition, as well as some material thickening at point of fracture, and to have shown the shadowgraph to the patient, in his then mental condition, would have, undoubtedly, precipitated litigation.[38]

While the power of the x-ray resonates in a number of different ways in *fin-de-siècle* American legal culture, my focus here is on the rules American courts developed for laying the proper evidentiary foundation for the admission of x-ray images. Especially given the extent of the social and cultural power that attended x-ray imaging techniques, it is imperative to assess whether the considerable epistemic power of analogical reasoning could stretch to cover such techniques. That is, there is little question that photographs featured extensive social and cultural power of their own. Yet, American judges were nevertheless able, via analogy, to assimilate the evidentiary use of the photograph into the relatively well-established usage of models, maps, and diagrams as mere visual compilations of verbal testimony. Could the same operation be used for the matter of the x-ray's evidentiary status? Could American judges preserve their authority as the epistemic arbiters of the x-ray? The answer to both of these questions is a qualified 'no'. And the reasons for such answers have great significance for contemporary neuroscientific evidence in American courts.

Regardless of this ultimate and critical difference, the legal discourse on x-rays shared much with that as to photographs. The two competing paradigms Mnookin identifies—concern with the deceptive possibilities of the photograph and faith in its ultimate veridical capacity—are evident in the early debate over the evidentiary status of x-rays. This debate played out most publicly in November 1896 when a plaintiff sought to introduce an x-ray image as evidence in a malpractice suit against his physician in Denver, Colorado. The plaintiff, James Smith, sued the physician, William Grant, for failing to diagnose Smith's fractured femur. Thus, the posture of the case captured the darkest fears of the medical profession: the aggrieved patient using the physician's own technical apparatus in malpractice litigation against that physician.

In the trial, Smith's lawyers proceeded to assemble an x-ray machine in the courtroom itself, and offered to show the jury x-ray images of their own hands. Grant's attorneys objected, and for three hours emphasized the subjective, deceptive nature of the images.[39] As argued above, this line of reasoning is one of the principal frames in which photographs were interpreted in American legal culture. Given the larger cultural discourse on objectivity, the role of science as a legitimizing institution in *fin-de-siècle* America, and the increasing focus on what the investigating eye could see in the body, it is unsurprising that the same concern

---

[38] O. C. Strickler, 'Dislocation of the Shoulder' (1900) 20 *Northwestern Lancet* 266.
[39] Note 37, above, 191.

and suspicion of the power of images with which some Americans viewed photographs is present as to x-rays. But the concern over the power of x-rays was not limited to its duplicitous potential. If it were, it is conceivable that American litigants, attorneys, witnesses, and judges might have succeeded in assimilating the evidentiary role of x-ray images into the more comfortable and conventional paradigm of models, maps, and diagrams.

This did not happen, and the principal reason it did not happen is that x-ray images simply could not be authenticated as models, maps, diagrams, and photographs could. In their argument to Judge Owen LeFevre, Grant's attorneys explained why: no witness could possibly testify as to the verity of 'an object unseen to the naked eye'.[40] This was the major impediment to the analogical categorization of x-ray images as simply the latest instantiation of models, maps, diagrams, and photographs. All of these latter items could be authenticated by a witness's testimony that the image fairly and accurately represented the truth of the matter asserted therein. This means of authentication buttressed the coherence of Wigmore's assessment that such images were nothing more than a witness's verbal testimony captured in visual form. That x-ray images could not be authenticated in this way threatened to expose, in stark terms, the tenuous fiction that the evidentiary treatment of photographs in American courts represented.

Although, by the end of the nineteenth century, American judges and courts had generally managed to harness the epistemic power of photographs by assimilating them into more familiar categories, I have noted that this process was hardly seamless. Daston and Galison emphasize in general that that '[d]espite all evidence to the contrary, the mythology of the [photographic] image as evidence untouched by human hands remained powerful'.[41] This is evident in various cases discussing the evidentiary status of photographs, in which the veridical power of photographs were emphasized and concomitant admissibility decisions made, thereby leaving no doubt that stakeholders in practice often regarded photographic images as evidentiary animals quite different from models, maps, and diagrams.[42] The perceived epistemic power of the x-ray image would push these tenuous boundaries even

---

[40] Note 37, above, 191.
[41] Note 21, above, 440, note 31.
[42] One of the most oft-cited cases for this proposition is *Franklin v State*, in which the Supreme Court of Georgia ruled that the trial court had correctly admitted a photograph as independent evidence of a murder, stating that 'we cannot conceive of a more impartial and truthful witness than the sun, as its light stamps and seals the similitude of the wound on the photograph put before the jury...' 69 Ga 36, 42 (1882). Similarly, in 1899, the Supreme Court of Missouri emphasized the veridical power of the photograph in reversing a negligence judgment against a railway company: 'But counsel for plaintiff, while endeavoring to establish that the path occasionally ran upon the ties, surely forgot one of his most important witnesses, a witness that could not be approached, coached nor bribed; a witness who could not be subpoenaed, although his deposition might be taken. That deposition has been taken in photograph number 2, and that witness the unimpeachable sun. His deposition, taken in indelible characters of heavenly light, reproduces the surroundings of the scene of the accident, the tracks, the space between them, the pathway and the end of the ties, as they existed at the time of the injury.' *Kreis v Mo. Pac. Ry. Co.* 49 SW 877, 880 (Mo 1899). The metaphor of light and the sun is common in discussions of scientific imaging in the mid-to-late nineteenth century: note 32, above.

further, at least in part by exposing the furthest limits of the analogy that already functioned imperfectly in legal practice.

Shorn, because of the authentication problem, of recourse to the argument that x-ray images were relevantly similar to models, maps, diagrams, and photographs, Smith's attorneys needed a different argument if they were to convince Judge LeFevre to admit the proffered x-ray image. They turned, not to a new rhetoric, but to an old one, a schema that Tal Golan traces to the very genesis of discourse on expert testimony in Anglo-American law. Namely, Smith's attorneys argued that the proper foundation for admitting the x-ray images existed in the reliability of the scientific method used to produce them. And while this argument was not in and of itself novel, its application would come to profoundly change American legal culture, in ways that have great significance for contemporary neuroimaging evidence.

Golan traces the tension between scientific testimony based on firsthand, personal familiarity and testimony by an expert witness lacking such familiarity to the famous 1782 decision of Lord Mansfield in *Folkes v Chadd*.[43] In that case, Lord Mansfield admitted the testimony of the illustrious John Smeaton, one of the founders of the profession of civil engineering, on a question regarding the propriety of draining marshlands in Norfolk to create a sluice that would enhance the safety of Wells Harbour. At issue was Smeaton's report regarding the cause of the decay of the natural harbour in Wells. Lord Mansfield ultimately admitted the report and relied on it in guiding his decision, which, according to Golan, illustrated 'the growing legal recognition by the end of the eighteenth century that there was a new class of persons, skilled in matters of science, who could give their opinion, even if was not based on the traditional trustworthiness of the senses'.[44]

Yet Lord Mansfield's decision to admit and rely upon Smeaton's report was in large part a function of Smeaton's expert qualifications. The extent of the expert witness's (scientific) qualifications remains a central part of the American law of evidence, and prefigures the use of expert testimony to support the admission of x-ray images as evidence. The continued relevance of the scientific expert in the common law tradition from 1782 to the present suggests the enduring power of a truth-to-nature notion of objectivity, in which subjective expertise in interpreting scientific evidence is critical to their veracity. While the idea that an expert witness could provide testimony even if not based on first-hand knowledge with the matter at issue was over a century old by the *fin-de-siècle*, the idea that scientific images could be offered into evidence without a witness who possessed first-hand knowledge was comparatively unprecedented—at least until the photograph.

Returning to *Smith v Grant*, Grant's attorneys recognized the authentication problem Smith faced, and argued, with the weight of tradition, that a witness with first-hand knowledge was the sole means of laying a proper foundation for the introduction of images into evidence. Smith's attorneys responded by conceding

---

[43] 3 Doug KB 157.   [44] Note 37, above, 45.

that such a means of authentication was the most common way of admitting visual images into evidence, but noted that it was not the exclusive method. Rather, in the discourse over the evidentiary role of photographs, a relatively new means of authentication had been advanced, one which posited the reliability of the method used to produce photographs as sufficient to lay the foundation for admissibility. Indeed, I have already noted that the reliability of the method was an argument used by expert witnesses in the Mumler hearing to shore up the veridical capacity of photographs, even while their testimony was offered to highlight the falsity of Mumler's own photographs.[45]

There is evidence in the relevant case law that this means of authentication was at least recognized contemporaneous with *Smith v Grant*.[46] However, relying too heavily on such a method risked vitiating the analogy between models, maps, and diagrams on the one hand, and photographs on the other. If the latter, unlike the former, could be authenticated by a witness without first-hand knowledge of the matter asserted therein, that suggested a critical difference between the two classes of images. Accordingly, Golan notes that this means of authentication was not regularly utilized in practice. But Smith's attorneys recognized its potential applicability to the question of the admissibility of x-ray images, and exploited it to great effect—Judge LeFevre was eventually won over by the argument:

In addition to these exhibits in evidence, we have nothing to do or say as to what they will purport to represent; that will, without doubt, be explained by eminent surgeons. These exhibits are only pictures or maps, to be used in explanation of a present condition, and therefore are secondary evidence and not primary. They may be shown to the jury as illustrating or making clear the testimony of experts... It must not be said of the law that it is wedded to precedent; that it will not lend a helping hand. Rather let the courts throw open the door to all well considered scientific discoveries. Modern science has made it possible to look beneath the tissues of the human body, and has aided surgery in telling of its hidden mysteries. We believe it to be our duty in this case to be the first, if you please, to so consider it, in admitting in evidence a process known and acknowledged as a determinate science.[47]

There is much in Judge LeFevre's language here that warrants scrutiny. His analysis begins with the rhetorical device Mnookin notes; namely, he uses the power of analogy to undermine the novelty of the question he is facing. X-ray images are 'only pictures or maps', which may be authenticated and explained by duly qualified experts just as Wigmore would come to note several years later. However, Judge LeFevre's language suggests his awareness that the analogy is strained for the reason that no witness can testify with first-hand knowledge as to the truth of the matter represented in the x-ray images. Even any expert witness's familiarity with the truth of the matter asserted therein is perforce second hand, acquired indirectly via the technical apparatus itself. Accordingly, Judge LeFevre goes on to admonish those beholden to the hoary hand of precedent, observing that

---

[45] Note 7, above, 37–9.
[46] See, e.g., *Cowley v People* 83 NY 464, 478–9 (1881); *Hampton v Norfolk & W.R. Co.* 27 SE 96, 98 (NC 1897).
[47] *Smith v Grant* (1895) *Chicago Legal News* 145.

law must welcome 'well considered scientific discoveries', thereby suggesting that Judge LeFevre is well aware that something distinctive and perhaps even unprecedented attends the decision on whether to admit x-ray images as evidence.[48] This distinction, of course, is the difficulty of laying a proper foundation for the images, and Judge LeFevre delivers the rule of the case in the final two sentences, both of which are crucial. In the first sentence, he emphasizes the social and cultural, not to mention epistemic and evidentiary, power available in illuminating the visible inner body. In the second sentence, he notes that because of such power, it is the 'duty' of the court to admit the images as the result of a 'process known and acknowledged as a determinate science'. It is the reliability and accuracy of the process that suffices to authenticate x-ray images as evidence, and Judge LeFevre's language suggests he is to some extent aware of the tension, if not the internal inconsistency, between regarding x-ray images as part of the same evidentiary taxonomy constituted by models, maps, diagrams, and photographs and the brute fact that x-ray images could not without difficulty be authenticated according to that taxonomy.

To be clear, my argument is not that the means of authentication Judge LeFevre relied on was fabricated entirely out of the need to categorize the epistemic and evidentiary valence of x-ray images in *fin-de-siècle* American legal culture. Rather, my argument is that while this potential method of authentication had been noted for at least several decades prior, and was itself part of an older tradition that substituted scientific reliability for first-hand knowledge as a criterion for admissibility, the matter of the x-ray dramatically accelerated the extent to which stakeholders relied on this means. Using the rigour of the particular scientific method may have been noted in the discourse over photographs, but there is little indication that stakeholders relied on it with any frequency. Indeed, why would they, if the evidentiary status of photographs could be fitted, albeit not with perfect comfort, into the category of models, maps, and diagrams? But x-ray images simply could not be so fitted, and so Judge LeFevre, and the vast majority of courts to follow, had to rely upon a different epistemic basis for proving up the evidentiary foundation for those images.

It is, however, unlikely that this secondary means of authentication would, as regards scientific and clinical images, quickly become the primary means if the epistemological core of the rule had no purchase in the larger social discourse. That is to say, unless there existed a prior conceptual scheme in which scientific images of the inner body were perceived to possess profound veridical power, the argument for authentication without first-hand knowledge could not have prevailed so easily.[49] Judge LeFevre's own language demonstrates this; he precedes his final ruling with a declaration on the power evident in modern science's capacity to see the hitherto dark spaces of the human body. Moreover, I have already touched on some of this larger social context regarding the power of illuminating the visible inner body in

---

[48] Note 37, above, 193.
[49] Note 32, above; see also Michael J. Sappol, *A Traffic of Dead Bodies: Anatomy and Embodied Social Identity in Nineteenth-Century America* (Princeton, Princeton University Press, 2002).

tandem with the increasing hold the concept of mechanical objectivity exerted by the *fin-de-siècle*.

Time and space do not permit further discussion of the true power of 'visibilizing' the inner body, but the evidentiary accommodation Judge LeFevre prescribed would take firm hold in American legal culture by the 1920s. However, even prior to this time, courts considering the admissibility of x-ray images cited and relied upon his analysis in *Smith v Grant*. Thus, as early as 1899, the Supreme Court of Minnesota declared in *dicta* that 'the utility and readability of [the x-ray's] results are already so well-established as scientific facts that courts ought to take judicial notice of them'.[50] Of course, a court's willingness to take judicial notice of x-ray images fully satisfies authentication requirements under the American law of evidence, which is a reflection of just how much epistemic power such images might carry at the time. However, even if the language in *Wittenberg* is deemed an outlier, American courts nevertheless quickly began to rely upon the reliability of the scientific process that produced the x-ray as a criterion for its authentication.

The first published case from an American court of last resort comes in 1897, from the Supreme Court of Tennessee. In the case styled *Bruce v Beall*, the court ruled that x-ray images offered by the injured plaintiff were admissible, noting that:

[n]ew as this process is, experiments made by scientific men, as shown by this record, have demonstrated its power to reveal to the natural eye the entire structure of the human body, and that its various parts can be photographed, as its exterior surface has been and now is.[51]

In other words, it is the process that suffices to authenticate the x-ray image, although the court hastened to emphasize that the testimony of properly qualified experts was required for admissibility. Moreover, as noted above, the fact that American courts insisted on expert testimony as a requirement for admissibility reflects a concern over the epistemic and rhetorical power of scientific and clinical images (as well as the truth-to-nature concept of objectivity). It is precisely because images of the inner body may have such persuasive power that courts took pains to circumscribe that power by requiring expert testimony.[52]

By the late 1920s, as a matter of ordinary practice, x-ray images could not be admitted into evidence without accompanying expert testimony. But even in these cases, the authentication requirements—an independent rule of evidence from that governing expert testimony—virtually all relied upon the reliability of the process itself. My argument is that x-rays catalysed this evidentiary regime, one that differed in important ways from the evidentiary regulation of photographs. The power of x-ray images was such that it accelerated the development of a framework of evidence law in which the perceived validity of the scientific process itself was sufficient to authenticate such images. In short, x-ray images changed American

---

[50] *Wittenberg v Onsgard* 81 NW 14, 14–5 (1899).
[51] *Bruce v Beall* 41 SW 445, 446 (1897).
[52] Courts ruling that expert testimony was required to admit the x-ray include *Marion v Coon Constr. Co.* 131 NYS 647, 651 (NY App Div 1915); *Lupton v S. Express Co.* 86 SE 614, 615 (NC 1915); and *Reinhold v Spencer* 26 P2d 796 (Id 1933).

evidence law. Of course, this is not to assert that such changes were entirely unprecedented, nor even that these changes might not have occurred in the absence of x-ray images. But in straining if not vitiating the analogy between maps, models, diagrams, and photographs on the one hand, and x-rays on the other, x-rays at the very least facilitated a fundamentally distinct means for evaluating the evidentiary status of scientific and clinical images.

In 1927, this regime would become even further cemented in American legal culture, in the case of *Frye v United States*.[53] *Frye* is typically regarded as a seminal case in the American law of scientific and expert evidence because it seems to play a pivotal role in instantiating the legal regime in which the reliability of the scientific process becomes a key criteria for its authentication (and hence its admissibility). Given how much ink has been spilled over the evidentiary status of neuroimaging-based lie-detection techniques, it is intriguing to note that *Frye* also involved a lie-detection test. In that case, Alphonso Frye, a young African-American, was accused of the murder of a wealthy African-American physician.[54] Frye confessed to the crime in August 1921 but subsequently sought to recant.[55] In an effort to shore up Frye's claims of innocence, Frye's attorney, Richard Mattingly, sought to introduce the results of William Marston's novel lie-detection test, through which, according to Golan, Mattingly himself had become convinced of Frye's innocence.[56] Both the trial court and the appellate court ruled that the test was inadmissible, but Golan observes that the doctrinal basis for rejecting the evidence was flimsy, since the results of the test were almost certainly relevant and probative, and Marston's qualifications as an experimental psychologist were indisputable.[57]

As such, the Court of Appeals reasoned that because lie-detection tests were not sufficiently reliable so as to enjoy general acceptance among the relevant scientific community of psychologists, it was inadmissible.[58] Admittedly, whether the reliability of the scientific process is established and whether the validity of the process is generally accepted among the relevant scientific community are not identical criteria. Yet it is plain from the *Daubert* standard itself that the overarching evidentiary criterion is whether the evidence proffered is produced via a valid scientific methodology.[59] The US Supreme Court in *Daubert* explained that whether the technique is generally accepted in the relevant scientific community

---

[53] 293 F 1013 (DC Cir 1923).
[54] Note 37, above, 245.
[55] Ibid. 245.
[56] Ibid. 246.
[57] Ibid. 251.
[58] Ibid. 251.
[59] In truth, the relationship between *Daubert* and *Frye* is intricate in the extreme. *Daubert* officially overruled *Frye* as to cases governed by the federal rules of evidence, but fifteen states and the District of Columbia (including seven of the largest jurisdictions: Arizona, California, Florida, Illinois, New York, Pennsylvania, and Washington) adhere to *Frye*. See, e.g., Simon A. Cole, 'Out of the Daubert Fire and Into the Fryeing Pan? Self-Validation, Meta-Expertise and the Admissibility of Latent Print Evidence in Frye Jurisdictions' (2008) 9 *Minnesota Journal of Law, Science & Technology* 453, 462; Alice B. Lustre, 'Post–Daubert Standards for Admissibility of Scientific and Other Expert Evidence in State Courts' (2001) 90 ALR 5th 453. Regardless of the particulars, there is no question that the reliability of the scientific process is a central evidentiary inquiry under both standards.

is only one of the factors used to assess the extent to which a particular technique is scientifically valid.[60] Thus, the principal evidentiary inquiry in weighing the admissibility of scientific and clinical images is whether the relevant image was produced *via* a reliable method. There are two key points here. First, while the rule announced in *Frye* refers to the 'general acceptance' standard, that criterion is only a smaller part of the larger inquiry the Court of Appeals weighed in *Frye*, *viz.* the reliability of the scientific process used to fabricate the lie detection measures. Second, this doctrine would come to dominate the American law of scientific evidence for at least the following seventy years. As Golan notes, '[i]t has survived many assaults on its exclusionary nature and blurred concepts, and was accepted for most of the twentieth century as the standard for the admissibility of new scientific evidence in practically all of America's courts'.[61] The rule is codified in Federal Rule of Evidence 702, and its counterpart (typically used as a model for state rules of evidence), Uniform Rule of Evidence 702. Anthony Benedetto notes:

A radiologist testifying as to the interpretation of a film-based radiograph is allowed to do so under Rule 702 because the process of making radiographs is standardized and has been accepted as being reliable for many years. The standardization of positioning, x-ray equipment, x-ray film, and x-ray film developing make it straightforward for the testifying radiologist to answer questions about the accuracy and reliability of the radiographs at issue in the trial.[62]

The literature on the relative merits of the *Frye* and *Daubert* standards is voluminous, and is far beyond the scope of this paper. The key point is simply that it is the epistemic power of a scientific and clinical image—the x-ray—that, while not creating a new evidentiary regime in American law, catalyses it. Prior to the discourse over the evidentiary status of the x-ray, American courts managed to assimilate such images under more conventional standards of American evidence law used to govern the admissibility of images in litigation. Because x-ray images simply could not be authenticated by a witness with first-hand knowledge, the x-ray stretched the plausibility of this process of assimilation. The tenuousness of categorizing x-rays as analogous to models, maps, diagrams, and photographs forced American courts to rely on different criteria for authentication and ultimate admissibility, including tying the authentication of the images to the reliability of the scientific process used to create them. In a sentence, then, the point of the entire analysis thus far is this: by the *fin-de-siècle*, scientific and clinical images possessed sufficient epistemic and evidentiary power to catalyse a relatively novel regime for managing scientific evidence in American legal culture. That scientific and clinical images are culturally powerful in the west is hardly news; that they played a significant role in altering the landscape of the American law of evidence is, I submit, comparatively less understood.

---

[60] *Daubert v Merrell Dow Pharmaceuticals* 509 US 579 (1993).
[61] Note 37, above, 253.
[62] Anthony R. Benedetto, 'Not Just 'X-Rays' Today: Recommendations for Admissibility of Modern Radiology Images' (2007) 49 *South Texas Law Review* 113, 130.

With these observations in mind, I can turn to assessing the implications for contemporary neuroscientific evidence, especially neuroimaging evidence.

27.3 The Power to Reshape Law: Implications for Law and Neuroscience

The argument is that the epistemic power of the x-ray in American legal culture was a significant factor in effecting fairly dramatic changes in the American law of scientific evidence. How is this relevant to contemporary discourse on the evidentiary status of neuroimaging evidence in American courts?

I suggested in section I that if analysis begins only with the assessment of whether neuroimaging techniques produce images that satisfy evidentiary standards in American law, such analysis necessarily lacks critical context in evaluating the power of scientific and clinical images. I have thus far endeavoured to provide a sketch of that context, with the overarching claim being that the power of these images catalysed the development of a distinct regime in American evidence law, one that in most of the essential respects continues to govern. Moreover, the fact that American courts have been, since the mid-nineteenth century, acutely aware of the epistemic and rhetorical power of scientific images supplements this larger picture regarding the concern over the potential use and abuse of neuroimaging evidence in American legal culture. If scientific and clinical images have social and cultural resonance sufficient to facilitate major changes in as conservative an institution as the American law of evidence, that has critical implications for the choices made in regulating the use of neuroimaging evidence in American courts.

I will unpack some of these implications momentarily, but it is first important to deepen the analysis by addressing the fact that neuroimaging evidence consists of images of the brain. This point is crucial because of the paramount role that the brain itself, as well as scientific images of the brain, exert in shaping enormous terrain in American culture. In his seminal history of the brain, Robert Martensen develops the concept of 'codes of signification built into [biomedicine's] conceptual foundations'.[63] Martensen's point is precisely that the epistemic significance of images of neural anatomy and function remains at the very centre of what it means to participate in American cultures of biomedicine. These codes are fundamentally tied to the power of scientific sight. This is at least in part why Daston and Galison insist that the scope of the investigating, scientific eye is central to defining the scientific self. Indeed, in the most significant western atlas of human anatomy of the last millennium, Vesalius spends a large portion of the early chapters discussing the anatomy of the head.[64] Similarly, in one of the cornerstone texts of modern neural anatomy, *Cerebri anatome* (1664), Thomas Willis noted the importance of 'ocular demonstrations' of neural anatomy.[65] And the preface to Walter Charleton's 1680

---

[63] Robert L. Martensen, *The Brain Takes Shape: An Early History* (New York, Oxford University Press, 2004) 203.
[64] Note 24, above.   [65] Note 63, above, 199.

treatise *Enquiries into Human Nature: Anatomic Praelections in the New Theatre of the Royal Colledge of Physicians in London* demonstrates the reverence with which early modern physicians treated the anatomy of the brain:

> Who can look into the Sanctum Sanctorum of this Temple, the Brain, and therein contemplate the pillars that support it, the arch'd roof that covers and defends it, the fretwork of the Ceiling, the double membrane that invests it, the resplendent partition that divides it, the four vaulted cells that drain away impurities, the intricate labyrinths of arteries that bring in from the heart rivulets of vital blood to heat and invigorate it, the Meanders of veins to export the same blood, the Aqueducts that preserve it from inundation, the infinite multitude of slender and scarce perceptible filaments that compose it, the delicate nerves or chords spun from those threds, the original of that silver chord (as Ecclesiastes calls it) or Spinal marrow, upon which the strength of back and limbs chiefly depends; and many other parts of the wonderful Engine: and not discern an infinite Wisdom in the design and construction of them?

The point, again, is that images of the brain are critical in shaping modern science and medicine in the West.

However, philosophers of science such as Evelyn Fox Keller and Paul Feyerabend have, along Kuhnian lines, warned of the error of seeing scientific cultures as somehow hermetically sealed from larger public discourse.[66] The centrality of neural anatomy in American biomedicine and science reflects and is reflected by its similar pride of place in American culture in general. For example, in his 2003 book, Joseph Dumit painstakingly documents how American illness sufferers literally construct their own biomedical identity on the basis of brain scans.[67] Similarly, José van Dijk explains how clinical imaging embodies the ideal of transparency in western culture, which is a significant part of wherefrom it derives its social and cultural power.[68]

These concerns are not merely academic. An elegant proof of the compelling rhetorical power of scientific images of the brain can be found in David McCabe and Alan Castel's study, aptly entitled 'Seeing is believing: The effect of brain images on judgments of scientific reasoning'. In the study, subjects read three brief scientific articles that contained various logical fallacies. The control group was only provided with the written text, while experimental groups were provided either the written text and a fMRI brain image, or the written text and a bar graph. Even though the written texts provided to the control and the experimental group were exactly the same, and hence contained identical fallacies, McCabe and Castel found that 'texts accompanied by a brain image were given the highest ratings of scientific reasoning, differing reliably from both the control ... and bar graph conditions'.[69]

---

[66] Evelyn Fox Keller, *Making Sense of Life: Explaining Biological Development with Models, Metaphors, and Machines* (Cambridge, MA, Harvard University Press, 2002); Paul Feyerabend, *Farewell to Reason* (New York, Verso Books, 1987).

[67] Note 2, above.

[68] *A Cultural Analysis of Medical Imaging* (Seattle, Washington, University of Washington Press, 2005).

[69] David P. McCabe and Alan D. Castel, 'Seeing is Believing: The Effect of Brain Images on Judgments of Scientific Reasoning' (2007) 107 *Cognition* 343, 346.

From these results, McCabe and Castel conclude 'that there is, indeed, something special about the brain images with respect to influencing judgments of scientific credibility'.[70] 'This tendency to interpret brain images as credible may be related to people's natural affinity for reductionistic explanations for cognitive phenomena...'.[71]

McCabe and Castel's findings are buttressed by the results of a recent study conducted by Weisberg, Keil, Goodstein, Rawson, and Gray. In the latter, the investigators performed three experiments on three groups of subjects (novices, neuroscience students, and neuroscience experts), and found that:

explanatorily irrelevant neuroscience information encouraged [novices and students] to judge [bad] explanations more favorably... That is, extraneous neuroscience information makes explanations look more satisfying than they actually are, or at least more satisfying than they otherwise would be judged to be. The students in the cognitive neuroscience class showed no benefit of training, demonstrating that only a semester's worth of instruction is not enough to dispel the effect of neuroscience information on judgments of explanations. Many people thus systematically misunderstand the role that neuroscience should and should not play in psychological explanations, revealing that logically irrelevant neuroscience information can be seductive—it can have much more of an impact on participants' judgments than it ought to.[72]

These findings are consistent with Racine, Bar-Ilan, and Illes's finding that what they term the fallacy of 'neuro-realism' is widespread.[73] This fallacy, and its prevalence, 'reflects the uncritical way in which an fMRI [investigation] can be taken as validation or invalidation of our ordinary view of the world'.[74] Moreover, several studies document that misunderstandings of the meaning of neuroimages are ubiquitous in the media.[75] 'Given the obviously powerful role of the media in both reflecting and shaping public discourse, this may be an additional reason to worry...'.[76]

Accordingly, there is ample evidence that neuroimages have compelling rhetorical power. However, there is comparatively less analysis explaining why it might be that images of the brain convey such power, and perhaps even less scholarship tracing the history of such power in American legal culture. In part, my point is that such power has, on at least one occasion, significantly contributed to the reshaping of the American law of scientific evidence itself. Scientific and clinical images are powerful and compelling; so powerful and compelling that they may not only

---

[70] Note 69, above, 346.
[71] Ibid. 344.
[72] Deena Skolnick Weisberg, Frank C. Keil, Joshua Goodstein, Elizabeth Rawson, and Jeremy R. Gray, 'The Seductive Allure of Neuroscience Explanations' (2008) 20(3) *Journal of Cognitive Neuroscience* 470, 475.
[73] Eric Racine, Ofek Bar-Ilan, and Judy Illes, 'fMRI in the Public Eye' (2005) 65(2) *Nature Reviews Neuroscience* 159.
[74] Note 73, above, 160.
[75] E.g., Eric Racine, Ofek Bar-Ilan, and Judy Illes, 'Brain Imaging: A Decade of Coverage in the Print Media' (2006) 28(1) *Science Communication* 122.
[76] Daniel S. Goldberg, 'The Detection of Constructed Memories and the Risks of Undue Prejudice' (2008) 8(1) *American Journal of Bioethics—Neuroscience* 23, 24.

possess rhetorical power that seems to far outstrip the truth value of the matter asserted, but also may, ironically, carry the power to mold the very legal practices and institutions used to regulate that power.

There are any number of ways in which this argument may contribute to specific evidentiary concerns in American legal culture. One particular example is the impact on an evidentiary objection lodged under Federal Rule of Evidence 403, and its state counterparts.[77] Rule 403 provides what is often referred to as a 'catch-all' objection. That is, it has wide applicability and can be used in a variety of circumstances in which evidence is deemed objectionable by a party. It provides: 'Although relevant, evidence may be excluded if its probative value is substantially outweighed by the danger of unfair prejudice, confusion of the issues, or misleading the jury, or by considerations of undue delay, waste of time, or needless presentation of cumulative evidence.'

As I have noted previously, the Rule imposes a high burden on the party opposing admission because the probative value of the evidence must be 'substantially outweighed' by the risk of undue prejudice. 'Thus, where the probative value of the evidence is significant, it is difficult to show that the quantum of probative value is *substantially* outweighed by the risks of undue prejudice or of misleading the jury.'[78] Thus, Rule 403 objections are typically deployed as a last resort, after all or most other relevant objections have been tried. However, this tactical consideration should not be confused with a lack of significance in the overall structure of the rules; Victor Gold noted that it is the 'cornerstone' of the federal rules.[79] When it comes to neuroimaging evidence, the larger context regarding the epistemic and rhetorical power of scientific and clinical images (especially of the brain) suggests that perhaps a Rule 403 objection may have greater use and impact than might otherwise be conjectured. If indeed scientific and clinical images are powerful enough to reshape the American law of evidence itself, and if neuroimages have demonstrated rhetorical and epistemic power that vastly surpasses their relative truth value, the argument that the risk of undue prejudice substantially outweighs its probative value may have greater rhetorical force of its own. This is especially true where many commentators have observed that the probative value of such images is not particularly large to begin with.[80]

But the merits of a 403 objection, and the importance of understanding the full evidentiary power of scientific and clinical images, provide an additional bite at the apple, so to speak. This is because the evidentiary standard promulgated in *Frye* prescribes the general acceptance test as a key criterion for scientific reliability (and hence admissibility in general). The US Supreme Court formally retained this test in *Daubert*, and there is general if not universal agreement that neuroimages likely do not satisfy the general acceptance test at the present time.[81] But, in concert with the balancing mechanism contained in Rule 403, comprehending the epistemic power of scientific and clinical images of the brain provides at least some reason for

---

[77] Note 3, above (citing sources). [78] Note 76, above, 23.
[79] Note 5 above, 497 and note 5. [80] Note 3, above.
[81] Ibid.

judging the admissibility of such images sceptically even in the event that the reliability and accuracy of the images becomes generally accepted in the relevant scientific community. Under this counterfactual, even where the probative value of neuroimages is relatively high (in so far as they are generally accepted as scientifically valid), the risk of unfair prejudice stemming from the vast epistemic and rhetorical power of such images may, in at least some cases, substantially outweigh the quantum of probative value.

27.4 Whither Scientific and Clinical Images of the Brain in American Legal Culture?

The scholarship addressing the evidentiary status of neuroscientific evidence is explosive, especially as to neuroimaging techniques. The majority of this scholarship warns of the potentially prejudicial effect of neuroimages, which, combined with the currently low probative value of such evidence, tilts the balance strongly against the admission of the evidence. However, a review of the literature demonstrates that while much of the analysis asserts the prejudicial effect of neuroimages, there is comparatively less discussion explaining precisely why neuroimages are likely to be so prejudicial, particularly compared with other forms of scientific and clinical evidence. There is a rich interdisciplinary literature crossing the medical humanities, the history of science, medicine, and technology, and science and technology studies (to name but a few) that delves deeply into the Western and American fascination with scientific and clinical images, and especially images of the brain.

To the best of my knowledge, there are relatively few attempts to link this corpus to the legal history of scientific images in American culture. Such an analysis is arguably a worthwhile contribution in its own right, but may also be, I hope, modestly useful in as much as it contextualizes the current legal debate over neuroscientific evidence in American courts and suggests implications for that debate. The lesson in this narrative is that scientific and clinical images are not simply persuasive to lay persons (although there is ample scholarship suggesting that the meaning-making power of scientific and clinical images, including neuroimages, is not limited to scientists and physicians). Daston and Galison's history by itself definitively shows the central role such images play in constructing the scientific self for over four centuries in the west. Consistent with this history, Martensen's account shows that the history of neuroimages is deeply and fundamentally linked with American cultures of biomedicine and science.

In terms of American legal history, the epistemic power of scientific images was sufficient to catalyse the development of a distinct regime through which American courts regulated scientific and clinical evidence. Both the photograph and the x-ray played pivotal roles in this process of development. The x-ray is of special importance, as it was largely the fact that even expert witnesses could not testify via first-hand knowledge that led courts to begin permitting the reliability of the scientific methodology used to produce the x-ray images stand as a primary means of authentication.

In the few decades following the initial decision in *Smith v Grant*, this particular standard would become formalized as the principal framework by which the admissibility of scientific evidence would be governed in American legal culture. With some changes, it remains as such.

In short, scientific and clinical images not only changed the Western world; they changed the American law of evidence. Such is their epistemic and rhetorical power.

28

Lost in Translation? An Essay on Law and Neuroscience

Stephen J. Morse[*]

28.1 Introduction

The rapid expansion in neuroscientific research fuelled by the advent of functional magnetic resonance imaging [fMRI] has been accompanied by popular and scholarly commentary suggesting that neuroscience may substantially alter, and perhaps will even revolutionize, both law and morality. This essay will attempt to put such claims in perspective and to consider how properly to think about the relation between law and neuroscience. The overarching thesis is that neuroscience may indeed make some contributions to legal doctrine, practice, and theory, but such contributions will be few and modest for the foreseeable future.

The first part of this essay describes the law's implicit folk psychological view of human behaviour and why any other model is not possible at present. It then turns to dangerous distractions that have bedevilled clear thinking about the relation between scientific explanations of human behaviour and law. Next, the essay considers how to translate the mechanistic findings of neuroscience into the folk psychological concepts the law employs. Finally, illustrative case studies of the legal relevance of neuroscience studies are presented. The discussion and all the examples focus on criminal law and on competence for the sake of simplicity and coherence, but the arguments are almost all generalizable to other legal contexts.

28.2 The Criminal Law's Implicit Psychology and Legal Criteria

Lawyers take the criminal law's implicit psychology for granted because there is seldom any need to identify or to question it. The new neuroscience may call the

---

[*] Ferdinand Wakeman Hubbell Professor of Law & Professor of Psychology and Law in Psychiatry, University of Pennsylvania Law School and School of Medicine. The author thanks Ed Greenlee for his invaluable assistance. His personal attorney, Jean Avnet Morse, provided sound, sober counsel and moral support.

law's psychology into question, however, so it is crucial consciously to recognize it and to understand what would be entailed if it were undermined.

Criminal law presupposes the 'folk psychological' view of the person and behaviour. This psychological theory causally explains behaviour in part by mental states such as desires, beliefs, intentions, willings and plans. Biological, other psychological, and sociological variables also play a role, but folk psychology considers mental states fundamental to a full explanation of human action. Human behaviour cannot be adequately understood if mental state causation is completely excluded or eliminated. Lawyers, philosophers, and scientists do of course argue about the definitions of mental states and theories of action, but that does not undermine the general claim that mental states are fundamental. Indeed, the arguments and evidence disputants use to convince others presuppose the folk psychological view of the person. Brains do not convince each other; people do.

For example, the folk psychological explanation for why you are reading this paper is, roughly, that you desire to understand the relation of neuroscience to law to improve your work, you believe that reading the paper will help fulfil that desire, and thus you formed the intention to read it. This is a 'practical' explanation rather than a deductive syllogism.

Folk psychology does not presuppose the truth of free will, it is perfectly consistent with the truth of determinism, it does not hold that we have minds that are independent of our bodies (although it, and ordinary speech, sound that way), and it presupposes no particular moral or political view. It does not claim that all mental states are necessarily conscious or that people go through a conscious decision-making process each time that they act. It allows for 'thoughtless', automatic, and habitual actions and for non-conscious intentions. It does presuppose that human action will at least be rationalizable by mental state explanations or that it will be responsive to reasons, including incentives, under the right conditions. The definition of folk psychology being used does not depend on any particular bit of folk wisdom about how people are motivated, feel, or act. Any of these bits, such as that excited utterances are reliable enough to be an exception to the hearsay rule, may be wrong, as neuroscience might help disclose. Moreover, neuroscience might help the law define and identify legally relevant mental states with more precision. The definition insists only that human action is in part causally explained by mental states.

To understand the importance of mental states, consider the criteria for criminal responsibility: the elements of the *prima facie* case—primarily acts and mental states— and the absence of an affirmative defence. All are infused with mental states. All crimes include a 'voluntary' act requirement, which is defined, roughly, as an *intentional* bodily movement (or omission in cases in which the person has a duty to act) done while the agent is in a reasonably integrated state of consciousness. Although the meaning of an intentional bodily movement is seldom specified, the best definition is a bodily movement that can be in principle understood according to the person's mental states. One can almost always ask of any act, 'Why did you do that?', and expect some explicit or implicit mental explanation. If there is none even implicitly possible, it is probable that the agent's bodily movement was not an act at all.

Other than crimes of strict liability, all crimes also require a culpable further mental state, such as purpose, knowledge, or recklessness. Some crimes are also defined with the *mens rea* of negligence, which appears to be the absence of a mental state. This is a controversial issue among legal scholars, but the best explanation is that the failure to pay attention when the agent was creating the substantial and unjustifiable level of risk that supports criminal liability is itself a type of culpable omission. On the other hand, some scholars believe that negligence is indistinguishable from strict liability or, even if it is distinguishable, it is not really a mental state at all.

All affirmative defences of justification and excuse involve an inquiry into the person's mental state, such as belief that self-defensive force was necessary or the lack of knowledge of right from wrong. Of course the person's mental state is influenced by biological, psychological, and sociological variables, and knowledge of these variables may help determine what the person's mental state was, but the law is ultimately concerned with the mental state itself, rather than the causes of it.

Brief reflection should indicate that the law's psychology must be a folk psychological theory, a view of the person as a conscious (and potentially self-conscious) creature who forms and acts on intentions that are the product of the person's other mental states such as desires, beliefs, willings, and plans. We are the sort of creatures that can act for and respond to reasons, including legal rules and standards that are expressed and understood linguistically. The law treats persons generally as intentional creatures and not as mechanical forces of nature.

Law is primarily action-guiding[1] and could not guide people *ex ante* and *ex post* unless people could use rules as premises in their reasoning about how they should behave. Otherwise, law as an action-guiding normative systems of rules and standards would be useless, and perhaps incoherent.[2] Law can directly and indirectly affect the world we inhabit only by its influence on human beings who can potentially use legal rules to guide conduct. Unless people were capable of understanding and then using legal rules to guide their conduct, law would be powerless to affect human behaviour.[3] As John Searle wrote:

Once we have the possibility of explaining particular forms of human behavior as following rules, we have a very rich explanatory apparatus that differs dramatically from the explanatory apparatus of the natural sciences. When we say we are following rules, we are accepting

---

[1] See, George Sher, *In Praise of Blame* (New York: Oxford University Press, 2006) 123 (stating that although philosophers disagree about the requirements and justifications of what morality requires, there is widespread agreement that 'the primary task of morality is to guide action'); John R. Searle, 'End of the Revolution' (2002) 43 *N.Y. Rev. of Books* at 33, 35.

[2] Scott J. Shapiro, 'Law, Morality, and the Guidance of Conduct' (2000) 6 *Legal Theory* 127, 131–2.

[3] Ibid, at 131–2. This view assumes that law is sufficiently knowable to guide conduct, but a contrary assumption is largely incoherent. As Shapiro (ibid.) writes: 'Legal skepticism is an absurd doctrine. It is absurd because the law cannot be the sort of thing that is unknowable. If a system of norms were unknowable, then that system would not be a legal system. One important reason why the law must be knowable is that its function is to guide conduct.' I do not assume that legal rules are always clear and thus capable of precise action guidance. If most rules in a legal system were not sufficiently clear most of the time, however, the system could not function. Further, the principle of legality dictates that criminal law rules should be especially clear.

the notion of mental causation and the attendant notions of rationality and existence of norms...

...The content of the rule does not just describe what is happening, but plays a part in *making it happen*.[4]

Legal rules are not simply mechanistic causes that produce 'reflex' compliance, although they can certainly help to inculcate law-abiding 'habits'. They operate within the domain of folk psychology. Legal rules are thus action-guiding primarily because they provide an agent with good moral or prudential reasons for forbearance or action. For example, no instinct governs how fast a person drives on the open highway. Among the various explanatory variables, however, the posted speed limit and the belief in the probability of suffering the consequences for exceeding it surely play a large role in the driver's choice of speed. Human behaviour can be modified by means other than influencing deliberation and human beings do not always deliberate before they act. Nonetheless, the law presupposes folk psychology, even when we most habitually follow the legal rules. The inculcation of law-abiding habits, for example, has an intentional component and we constantly act in the 'shadow of the law', especially when criminal conduct is at stake.

The legal view of the person does not hold that people must always reason or consistently behave rationally according to some preordained, normative notion of rationality. Rather the law's view is that people are capable of acting for reasons and are capable of minimal rationality according to predominantly conventional, socially constructed standards. The type of rationality the law requires is the ordinary person's common-sense view of rationality, not the technical notion that might be acceptable within the disciplines of economics, philosophy, psychology, computer science, and the like.

Virtually all actions for which agents deserve to be praised, blamed, rewarded, or punished are the product of mental causation[5] and, in principle, responsive to reason. Machines may cause harm, but they cannot do wrong and they cannot violate expectations about how people ought to live together. Machines do not deserve praise, blame, reward, punishment, concern, or respect because they exist or because of the results they cause. Only people, intentional agents with the potential to act, can violate expectations of what they owe each other and only people can do wrong.

Many scientists and some philosophers of mind and action consider folk psychology to be a primitive or pre-scientific view of human behaviour and the next section of this paper considers such views. For the foreseeable future, however, the law will be based on the folk psychological model of the person and behaviour described and this paper will proceed on that premise. Until and unless scientific discoveries convince us that our view of ourselves is radically wrong, the basic explanatory apparatus of folk psychology will remain central. It is vital that lawyers

---

[4] Searle, note 1, above, at 35.
[5] I do not mean to imply dualism here. I am simply accepting the folk-psychological view that mental states—which are fully produced by and realizable in the brain—play a genuinely causal role in explaining human behaviour.

and legal policy makers not lose sight of this model lest they fall into confusion when various claims based on neuroscience are made. Once again, any neuroscientific data or evidence must always be relevant to the law's folk psychological criteria. If neuroscience is to have any influence on current law and legal decision-making, it must be through this framework.

This author's view is that a primary task for neuroscience is to explain agency, not to explain it away reductively. That is, neuroscience should attempt to explain how a two and a half pound hunk of organic matter can produce consciousness, intentionality, reasons responsiveness, mental causation, and all the other mental apparatus that is so central to our lives.

28.3 Potential Distractions and Confusions

This section of the paper considers a number of underlying and related issues that are often thought to be relevant to criminal responsibility and competence but that are irrelevant or confusions and distractions: free will, causation as an excuse, causation as compulsion, prediction as an excuse, dualism, and the non-efficacy of mental states (alluded to just above).

Contrary to what many people believe and what judges and others sometimes say, free will is not a legal criterion that is part of any doctrine and it is not even foundational for criminal responsibility. Criminal law doctrines are fully consistent with the truth of determinism or universal causation that allegedly undermines the foundations of responsibility. Even if determinism is true, some people act and some people do not. Some people form prohibited mental states and some do not. Some people are legally insane or act under duress when they commit crimes, but most defendants are not legally insane or acting under duress. Moreover, these distinctions matter to moral and legal theories of responsibility and fairness that we have reason to endorse. Thus, law addresses problems genuinely related to responsibility, including consciousness, the formation of mental states such as intention and knowledge, the capacity for rationality, and compulsion, but it never addresses the presence or absence of free will.

When most people use the term free will or its lack in the context of legal responsibility, they are typically using this term loosely as a synonym for the conclusion that the defendant was or was not criminally responsible. They typically have reached this conclusion for reasons that do not involve free will, such as that the defendant was legally insane or acted under duress, but such usage of free will only perpetuates misunderstanding and confusion. Once the legal criteria for excuse have been met, for example—and none includes lack of free will as a criterion—the defendant will be excused without any reference whatsoever to free will as an independent ground for excuse.

There is a genuine metaphysical problem about free will, which is whether human beings have the capacity to act uncaused by anything other than themselves and whether this capacity is a necessary foundation for holding anyone legally or morally accountable for criminal conduct. Philosophers and others have debated these issues in various forms for millennia and there is no resolution in sight. Indeed, some people think the problem is not resolvable. This is a real philosophical issue, but, it is not a problem for the law, and neuroscience raises no new challenge to this conclusion. Solving the free will problem would have profound implications for responsibility doctrines and practices, such as blame and punishment, but, at present, having or lacking libertarian freedom is not a criterion of any civil or criminal law doctrine.

Neuroscience is simply the most recent mechanistic causal science that appears deterministically to explain behaviour. It thus joins social structural variables, behaviourism, genetics, and other scientific explanations that have also been deterministic explanations for behaviour. In principle, however, neuroscience adds nothing new, even if it is better, more persuasive science than some of its predecessors. As long as free will in the strong sense is not foundational for just blame and punishment and is not a criterion at the doctrinal level—which it is not—the truth of determinism or universal causation poses no threat to legal responsibility. Neuroscience may help shed light on folk psychological excusing conditions, such as automatism or insanity, for example, but the truth of determinism is not an excusing condition. The law will be fundamentally challenged only if neuroscience or any other science can conclusively demonstrate that the law's psychology is wrong and we are not the type of creatures for whom mental states are causally effective. This is a different question from whether determinism undermines responsibility, however, and this paper returns to it below.

A related confusion is that behaviour is excused if it is caused, but causation per se is not a legal or moral mitigating or excusing condition. At most, causal explanations can only provide evidence concerning whether a genuine excusing condition, such as lack of rational capacity, was present. For example, suppose a life history marked by poverty and abuse played a predisposing causal role in a defendant's criminal behaviour. Or suppose that an alleged new mental syndrome played a causal role in explaining criminal conduct. The claim is often made that such causes, which are not within the actor's capacity to control rationally, should be an excusing or mitigating position per se, but this claim is false.

All behaviour is the product of the necessary and sufficient causal conditions without which the behaviour would not have occurred, including brain causation, which is always part of the causal explanation for any behaviour. If causation were an excusing condition per se, then no one would be responsible for any behaviour. Some people welcome such a conclusion and believe that responsibility is impossible, but this is not the legal and moral world we inhabit. The law holds most adults responsible for most of their conduct and genuine excusing conditions are limited. Thus, unless the person's history or mental condition, for example, provides evidence of an existing excusing or mitigating condition, such as lack of rational capacity, there is no reason for excuse or mitigation.

Even a genuinely abnormal cause is not an excusing condition. For example, imagine a person with paranoid suspiciousness who constantly and hypervigilantly scans his environment for cues of an impending threat. Suppose our person with paranoia now spots a genuine threat that no normal person would have recognized and responds with proportionate defensive force. The paranoia played a causal role in explaining the behaviour, but no excusing condition obtained. If the paranoia produced a delusional belief that an attack was imminent, then a genuine excuse, legal insanity—an irrationality-based defence—might be appropriate.

In short, a neuroscientific causal explanation for criminal conduct, like any other type of causal explanation, does not per se mitigate or excuse. It provides only evidence that might help the law resolve whether a genuine excuse existed or data that might be a guide to prophylactic or rehabilitative measures.

Compulsion is a genuine mitigating or excusing condition, but causation, including brain causation, is not the equivalent of compulsion. Compulsion may be either literal or metaphorical and normative. If compulsion is literal, say, a person's arm moves because the person had a neuromuscular spasm or because a much stronger person pushed the arm, the person has not acted at all. Metaphorical compulsion is more difficult to understand, but it includes cases in which someone acts in response to a do-it-or-else threat (such as the excuse of duress) or acts in response to strong internal urges or desires (such as the control test for legal insanity). In all metaphorical compulsion cases the person acts, however, and deciding when to mitigate or excuse in such cases is a normative legal question.

It is crucial to recognize that most human action is not plausibly the result of either type of compulsion, but all human behaviour is caused by its necessary and sufficient causes, including brain causation. Even abnormal causes are not compelling. Suppose, for example, that a person with paedophilic urges has them weakly and is weakly sexed in general. If the person molested a child there would be no ground for a compulsion excuse. If causation were per se the equivalent of compulsion, all behaviour would be compelled and no one would be responsible. Once again, this is not a plausible account of the law's responsibility conditions. Causal information from neuroscience might help us resolve questions concerning whether legal compulsion existed or it might be a guide to prophylactic or rehabilitative measures when dealing with plausible legal compulsion. But causation is not per se compulsion.

Causal knowledge, whether from neuroscience or any other science, can enhance the accuracy of behavioural predictions, but predictability is also not per se an excusing or mitigating condition, even if the predictability of the behaviour is perfect. To understand this, just consider how many things each of us does that are perfectly predictable for which there is no plausible excusing or mitigating condition. Even if the explanatory variables that enhance prediction are abnormal, excuse or mitigation is warranted only if a genuine excusing or mitigating condition is present. For example, recent research demonstrates that a history of childhood abuse coupled with a specific genetically produced neurotransmitter abnormality vastly increases the risk that a person will behave antisocially as an adolescent or young adult.[6] Again, such information may be of prophylactic or rehabilitative use for people affected, but no excuse or mitigation is applicable just because these

---

[6] Avshalom Caspi et al., 'Role of Genotype in the Cycle of Violence in Maltreated Children' (2002). 297 *Science* 851.

variables make antisocial behaviour far more predictable. If the variables that enhance prediction also produce a genuine excusing or mitigating condition, then excuse or mitigation is justified for the latter reason and independent of the prediction.

Most informed people are not 'dualists' about the relation between the mind and the brain. That is, they no longer think that our minds (or souls) are independent of our brains (and bodies more generally) and can somehow exert a causal influence over our bodies. It may seem, therefore, as if law's emphasis on the importance of mental states as causing behaviour is based on a pre-scientific, outmoded form of dualism, but this is not the case. Although the brain enables the mind, we have no idea how this occurs and have no idea how action is possible. It is clear that, at the least, mental states are dependent upon or supervene on brain states, but neither neuroscience nor any other science has demonstrated that mental states play no independent and partial causal role.

Despite the lack of understanding of the mind–brain-behaviour relation, some scientists and philosophers question whether mental states have any causal effect, treating mental states as psychic appendixes that evolution has created but that have no genuine function. These claims are not strawpersons. They are seriously made by serious, thoughtful people. If accepted, they would create a complete and revolutionary paradigm shift in the law of criminal responsibility and competence (and more widely). They are treated in a later section.

In conclusion, legal actors must always keep the folk psychological view present to their minds when considering claims or evidence from neuroscience and must always question how the science is legally relevant to the law's action and mental states criteria. The truth of determinism, causation and predictability do not in themselves answer any doctrinal or policy issue.

28.4 Legal Relevance and the Need for Translation

What in principle is the possible relation of neuroscience to law? We must begin with a distinction between internal relevance and external relevance. An internal contribution or critique accepts the general coherence and legitimacy of a set of legal doctrines, practices or institutions and attempts to explain or alter them. For example, an internal contribution of criminal responsibility may suggest the need for doctrinal reform, of, say, the insanity defence, but it would not suggest that the notion of criminal responsibility is itself incoherent or illegitimate. By contrast, an externally relevant critique suggests the doctrines, practices or institutions are incoherent, illegitimate, or unjustified.

This section will consider the internal potential relevance of neuroscience. It begins with general considerations concerning the relevance of neuroscience to law, using criminal law and competence determinations as its primary examples. Then it turns to a taxonomy of types of internal contributions. Finally, it considers more radical, external challenges to law from neuroscience.

28.4.1 General considerations concerning translation

The law's criteria for responsibility and competence are essentially behavioural—acts and mental states. The criteria of neuroscience are mechanistic—neural structure and function. Is the apparent chasm between those two types of discourse bridgeable? This is a familiar question in the field of mental health law,[7] but there is even greater dissonance in neurolaw. Psychiatry and psychology sometimes treat behaviour mechanistically, sometimes treat it folk psychologically, and sometimes blend the two. In many cases, the psychological sciences are quite close in approach to folk psychology. Neuroscience, in contrast, is purely mechanistic and eschews folk psychological concepts and discourse. Thus, the gap will be harder to bridge.

The brain enables the mind. If your brain is dead, you are dead, you have no mind, and you do not behave at all. Therefore, facts we learn about brains in general or about a specific brain in principle could provide useful information about mental states and human capacities in general and in specific cases. Some believe that this conclusion is a category error.[8] This is a plausible view and perhaps it is correct. If it is, then the whole subject of neurolaw is empty and there was no point to writing this paper in the first place. Let us therefore bracket this pessimistic view and determine what follows from the more optimistic position that what we learn about the brain and nervous system can be potentially helpful to resolving questions of criminal responsibility if the findings are properly translated into the law's psychological framework. Then, the question is whether the new neuroscience is legally relevant because it makes a proposition about responsibility or competence more or less likely to be true. At present, few such data exist, but neuroscience is advancing so rapidly that such data may exist in the near or medium term. Moreover, the argument is conceptual and does not depend on any particular neuroscience findings.

Some preliminary points of general applicability must be addressed first, however. The most important is simply to repeat the message of the prior section of this paper. Causation by biological variables, including abnormal biological variables, does not per se create an excusing or mitigating condition. Any excusing condition must be established independently. The goal is always to translate the biological evidence into the criminal law's folk psychological criteria.

Assessing criminal responsibility involves a retrospective evaluation of the defendant's mental states at the time of the crime. No criminal wears a portable scanner or other neurodetection device that provides a measurement at the time of the crime. Further, neuroscience is insufficiently developed to detect specific, legally relevant mental content or to provide a sufficiently accurate diagnostic marker for even severe mental disorder. Nonetheless, certain aspects of neural

---

[7] Alan A. Stone, *Law, Psychiatry, and Morality* (New York: Cambridge University Press, 1984) 95–6.
[8] M. R. Bennett and P. M. S. Hacker, *Philosophical Foundations of Neuroscience* (Malden, MA: Blackwell Publishing, 2003); Michael Pardo and Dennis Patterson, 'Philosophical Foundations of Neuroscience' (2010) *U.Ill.L. Rev.* 1211.

structure and function that bear on legally relevant capacities, such as the capacity for rationality and control, may be temporally stable in general or in individual cases. If they are, neuroevidence may permit a reasonably valid retrospective inference about the defendant's rational and control capacities and their impact on criminal behaviour. This will of course depend on the existence of adequate science to do this. We now lack such science, but future research may remedy this.

Questions concerning competence or predictions of future behaviour address a subject's present condition. Thus, the retrospective problems besetting retrospective responsibility analysis do not apply to such questions. The criteria for competence are functional. They ask whether the subject can perform some task, such as understanding the nature of a criminal proceeding or understanding a treatment option that is being offered, at a level the law considers normatively acceptable to warrant respecting the subject's choice and autonomy.

At present, most neuroscience studies on human beings involve very small numbers of subjects. Further, most studies average the neurodata over the subjects and the average finding may not accurately describe the brain structure or function of an actual subject in the study. Finally, the neuroscience of cognition and interpersonal behaviour is largely in its infancy and what is known is quite coarse-grained and correlational rather than fine-grained and causal. Over time, however, these problems may ease as imaging and other techniques become less expensive and more accurate, as research designs become more sophisticated, and as the sophistication of the science increases generally. It is also an open question whether accurate inferences or predictions about individuals are possible using group data for a group that includes the individual. This is a very controversial topic, but even if it is difficult or impossible now, it may become easier in the future.

Virtually all neuroscience studies of potential interest to the law involve some behaviour that has already been identified as of interest and the point of the study is to identify that behaviour's neural correlates. Neuroscientists do not go on general 'fishing' expeditions. There is usually some bit of behaviour, such as addiction, schizophrenia, or impulsivity, that they would like to understand better by investigating its neural correlates. To do this properly presupposes that they have identified and validated the behaviour under neuroscientific investigation. On occasion, the neuroscience might suggest that the behaviour is not well-characterized or is neurally indistinguishable from other, seemingly different behaviour. In general, however, the existence of legally relevant behaviour will already be apparent. For example, some people are grossly out of touch with reality. If, as a result, they do not understand right from wrong, we excuse them because they lack such knowledge. We might learn a great deal about the neural correlates of such psychological abnormalities, but we already knew without neuroscientic data that these abnormalities existed and we had a firm view of their normative significance. In the future, however, we may learn more about the causal link between the brain and behaviour and studies may be devised that are more directly legally relevant.

Whatever neuroevidence is adduced must be translated into the folk psychological criteria the law employs. That is, the expert must be able to explain precisely how, for example, the neuroevidence bears on whether the agent acted, formed a required *mens rea*, or met the criteria for an excusing condition. If the evidence is not directly relevant, the expert should be able to explain the chain of inference from the indirect evidence to the law's criteria. In addition, over time there will be feedback between the folk psychological criteria and the neuroscientific data. Each might inform the other. Conceptual work on mental states might suggest new neuroscientific studies, for example, and the neuroscientific studies might help refine the folk psychological categories. The ultimate goal would be a reflective conceptual-empirical equilibrium.

At present, we lack the ability neurally to identify the content of a person's legally relevant mental states, such as whether the defendant acted intentionally or knowingly, but we are increasingly learning about the relation between brain structure and function and behavioural capacities, such as executive functioning, that are apparently relevant to broader judgements about responsibility and competence. We are unlikely to make substantial progress with neural assessment of mental content, but we are likely to learn more about capacities that will bear on excuse or mitigation.

Finally, and most importantly, because the responsibility and competence criteria are behavioural, actions speak louder than images. This is a truism for all criminal responsibility assessments. If the finding of any test or measurement of behaviour is contradicted by actual behavioural evidence, then we must believe the behavioural evidence because it is more direct and probative of the law's behavioural criteria. For example, if the person behaves rationally in a wide variety of circumstances, the agent is rational even if the brain appears structurally or functionally abnormal. And we confidently knew that some people were behaviourally abnormal, such as being psychotic, long before there were any psychological or neurological tests for such abnormalities. An analogy from physical medicine may be instructive. Suppose someone complains about back pain, a subjective symptom, and the question is whether the subject actually does have back pain. We know that many people with abnormal spines do not experience back pain, and many people who complain of back pain have normal spines. If the person is claiming a disability and the spine looks dreadful, evidence that the person regularly exercises on a trampoline without difficulty clearly indicates that there is no disability caused by back pain. If there is reason to suspect malingering, however, and there is not clear behavioural evidence of lack of pain, then a completely normal spine might be of use in deciding whether the claimant is malingering. Unless the correlation between the image and the legally relevant behaviour is very powerful, such evidence will be of limited help, however.

If actions speak louder than images, however, what room is there for using neuroevidence? Let us begin with cases in which the behavioural evidence is clear and permits an equally clear inference about the defendant's mental state. For example, lay people may not know the technical term to apply to people who are manifestly out of touch with reality, but they will readily recognize this unfortunate condition. No further tests of any sort will be necessary to prove this. In such cases, neuroevidence will be at most convergent and increase our confidence in what we already had confidently concluded. Whether it is worth collecting the neuroevidence will depend on how cost–benefit justified obtaining convergent evidence will be.

The most striking example of just such a case was the US Supreme Court's decision, *Roper v Simmons*,[9] which categorically excluded the death penalty for capital murderers who killed when they were sixteen or seventeen years old because such killers do not deserve the death penalty. The *amicus* briefs were replete with neuroscience data showing that the brains of late adolescents are not fully biologically mature, and advocates used such data to suggest that the adolescent killers could therefore not fairly be put to death. Now, we already knew from common-sense observation and rigorous behavioural studies that juveniles are on average less rational than adults. What did the neuroscientific evidence about the juvenile brain add? It was consistent with the undeniable behavioural data, and perhaps provided a partial causal explanation of the behavioural differences. The neuroscience data was therefore merely additive and only indirectly relevant.

Whether adolescents are sufficiently less rational on average than adults to exclude them categorically from the death penalty is of course a normative legal question and not a scientific or psychological question. Advocates claimed, however, that the neuroscience confirmed that adolescents are insufficiently responsible to be executed, thus confusing the positive and the normative. The neuroscience evidence in no way independently confirms that adolescents are less responsible. If the behavioural differences between adolescents and adults were slight, it would not matter if their brains are quite different. Similarly, if the behavioural differences were sufficient for moral and constitutional differential treatment, then it would not matter if the brains were essentially indistinguishable.

If the behavioural data are not clear, then the potential contribution of neuroscience is large. Unfortunately, it is in just such cases that the neuroscience at present is not likely to be of much help. I term this the 'clear cut' problem. Recall that neuroscientific studies usually start with clear cases of well-characterized behaviour. In such cases, the neural markers might be quite sensitive to the already clearly identified behaviours precisely because the behaviour is so clear. Less clear behaviour is simply not studied or the overlap between less clear behaviour and controls is greater. Thus the neural markers of clear cases will provide little guidance to resolve behaviourally ambiguous cases of legally relevant behaviour. For example, suppose in an insanity defence case the question is whether the defendant suffers from a major mental disorder such as schizophrenia. In extreme cases, the behaviour will be clear and no neurodata will be necessary. Now, small but statistically significant differences in neural structure or function between people who are clearly suffering from schizophrenia and those who are not have been discovered. Nonetheless, in a behaviourally unclear case, the overlap between data on the brains of people with schizophrenia and people without the disorder is so great that a scan is insufficiently sensitive to be used for diagnostic purposes.

With these general considerations in mind, let us now turn to the specific types of contributions neuroscience might make to the law as it now exists.

---

[9] *Roper v Simmons* 543 US 551 (2005).

28.4.2 Potential internal contributions of neuroscience

28.4.2.1 *Help demonstrate that the particular bit of folk wisdom or apparent truth about the world underlying a particular doctrine or legal practice is wrong and thus suggest that the doctrine or practice should change or be abandoned*

The proponent must show how the evidence confirms or challenges the folk psychological suppositions that underlie the doctrine in question. Showing that brain activation, even abnormal activation, is present or absent and played a causal role is virtually never per se legally relevant. The brain is always actively playing a role. The proponent must show specifically how the neurodata confirm or challenge the underlying folk psychological assumptions.

For example, excited utterances are an exception to the hearsay doctrine because the law assumes that they are likely to be true. Neuroscience might demonstrate that when people are excited in the way the law requires for this exception, the parts of the brain activated or deactivated are consistent with decreased accuracy of a subject's recollections. (One would have to induce experimentally excitement in the subject and keep him or her still enough to be validly scanned during the experimental task.) But the gold standard demonstration would be behavioural because if it turns out that excited utterances are in fact more likely to be accurate, the exception makes sense even though accompanying brain states seem inconsistent. In any case, jettisoning such a doctrine would not cast doubt on the folk psychological model. It presupposes it.

For another example, Terry Maroney has demonstrated that Supreme Court justices use incorrect emotional common sense as a basis for deciding certain cases.[10] This is clearly poor judicial reasoning and incorrect behavioural premises should be corrected by the science. Note that the science must begin with correctly identifying the statistically normative emotional response and neuroscience will do little more than confirm this.

28.4.2.2 *Suggest the need for new doctrine or practices*

One might claim that neuroscientific findings demonstrate that the law needs new doctrines and that injustice is created by their absence. For example, only a minority of states now have a control test for legal insanity. I believe that virtually all cases in which justice might demand a control test for criminal responsibility can be explained according to a theory of impaired rationality,[11] but one can certainly imagine neuroscience data that might be consistent with a control problem entirely independently of any rationality defect. How it would be consistent would have to be demonstrated. The psychological intermediary variables would have to be

---

[10] See Terry A. Maroney, 'Adolescent Brain Science and Juvenile Justice', above, 255.
[11] Stephen J. Morse, 'Uncontrollable Urges and Irrational People' (2002) 88 *Va. L. Rev.* 1025.

identified. If all this were done, it might be strong evidence to adopt a control test if one believes that people with such problems deserve some type of excuse or that they cannot be deterred. Note that the capacity to control oneself is a folk psychological concept. Machines may have internal regulatory mechanisms built into their design, but only people control themselves by suppressing their desires and intentions to act in prohibited ways. Folk psychology is again presupposed.

Permit me a bit of special pleading for my next example. Anglo-American criminal law does not have a generic mitigating doctrine that is applicable at trial. Unless a defendant meets the criteria for a complete excuse such as legal insanity or duress, the law expects the defendant to control himself despite temptations and provocations. Mitigation is then left to the discretion of the sentencing judge. I have argued, in contrast, that criminal law should provide defendants with a partial mitigating doctrine applicable at trial that would in appropriate cases reflect a defendant's substantially diminished rational or control capacities.[12] Neuroscience studies might help confirm that some defendants have much greater difficulty obeying the law in cases of provocation and temptation than criminal law and common sense presuppose. If this were true, in the interest of fairness, of accurately responding to a defendant's culpability, perhaps the criminal law would be justified in adopting a generic partial mitigating doctrine.

28.4.2.3 Help adjudicate an individual case by providing evidence of the folk psychological criteria

Again, the proponent of using the evidence would have to show that the neurodata confirm or challenge the presence of folk psychological criterial mental states and actions.

For example, suppose there was clear evidence that a person lacked the neural capacity that supports the folk psychological process of controlling one's intention to act in a prohibited way. This would be relevant to criminal responsibility for the insanity defence or sentencing purposes in some jurisdictions. How much lack of capacity would be required to excuse or mitigate could not be answered by neuroscience. It is a normative issue that folk psychological lawmakers and law appliers would have to address. Now, most cases will not be so clear and actions speak louder than images. Again, however, the use of neuroscience evidence to help adjudicate individual cases according to folk psychological legal criteria presupposes folk psychology.

28.4.2.4 Help to implement a current policy more effectively

The question in such cases is whether using neurodata would be cost–benefit justified. For example, criminal sentencing and many civil and quasi-criminal

---

[12] Stephen J. Morse, 'Diminished Rationality, Diminished Responsibility' (2003) 1 *Ohio St. J. Crim. L.* 289.

commitment laws contain a prediction of future behaviour criterion. This would provide the easiest and most straightforward case for relevance because, in such cases, mental state analysis is not crucial and neural markers might increase predictive accuracy. Notice that such doctrines address not the person being predicted, but the people who must decide how to respond to a particular prediction, a folk psychological decider performing the folk psychological process of deciding. Neuroscience might also enhance the accuracy of predicting who will benefit from diversion and from treatment programmes for those sentenced or civilly committed. Again, even if neurodata are useful in principle, is there sufficient value-added to justify the cost of collecting it?

28.4.3 External challenges from neuroscience

Recall that any external challenge to doctrines, institutions or practices suggests that they are incoherent and unjustifiable *ab initio*. This part addresses the two most radical external challenges from neuroscience: the challenge from neurodeterminism and the threat to the law's concept of the person, which grounds its concept of responsibility. The first challenge is familiar and has a good compatibilist response. The second is more radical and threatens the nature of law itself. The second fails on conceptual and empirical grounds, however. The coherence of responsibility and competence doctrines and practices are safe. At least for now.

28.4.3.1 The challenge from neuroscientific determinism

Many think that the proof of neuroscientific determinism undermines the legitimacy of all blaming practices, which are crucial to the criminal law. As the earlier section on distractions indicated, this is a mistake, but it is an external claim because it challenges the legitimacy of present criminal law as a whole and not just specific aspects of it. There is a profound metaphysical question about the truth of determinism, but this question is not itself resolvable by neuroscience. Further, as we have seen, the truth of determinism is fully consistent with the law's view of the person and all legal doctrines. The truth of determinism is not inconsistent with the view that mental states matter to the causation of behaviour and that human beings are capable of being guided by reason, including the law's commands. Let us therefore turn to a genuine challenge.

28.4.3.2 The challenge to personhood

The claim advanced by many that mental states are epiphenomenal and do no explanatory work presents a challenge to the coherence of all law, and not just to responsibility practices. If the concept of mental causation that underlies folk psychology and current conceptions of responsibility is false, our responsibility practices, and many others, would appear unjustifiable.

Such extreme claims are not strawpersons. Here is a lengthy quote from a widely quoted article by neuroscientists Joshua Greene and Jonathan Cohen that expresses the mechanistic conception:[13]

[A]s more and more scientific facts come in, providing increasingly vivid illustrations of what the human mind is really like, more and more people will develop moral intuitions that are at odds with our current social practices... Neuroscience has a special role to play in this process for the following reason. As long as the mind remains a black box, there will always be a donkey on which to pin dualist and libertarian positions... What neuroscience does, and will continue to do at an accelerated pace, is elucidate the 'when,' 'where' and 'how' of the mechanical processes that cause behaviour. It is one thing to deny that human behaviour is purely mechanical when your opponent offers only a general philosophical argument. It is quite another to hold your ground when your opponent can make detailed predictions about how these mechanical processes work, complete with images of the brain structures involved and equations that describe their function... At some further point... [p]eople may grow up completely used to the idea that every decision is a thoroughly mechanical process, the outcome of which is completely determined by the results of prior mechanical processes. What will such people think as they sit in their jury boxes? Will jurors of the future wonder whether the defendant... *could have done otherwise*? Whether he really *deserves* to be punished...? We submit that these questions, which seem so important today will lose their grip in an age when the mechanical nature of human decision-making is fully appreciated. The law will continue to punish misdeeds, as it must for practical reasons, but the idea of distinguishing the truly, deeply guilty from those who are merely victims of neuronal circumstances will, we submit, seem pointless.

Alternatively, to use another of Greene and Cohen's arguments, suppose that 'neuroscience holds the promise of turning the black box of the mind into a *transparent bottleneck*'.[14] They mean that the brain is the final mechanistic pathway through which all types of explanations of behaviour must ultimately operate and that neuroscience will be able to demonstrate that brain mechanisms, not mental states, are doing all the work.[15] They speculate that we may some day possess 'extremely high-resolution scanners that can simultaneously track the neural activity and connectivity of every neuron in the human brain...' and that, with the help of computers and software, can help people see the neural events that are alone causally responsible for their behaviour.[16] If such mechanistic understanding and knowledge were available and widespread, Greene and Cohen are probably correct that notions of responsibility would wither away because most would believe that

---

[13] Joshua Greene and Jonathan Cohen, 'For the Law, Neuroscience Changes Nothing and Everything' in S. Zeki and O. Goodenough (eds.), *Law & the Brain* (New York: Oxford University Press, 2006) 207, 217–8.
[14] Ibid. at 218.
[15] Ibid.
[16] I will assume that the scanning and computing abilities that the argument employs are possible, although the brain has 10^{11} cells and at least 10^{15} connections. Is it really likely, however, that the computer would predict what precise sentences we would speak? At present, of course, the speculation is pure science fiction and, in my opinion, is likely to remain so. The real problem with the argument is not that it assumes a (barely) plausible computational ability, but that it assumes that mental states can be reduced simply to brain states, an assumption that this subsection addresses.

it was the brain that 'did it', not the agent, and we do not hold brains morally responsible.

This picture of human activity exerts a strong pull on the popular, educated imagination as well as on the theorizing of scientists. Consider the following example. In an ingenious recent study,[17] investigators were able to predict accurately based on which part of the brain was physiologically active whether a shopper-subject would or would not make a purchase. Activity in these regions predicted immediately subsequent purchases 'above and beyond self-report variables'.[18] As we shall see in this subsection, this does not mean that the person's weighing of preferences and prices and the final decision played no role. Activity in the nucleus accumbens, the insula, and the mesial prefrontal cortex is not 'weighing' and 'deciding'. The latter are the activities of people, not brains.[19] The findings interestingly, although unsurprisingly, suggest, however, that specific brain regions play a crucial role in particular types of psychological processes.

This study was reported in the 'Science Times' section of the *New York Times* by John Tierney.[20] Here is how the story was 'spun', beginning with its title: 'Findings: The Voices in My Head Say "Buy It!" Why Argue?'. The shopper is simply the hapless puppet of brain processes and plays no role as an agent in the purchase process. The decision is not up to the shopper; it is up to their brain. The conclusion considers how the study might help us deal with feckless consumerism:

> You might remove the pleasure of shopping by somehow dulling the brain's dopamine receptors so that not even the new Apple iPhone would get a rise in the nucleus accumbens, but try getting anyone to stay on that medication. Better the occasional jolt of pain. Charge it to the insula.[21]

In addition to getting the study wrong—insula activation was associated with excessive prices and the decision not to purchase[22]—it betrays once again the mechanistic view of human activity. What people do is simply a product of brain regions and neurotransmitters. The person disappears. There is no shopper. There is only a brain in a mall.

What if all these thinkers who claim that we are just victims of neuronal circumstances (VNCs) are correct? Suppose neuroscience convinces us that agency and folk psychology are an illusion, that intentional bodily movements and reflexes are morally indistinguishable because both are simply the outcomes of mechanistic

---

[17] Brian Knutson et al., 'Neural Predictors of Purchases' (2007) 53 *Neuron* 147.
[18] Ibid. at 147.
[19] M. R. Bennett and P. M. S. Hacker, 'Philosophical Foundations of Neuroscience: An Excerpt from Chapter 3', in Maxwell Bennett et al. (eds.), *Neuroscience & Philosophy: Brain, Mind & Language* (New York: Columbia University Press, 2007) 15, 18–23 (describing ascription of psychological attributes to the brain as 'senseless'). But see Daniel Dennett, 'Philosophy as Naive Anthropology: Comment on Bennett and Hacker' in ibid. at 73, 86–8 (claiming that it makes sense to attribute 'attenuated' sorts of psychological attributes to parts of the brain).
[20] John Tierney, 'Findings: The Voices in My Head Say "Buy It!" Why Argue?' (2007) *New York Times*, 16 January, F1.
[21] Ibid. at F6. [22] Ibid.

biophysical processes? What if all the contending conceptions about responsibility depend on a mistake about human activity? What if, for example, mental states do not explain actions but are simply *post hoc* rationalizations the brain creates to 'make sense of' the bodily motions or non-motions that brains produce? We are just mechanisms, although the illusion of conscious will may play a positive role in our lives.[23] Some people, including many psychologists and neuroscientists, think that new discoveries about the causation of behaviour are leading inexorably to a purely mechanistic view of the link between the brain and behaviour, and thus to a purely mechanistic view of human behaviour. Will the agentic person disappear and be replaced by the biological victim of neuronal circumstances?

If Greene and Cohen are right, we are all allegedly 'merely victims of neuronal circumstances.' But are we? And will criminal justice system as we know it, which includes robust notions of personhood and desert, wither away as an outmoded relic of a prescientific and cruel age? And not only criminal law is in peril. What will be the fate of contracts, for example, when a biological machine that was formerly called a person claims that it should not be bound because it did not make a contract. The contract was simply the outcome of various 'neuronal circumstances'. Although I predict that we will see far more numerous attempts to introduce neuroevidence in the future, the dystopia that Greene and Cohen predict is not likely to come to pass.

It is important to note from the outset, however, that compatibilism or other responses to the determinist challenges to responsibility will not save the disappearing person. Determinism is consistent with either of two inconsistent views of human behaviour. The truth of determinism is consistent with the existence or non-existence of agency, with the causal role or non-causal role of mental states in explaining behaviour. Compatibilism presupposes that a folk psychological account of action is accurate and that distinctions based on it, such as the difference between actions and non-actions, should make a moral and legal difference. The new, VNC claims deny precisely this. The person and responsibility can only be saved if VNC is false or, if it is true, we learn to live with the illusion that it is false. Otherwise, all agency-based conceptions of responsibility must be abandoned.

At present, no such radical, external challenge from neuroscience even remotely approaches plausibility. It is true that the law's fundamental presuppositions about personhood and action are open to profound objection. Most fundamentally, action and consciousness are scientific and conceptual mysteries.[24] We do not

---

[23] This claim should not be confused with the apparently similar claim that 'personhood' is an illusion. See Martha J. Farah and Andrea S. Heberlein, 'Personhood and Neuroscience: Naturalizing or Nihilating?' (2007) 7 *Am. J. Bioethics* 37, 40 (claiming that our construct of 'personhood' is simply the illusory product of innate and automatic brain systems that is 'projected' onto the world). There are many problems with the logic of this claim, but even if it is correct, it does not deny that creatures like us have mental states, such as desires and beliefs, that can be causally explanatory. Most charitably interpreted, it simply denies the explanatory usefulness of the normative concept of a 'person'.

[24] See Robert Audi, *Action, Intention and Reason* (Ithaca: Cornell University Press, 1993) 1–4 (describing the 'basic philosophical divisions' in each of the four major problem areas in action theory); Colin McGinn, *The Mysterious Flame: Conscious Minds in a Material World* (New York: Basic Books, 1999) (describing the immense difficulty of explaining consciousness and doubting the ability of human beings to do so).

know how the brain enables the mind[25] and we do not know how action is possible. At most we have hypotheses or *a priori* arguments. Moreover, causation by mental states seems to depend on now largely discredited mind–brain dualism that treats minds and brains as separate entities that are somehow in communication with one another.[26] How can such tenuously understood concepts be justifiable premises for legal practices such as blaming and punishing? If our picture of ourselves is wrong, as many neuroscientists claim, then our responsibility practices are morally unjustified according to any moral theory we currently embrace. On the other hand, given how little we know about the brain–mind and brain–action connection, to claim based on neuroscience that we should radically change our picture of ourselves and our practices is a form of neuroarrogance.

To see in more specific detail why we need not abandon our robust conception of agency despite such claims, let us turn to the indirect and allegedly direct evidence for them. The real question is whether scientific and clinical investigations have shown that agency is rare or non-existent; that conscious will is largely or entirely an illusion. Four kinds of indirect evidence are often adduced: first, demonstrations that a very large part of our activity is undeniably caused by variables we are not in the slightest aware of; second, studies indicating that more activity than we think takes place when our consciousness is divided or diminished; third, laboratory studies that show that people can be experimentally misled about their causal contribution to their apparent behaviour; and, fourth, evidence that particular types of psychological processes seem to have associated neurophysiological activity in specific regions of the brain. None of these types of evidence offers logical support to VNC, however.

Just because a person may not be aware of all the causes for why he formed an intention does not mean that he did not form an intention, that he was not a fully conscious agent when he did so, and that his intention played no causal role in explaining the person's behaviour. Even if human beings were never aware of the causes of their intentions to act and of their actions, it would not necessarily follow that they were not acting consciously, intentionally and for reasons that make eminent sense to anyone under the circumstances.

Human consciousness can undeniably be divided or diminished by a wide variety of normal and abnormal causes.[27] We have known this long before contemporary scientific discoveries of what causes such states and how they correlate with brain structure and processes. Law and morality agree that if an agent's

---

[25] Paul R. McHugh and Philip R. Slavney, *The Perspectives of Psychiatry*, 2nd edn (Baltimore: Johns Hopkins University Press, 1998) 11–12.

[26] It is almost impossible not to talk 'dualistically' in ordinary speech and writing. Every time a monist neuroscientist uses a personal pronoun in speaking or writing, for example, he seems to imply that there is a genuine 'him' or 'her' that is somehow distinguishable from his brain activity. This does not mean, however, that the neuroscientist (or anyone else) is really a crypto-dualist. It is simply an inevitable feature of current language, and perhaps it always will be.

[27] See Jeffrey L. Cummings and Michael S. Mega, *Neuropsychiatry and Behavioral Neuroscience* (New York, Oxford University Press,, 2003) 333–43 (description of dissociative and related states and their causes and treatments); D. Vaitl, N. Birbaumer et al., 'Psychobiology of Altered States of Consciousness' (2005) 131 *Psychol. Bull.* 98.

capacity for consciousness is non-culpably diminished, responsibility is likewise diminished. Some suggest that it is diminished because bodily movements in the absence of fully integrated consciousness are not 'actions'.[28] Others believe that apparently goal-directed behaviour that is responsive to the environment, such as sleepwalking, is action, but that it should be excused because diminished consciousness reduces the capacity for rationality.[29] Let us assume that the former view is correct, because it offers more direct support to VNC and therefore the greatest challenge to traditional notions of individual responsibility. Let us also assume that divided or diminished consciousness is more common than it appears to be. Nevertheless, neither of these assumptions supports the more radical, general VNC thesis and the arguments for automatistic imperialism have been termed the 'automaticity juggernaut'.[30]

Demonstrating that divided or partial consciousness is more common than it appears certainly extends the range of cases in which people are not responsible or have diminished responsibility. Such studies do not demonstrate, however, that most human bodily movements that appear intentional and rational (apparently rational actions) occur when the person has altered consciousness. One cannot generalize to all human behaviour from genuinely deviant cases or from cases in which a known abnormality is present. A model of action (or, we should say, non-action) built on sleepwalking, for example, is hardly a threat to orthodox notions of individual responsibility.

There is substantial empirical evidence to suggest that laboratory manipulations of unsuspecting subjects can cause the subjects to believe that their intentions were producing action when this was not the case.[31] That subjects can be cleverly misled by experimental manipulations hardly indicates that intentions generally play no role in explaining our behaviour. Self-deception under laboratory conditions of deceit does not entail that intentions generally do not causally explain action. Universal deception about personal causal efficacy hardly seems to be a plausible evolutionary outcome.

Finally, there is accumulating evidence that various psychological processes have associated neural activity in localized regions of the brain. We have long known that many behavioural activities are biologically based in highly specific regions. For example, there is substantial evidence that ability to recognize faces is localized in a region of the temporal lobe of the right hemisphere referred to as the 'fusiform face area'. Should this area become lesioned, the subject loses the ability to recognize

---

[28] See, e.g., Michael S. Moore, *Act and Crime* (New York, Oxford University Press, 1993) 49–52, 135–55, 257–8 (arguing that cases of compromised consciousness should be treated as non-action); see also Michael S. Moore, 'More on Act and Crime' (1994) 142 *U. Pa. L. Rev.* 1749, 1804–20.

[29] Stephen J. Morse, 'Culpability and Control' (1994) 142 *U. Pa. L. Rev.* 1587, 1641–52 (arguing that clouded consciousness should be treated as an affirmative defence); see also Bernard Williams, 'The Actus Reus of Dr. Caligari' (1994) 142 *U. Pa. L. Rev.* 1661 (arguing that human activity with clouded consciousness is action).

[30] John F. Kihlstrom, 'The Automaticity Juggernaut—or, Are We Automatons After All?' in John Baer, James C. Kaufman, and Roy F. Baumeister (eds.), *Are We Free? Psychology and Free Will* (New York: Oxford University Press, 2008) 155–73.

[31] See John A. Bargh, 'Bypassing the Will: Toward Demystifying the Nonconscious Control of Social Behavior' in Ran R. Hassin, James S. Uleman, and John A. Bargh (eds.), *The New Unconscious* (New York: Oxford University Press, 2005) at 37, 51–4 (reviewing the evidence and concluding that the 'will' is not primarily responsible for action).

faces, a condition called prosopagnosia.[32] Now, however, functional neuroimaging techniques permit the exploration of brain activity during more complicated psychological processes and can identify associated neurophysiological activation for the processes. I have already discussed the example of brain regions associated with decisions to purchase an object.[33] For another example, a recent study demonstrated that investigators could determine from the region of brain activity which mental process—adding or subtracting—a subject had covertly intended to, but had not yet, performed.[34]

The localization evidence is immensely interesting and suggestive, but it does not indicate that mental states play no role in causally explaining behaviour. There must be a biological substrate in the brain for all human behaviour. If your brain is dead, you are dead and not behaving at all. Nor is it surprising that particular regions of the brain are associated with particular psychological processes. For example, a leading, albeit controversial, theory of how the mind works suggests that it is composed of different systems that perform different functions.[35] Although we do not know how the brain enables the mind, it makes sense to assume that specific psychological processes would have brain substrates specific to each individual process. Based on what we already know about localization and on the reasonable assumption that it would be inefficient if all regions of the brain needed equal activation to support all psychological processes, localization is most likely to be true. Even if all this is correct, however, it does not follow that mental states do no causal explanatory work. It demonstrates at most that the neural network substrates for specific mental functions may be located in specific regions of the brain.

What is needed to support VNC is a general, direct demonstration that causal intentionality is an illusion *tout court*, but no such general demonstration has yet been produced by scientific study. The most interesting evidence has arisen from studies done by neuroscientist, Benjamin Libet,[36] which have generated an

---

[32] Elinore McKone, Kate Crooke, and Nancy Kanwisher, 'The Cognitive and Neural Development of Face Recognition in Humans' in Michael S. Gazzaniga (ed.), *The Cognitive Neurosciences,* 4th edn (Cambridge, MA: MIT Press, 2009) 467, 468–9; Daniel Trench, 'Functional Neuroanatomy: Neuropsychosis Correlates of Cortical and Subcortical Damage' in Stuart C. Yudovsky and Robert Hales (eds.), *Neuropsychiatry and Clinical Neuroscience*, 4th edn (Washington, DC: American Psychiatric Publishing, 2002) 71, 80–2.

[33] See notes 17–19, above, and accompanying text.

[34] John-Dylan Haynes et al., 'Reading Hidden Intentions in the Human Brain' (2007) 17 *Current Biology* 323, 323–8. It is important to recognize that the brain activity accurately predicted only which *type* of process the subject had covertly formed the intention to perform. It did not identify the specific content of the intention, such as which two numbers the subject intended to add or subtract. Despite the enormous advances in cognitive neuroscience, we do not know how to read minds using neuroimaging or any other technique. Cf. Martha J. Farah, 'Bioethical Issues in the Cognitive Neurosciences' in *The Cognitive Neurosciences* note 32, above, at III, 1309, 1309–10 (referring to the ability to identify traits and states as 'a crude form of mindreading').

[35] See, e.g., Jerry A. Fodor, *The Modularity of Mind* (Cambridge, MA: MIT Press, 1983) (providing a modular theory).

[36] Benjamin Libet, 'Do We Have Free Will' in Benjamin Libet et al. (eds.), *The Volitional Brain: Towards a Neuroscience of Free Will* (Exeter: Imprint Academic, 1999) 47 (summarizing the findings and speculating about their implications). For a more recent, powerful demonstration of a similar finding, see C. S. Soon et al., 'Unconscious Determinants of Free Decisions in the Human Brain' (2008) 11 *Nature Neuroscience* 543.

immense amount of comment.[37] Indeed, many claim that Libet's work and later, similar studies are the first direct neurophysiological evidence of VNC.[38] Libet's exceptionally creative and careful studies demonstrate that measurable electrical brain activity associated with intentional actions occurs in the relevant motor area of the brain about 550 milliseconds before the subject actually acts and for about 350–400 milliseconds before the subject is consciously aware of the intention to act.

Let us assume, with cautious reservations,[39] the basic scientific methodological validity of these studies.[40] The crucial question then becomes whether the interpretation of these findings as supporting VNC is valid. Michael Moore has usefully shown that the Libetian conception of the role of brain events in causing behaviour is confused.[41] Indeed, it is not clear precisely what the claim is, but the most profound challenge would be that mental states are epiphenomenal. If this is true, the folk psychological basis for all law is incoherent. Alfred Mele has shown that Libet's work does not establish VNC, has exposed numerous confusions, and has usefully described the type of experiment that might achieve this result.[42] Rather than repeat their analyses, which bear close reading, this section will instead offer a more empirical and common-sense critique.

It does not follow from Libet's discovery of the temporal ordering that conscious intentionality does no causal work. It simply demonstrates that non-conscious brain events precede conscious experience. Once again, we have no idea how the brain enables the mind, but this seems precisely what one would expect of the mind-brain. Electrical impulses move quickly among neurons, but some lag between brain activity and conscious experience seems unsurprising. Once again, if the brain is dead, the person is dead. Prior electrical activity does not mean that intentionality played no causal role. Electrical activity in the brain is precisely that: electrical activity in the brain and not a mental state such as a decision or an intention. A readiness potential is not a decision.[43] A perfectly plausible reading

---

[37] Daniel Wegner, *The Illusion of Conscious Will* (Cambridge, MA: The MIT Press, 2002) 54–5 (characterizing the recounting of Libet's results as a 'cottage industry' and noting the large and contentious body of commentary).

[38] William P. Banks and Susan Pocket, 'Benjamin Libet's Work on the Neuroscience of Free Will' in Max Velmans and Susan Schneider (eds.), *The Blackwell Companion to Consciousness* (Malden, MA: Blackwell, Publishing, 2007) 657, at 658.

[39] See, e.g., Henrik Walter, *Neurophilosophy of Free Will: From Libertarian Illusions to a Concept of Natural Autonomy* (Cynthia Klor trans.) (Cambridge, MA: MIT Press, 2001) 250–2; Jing Zhu, 'Reclaiming Volition: An Alternative Interpretation of Libet's Experiment' (Nov. 2003) 61 *J. Consciousness Stud.* 1–77.

[40] Banks and Pocket, note 38, above, at 659–62 (concluding after a careful review of possible artefacts that 'readiness potentials do start before the subject consciously 'decides' to move').

[41] Michael S. Moore, 'Libet's Challenges to Responsible Human Agency' in Walter Sinnott-Armstrong and Lynn Nadel (eds.), *Conscious Will And Responsibility: A Tribute to Benjamin Libet* (New York: Oxford University Press, 2010), ch. 18.

[42] Alfred R. Mele, *Effective Intentions: The Power of the Conscious Will* (New York: Oxford University Press, 2009). See also, M. R. Bennett and P. M. S. Hacker, *Philosophical Foundations of Neuroscience* (Malden, MA: Blackwell Publishing, 2003) 228–31 (criticizing Libet's account of action).

[43] Moreover, Libet does not carefully distinguish between urges or wants on the one hand and decisions and intentions on the other. Indeed, Alfred Mele argues that the experimental evidence is

of Libet's work is that various non-conscious causal variables, including non-conscious urges, precede action—who would have thought otherwise?—but intentionality is nonetheless necessary for action.

Libet also suggests that people can 'veto' the act during the delay between becoming aware of the intention and performing the intended action, which he surprisingly conceives of as an undetermined act. Other researchers appear to have localized the part of the brain that is associated with the activity of vetoing.[44] But, in addition to the implausibility of the veto being undetermined,[45] the conceptual foundations of the interpretation that the subjects were exercising a genuine veto are shaky at best.[46] This suggestion undermines the claim that the brain is doing all the work because it is an agent's mental state, a newly formed intention to veto, that causes the agent not to perform the act. In short, Libet's work presupposes agency at every step in the process.

Libet's task involved 'random' finger movements that involved no deliberation whatsoever and no rational motivation for the specific movements involved.[47] This is a far cry from the behavioural concerns of the criminal law or morality, which address intentional conduct when there is always good reason to refrain from harming another or to act beneficently. In fact, it is at present an open question whether Libet's paradigm is representative of intentional actions in general because Libet used such trivial behaviour.[48]

In addition to direct problems with the alleged implications of Libet's work, there are also good reasons to reject it. Answers to the possibility of VNC are rooted in common sense, a plausible theory of mind, our evolutionary history, and practical necessity. Virtually every neurologically intact person consistently has the experience of first person agency, the experience that one's intentions flow from one's desires and beliefs and result in action. Indeed, this folk-psychological experience is so central to human life and so apparently explanatory that it is difficult to imagine giving it up or a good reason to do so, even if it were possible to give it up. As the eminent philosopher of mind, Jerry Fodor, has written:

[I]f commonsense intentional psychology were really to collapse, that would be, beyond comparison, the greatest intellectual catastrophe in the history of our species; if we're that wrong about the mind, then that's the wrongest we've ever been about anything. The collapse of the supernatural, for example, didn't compare... Nothing except, perhaps, our

---

much more consistent with the RP being associated with an urge rather than an intention or a decision: Alfred R. Mele, *Free Will and Luck* (New York: Oxford University Press, 2006) 33,40. I am not convinced that this problem is major, but associating the RP with 'desire' rather than intention perhaps weakens the case that Libet's work establishes NAT.

[44] See, Maurice Brass and Patrick Haggard, 'To Do or Not to Do: The Neural Signature of Self-control' (2007) 27 J. *Neurosci.* 9141 (identifying the part of the brain that is activated when the 'veto' is exercised).

[45] Banks and Pockett, note 38 above, at 667.

[46] Mele, note 42 above, at 34–5.

[47] Participating in the study and cooperating with the investigator can be rationally motivated, of course. But the experimental task was to move one's finger randomly, for no good reason.

[48] Banks and Pockett, note 38 above, at 662–3.

commonsense physics... comes as near our cognitive core as intentional explanation does. We'll be in deep, deep trouble if we have to give it up...
... But be of good cheer; everything is going to be all right.[49]

Folk psychology has much explanatory power and is capable of scientific investigation.[50] There is compelling psychological evidence that intentions play a causal role in explaining behaviour.[51] Finally, despite Mele's attempt, it is hard to imagine the nature of a scientific study that would prove conclusively that mental states do no work to creatures who have created and will assess that study with mental states.

The plausible theory of mind that might support mental state explanations is thoroughly material, but non-reductive and non-dualist. It hypothesizes that all mental and behavioural activity is the causal product of lawful physical events in the brain, that mental states are real, that they are caused by lower level biological processes in the brain, that they are realized in the brain—the mind-brain—but not at the level of neurons, and that mental states can be causally efficacious.[52] It accepts that a fully causal story about behaviour will be multifield and multilevel.[53]

There is a perfectly plausible evolutionary story about why folk psychology is causally explanatory and why human beings need rules such as those provided by law. We have evolved to be self-conscious creatures that act for reasons and are reasons responsive. Acting for reasons is inescapable for creatures like ourselves who inevitably care about the ends they pursue and about what reason they have to act in one way rather than another.[54] Because we are social creatures whose interactions are not governed primarily by innate repertoires, it is inevitable that rules will be necessary to help order our interactions in any minimally complex social group.[55] As a profoundly social species, it seems apparent that our ancestors would have been much less successful, and therefore much less likely to be our ancestors, if they were unable to understand the intentions of others, not sure they could convert their intentions into action, and were not also equipped with powerful assumptions that

---

[49] Jerry A. Fodor, *Psychosemantics: The Problem of Meaning in the Philosophy of Mind* (Cambridge, MA: MIT Press, 1987) xii.

[50] See, e.g., Bertram F. Malle, *How The Mind Explains Behavior: Folk Explanations, Meaning and Social Interaction* (Cambridge, MA: MIT Press, 2004) (providing a full theoretical account and empirical support). There is also growing recognition within psychology that 'mental-state inference is one of the most fundamental tools of social cognition'. Bertram F. Malle, 'Folk Theory of Mind: Conceptual Foundations of Human Social Cognition' in Ran R. Hassin, James S. Uleman, and John A. Bargh (eds.), *The New Unconscious* (New York: Oxford University Press, 2005) 225, 229.

[51] See, Richard Holton, *Willing, Wanting, Waiting* (New York: Oxford University Press, 2009) 5–9 (reviewing psychologist Peter Gollwitzer's work and explaining how it supports the role of a distinct psychological kind, intention, as playing a causal role in behaviour); Mele, note 42 above, at 134–6.

[52] See, e.g., John R. Searle, *Mind: A Brief Introduction* (New York: Oxford University Press, 2004) at 113–4 (terming his position 'biological naturalism' about consciousness).

[53] See, Carl Craver, *Explaining the Brain: Mechanisms and the Mosaic of Neuroscience* (New York: Oxford University Press, 2007)(providing an account of how neuroscience fits into a multifield, multilevel framework for explaining behaviour).

[54] Hilary Bok, *Freedom and Responsibility* (Princeton: Princeton University Press, 1998) at 75–91, 129–31, 146–51.

[55] Larry Alexander et al., *The Rule of Rules: Morality, Rules and the Dilemmas of Law* (Durham, NC: Duke University Press, 2001) 11–25 (explaining why rules are necessary in a complex society and contrasting their account with H. L. A. Hart's theory).

that stranger coming over the hill is equipped with the same capacity for harmful intentions as they are.[56]

One of the qualities that makes us most human is the ability to infer that others have independent mental states and then to use that information to understand and predict the behaviour of others. Psychologists call this having a 'theory of mind'.[57] Human beings who do not develop an adequate 'theory of mind', such as those with autism, experience profound difficulties in their interpersonal lives. The lengthy conservation of mental states and their ubiquitousness and centrality suggest that they do play an important causal role and that they are very evolutionarily expensive if they do not. This is of course not an incontrovertible analytic argument against VNC, but surely the burden of persuasion is on those who argue to the contrary. At the very least, we remain entitled to presume that conscious intentions are causal until the burden is met.

Libet's work is fascinating, but it does not prove that humans are generally not conscious, intentional agents or capable of employing their conscious intentionality when they have good reason to do so.[58] Even if the work is methodologically valid, various conceptual and interpretive arguments undermine the claim that Libet has demonstrated that VNC is true.

In short, despite the often astonishing findings and impressive advances in neuroscience and allied disciplines, there is no compelling evidence yet that VNC is generally true. Do any of the foregoing arguments entail that mental states are not epiphenomenal? Of course not. But we are nowhere close to demonstrating epiphenomenalism experimentally (indeed, it would be hard to imagine what that study would look like), the best theories about reduction suggest that it is probably false, and evolution and common sense suggest that it is false. Future discoveries may undermine this conclusion, however, so I now turn to the implications of VNC.

VNC alas can provide no guidance about what people should do next and, in any event, degenerates into self-referential incoherence. Suppose that you were convinced by the mechanistic view that you were not an intentional, rational agent after all. (Of course, the notion of being 'convinced' would be an illusion, too.[59] Being convinced means that you were persuaded by evidence or argument, but a

---

[56] See Justin N. Wood et al., 'The Perception of Rational, Goal-Directed Action in Nonhuman Primates' (2007) 317 *Science* 1402, 1405 (demonstrating that the ability to understand the intentions of other creatures evolved in primates 40 million years ago); see also Esther Herrmann et al., 'Humans Have Developed Specialized Skills of Social Cognition: The Cultural Intelligence Hypothesis' (2007) 317 *Science* 1360 (comparing chimpanzees and orang-utans to two-and-a-half-year-old humans and discovering that they have approximately equal cognitive skills concerning the physical world, but that humans have superior cognitive skills for understanding social interaction).

[57] Geraldine Dawon and Karen Toth, 'Autism Spectrum Disorders' in Dante Cicchetti and Donald J. Cohen (eds.), *Developmental Psychopathology*, 2nd edn, *Risk, Disorder and Adaptation* (Hoboken, NJ: John Wiley & Sons, 2006) vol. 3, 327.

[58] See Jerry Fodor, 'Making the Connection' (2002) *Times Literary Supplement*, 17 May, at 4 (arguing that the new neuroscience rarely has much to contribute when the phenomenon in question is complex social behaviour).

[59] See Daniel C. Dennett, 'Calling in the Cartesian Loans' (2004) 27 *Behav. Brain Scis.* 661, 661 (wondering, in response to Professor Wegner, who is this 'we' that inhabits the brain).

mechanism is not persuaded by anything. It is simply neurophysically transformed.) What should you do now? You know that it is an illusion to think that your deliberations and intentions have any causal efficacy in the world. (Again, what does it mean according to the purely mechanistic view to 'know' something? But enough.) You also know, however, that you experience sensations such as pleasure and pain and that you care about what happens to you and to the world. You cannot just sit quietly and wait for your neurotransmitters to fire. You cannot wait for determinism to happen. You must, and will of course, deliberate and act.

If one still thought that VNC was correct and that standard notions of genuine moral responsibility and desert are therefore impossible, one might nevertheless continue to believe that the law would not necessarily have to give up the concept of incentives. Indeed, Greene and Cohen concede that we would have to keep punishing people for practical purposes.[60] Through poorly understood automatic processes, it is possible that various potential rewards and punishments would shape behaviour even if they did not do so as premises in practical reasoning. Such an account would be consistent with 'black box' accounts of economic incentives. For those who believe that a thoroughly naturalized account of human behaviour entails complete consequentialism, such a conclusion might not be unwelcome.

On the other hand, this view seems to entail the same internal contradiction just explored. What is the nature of the 'agent' that is discovering the laws governing how incentives shape behaviour? Could understanding and providing incentives via social norms and legal rules simply be epiphenomenal interpretations of what the brain has already done? How do 'we' 'decide' which behaviours to reward or punish? What role does 'reason'—a property of thought and agents, not a property of brains—play in this 'decision'? Once again, the VNC account seems to swallow itself. Moreover, VNC proponents of consequentialism could hardly complain about those who refuse to 'accept' what the proponents think rationality requires. The allegedly misguided people who resist are simply the victims of their automatic brain states. They cannot be expected intentionally to use their capacity for reason to accept what the consequentialists believe reason demands. Indeed, the consequentialist's belief is also an illusory mental state or it exists but plays no role in explaining behaviour.

Even if our mental states play no genuinely causal role (about which we will never be certain until we solve the mind–body problem) human beings will find it almost impossible not to treat themselves as rational, intentional agents unless there are major changes in the way our brains work. Moreover, if one uses the truth of pure mechanism as a premise in deciding what to do, this premise yields no particular moral, legal, or political conclusions. It will provide no guide to how one should live or how one should respond to the truth of VNC. If reasons, which are mental states, are epiphenomenal and normativity depends on reason, VNC is normatively inert.[61]

---

[60] Greene and Cohen, note 13, above, at 218.
[61] I was first prompted to this line of thought by Mitchell Berman's discussion of determinism and normativity. See, Mitchell Berman, 'Punishment and Justification' (2008). 118 *Ethics* 258, 271 n. 34.

28.5 Case Studies

This section considers case studies that will illuminate the conceptual and abstract theses of the preceding sections. It begins with examination of the legal implications of two recent, widely-noticed and excellent neuroscience studies that appear to have legal relevance. Then it considers a body of neuroscience research that many people think is useful to the law. Finally, it addresses the criminal responsibility of an individual paedophile for whom there was compelling evidence that a brain tumour was the source of the defendant's paedophilic urges and behaviour.

28.5.1 The neural corrrelates of third-party punishment[62]

In scenarios involving third-party criminal punishment, the right dorsolateral prefrontal cortex (rDLPFC) was activated when the sixteen subjects decided that the harmdoer should be punished because there was culpability—a result consistent with previous findings that rDLPFC is activated when people make punishment decisions in two-party games. It also found that the amygdala, a region in an older portion of the brain that is associated with the expression of emotion and affective processing, was activated when the subjects decided how much to punish.

The study is compelling, but one can fairly raise objections to various aspects of the language of the study that are common but which overclaim or mislead. For example, it claims that it 'elucidated the neural dynamics that underlie human altruistic punishment'; that 'prefrontal and parietal activity is modulated by a punishment-related decisional process'; that 'the two fundamental components of third-party legal decision making... are not supported by a single neural system'; and that 'these findings seem to highlight an important conceptual overlap between moral reasoning and legal reasoning...'. The study does show which brain regions are apparently activated when subjects perform specific tasks, but this is different from 'elucidating neural dynamics'. If this means that the study demonstrates the causally explanatory interactions between neurons or neural networks, it is false. If it means simply that there is a correlation between activation in some regions and specific tasks—as the title of the study properly indicates—it is true but the language appears to make a more expansive claim. 'Modulated' suggests causal understanding that we do not possess. 'Supported by a single neural system' again goes beyond our understanding. It may well be true, but the language suggests an understanding of causation, of necessary and perhaps sufficient neural conditions for the behaviour, that we in fact lack. Finally, the suggestion that these neural correlates 'highlight an important conceptual overlap' between two types of reasoning is a category mistake because it confuses the positive—the brain

---

[62] J. W. Buckholtz et al., 'The Neural Correlates of Third-Party Punishment' (2008) 60 *Neuron* 930. The study deservedly received an admiring comment in the same issue of the journal in which the study appeared. J. Haushofer and E. Fehr, 'You Shouldn't Have: Your Brain on Others' Crimes' (2008) 60 *Neuron* 738.

findings—with the normative—the concepts of moral and legal reasoning. Even if precisely the same brain regions were activated during moral and legal-reasoning tasks, it would not follow that they are not conceptually distinct.

The interesting question is the legal relevance of the study. It is unsurprising that two different types of decisions are associated with activation in different brain regions. This study does help confirm the apparent similarities between two-party and three-party norm-violation punishment decisions, and it is thus a contribution to understanding the evolution of punishment practices. But what is its current legal relevance? It does not tell us who to punish or how much. In other words, it does not contribute to the substance of crucial normative questions in criminal law.[63]

The prepotent rDLPFC response to punish and the role of the amygdala in such decisions are also not surprising because everyone understands that virtually all human beings will experience emotional responses when making decisions about whether punishment is warranted and how much to punish norm violators. This might create concerns about whether judges may sometimes be influenced by emotional factors and then decide unfairly or with implicit bias. But we already knew they sometimes do this and deciding which decisions are not fair or impartial is a normative question of folk psychological practical reasoning that cannot be read off from the brain (unless we found a perfect neural marker for those decisions that we had already decided on normative grounds were not fair, which is a fantasy at present and perhaps for the future). Assume that a particular judge showed particularly active amygdala activity when sentencing. What follows?

In short, this is first-rate cognitive neuroscience, but it is legally inert.

28.5.2 The neural correlates of moral decision-making in psychopathy[64]

In a task involving emotional decision-making with seventeen subjects, those with higher psychopathy scores showed decreased amygdala activation and those scoring particularly high on the interpersonal factor for psychopathy (manipulation, conning, superficiality, and deceitfulness) also exhibited reduced activity in what is called the general 'neural moral circuit'. Decreased activation in these areas is associated with deficits in the ability to envision harming others and more generally with deficits involving the proper use of emotions and understanding of interpersonal processes. We already possess a well-characterized understanding and operationalized measure of the emotional and interpersonal deficits of psychopaths. We understand that they cannot use conscience, empathy, and similar qualities to guide

---

[63] See Selim Berker, 'The Normative Insignificance of Neuroscience' (2009) 37 *Philosophy & Public Affairs* 293; Frances Kamm, 'Neuroscience and Moral Reasoning: A Note on Recent Research' (2009) 37 *Philosophy & Public Affairs* 330.

[64] A. L. Glenn, A. Raine, and R. A. Schug, 'The Moral Correlates of Moral Decision-making in Psychopathy' (2009) 14 *Molecular Psychiatry* 5. Publication of this article was followed by a useful exchange about it. See, S. Tassy et al., 'Do Psychopathic Patients Use their DLPFC when Making Decisions in Moral Dilemmas?'. (2009) 14 *Molecular Psychiatry* 908; A. L. Glenn et al., 'Increased DLPFC Activity during Moral Decision-making in Psychopathy' (2009) 14 *Molecular Psychiatry* 909.

their conduct. Nevertheless, the criminal law provides no excuse or mitigation for this condition. Do these findings imply that we should?

First, it is again unsurprising that the brain activity of psychopaths doing moral judgement tasks differs from the activity of non-psychopathic controls. Their behaviour is different and different brain activity is consistent with different behaviours. Even assuming that the brain findings were causal and not correlational, should these data cause us to excuse or mitigate the criminal conduct of psychopaths?

All behaviour has associated brain activity and even abnormal brain activity is not per se an excusing condition. Nor is brain causation the equivalent of compulsion or lack of self-control. The criteria for excuse—roughly, lack of rational capacity or the capacity for self-control—are folk psychological and we already understood the folk-psychological differences between people with and without psychopathy. (Whether we consider the psychopathic characteristics 'deficits' is a normative question.) Moreover, most informed observers already concluded that these deficits are not the psychopath's own fault. It is how they are constituted.[65] Thus, even without the brain findings, we knew everything we needed to know to argue normatively about whether psychopaths ought to be excused or mitigated. We already knew what we had to know about psychopaths' rational and control capacities, which are properties of folk psychological people, not brain criteria. The study does provide weakly convergent evidence for the validity of psychopathy as an identifiable condition, but that only becomes important if we had already decided on normative grounds that psychopathy should have legal relevance. In short, this is excellent cognitive neuroscience, but it is legally inert once again.

28.5.3 Executive functioning

Executive function is the term applied to a wide range of abilities that enable purposive, goal-oriented, successful behaviour.[66] These include the capacities to initiate and plan behaviour, to focus attention, and to self-monitor and self-regulate, including inhibition of inappropriate desires. Defects in these functions, which stem from lateral prefrontal cortex (LPFC) malfunction, tend to be global and to affect the person's behaviour generally. Not all functions need be impaired, however. The extent of impairment depends on the specific pathology involved. A person with such defects may be excitable, impulsive, and erratic, or, in the alternative, avolitional, perseverating, and with flattened affect. Narrowly construed

---

[65] I will not address whether psychopathy is a disorder that exists and can be properly diagnosed among children and adolescents. I simply assume that the characteristics adult psychopaths exhibit are not their fault but are the product of causal variables beyond their control.

[66] Jeffery L. Cummings and Bruce L. Miller, 'Conceptual and Clinical Aspects of the Frontal Lobes' in Jeffrey L. Cummings and Bruce L. Miller (eds.), *The Human Frontal Lobes: Functions and Disorders*, 2nd edn (New York: Guilford Press, 2007) 12, 15–18; Joaquin M. Fuster, *The Prefrontal Cortex*, 4th edn (Boston: Academic Press, 2008) 178ff.; Joel H. Kramer and Lovingly Quitania, 'Bedside Frontal Lobe Testing' ibid. at 2279, 279–85; Muriel D. Lezak et al., *Neuropsychological Assessment*, 4th edn (New York: Oxford University Press, 2004) 35–7. There is some variation among writers concerning the characterization of executive functions, but the description I will give is common and sufficient for our purposes.

cognitive functioning may not be impaired, but, for example, the person's ability to use the intelligence and knowledge he possesses is diminished. These problems can impair the capacity for normal, independent life, including the ability to have successful interpersonal relations and to avoid inappropriate and unlawful behaviour.

One might well expect to find defects in such functions, whether or not associated with clear organic pathology, in large numbers of people who violate the criminal law or who have trouble meeting various competence criteria. Such defects, especially of the type involving disinhibition and poor planning, surely might play a causal role because they make it more difficult to behave well. Nevertheless, how are these impairments legally relevant. Simply having defects in rational or control capacities, no matter how they are caused, is not per se an excusing or mitigating condition. Indeed, there is no criminal law excuse for having control difficulties generally, but such difficulties may be relevant to legal insanity if they are a product of mental illness and to various forms of involuntary civil or quasi-criminal commitment.[67] Such defects may also be relevant to sentencing.

All these cases must be decided behaviourally, however, so what does localization in the LPFC of many of these defects add? In cases in which the defect is arguably sufficient to warrant a conclusion concerning mitigation, excuse or incompetence, the behavioural evidence will typically be so manifest that no neuroscience evidence will be needed.[68] Moreover, there is empirical reason to question whether tests of executive function explain much of the variance in the real world expression of executive function.[69] Therefore, neuroscience evidence concerning executive function is not likely to help resolve close cases for the general reasons discussed above concerning how neuroscience data are obtained, but that may change in the future if neuromarkers become more sensitive. Similarly, if the neuroscience becomes more sensitive, it might help decide cases in which there is concern about malingering. Most important, criminal responsibility and competence are normative legal criteria. Neurodata must therefore be considered in light of the behavioural data and normative considerations. It cannot resolve the legal question to which it may potentially be relevant.[70]

It also does not seem that the neuroscience evidence presents a challenge to existing doctrine. It could, however, motivate expanding the categories of people who might be mitigated or excused if the neuroscience taught us that larger numbers of people have serious difficulty controlling their behaviour than we thought previously or if it showed that the behavioural evidence is misleading concerning control capacities. We do not have such data at present, but we might in

---

[67] E.g., *Kansas v Crane* 534 US 407 (2002) (requiring for mentally abnormal sexually violent predator commitments that the subject has 'serious difficulty controlling himself').

[68] See Lezak et al., note 66, above, at 36 ('Many of the behavior problems arising from impaired executive functions are apparent even to casual or naïve observers.')

[69] See, Russell A. Barkley and Kevin R. Murphy, 'Impairment in Occupational Functioning and Adult ADHD: The Predictive Utility of Executive Function (EF) Ratings Versus EF Tests' (2010) 25 *Archives of Clinical Neuropsychology* 157.

[70] Dean Mobbs et al., 'Law, Responsibility and the Brain' (2007) 5 *PLoS Biology* 0693, 96–7.

the future. Finally, neuromarkers of executive function defects may have marginal utility in helping to make prediction decisions for sentencing, parole and the like. Whether such markers are sufficiently helpful beyond behavioural methods to justify the cost of collecting the data is a normative question. For now, no such data exist, but it may in the future and it is possible that obtaining that data will be cost–benefit justified.

Understanding the role of brain abnormalities in producing executive function deficits may of course lead to treatments or other interventions that would prevent crime and other untoward behaviours. For now, however, neuroscience evidence of executive function deficits simply provides some hope for evidentiary assistance in the future. It suggests no major reforms of doctrine or practice.

28.5.4 The case of Mr Oft

This case was first reported in a neurology journal.[71] Oft was a forty-year-old school teacher who was married and had a step daughter. He had an interest in pornography dating to his adolescence, but at the time in question he experienced a growing sexual interest in children and he collected child pornography and visited child pornographic Internet sites. He also solicited prostitution at 'massage parlours', which he had not previously done. Oft tried to conceal his activities because he knew that they were unacceptable. Nevertheless, he continued to act on his sexual impulses because, he said, the 'pleasure principle overrode' his restraint. Oft began to make subtle sexual advances to his prepubescent stepdaughter, who informed her mother.

Oft was convicted of child molestation and ordered to undergo an inpatient rehabilitation programme instead of prison. Despite his desire to avoid prison, he solicited sexual favours from staff and other patients in his programme and he was expelled.

The evening before his prison sentencing, Oft was admitted to a hospital emergency room complaining of headache. Although no physiologic cause was suspected, he was admitted on psychiatric grounds with a diagnosis of paedophilia. He expressed suicidal ideation and a fear that he would rape his landlady. During neurologic examination he solicited female staff for sexual favours and was unconcerned that he had urinated on himself. He had various neurological signs, including problems with his gait. Oft was alert and completely oriented. His memory was intact, his speaking and reading skills were unimpaired, and he was able to inhibit motor responses on a standard test of this ability. Word generation was somewhat impaired. He did suffer from constructional apraxia, the inability to assemble a coherent whole from its constituent elements, as demonstrated by his inability to draw a clock or to copy figures. He also could not write a legible sentence. A magnetic resonance imaging (MRI) test was performed.

---

[71] Jeffrey M. Burns and Russell H. Swerdlow, 'Right Orbitofrontal Tumor with Pedophilia Symptom and Constructional Apraxia Sign' (2003) 60 *Arch. Neurology* 437. The name Mr Oft is a pseudonym.

Oft had a large orbitofrontal tumour. The orbitofrontal cortex is involved in the regulation of social behaviour. Lesions acquired in this region later in life are associated with impulse control problems and antisocial conduct, but previously established moral judgement is preserved. The tumour was surgically removed and Oft quickly recovered bladder control and normal walking activity. Two days post surgery, his neurologic examination was essentially normal. Oft then successfully completed an outpatient treatment programme for his sexual disorder. He was no longer considered a threat and returned home. About a year later, he experienced a persistent headache and again began secretly collecting pornography. MRI showed tumour regrowth and the new tumour was successfully removed.

In their discussion of Oft's case, the authors said that Oft 'could not refrain from acting on his pedophilia despite the awareness that the behavior was inappropriate'.[72] They hypothesized that the problem was caused by a disruption of his somatic marker system, which lead to a preference for short-term reward and thus impaired the 'subject's ability to appropriately navigate social situations'.[73]

Although paedophilia is not a sufficient mental disorder to support an insanity defence, it is not absurd to think that perhaps Oft deserved mitigation or excuse for his sexual deviance on the ground that he could not control himself. With respect, however, we do not know whether Oft could not—that is, lacked the ability to—control his sexual behaviour, or whether he simply did not. Given the timing of the appearance of the sexual deviance and the tumour growth, we can be quite confident that the tumour played a causal role in producing and heightening his sexually deviant urges and in undermining his inhibitory processes.

The general legal question is how Oft is relevantly different from any other paedophile with similar urges and similar inhibitory controls? One assumption is that the sexual behaviour is a mechanistic product of the tumour and is thus just like the mechanistic sign of any other disease. This assumption begs the question of responsibility, however. Oft's desires may have been mechanistically caused, but acting on them was intentional action. An abnormal cause for his behaviour does not mean that he could not control his actions. This must be shown independently. We can reasonably infer that Oft had difficulty controlling behaviour that harmed himself because he acted in ways he knew would have negative consequences. But this is true of all paedophiles and we do not excuse them. He may have had impaired executive function, but this may also be true of many paedophiles and would again need to be established independently. Although there is reason to question whether Oft differs substantially from paedophiles generally, the temptation to respond to Oft differently is strongly influenced by the lure of mechanism.

Now let us be highly specific. Recall that the neurologists who examined him concluded that Oft could not refrain from acting on his paedophilic urges. Oft was clearly not responsible for having those urges in the first place. We do not know how strong his paedophilic urges were, however, nor do we know which inhibitory functions, if any, were compromised. We do know that Oft did not control his

---

[72] Ibid. at 440.   [73] Ibid.

paedophilic and other sexual urges, including in circumstances in which it was unlawful or completely inappropriate to express them. Moreover, Oft understood that his behaviour was unacceptable and he reported that the pleasure principle overrode his inhibitions. It is reasonable to conclude based on common-sense inferences that Oft experienced substantial difficulty controlling himself, but how do we know that he lacked sufficient control capacity to deserve mitigation or excuse? Oft was firmly in touch with reality and fully understood the moral and legal rules. He understood that important interests were at stake and that he should not violate them. We do not know how firmly Oft resolved not to yield to his impulses or whether he took steps to restrain them. There is a hint in his comment about the pleasure principle that he took no such steps.

At least in the beginning when the paedophilic urges emerged, Oft was able to conceal many of his unacceptable activities, which often requires inhibiting one's response to desires, and his spouse evidently noted nothing amiss in his behaviour. Oft's paedophilia and pornographic interests were discovered only when his stepdaughter reported his sexual advances. Later on, however, the picture becomes more mixed. Most of the neuropsychological testing done in the hospital prior to surgery that bears on his executive functioning was normal. His memory was intact, and Oft 'verbally shifted between letter and number sets, conceptualized, performed sequential hand movements, and inhibited motor responses on the Luria go-no go test'.[74] On the other hand, complete neuropsychological testing of frontal lobe function was not performed because prompt surgical intervention was necessary. Such testing might have disclosed other executive function deficits. Further, during his neurological examination, Oft appeared unconcerned that he had urinated on himself, which suggests some type of irrational disconnection from his situation, and he solicited hospital personnel for sexual favours. Perhaps most importantly, Oft's sexual behaviour in his treatment programme threatened his ability to avoid prison. In short, Oft's ability to control himself apparently deteriorated quite rapidly and became markedly worse by the time of hospitalization compared to the time of the offence.

We know a great deal about the cause of Oft's paedophilic desires and have a plausible causal account of how his executive function might have been undermined. Brain causation, even by such a manifest abnormality, still does not answer our question about how difficult it was for him to control himself. The case study authors proposed a 'somatic marker' hypothesis to explain behaviour that is similar to acquired sociopathy. If true, this helps explain why a hitherto continent agent began to act for immediate gratification and with insufficient regard for future consequences. Put another way, it helps explain why his judgement was impaired and poor judgement makes controlling oneself more difficult. Nevertheless, it does not tell us how impaired Oft's judgement was or the role such impairment played in explaining his inappropriate and criminal behaviour.

---

[74] Burns and Swerdlow, note 71, above, at 438. The Luria test examines how well the subject is able to inhibit a prepotent response.

Finally to decide whether Oft deserves mitigation or excuse we must use all the considerations just reviewed and come to an all-things-considered normative evaluation about his capacities at the time of the crime. This is a standard question of deciding when a person had sufficient capacity to resist temptation. To use the capacious Model Penal Code language, did Oft lack substantial capacity to understand the wrongness of his actions or to conform his conduct to moral and legal norms? These capacities range along a continuum. At the time of the crime, Oft was clearly in the grey area between the kinds of situation in which virtually every person would have trouble controlling undesirable behaviour and those in which virtually no one would fail to exercise control. I will leave readers to decide the question for themselves, but the answer cannot be based on the presence of abnormal brain causation per se. Oft's case might elicit sympathy for his plight, which was terrible luck, and it certainly suggests that a medical rather than punitive response might be cost–benefit justified. But these considerations are distinct from whether Oft deserved mitigation or excuse.

28.6 Conclusion

The relation between neuroscience and legal doctrine and practice is conceptually fraught. Neuroscience has the potential to make internal contributions to legal doctrine and practice if the relation is properly understood. For now, however, such contributions are modest at best and neuroscience poses no genuine, radical challenges to concepts of personhood, responsibility, and competence.

Index of Names

Alcmaeon of Croton 1
Alexander, L. 552
Appelbaum, P.S. 302–3, 306
Aristotle 182
Aronson, J.D. 236, 242, 277
Ashworth, A. 159
Astrow, A.B. 294
Audi, R. 546

Baird, A. 233–6, 246, 257, 277
Bargh, J.A. 548
Bender, L. 229
Bennett, M.R. 544, 545, 550
Berker, S. 556
Berman, M. 554
Beschle, D.L. 276
Blair, J. 21–2, 24, 440, 441
Blakemore, C. 146
Bland case 10
Blumoff, T. 9
Boella, L. 198, 199
Bok, H. 552
Boudreau, C. 11
Boyd, R. 436, 437
Braman, D. 239
Brass, M. 551
Breidbach, O. 511
Brizendine, L. 275
Brodie, J.D. 236–7, 240
Brown, G. 155–6
Brunner, H.G. 65, 180
Buckholtz, J. 8, 555
Buren, A.M. 376
Burns, S.M. 238, 559, 561
Buss, E. 251, 282
Byrne, R.W. 436

Cáceres, E. 11
Cajal, S.R. y. 3
Caparo, L. 11
Carbone, J. 9
Carlson, R. 453–4
Caspi, A. 66, 179–81, 184, 533
Castel, A.D. 102, 280, 524, 525
Chamallas, M. 229
Charland, L.C. 303, 306
Chorvat, T. 197
Churchland, P.M. 154
Claydon, L. 8–9
Cohen, G.B. 253
Cohen, J. 239, 248, 253–4, 544, 546, 554
Coles, M.G.H. 404
Colombetti, G. 439

Craver, C. 552
Crawford, V. 396
Crespo, A. 472
Cummings, J.L. 547, 557

Daguerre, L–J.M. 508
Dalai Lama 11, 444–5
Damasio, A.R. 179, 305–6, 411–14
Darwin, C. 435
Daston, L. 510, 515, 527
Daubert rule 98, 521–2, 526
da Vinci, Leonardo 2
Dawson, G. 553
Decker, E. 453–4
Dennett, D. 545, 553
Denno, D. 242
Descartes, R. 2, 336, 411–12
Devlin, P.D. 407
Dobbs, D. 203, 207, 209, 210
Donchin, E. 404
Dresser, R. 28
Du Bois-Reymond, E. 2
Duff, R. 302
Duffy, J. 11

Eddersheim, J.G. 226
Eisenberger, N. 212
Enslinger, P.J. 295
Etkia, A. 218

Farah, M.J. 546, 549
Fehr, E. 555
Feigenson, N. 99, 408
Feld, B. 251
Feyerabend, P. 524
Fins, J. 10
Fischer, J.M. 7
Fodor, J.A. 549, 551–3
Foster, C. 10
Fox, D. 10
Frank, J. 408
Frankfurt, H. 32, 45–6, 52, 53, 56, 57
Frye v U.S. 97, 521
Fugelsang, J. 233
Fukuyama, F. 484
Fuster, J. 557

Gage, Phineas 4–5, 90
Galen 1–2
Galison, P. 512, 516, 527
Gall, F.J. 2
Galvani, L. 2
Garland, B. 417

Index of Names

Gauguin, P. 175
Gazzaniga, M. 6, 417, 449
Gibbard, A. 183
Gilligan, C. 245
Glannon, W. 6, 7
Glenn, A.L. 556
Golan, T. 514, 517
Gold, V. 526
Goldberg, D. 11–12
Goldberg, S. 11
Golgi, C. 3
Goodenough, O. 200
Graham v Florida 253, 263, 264
Greely, H. 7–8, 77, 432
Greene, J. 239, 248, 249, 544, 546, 253–4
Greenfield, S. 141, 156, 165, 169
Grey, B. 9
Griffin, D.R. 453
Grisso, T. 239, 302–303
Gurley, J.R. 102
Gut, R. 235

Hacker, P.M.S. 544, 545, 550
Haggard, P. 551
Hamer, D. 453
Hare, R. 22–3, 491
Harlow, J. 4–5
Hart, H.L.A. 13, 118, 460, 552
Haushofer, J. 555
Haynes, J-D. 549
Heberlein, A.S. 546
Heinemann case 267–8
Hermann, E. 553
Herzig, R. 514
Hinckley case 71, 73
Hinntikke, J. 39
Hippocrates 1
Hirasawa, K. 454
Hobbes, T. 11, 433, 434, 439
Hoffman, M. 11
Holmes, O.W. 118, 497
Holton, R. 552
Holtz, R.L. 141–2
Horgan, T. 44
Hornsby, J. 145
Houser, S. 185
Hounsfield, Sir G. 4
Hudson, B. 159
Hume, D. 183, 483, 484, 485
Hutcherson, C.A. 446

James, W. 41, 446, 451–2
Jennett, B. 368
Jones, N. 220

Kahan, D.M. 239, 280
Kalven, H. 408
Kamm, F. 556
Kane, R. 54

Kant, I. 175, 182, 185, 409, 410
Keller, E.F. 524
Kihlstrom, J.F. 548
Knutson, B. 545
Koch, R. 63–4
Kohlberg, L. 245
Korsgaard, C. 175
Kulynych, J. 4, 238, 241

Lancaster, J. 487
Laudan, L. 460
Lazar, S.W. 443
Le Doux, J. 212, 423
Lehrer, K. 44
Leibniz. G. 31
Lesak, M.D. 557, 558
Lewis, D. 44
Libet, B. 6, 549, 550–1, 553
Lief, J. 446–7
Lombroso, C. 2
Luck, S. 402
Lupia, A. 395, 396, 397
Lutz, A. 443

M'Naghten case 16, 84
Macadam, J. 300
MacIntyre, A. 484
Mackenzie, R. 10
Maibom, H. 8, 79, 80, 86–91, 93–5
Malle, B.F. 552
Marcus, D.K. 102
Maroney, T. 9, 204, 301, 541
Martinsen, R. 523, 527
Matthews, M. 220
McCabe, D.P. 102, 280, 524, 525
McCabe, K. 199
McClelland, J.L. 466
McCubbins, M.D. 395–7
McGinn, C. 546
McHuff, P.R. 547
Mega, M.S. 547
Mele, A. 35, 550–2
Mendez, M.F. 293, 294, 296
Metzgor, E. 507
Mille, B.L. 557
Mitsumoto, H. 295
Mnookin, J. 507, 508, 510, 512
Mobbs, D. 24, 558
Moffitt, T. 179, 181
Moore, M.S. 17, 186–8, 550
Moore, G.E. 483, 485
Morris, J.P. 406
Morse, S. 6, 12, 28, 62, 90, 186–8, 237, 239, 240–1, 302, 431–2

Nagel, T. 175–7
Nemeroff, C.B. 336
Norrie, A. 145, 147

Index of Names

O'Connor, T. 42
O'Shea, M. 3

Perry, J. 44
Persinger, M. 449, 451
Pfaff, D. 11, 389, 390, 392
Piaget, J. 236, 245
Piffman case 265–6
Pinker, S. 279
Pizzaro, D. 413
Plum, F. 368, 369, 373
Popper, K. 450–1
Posner, R. 280
Potter, N.N. 89
Prehn, K. 200

Quine, W.V. 35
Quinlan case 10, 367–72
Quitania, L. 557

Rabkin, J.C. 295
Racine, E. 525
Rawls, J. 413
Raz, J. 146
Redding, R.E. 240
Reimer, M. 8, 80, 85–7, 93–5
Richerson, P.J. 436–7
Riedijk, S. 298
Rizzulatti, G. 440
Robinson, P. 489–90, 493–4, 497–9
Roentgen, W.C. 4
Roper v Simmons case 9, 20–1, 242–53, 255, 259–60, 264, 431, 456–7, 540
Rose, N. 77, 141, 169
Rose, S. 144, 153–4
Roskies, A. 8
Russell, B. 34

Sabbagh, C. 293
Sadler, J.Z. 89
Sapolsky, R. 62, 63, 89
Schäfer, B. 37
Scott, E. 235, 239, 247, 249, 256, 257, 260
Searle, J. 521, 522
Sexton case 5, 16–17
Shakespeare, W. 419
Shapira, J. 293, 294
Sher, G. 521
Shermer, M. 487
Simon, D. 463, 464
Sinnott-Armstrong, W. 8
Siraisi, N.G. 510
Slavney, P. 547
Smith, R. 253

Snead, C. 239, 240–1, 245
Sobel, J. 396
Spear, L.P. 258
Spencer, H. 486
Steinberg, L. 235, 239, 247, 256, 260, 273, 281
Stone, A.A. 537
Strickler, O.C. 514–15
Sunstein, C. 236
Swerdlow, R.H. 238, 559, 561

Tanaka, J.W. 549
Tancredi, L.R. 90, 236–7, 240
Taruffo, M. 460
Tassy, S. 556
Terracina, D. 9
Teter, M. 507
Thompson, C. 439
Tierney, J. 545
Tomasso, J. 450
Toth, K. 553
Tucker, J. 508, 510

Uaitl, D. 547
Utab, W.R. 226

Van Dijk, J. 524
Van Inwagen, P. 43, 47–8, 53
Vesalius 523
Viens, A. 10–11, 227
Vincent, N. 8, 75

Wager, T.D. 218
Walker, M.U. 176
Walton, D. 461
Watson, G. 87, 88, 92
Wegner, D. 550
Wernicke, C. 2–3
Whiten, A. 436
Wiesberg, D.S. 525
Wigmore, J.H. 512
Wilkinson, D. 10
Williams, B. 174, 185, 548
Willis, T. 2
Wolf, S. 87
Wood, J.N. 553

Yang, Y. 442

Zachar, P. 89
Zau, 427
Zeisel, H. 408
Zhu, J. 550
Zimmermann, M.J. 184
Zimring, F.E. 257

Index of Subjects

A, Re 324
abnormality
  assessing 106–7
abuse of children 9, 87, 176, 179–4, 535
adolescents 231–54, 255–82
  death penalty and 241–53, 259–61, 540
adolescent brain 430
  neuroscience and 232–79
  traditional social science research and 235–6
advance decisions refusing treatment
  (ADRT) 300–7
aggression
  neurobiology of 7, 386, 428–9
altruism 435
Alzheimer's disease 64–5, 285–6, 289–90, 292, 300
amnesia 25–7
Arthur's case 316–17
artificial intelligence 11, 29–39, 461–2
assisted dying 300
atheism, militant 449
autism 23
autonomy 433, 438

B, Re, ('Baby Alexandra') 316–17, 321–2
brain
  Descartes on 2
  Galen on 1
  Golgi on 3
  Harlow on 4–5
  Willis on 2
brain scans
  legal evidence, as 97–114
British Medical Association 314–15
Brunner Syndrome 65, 67–8, 75–6
bystander claims 210

C, Re 315, 322–3
'Can't Help Himself'
  physical evidence 63–8
  behavioural evidence 68–70
  individual's proof 70
Carmentis Machine 311–12, 325–7
causation 110
Children Act 1989 316
civil law tradition 459–82
Clark, Sally 163–5
compassion 1, 435–7, 447
contemplation 445
contract law 421–2, 488–9
  good faith in 422
correctionism 466–9
cranioscopy 2–3

criminal intent 14–17, 145–46
Criminal Justice Act 2003 152
criminal law 439–432
  relative blameworthiness 489–90, 494
  implicit psychology of 529–33
criminal negligence 25–27
criminal responsibility
  brain scans, and, 104

death penalty 9
  juveniles, for 20–1
determinism
  and luck 41–59
  dilemma of 41–2

emotional contagion 434, 435
emotional harm 203–29
emotions 200
  evidence and 414–16
  damages cases and 420–2
empathy 11, 434, 435–7
  medication and 442–4
end-of-life decisions 10, 300–7
England
  criminal law and 142–69
  emotional harm and 220–4
  felony murder rule in 497
evidence
  brain images as 8, 97–114, 148–58
  expert evidence
    Law Commission and 165–6, 169
  photographs, power of 100
  prejudicial effect of 99–103
  scientific images 508
evolutionary jurisprudence 483–503

falsifiability
  Popper on 450–1
fear 2, 23–4, 214–17, 423–6, 445, 447
felony murder rule 496–8
  abolish 497, 500–1
  exceptions to 499–500
Fifth Amendment 335–6
fMRI scan 17–18, 21
free will 6, 185, 239, 250
frontotemporal dementia (FTD) 283–304

Gage, Phineas 4–5, 67, 90
gender bias
  emotional harm and 229
General Medical Council 314

Germany
  legal system in 29, 39
God 449–51
  Darwin and 453
  gene 452–3
golden rule 419–20, 422–4, 426, 435, 439

Hasan v R 151–2
health *see* public health
Hill v R 148–51
history
  knowledge of brain 1–5, 513, 523–24
  scientific images, of 505–28
Hobbes, Thomas
  human nature, and 433, 437
Homicide Act 1957 148–9
human genome project 153
Huntington's disease 64

illness of children
  non-treatment of, 454–456
'is' and 'ought' 483–487
images *see* brain scans
immunisation law, 386
impulse control 17–21
incrimination 336–40
injury
  sense brain 367
  P.V.S. and 367–72
insanity defence 16, 74, 92, 455
instructions
  jury, 410–411
intent, criminal *see* criminal intent
intolerability
  end of life decisions and 326–327
Italy 409
  legal system in 410

J, Re 317, 323
juries 11, 307–17, 395–406, 464, 494
  Devlin, Lord on 407
  emotions and 408, 410–17
  Frank on 408
  Kalven and Zeisel on 408
  Simon on 463–4
juveniles
  competence and 261
  death penalty and 241–53, 259–61
  lengthy sentences 263–64
  mens rea and 261
  punishment, evidence, 261–3
  reasonableness 261–8
  transfer to adult court 261, 265–6
  waiver of rights 261, 268–9

kleptomania 75

lie-detector test 83, 521
logic 30–9
  and *mens rea* 32–4
  mens rea and scope distinctions 36–7
  quantifiers and scope distinctions 34–6
luck
  determinism and 41, 59
  moral 9, 172–7

Mccann's case 152–3, 158
MAOA 65–7, 171, 179–81, 183, 189–91
  the environment and causation 182–3
maternal behaviour 427
MB, Re 324–5
meditation 442–4
memory 214–17
mens rea 7, 14–17
  logic and the brain 29–39
Mental Capacity Act 2005 300–1, 303, 316
Mexico 469, 482
minimally conscious state 371–3, 375–6, 383–4
moral luck 174–177 *see also* luck
  objections 184–9
moral misbehaviour syndrome 305–6
moral 11, 387–389, 419–432
  neurobiological research and, 389–390
  qualitative research and, 390–391
morality 438–39
  neural correlate of 440–1
  care-based 441–2, 445
  social convention and 441, 445
motor neuron disease (MND/ALS) 283–6, 290–1, 294–6, 298–300, 304

naturalistic fallacy 484–7
  Hume and 484–5
Neanderthals 436
neonates *see* newborn infants
neuroethics 198–9
neuroethics society 5
neuroimaging 4 *see also* brain scans
  and brain function 13
  interpretation of 30
newborn infants
  damaged 10
  treatment withdrawal 309–33
NICE 296
North Dakota 410
Nuffield Council on Bioethics 315–16

objectivity 515
  Daston and Galison on 510, 511, 516
opiates 451

Pagett's case 160–1
parole 386
persistent vegetative state (PVS) 10, 367–72
  civil rights issues 376–84
  diagnostic assessment of 372–74
persuasion
  neuroscience to study 398–400

Index of Subjects

PET scans 15–18, 21, 24, 217–18
phrenology 2–3
photographs 100, 415, 508–14
placebos 451
post-traumatic stress disorder 204–6, 213–17
  neuroscience studies of 217–19
property law
  origins of 196–8
  copyright 199–200
psychiatric illness
  English law and 220–224
psychology 21–4, 446, 490–92, 556–7
  Glenn, Raine and Schuj on 22
  Hare on 23–4
  disorder of emotional empathy and 441–2
public health
  history of 383–7
  neuroscience and 10–11, 385–93
public health law
  neuroscience and 385–7, 391–3
  reciprocity and 387–9
punishment
  neural correlates of third party punishment 115–39, 555–6
  harsher 240–1
  life without parole 263–4

quarantines 387, 389, 391

reciprocity *see* moral reciprocity
religion 11
  Christian scientists and 454–8
  establishment of 450
  Gazzaniga and 451
  James and 451–54
  neuroscience and free exercise of 449–58
  theories of evolution and 454–6
Roman law 507
Royal College of Paediatrics and Child Health guidelines 313–14

SARS 392
sentencing practices
  in United States 493–4
sexual behaviour 426–7
silence, right to 10
slavery
  religion and 451
sterilisation
  male 324
sympathy 434–5

Thomas's Case 161–3
Tibet 442–4
tort, law of 9, 203–29, 386
  emotional harm 203–29
Tourettes Syndrome 7, 8, 62, 68, 70, 75–6
trust 396–7, 400, 405

Vietnam veterans 217
violent impulses 17–21
voluntariness 158–61

X-ray 4, 12, 514–23, 527
  power and 514